Tumor Dormancy, Quiescence, and Senescence
Volume 2

Tumor Dormancy and Cellular Quiescence and Senescence
Volume 2

For further volumes:
http://www.springer.com/series/10582

Tumor Dormancy, Quiescence, and Senescence
Volume 2

Tumor Dormancy, Quiescence, and Senescence

Aging, Cancer, and Noncancer Pathologies

Edited by

M.A. Hayat
Distinguished Professor
Department of Biological Sciences
Kean University, Union, NJ, USA

Editor
M.A. Hayat
Department of Biological Sciences
Kean University
Room 213, Library building
Morris Avenue 1000
Union, NJ 07083, USA

ISBN 978-94-007-7725-5   ISBN 978-94-007-7726-2 (eBook)
DOI 10.1007/978-94-007-7726-2
Springer Dordrecht Heidelberg New York London

Library of Congress Control Number: 2013931982

© Springer Science+Business Media Dordrecht 2014
This work is subject to copyright. All rights are reserved by the Publisher, whether the whole or part of the material is concerned, specifically the rights of translation, reprinting, reuse of illustrations, recitation, broadcasting, reproduction on microfilms or in any other physical way, and transmission or information storage and retrieval, electronic adaptation, computer software, or by similar or dissimilar methodology now known or hereafter developed. Exempted from this legal reservation are brief excerpts in connection with reviews or scholarly analysis or material supplied specifically for the purpose of being entered and executed on a computer system, for exclusive use by the purchaser of the work. Duplication of this publication or parts thereof is permitted only under the provisions of the Copyright Law of the Publisher's location, in its current version, and permission for use must always be obtained from Springer. Permissions for use may be obtained through RightsLink at the Copyright Clearance Center. Violations are liable to prosecution under the respective Copyright Law.
The use of general descriptive names, registered names, trademarks, service marks, etc. in this publication does not imply, even in the absence of a specific statement, that such names are exempt from the relevant protective laws and regulations and therefore free for general use.
While the advice and information in this book are believed to be true and accurate at the date of publication, neither the authors nor the editors nor the publisher can accept any legal responsibility for any errors or omissions that may be made. The publisher makes no warranty, express or implied, with respect to the material contained herein.

Printed on acid-free paper

Springer is part of Springer Science+Business Media (www.springer.com)

Although touched by technology, surgical pathology always has been, and remains, an art. Surgical pathologists, like all artists, depict in their artwork (surgical pathology reports) their interactions with nature: emotions, observations, and knowledge are all integrated. The resulting artwork is a poor record of complex phenomena.

Richard J. Reed, MD

One Point of View

All small tumors do not always keep growing, especially small breast tumors, testicular tumors, and prostate tumors. Some small tumors may even disappear without a treatment. Indeed, because prostate tumor grows slowly, it is not unusual that a patient may die at an advanced age of some other causes, but prostate tumor is discovered in an autopsy study. In some cases of prostate tumors, the patient should be offered the option of active surveillance followed by PSA test or biopsies. Similarly, every small kidney tumor may not change or may even regress. Another example of cancer or precancer reversal is cervical cancer. Precancerous cervical cells found with Pap test, may revert to normal cells. Tumor shrinkage, regression, dormancy, senescence, reversal, or stabilization is not impossible. Can prosenescence therapy be an efficient alternative strategy to standard therapies for cancer prevention and treatment?

Another known example of cancer regression is found in pediatric neuroblastoma patients. Neuroblastoma shows one of the highest rates of spontaneous regression among malignant tumors. In addition to the well-known spontaneous regression in stage 4S disease, the high incidence of neuroblastoma remnants found during autopsy of newborns suggest that localized lesions may undergo a similar regression (Guin et al. 1969). Later studies also indicate that spontaneous regression is regularly seen in infants with localized neuroblastoma and is not limited to the first year of life (Hero et al. 2008). These and other studies justify the "wait and see" strategy, avoiding chemotherapy and radiotherapy in infants with localized neuroblastoma, unless *MYCN* gene is amplified. Infants with nonamplified *MYCN* and hyperdiploidy can be effectively treated with less intensive therapy. Infants with disseminated disease without *MYCN* have excellent survival with minimal or no treatment. Another example of spontaneous shrinkage and loss of tumors without any treatment is an intradural lipoma (Endoh et al. 1998).

Although cancers grow progressively, various lesions such as cysts and thyroid adenomas show self-limiting growth. Probably, cellular senescence occurs in many organ types following initial mutations. Cellular senescence, the growth arrest seen in normal mammalian cells after a limited number of divisions, is controlled by tumor suppressors, including p53 and p16, and so this phenomenon is believed to be a crucial barrier to tumor development. It is well-established that cell proliferation and transformation induced by oncogene activation are restrained by cellular senescence.

Metastasis is the main cause of death from cancer. Fortunately, metastasis is an inefficient process. Only a few of the many cancer cells detached from the primary tumor succeed in forming secondary tumors. Metastatic inefficiency varies depending on the location within an organ, but the malignancy may continue to grow preferentially in a specific tissue environment. Some of the cancer cells shed from the primary tumor are lost in the circulation due to hemodynamic forces or the immune system, macrophages, and natural killer cells.

Periodic rejection of a drug by FDA, which was previously approved by the FDA, is not uncommon. Most recently, the FDA ruled that Avastin should not be used to treat advanced breast cancer, although it remains on the market to treat other cancers, including colon and lung malignancies. Side-effects of Avastin include high blood pressure, massive bleeding, heart attack, and damage to the stomach and intestines.

Unwanted side effects of some drug excipients (e.g., propylene glycol, menthol) may also pose safety concerns in some patients. Excipients are defined as the constituents of the pharmaceutical formulation used to guarantee stability, and physicochemical, organoleptic and biopharmaceutical properties. Excipients frequently make up the majority of the volume of oral and parenteral drugs. Not all excipients are inert from the biological point of view. Although adverse drug reactions caused by the excipients are a minority of all adverse effects of medicinal products, the lack of awareness of the possible risk from excipients should be a concern for regulatory agencies, physicians, and patients (Ursino et al. 2011). Knowledge of the potential side effects of excipients is important in clinical practice.

It is known that chemotherapy can cause very serious side-effects. One most recent example of such side-effects was reported by Rubsam et al. (2011). Advanced hepatocellular carcinoma (HCC) induced by hepatitis C virus was treated with Sorafenib. It is an oral multikinase inhibitor that interferes with the serine/threonine kinases RAF-1 and B-Raf and the receptor tyrosine kinases of the vascular endothelial growth factor receptors and the platelet-derived growth factor receptor-beta. Although sorafenib is effective in regressing HCC, it shows serious side-effects including increasingly pruritic and painful skin changes (cutaneous eruption).

An example of unnecessary surgery is the removal of all the armpit lymph nodes after a biopsy when a sentinel node shows early stage breast cancer; removal of only the sentinel node may be needed. Limiting the surgery to the sentinel node avoids painful surgery of the armpit lymph nodes, which can have complications such as swelling and infection (such limited surgery is already being practiced at the Memorial Sloan-Kettering Cancer Research Center). Radiation-induced second cerebral tumors constitute a significant risk for persons undergoing radiotherapy for the management of cerebral neoplasms. High-grade gliomas are the most common radiation-induced tumors in children (Pettorini et al. 2008). The actual incidence of this complication is not known, although it is thought to be generally low.

Medical Radiation

Chromosome aberrations induced by ionizing radiation are well-known. Medical radiation-induced tumors are well-documented. For example, several types of tumors (sarcomas, meningiomas) can develop in the CNS after irradiation of the head and neck region (Parent 1990). Tumorigenic mechanisms underlying the radiation therapy of the CNS are discussed by Amirjamshidi and Abbassioun (2000) (see below).

Radiation therapy is commonly used to treat, for example, patients with primary and secondary brain tumors. Unfortunately, ionizing radiation has limited tissue specificity, and tends to damage both neoplastic and normal brain tissues. Radiation-induced brain injury, in fact, is a potential, insidious later cerebral side-effect of radiotherapy. Most commonly it consists of damage in small arteries and capillaries, resulting in secondary processes of ischemia.

After radiation therapy, imaging techniques (CT, MRI, SPECT) can be used to assess treatment response and detect radiation-induced lesions and recurrent tumors. Optical spectroscopy has also been used for detecting radiation damage (Lin et al. 2005). The F_{500} nm spectral peak allows accurate selection of tissues for biopsy in evaluating patients with new, contrast enhancing lesions in the setting of previous irradiation. This peak is highly correlated with a histological pattern of radiation injury. Deep lesions require a stereotactic biopsy to be conclusive. Also, much of the radiation effect is mediated by acute and chronic inflammatory cellular reactions. Biopsy samples supplement pathological differentiation of radiation effect from tumor progression. It should be noted that most of the biopsies show radionecrosis as well as scattered tumor cells.

Women treated with therapeutic chest radiation may develop cancer. This possibility becomes exceedingly serious considering that 50,000–55,000 women in the United States have been treated with moderate to high-dose chest radiation (~20 Gy). This possibility is much more serious for pediatric or young adult cancer patients, because these women are at a significantly increased risk of breast cancer and breast cancer mortality following cure of their primary malignancy (Martens et al. 2008). A recent study also indicates that such young women develop breast cancer at a young age, which does not appear to plateau (Henderson et al. 2010). In this high-risk population, ironically there is a benefit associated with early detection. In other words, young women with early stage breast cancer following chest radiation have a high likelihood for favorable outcome, although life-long surveillance is needed.

Presently, although approximately 80 % of the children with cancer are cured, the curative therapy could damage a child's developing organ system; for example, cognitive deficits following cranial radiotherapy are well known. Childhood survivors of malignant diseases are also at an increased risk of primary thyroid cancer (Sigurdson et al. 2005). The risk of this cancer increases with radiation doses up to 20–29 Gy. In fact, exposure to radiation therapy is the most important risk factor for the development of a new CNS tumor in survivors of childhood cancer, including leukemia and brain tumors. The higher risk of subsequent glioma in children subjected to medical radiation at a very young age reflects greater susceptibility of the developing brain to radiation. The details of the dose-response relationships, the expression of excess risk over time, and the modifying effects of other host and treatment factors have not been well defined (Neglia et al. 2006).

A recent study indicates that childhood brain tumor survivors are at an increased risk of late endocrine effects, particularly the patients treated with cranial radiation and diagnosed at a younger age (Shalitin et al. 2011). Among children with cancer, the application of radiotherapy, therefore, should not be taken lightly, and it should be administered only when absolutely necessary to successfully treat the primary tumor. When radiotherapy is administered, use of the minimum effective dose tends to minimize the risk of second CNS neoplasms (late effect). Prolonged follow-up of childhood cancer survivors (particularly those treated with radiation) is necessary because of the long period between treatment and the development of malignancy. This practice should be a part of the effective therapy of the primary disease.

It is well established that radiation doses are related to risk for subsequent malignant neoplasms in children with Hodgkin's disease. It has been reported that increasing radiation dose was associated with increasing standardized incidence ratio ($p=0.0085$) in survivors of childhood Hodgkin's disease (Constine et al. 2008). Approximately, 75 % of subsequent malignancies occurred within the radiation field. Although subsequent malignancies occur, for example, in breast cancer survivors in the absence of radiotherapy, the rise increases with radiation dose.

The pertinent question is: Is it always necessary to practice tumor surgery, radiotherapy, chemotherapy or hormonal therapy or a combination of these therapies? Although the conventional belief is that cancer represents an "arrow that advances unidirectionally", it is becoming clear that for cancer to progress, it requires cooperative microenvironment (niche), including immune system and hormone levels. However, it is emphasized that advanced (malignant) cancers do not show regression, and require therapy. In the light of the inadequacy of standard treatments of malignancy, clinical applications of the stem cell technology need to be expedited.

Prostate Cancer

There were an estimated 217,730 new cases of prostate cancer in the United States in 2010 with 32,050 deaths, making it the second leading cause of cancer deaths in men. Currently, there are more than 2,000,000 men in the United States who have had radical or partial prostate surgery performed. Considering this huge number of prostate surgeries and the absence of a cumulative outcome data, it seems appropriate to carefully examine the benefits of radical surgery, especially in younger men.

Clinical prostate cancer is very rare in men of the ages younger than 40 years. In this age group the frequency of prostate malignancy is 1 in 10,000 individuals. Unfortunately, the incidence of malignancy increases over the ensuing decades, that is, the chance of prostate malignancy may reach to one in seven in men between the ages of 60 and 79 years. Reactive or aging-related alterations in the tumor microenvironment provide sufficient influence, promoting tumor cell invasion and metastasis. It has been shown that nontumorigenic prostate epithelial cells can become tumorigenic when cocultured with fibroblasts obtained from regions near tumors (Olumi et al. 1999).

Prostate cancer treatment is one of the worst examples of overtreatment. Serum prostate specific antigen (PSA) testing for the early detection of prostate cancer is in wide use. However, the benefit of this testing has become controversial. The normal cut-off for serum levels of PSA is 4 ng/ml, so a man presenting with a PSA above this level is likely to require a rectal biopsy, but only 25 % of men with serum levels of PSA between 4 ng and 10 ng/ml have cancer (Masters 2007). The PSA threshold currently being used for biopsy ranges between 2.5 and 3.4 ng/ml. Up to 50 % of men presenting with prostate cancer have PSA levels within the normal range. It is apparent that screening of prostate cancer using PSA has a low specificity, resulting in many unnecessary biopsies, particularly for gray zone values (4 ng–10 ng/ml). According to one point of view, the risks of prostate cancer overdetection are substantial. In this context, overdetection means treating a cancer that otherwise would not progress to clinically significant disease during the lifetime of the individual. Overdetection results in overtreatment. The advantages and limitations of PSA test in diagnosing prostate cancer were reviewed by Hayat (2005, 2008).

Androgen deprivation therapy (ADT) is an important treatment for patients with advanced stage prostate cancer. This therapy is carried out by blocking androgen receptor or medical or surgical castration. Although ADT is initially very effective, treated tumors inevitably progress to androgen-independent prostate cancer (AIPC), which is incurable. One possible mechanism responsible for the development of AIPC is modulation of the tissue microenvironment by neuroendocrine-like cancer cells, which emerge after ADT (Nelson et al. 2007).

Recently, Pernicova et al. (2011) have further clarified the role of androgen deprivation in promoting the clonal expansion of androgen-independent prostate cancer. They reported a novel linkage between the inhibition of the androgen receptor activity, down-regulation of S-phase kinase-associated protein 2, and the formation of secretory, senescent cells in prostate tumor cells. It is known that several components of the SASP secretome, such as IL-6, IL-8, KGH, and epidermal growth factor, are capable of transactivating androgen receptor under androgen-depleted conditions (Seaton et al. 2008). It needs to be pointed out that androgen deprivation therapy, used in high-risk patients with prostate cancer, may cause reduced libido, erectile dysfunction, fatigue, and muscle loss; osteoporosis is also a late complication. Therefore, periodic bone density scanning needs to be considered.

Recently, the FDA cleared the use of NADiA (nucleic acid detection immunoassay) ProsVue prognostic cancer test. This proprietary nucleic acid detection immunoassay technology identifies extremely low concentrations of proteins that have not been routinely used as a diagnostic or prognostic aid. It is an <u>in vitro</u> diagnostic assay for determining the rate of change of serum total PSA over a period of time. The assay can quantitate PSA at levels <1 ng/ml.

This technique can be used as a prognostic marker, in conjunction with clinical evaluation, to help identify patients at reduced risk for recurrence of prostate cancer for years following prostatectomy. It targets the early detection of proteins associated with cancer and infectious diseases. This technique combines immunoassay and real-time PCR methodologies with the potential to detect proteins with femtogram/ml sensitivity (10–15 g/ml). Additional clinical information is needed regarding its usefulness in predicting the recurrence.

A significant decrease in the risk of prostate cancer-specific mortality is observed in men with few or no comorbidities. Indeed, active surveillance in lieu of immediate treatment (surgery or radiation, or both) is gaining acceptance. Most men with prostate cancer, even those with high-risk disease, ultimately die as a result of other causes (Lu-Yao et al. 2009). Debate on this controversy is welcome, but narrow opinions and facile guidelines will not lead to facts and new information; men worldwide deserve it (Carroll et al. 2011). Automatic linking positive diagnosis with treatment, unfortunately, is a common clinical practice. Unfortunately, even men who are excellent candidates for active surveillance in the United States often undergo some treatment. Deferment of treatment is advised in men with low-risk disease, especially of a younger age.

Active surveillance is proposed for patients with low-risk prostate cancer in order to reduce the undesirable effects of overdiagnosis. Prostate specific antigen serum level lower than 10 ng/L and Gleason score lower than 7 are the main criteria to select patients for active surveillance. The correct use of these two criteria is essential to differentiate between aggressive and non-aggressive prostate cancer. Autopsy studies indicate that approximately one out of three men older than 50 years show histological evidence of prostate cancer (Klotz 2008). Thus, a large proportion of prostate cancers are latent, never destined to progress, or affect the life of the patient. It is estimated that the percentage of low-risk prostate cancer is between 50 and 60 % of newly diagnosed cases. A large number of patients die having prostate cancer, but not because of this cancer (Filella et al. 2011).

First whole genome sequences of prostate tumors were recently published online in *Nature* journal (vol. 470: 214–220, 2011). This study revealed that rather than single spelling errors, the tumor has long "paragraphs" of DNA that seem to have broken off and moved to another part of the genome (rearrangement of genes), where they are most active. These portions of DNA contain genes that help drive cancer progression. The mutated genes involved include *PTEN, CADM2, MAG12, SPOP,* and *SPTA1.* This information may lead to the development of more efficient, less invasive ways to diagnose and treat this cancer. Such information, in addition, should lead to personalized therapeutics according to sequencing results of different gene mutations or chromosomal rearrangements. The urgent need of such studies becomes apparent considering the huge number of new cases of prostate problems reported every year.

In contrast to prostate cancer, cardiovascular disorders take the heavier toll of life. In other words, the risk of death for men in the United States between the ages of 55 and 74 years due to cardiovascular disease surpasses that of prostate cancer. Cardiovascular disease is the most common of the chronic non-communicable diseases that impact global mortality. Approximately, 30 % of all deaths worldwide and 10 % of all healthy life lost to disease are accounted for by cardiovascular disease alone.

In conclusion, initial treatment with standard surgery, irradiation, chemotherapy, or hormonal therapy, or combination of these protocols can result in both local and systemic sequelae. Therefore, surveillance for late recurrence and secondary primary malignancies is recommended for most cancer patients. Patients with breast, lung, prostate, colorectal, and head and neck cancers constitute the largest groups requiring long-term monitoring and follow-up care.

Eric Hayat

References

Amirjamshidi A, Abbassioun K (2000) Radiation-induced tumors of the central nervous system occurring in childhood and adolescence. Child's Nerv Syst 16:390–397

Carroll PR, Whitson JH, Cooperberg MR (2011) Serum prostate-specific antigen for the early detection of prostate cancer; always, never, or only sometimes? J Clin Oncol 29:345–346

Constine LS, Tarbell N, Hudson MM et al (2008) Subsequent malignancies in children treated for Hodgkin's disease; associations with gender and radiation dose. Int J Rad Oncol Biol Physol 72:24–33

Endoh M, Iwasaki Y, Koyanagi I, Hida K, Abe H (1998) Spontaneous shrinkage of lumbosacral lipoma in conjunction with a general decrease in body fat: case report. Neurosurgery 43(1):150–151, discussion 151–152

Filella X, Alcover J, Molina R (2011) Active surveillance in prostate cancer: the need to standardize. Tumor Biol 32:839–843

Guin P, Gilbert E, Jones B (1969) Incidental neuroblastoma in infants. Am J Clin Pathol 51:126–136

Hayat MA (2005) Prostate carcinoma: an introduction. In: Immunohistochemistry and in situ hybridization of human carcinomas, vol 2. Elsevier, San Francisco, pp 279–297

Hayat MA (2008) Prostate carcinoma. In: Methods of cancer diagnosis, therapy, and prognosis, vol 2. Springer, New York, pp 391–396

Henderson TO, Amsterdam A et al (2010) Surveillance for breast cancer in women treated with chest radiation for a childhood, adolescent or young adult cancer: a report from Children's Oncology Group. Ann Intern Med 152:1–22

Hero S, Simon T, Spitz R, Ernestus K, Gnekow A, Scheel-Walter H, Schwabe D, Schilling F, Benz-Bohm G, Berthold F (2008) Localized infant neuroblastomas often show spontaneous regression: results of the prospective trials NB95-S and NB 97. J Clin Oncol 26:1504–1510

Klotz L (2008) Low-risk prostate cancer can and should often be managed with active surveillance and selective delayed intervention. Nat Clin Pract Urol 5:2–3

Lin W-C, Mahadevan-Jansen A, Johnson MD, Weil R, Toms SA (2005) In vivo optical spectroscopy detects radiation damage in brain tissue. Neurosurgery 57:518–525

Lu-Yao GL, Albertsen PC, Moore DF et al (2009) Outcomes of localized prostate cancer following conservative management. JAMA 302:1202–1209

Masters JR (2007) Clinical applications of expression profiling and proteomics in prostate cancer. Anticancer Res 27:1273–1276

Mertens AC, Liu Q, Neglia JP et al (2008) Cause-specific late mortality among 5-year survivors of childhood cancer: the Childhood Cancer Survivor Study. J Natl Cancer Inst 100:1368–1379

Neglia JP, Robison LL, Stovall M, Liu Y, Packer RJ et al (2006) New primary neoplasms of the central nervous system in survivors of childhood cancer: a report from the childhood cancer survivor study. J Natl Cancer Inst 98:1528–1537

Nelson EC, Cambio AJ, Ok JH, Lara PN Jr, Evans CP (2007) Clinical implications of nueroendocrine differentiation in prostate cancer. Prostate Cancer Prostatic Dis 10:6–14

Olumi AF, Grossfeld GD, Hayward SW, Carroll PR, Tlsty TD, Cunha GR (1999) Carcinoma-associated fibroblasts direct tumor progression of initiated human prostatic epithelium. Cancer Res 59:5002–5011

Parent AD (1990) Late complications of radiation-induced neoplasms. Neurosurgery 26:1090–1091

Pernicova Z, Slabakova E, Kharaishvill G, Bouchal J, Kral M, Kunicka Z, Machalam M, Kozubik A, Soucek K (2011) Androgen depletion induces senescence in prostate cancer cells through down-regulation of SKp2. Neoplasia 13:526–536

Pettorini BL, Park Y-S, Caldarelli M, Massimi L, Tamburrini G, DiRocco C (2008) Radiation induced brain tumors after central nervous system irradiation in childhood: a review. Child's Nervous Syst 24:793–805

Rubsam K, Flaig MJ, Ruzicka T, Prinz JC (2011) Erythema marginatum hemorrhagicum: a unique cutaneous side effect of sorafenib. J Am Acad Dermatol 64:1194–1196

Seaton A, Scullin P, Maxwell PJ, Wilson C, Pettigrew J, Gallagher R, O'Sullivan JM, Johnston PG, Waugh DJ (2008) Interleukin-8 signaling promotes androgen-independent proliferation of prostate cancer cells via induction of androgen receptor expression and activation. Carcinogenesis 6:1148–1156

Shalitin S, Gal M, Goshen Y, Cohen I, Yaniv I, Philip M (2011) Endocrine outcome in long-term survivors of childhood brain tumors. Horm Res Paediatr 76:113–122

Sigurdson AJ, Ronckers CM, Mertens AC et al (2005) Primary thyroid cancer after a first tumor in childhood (the childhood cancer survivor study): a nested case–control study. Lancet 365:2014–2023

Ursino MG, Poluzzi E, Caramella C, DePonti F (2011) Excipients in medicinal products used in gastroenterology as a possible cause of side effects. Regul Toxicol 60:93–105

Preface

Little is known regarding the factors that regulate entry of residual cancer into a dormant state or the subsequent reinitation of growth. The prognostic factors present in the primary tumor are imprecise in predicting which patients will be cured by local treatment and which patients will have metastatic recurrence.

Although much progress has been made in identifying many of the genetic factors that contribute to cancer development, much remains to be learned about genetic and epigenetic factors that influence both tumor dormancy and the growth of metastasis. A majority of us have in situ tumors that may remain dormant or may progress into a lethal form of cancer; the former are prevented from recruiting their own blood supply.

This is volume 2 of the multivolume series discussing Tumor Dormancy, Quiescence, and Cellular Senescence. The role of tumor dormancy in a number of diseases, including breast cancer, melanoma, prostate cancer, liver cancer, and lung cancer is discussed. It is also pointed out that quiescent state regulates hematopoietic stem cells and muscle stem cells. The mediation of reversible quiescent state in a subset of ovarian, pancreatic, and colon cancers by the kinase is detailed. Molecular mechanisms underlying stress-induced cellular senescence and accumulation of reactive oxygen species and induction of premature senescence are presented. The importance of the role of microRNASE in oxidative stress-induced apoptosis and senescence and the effect of microRNA as a modulator of cell proliferation in lung cancer are detailed. Suppression of cellular senescence in glioblastoma brain tumor is also explained.

By bringing together a large number of experts (oncologists, neurosurgeons, physicians, research scientists, and pathologists) in various aspects of this medical field, it is my hope that substantial progress will be made against a terrible human disease and injury. It is difficult for a single author to discuss effectively the complexity of diagnosis, therapy, including tissue regeneration. Another advantage of involving more than one author is to present different points of view on a specific controversial aspect of cancer cure and tissue regeneration. I hope these goals will be fulfilled in this and other volumes of the series. This volume was written by 70 contributors representing 11 countries. I am grateful to them for their promptness in accepting my suggestions. Their practical experience highlights their writings, which should build and further the endeavors of the readers in these important areas of disease and injury. I respect and appreciate the hard work and exceptional insight into the role of dormancy, quiescence, and cellular senescence in various diseases and stem cell functions provided by these contributors. The contents of the volume are divided into three parts: Dormancy, Quiescence, and Cellular Senescence for the convenience of the readers.

It is my hope that subsequent volumes of the series will join this volume in assisting in the more complete understanding of the major human diseases and their treatments. There exists a tremendous, urgent demand by the public and the scientific community to address to cancer diagnosis, treatment, cure and hopefully prevention. In the light of existing cancer calamity, government funding must give priority to eradicating deadly malignancies over military superiority.

I am thankful to Dr. Dawood Farahi and Mr. Phil Connelly for recognizing the importance of medical research and publishing through an institution of higher education.

Union, NJ, USA
July, 2013

M.A. Hayat

Contents

Part I Molecular Mechanisms

1 **Asymmetric Dimethylarginine Accelerates Cellular Senescence**... 3
Fotunato Scalera and Stefanie M. Bode-Böger

2 **Membrane-Derived Extracellular Vesicles from Endothelial Progenitor Cells Activate Angiogenesis** 17
Vincenzo Cantaluppi, Federico Figliolini, Maria Chiara Deregibus, and Giovanni Camussi

3 **Induction of P21-Dependent Senescence: Role of NAE Inhibitor MLN4924** 27
Yongfu Pan, Yi Sun, and Lijun Jia

4 **Regulation of the Novel Senescence Pathway by SKP2 E3 Ligase** 33
Guocan Wang, Yuan Gao, Li Chen, Ying-Jan Wang, and Hui-Kuan Lin

5 **Oncogene-Induced Senescence: Role of Mitochondrial Dysfunction**...................................... 45
Olga Moiseeva and Gerardo Ferbeyre

6 **Interleukin-6 Induces Premature Senescence Involving Signal Transducer and Activator of Transcription 3 and Insulin-Like Growth Factor-Binding Protein 5** 53
Hirotada Kojima, Hiroyuki Kunimoto, Toshiaki Inoue, and Koichi Nakajima

7 **A Role for the Nuclear Lamina Shape in Cell Senescence and Aging**..................................... 61
Christiaan H. Righolt and Vered Raz

8 **Upregulation of Alpha-2-Macroglobulin in Replicative Senescence** 71
Li Wei Ma, Guo Dong Li, and Tan Jun Tong

9	**Elevation of Ceramide in Senescence: Role of Sphingolipid Metabolism** Mark E. Venable	81
10	**Molecular Signals Underlying Hair Follicle Morphogenesis and Cutaneous Regeneration** Xusheng Wang and Yaojiong Wu	89
11	**Role of Chromatin-Remodeling Factor Jun Dimerization Protein 2 (JDP2) in Cellular Senescence** Kazunari K. Yokoyama and Kung-Kai Kuo	101
12	**Induction of Cellular Senescence: Role of Mitogen-Activated Protein Kinase-Interacting Kinase 1** Samira Ziaei and Naoko Shimada	111
13	**Mechanisms of Premature Cell Senescence** Julien Maizel, Jun Chen, and Michael S. Goligorsky	121

Part II Tumor and Cancer

14	**Nuclear Protein Pirin Negates the Cellular Senescence Barrier Against Cancer Development** Silvia Licciulli and Myriam Alcalay	131
15	**Defects in Chromatin Structure and Diseases** Umberto Galderisi and Gianfranco Peluso	143
16	**The Role of Fibrosis in Tumor Progression and the Dormant to Proliferative Switch** Lara H. El Touny, Dalit Barkan, and Jeffrey E. Green	155
17	**Diagnosis of Branchial Cyst Carcinoma: Role of Stem Cells and Dormancy** Athanassios Kyrgidis	165
18	**Function of the ING Proteins in Cancer and Senescence** Uyen M. Tran, Uma Rajarajacholan, and Karl Riabowol	179
19	**Premalignancy and Cellular Senescence** Hussein A. Abbas and Raya Saab	195
20	**Loss of Cdh1 Triggers Premature Senescence in Part via Activation of Both the RB/E2F1 and the CLASPIN/CHK1/P53 Tumor Suppressor Pathways** Shavali Shaik, Pengda Liu, Zhiwei Wang, and Wenyi Wei	207
21	**Suppression of Premature Senescence and Promotion of Metastatic Transformation: Role of Reduced TGF-Beta Signaling in Human Cancer Progression** Shu Lin and Lu-Zhe Sun	219

22	**Senescence Escape in Melanoma: Role of Spleen Tyrosine Kinase SYK**..	227
	Marcel Deckert and Sophie Tartare-Deckert	
23	**Micrometastatic Cancer Cells: Role of Tumor Dormancy in Non-small Cell Lung Cancer (NSCLC)**...............	239
	Stefan Werner, Michaela Wrage, and Harriet Wikman	
24	**Quiescent CD4⁺ T Cells Inhibit Multiple Stages of HIV Infection**	253
	Jerome A. Zack and Dimitrios N. Vatakis	

Part III Stem Cells and Cancer Stem Cells

25	**Senescent-Derived Pluripotent Stem Cells Are Able to Redifferentiate into Fully Rejuvenated Cells**..........................	265
	Ollivier Milhavet and Jean-Marc Lemaitre	
26	**The Transcription Factor GATA2 Regulates Quiescence in Haematopoietic Stem and Progenitor Cells**	277
	Neil P. Rodrigues and Alex J. Tipping	
27	**Dormancy and Recurrence of Cancer Stem Cells in Bone: Role of Bone Morphogenetic Proteins**................	289
	Sambad Sharma, Fei Xing, and Kounosuke Watabe	
28	**Role of Microenvironment in Regulating Stem Cell and Tumor Initiating Cancer Cell Behavior and Its Potential Therapeutic Implications**...............................	301
	Ana Krtolica	

Index... 313

Contents of Volume 1

Part I Tumor Dormancy

1 Dormancy, Quiescence, and Cellular Senescence

2 Is Tumor Dormancy Clinically Relevant?

3 Microenvironmental Influence on Breast Cancer Dormancy and Metastasis

4 Determination of Breast Cancer Dormancy: Analysis of Circulating Free DNA Using SNP 6.0 Arrays

5 Clonogenicity of Cultured Prostate Cancer Cells Is Controlled by Dormancy: Significance and Comparison with Cell Culture Models of Breast Cancer Cell Dormancy

6 Dormancy and Metastasis of Melanoma Cells to Lymph Nodes, Lung, and Liver

7 Late Recurrence Is a Sign of Melanoma Dormancy: Need of Life-Long Follow-Up of Melanoma Patients

Part II Quiescence

8 Hematopoietic Stem Cell Quiescence and Long Term Maintenance: Role of SCL/TAL1

9 Regulation of Muscle Stem Cell Quiescent and Undifferentiated State: Roles of Hesr1 And Hesr3 Genes

10 The Kinase MIRK/DYRK1B Mediates a Reversible Quiescent State in a Subsetof Ovarian, Pancreatic, and Colon Cancers

Part III Cellular Senescence

11 Stress -Induced Senescence: Molecular Pathways

12 Accumulation of Reactive Oxygen Species and Induction of Premature Senescence: Role of DDB2

13 p21 Mediates Senescence by a Mechanism Involving Accumulation of Reactive Oxygen Species

14 Role of MicroRNAs and ZEB1 Downmodulation in Oxidative Stress-Induced Apoptosis and Senescence

15 Hypoxic Response in Senescent Brain Is Impaired: Possible Contribution to Neurodegeneration

16 Enhancing Reprogramming to Pluripotency by Controlling Senescence

17 Histone Deacetylase Inhibitor Induces Replicative Senescence of Mesenchymal Stem Cells

18 Senescence Arrest of Endopolyploid Cells Renders Senescence Into One Mechanism for Positive Tumorigenesis

19 The Two Faces of Senescence-Associated Epigenetic Alterations: Tumor Suppressors and Oncogenic Drivers

20 Chemotherapy- and Radiation-Induced Accelerated Senescence: Implications for Treatment Response, Tumor Progression, and Cancer Survivorship

21 Suppression of Cellular Senescence in Glioblastoma: Role of Src Homology Domain-Containing Phosphatase 2

22 Chemotherapy of Malignant Pleural Mesothelioma Induces both Senescence and Apoptosis

23 MicroRNA as a Modulator of Cell Proliferation and Senescence: Role in Lung Cancer Cells

24 Role of Senescence Induction in Cancer Therapy

25 Cellular Senescence Limits the Extent of Fibrosis Following Liver Damage

26 Formation of Secretory Senescent Cells in Prostate Tumors: The Role of Androgen Receptor Activity and Cell Cycle Regulation

Contributors

Hussein A. Abbas Children's Cancer Center of Lebanon, American University of Beirut, Beirut, Lebanon

Myriam Alcalay Department of Experimental Oncology, Istituto Europeo di Oncologia, Milan, Italy

Dalit Barkan Department of Biology, Faculty of Sciences, Haifa University, Haifa, Israel

Stefanie M. Bode-Böger Institute of Clinical Pharmacology, University Hospital, Otto-von-Guericke University, Magdeburg, Germany

Giovanni Camussi Molecular Biotechnology Center, Department of Medical Sciences, University of Turin, Turin, Italy

Vincenzo Cantaluppi Molecular Biotechnology Center, Department of Internal Medicine, University of Turin, Turin, Italy

Li Chen Department of Molecular and Cellular Oncology, MD Anderson Cancer Center, University of Texas, Houston, TX, USA

Jun Chen Department of Medicine, Renal Research Institute, New York Medical College, Valhalla, NY, USA

Marcel Deckert C3M, Team Microenvironment, Signaling and Cancer, INSERM, Nice, France

Maria Chiara Deregibus Molecular Biotechnology Center, Department of Medical Sciences, University of Turin, Turin, Italy

Lara H. El Touny Laboratory of Cancer Biology and Genetics, National Cancer Institute, Bethesda, MD, USA

Gerardo Ferbeyre Departmente de Biochimie, Universite de Montreal, Montreal, QC, Canada

Federico Figliolini Molecular Biotechnology Center, Department of Medical Sciences, University of Turin, Turin, Italy

Umberto Galderisi Department of Experimental Medicine, Section of Biotechnology and Molecular Biology, Second University of Naples, Naples, Italy

Yuan Gao Department of Molecular and Cellular Oncology, MD Anderson Cancer Center, University of Texas, Houston, TX, USA

Michael S. Goligorsky Department of Medicine, Pharmaclogy and Physiology, Renal Research Institute, New York Medical College, Valhalla, NY, USA

Jeffrey E. Green Laboratory of Cancer Biology and Genetics, National Cancer Institute, Bethesda, MD, USA

Toshiaki Inoue Division of Human Genome Science, Department of Molecular and Cellular Biology, School of Life Sciences, Faculty of Medicine, Tottori University, Yonago, Japan

Lijun Jia Cancer Institute, Fudan University Shanghai Cancer Center, Shanghai, China

Hirotada Kojima Department of Immunology, Graduate School of Medicine, Osaka City University, Osaka, Japan

Ana Krtolica Life Sciences Division, Lawrence Berkeley National Laboratory, Berkeley, CA, USA

Hiroyuki Kunimoto Department of Immunology, Graduate School of Medicine, Osaka City University, Osaka, Japan

Kung-Kai Kuo Division of Hepatobiliopancreatic Surgery, Department of Surgery, Kaohsiung Medical University Hospital, Kaohsiung, Taiwan

Athanassios Kyrgidis Department of Oral Maxillofacial Surgery, Aristotle University of Thessaloniki, Thessaloniki, Greece

Jen-Marc Lemaitre Laboratory of Plasticity of the Genome and Aging, Institute of Functional Genomics, Montpellier Cedex 05, France

Guo Dong Li Research Center on Aging, Department of Biochemistry and Molecular Biology, Peking University Health Science Center, Beijing, People's Republic of China

Silvia Licciulli Kissil Lab, Department of Cancer Biology, The Scripps Research Institute, Jupiter, FL, USA

Hui-Kuan Lin Department of Molecular and Cellular Oncology, MD Anderson Cancer Center, The University of Texas, Houston, TX, USA

Graduate Institute of Basic Medical Science, China Medical University, Taichung, Taiwan

Department of Biotechnology, Asia University, Taichung, Taiwan

Shu Lin Department of Cellular and Structural Biology, University of Texas Health Science Center, San Antonio, TX, USA

Pengda Liu Department of Pathology, Beth Israel Deaconess Medical Center, Harvard Medical School, Boston, MA, USA

Li Wei Ma Research Center on Aging, Department of Biochemistry and Molecular Biology, Peking University Health Science Center, Beijing, People's Republic of China

Julien Maizel Department of Medicine, Renal Research Institute, New York Medical College, Valhalla, NY, USA

Ollivier Milhavet Laboratory of Plasticity of the Genome and Aging, Institute of Functional Genomics, Montpellier Cedex 05, France

Olga Moiseeva Departmente de Biochimie, Universite de Montreal, Montreal, QC, Canada

Koichi Nakajima Department of Immunology, Graduate School of Medicine, Osaka City University, Osaka, Japan

Yongfu Pan Cancer Institute, Fudan University Shanghai Cancer Center, Shanghai, China

Gianfranco Peluso Institute of Biochemistry of Proteins and Enzymology, C.N.R., Naples, Italy

Uma Rajarajacholan Department of Biochemistry and Molecular Biology and Oncology, University of Calgary, Calgary, AB, Canada

Vered Raz Department of Human Genetics, Leiden University Medical Center, Leiden, The Netherlands

Karl Riabowol Department of Biochemistry and Molecular Biology and Oncology, University of Calgary, Calgary, AB, Canada

Christiaan H. Righolt Department of Imaging Science and Technology, TU Delft, Delft, The Netherlands

Manitoba Institute of Cell Biology, University of Manitoba, Winnipeg, Canada

Neil P. Rodrigues National Institutes of Health, Center for Biomedical research Excellence in Stem Cell Biology, Roger Williams Medical Center, Boston University of Medicine, Providence, RI, USA

Department of Dermatology, Boston University School of Medicine, Boston, MA, USA

Center for Regenerative Medicine, Boston University School of Medicine, Boston, MA, USA

Raya Saab Children's Cancer Center of Lebanon, American University of Beirut, Beirut, Lebanon

Fotunato Scalera Institute of Clinical Pharmacology, University Hospital, Otto-von-Guericke University, Magdeburg, Germany

Shavali Shaik Department of Pathology, Beth Israel Deaconess Medical Center, Harvard Medical School, Boston, MA, USA

Sambad Sharma Cancer Institute, University of Mississippi Medical Center, Jackson, MS, USA

Naoko Shimada Department of Biology, The City College of New York, New York, NY, USA

Lu-Zhe Sun Department of Cellular and Structural Biology, University of Texas Health Science Center, San Antonio, TX, USA

Yi Sun Division of Radiation and Cancer Biology, Department of Radiation Oncology, University of Michigan, Ann Arbor, MN, USA

Sophie Tartare-Deckert C3M, Team Microenvironment, Signaling and Cancer, INSERM, Nice, France

Alex J. Tipping University College London (UCL) Cancer Institute, London, UK

Tan Jun Tong Research Center on Aging, Department of Biochemistry and Molecular Biology, Peking University Health Science Center, Beijing, People's Republic of China

Uyen M. Tran Department of Biochemistry and Molecular Biology and Oncology, University of Calgary, Calgary, AB, Canada

Dimitrios N. Vatakis Division of Hematology-Oncology, Department of Medicine, David Geffen School of Medicine at UCLA, Los Angeles, CA, USA

Mark E. Venable Biology Department, Appalachian State University, Boone, NC, USA

Guocan Wang Department of Genomic Medicine, MD Anderson Cancer Center, University of Texas, Houston, TX, USA

Xusheng Wang Life Science Division, Tsinghua University Graduate School at Shenzhun, Shenzhen, China

Ying-Jan Wang Department of Molecular and Cellular Oncology, MD Anderson Cancer Center, The University of Texas, Houston, TX, USA

Department of Environmental and Occupational Health, National Cheng Kung University, Medical College, Tainan, Taiwan

Zhiwei Wang Department of Pathology, Beth Israel Deaconess Medical Center, Harvard Medical School, Boston, MA, USA

Kounosuke Watabe Cancer Institute, University of Mississippi Medical Center, Jackson, MS, USA

Wenyi Wei Department of Pathology, Beth Israel Deaconess Medical Center, Harvard Medical School, Boston, MA, USA

Stefan Werner Center of Experimental Medicine, Department of Tumor Biology, University Medical Center Hamburg-Eppendorf, Hamburg, Germany

Harriet Wikman Center of Experimental Medicine, Department of Tumor Biology, University Medical Center Hamburg-Eppendorf, Hamburg, Germany

Michaela Wrage Center of Experimental Medicine, Department of Tumor Biology, University Medical Center Hamburg-Eppendorf, Hamburg, Germany

Yaojiong Wu Life Science Division, Tsinghua University Graduate School at Shenzhun, Shenzhen, China

Fei Xing Cancer Institute, University of Mississippi Medical Center, Jackson, MS, USA

Kazunari K. Yokoyama Graduate Institute of Medicine, Kaohsiung Medical University, Kaohsiung, Taiwan

Jerome A. Zack Division of Hematology-Oncology, David Geffen School of Medicine at UCLA, Los Angeles, CA, USA

UCLA, AIDS Institute, Los Angeles, CA, USA

Samira Ziaei Department of Biology, The City College of New York, New York, NY, USA

Part I
Molecular Mechanisms

Asymmetric Dimethylarginine Accelerates Cellular Senescence

Fotunato Scalera and Stefanie M. Bode-Böger

Contents

Abstract 3
Introduction 4
Asymmetric Dimethylarginine 5
ADMA in Cardiovascular Diseases 5
ADMA Synthesis 5
ADMA Metabolism 6
ADMA in Cellular Senescence
and in the Elderly 8
**Asymmetric Dimethylarginine Accelerated
Cellular Senescence** 8
ADMA Accelerates Cellular Senescence
by Inhibited Telomerase Activity
and Shortened Telomere Length 8
ADMA-Inhibited NO Synthesis in Senescence
Is Associated with Enhanced ROS Formation
and Downregulated DDAH Activity 10
Discussion 12
References 15

F. Scalera (✉) • S.M. Bode-Böger
Institute of Clinical Pharmacology, University
Hospital, Otto-von-Guericke University, Magdeburg,
Germany
e-mail: fortunato.scalera@med.ovgu.de;
stefanie.bode-boeger@med.ovgu.de

Abstract

Cellular senescence, a physiological state of irreversible growth arrest, might contribute to cardiovascular diseases in the elderly. Cellular senescence can be modulated by several different endogenous and exogenous factors that contribute to delay or accelerate the process of senescence. Nitric oxide (NO), formed from L-arginine by nitric oxide synthase (NOS), has antiatherogenic properties and is one of endogenous factors leading to delayed cellular senescence. Intracellular factors that decrease NO synthesis may therefore represent important targets in the accelerated cellular senescence and consequently in the development of cardiovascular diseases. Long-term treatment of human endothelial cells with asymmetric dimethylarginine (ADMA), an endogenous competitive inhibitor of NOS and regarded as a novel cardiovascular risk factor, accelerates the process of cellular senescence by inhibiting NO synthesis. Additionally, ADMA accelerates the shortening of telomere length in a dose-dependent manner and inhibits the telomerase activity. Our results suggest a new mechanism through which elevated ADMA levels observed in the elderly might accelerate endothelial senescence and consequently might promote atherogenesis.

Keywords

Arginine methylation • Asymmetric dimethylarginine (ADMA) • Cardiovascular diseases

- Cellular senescence • Dimethylarginine dimethylaminohydrolase (DDAH) enzymes
- Human endothelial cells • Metabolism
- Symmetrical dimethylarginine (SDMA)
- Telomerase reverse transcriptase (TERT)

Introduction

Aging is a major independent risk factor for the development of cardiovascular diseases such as atherosclerosis, hypertension and coronary heart disease (Lakatta and Levy 2003). Aging is associated with complex structural and functional alterations in the endothelium. The endothelium is situated at the interface between blood and vascular wall/tissue and is a dynamic, heterogeneous, disseminated organ that has vital synthesis, secretory, metabolic and immunological functions. The functional integrity of the endothelium is therefore essential to prevent vascular leakage and the formation of atherosclerotic lesions. As endothelial cells are a critical component of the endothelium, their senescent status can have a significant impact in integrity and function of endothelium and consequently may promote endothelial dysfunction and the formation of atherosclerotic lesions. This hypothesis is based on recent evidence that strongly suggests a causative link between alterations seen in the structure and function of senescent human endothelial cells in culture and changes seen in age-related vascular diseases (Minamino and Komuro 2007).

The phenomenon of cellular senescence was first described by Hayflick and Moorhead (1961). The authors observed that human somatic cells isolated from tissue and grown in culture ceased to proliferate after a certain number of cell divisions. On the basis of these findings, the authors proposed that normal somatic cells have an intrinsic limited proliferative capacity and that this propriety represented a manifestation of aging at cellular level. Subsequently, Harley et al. (1990) showed that the proliferative arrest results from progressive cell division-dependent shortening of telomeres.

Telomeres are nucleoprotein complexes located at both ends of eukaryotic chromosomes. These complexes consist of a six-base-pair DNA repeat $(TTAGGG)_n$ that extends over a length of several thousand base pairs and a growing list of associated proteins that serve to protect telomeric ends. Telomeres are important structures involved in cell cycle control, and essential for maintaining genome stability and integrity and for an extended proliferative lifespan in both cultured cells and the whole organism. As a consequence of semi-conservative DNA replication, the extreme terminals of chromosomes are not duplicated completely, resulting in successive shortening of telomere with each cell division. When telomeres shorten below a critical length, eroded telomeres generate a persistent DNA damage response, which initiates and maintains the senescence growth arrest by activating signaling pathway of tumor suppressor protein p53 (Deng et al. 2008).

The shortening of telomeres is prevented by the enzyme telomerase. The telomerase is a large ribonucleoprotein complex and consists of two core compounds, the telomerase reverse transcriptase (TERT), which contains the catalytic activity of the enzyme, and the telomerase RNA component (TERC), which serves as the template for the synthesis of the new telomeric sequence. Telomerase adds telomeric repeats on the ends of chromosomes, thus maintaining their length despite cell division. However, most normal somatic cells do not express sufficient telomerase to maintain telomere length indefinitely, and therefore undergo telomere attrition with each mitotic cycle, both in cell culture as a function of population doublings (PDs) and during aging of the whole organism (Harley et al. 1990; Collins and Mitchell 2002).

Human endothelial cells are regarded as senescent after 30–40 PDs under normal culture conditions. In senescent status, endothelial cells are viable and metabolically active, but fail to grow despite optimal cell culture conditions. This condition is associated with a reduction of telomerase activity and a shortening of telomere length. Senescent cells show also change in gene expression, function and morphology such as an increase in size, polymorphic nuclei, flattening and vacuolisation compared to proliferating cells (Vasa et al. 2000; Minamino and Komuro 2007).

In addition, senescent cells express senescence-associated β-galactosidase (SA β-gal), the most widely used biomarker for determining cellular senescence originally discovered by Dimri et al. (1995). SA β–gal activity is absent in the majority of endothelial cells in early passage cultures, but present in a large proportion of cells in late passage cultures (>30 PDs). Moreover, one of the most important changes of senescent endothelial cells that may promote endothelial dysfunction and consequently cardiovascular diseases is the reduction of nitric oxide (NO) bioavailability (Minamino and Komuro 2007).

NO is a short-lived free radical gas molecule and responsible for the integrity and function of endothelium. NO exerts vasculoprotective and cardioprotective effects, such as maintenance of normal coronary blood flow, inhibition of platelet aggregation and inflammatory cell adhesion to endothelial cells, and disruption of proinflammatory cytokine-induced signaling pathways. NO is synthesized by NO synthase (NOS) enzyme, using L-Arginine as substrate. There are three NOS isoenzymes described: endothelial (eNOS), neuronal (nNOS) and inducible (iNOS) NOS.

In senescent cells, a reduced NO synthesis may be associated with an increase of asymmetrical dimethylarginine (ADMA), an endogenous competitive inhibitor of all NOS isoforms and regarded as a risk factor for cardiovascular diseases (Cooke 2004).

Asymmetric Dimethylarginine

ADMA in Cardiovascular Diseases

ADMA is an endogenous methylated form of the amino acid L-arginine that exerts its biological activity by competing with L-arginine for binding to the active site of NOS enzymes. This was first reported by Vallance et al. (1992) showing that this endogenous compound inhibits the synthesis and signaling pathway of NO *in vivo* and *in vitro*, and that the addition of L-arginine attenuates the inhibitory effect of ADMA on NO synthesis. At that time, Vallance et al. (1992) postulated that ADMA by impairing NO bioavailability might contribute to endothelial dysfunction and cardiovascular disease. On the basis of this hypothesis, increased plasma concentrations of ADMA have been observed in a large number of diseases that are correlated with an increased risk of endothelial dysfunction and atherosclerosis. Elevations in plasma ADMA have a high prevalence in hypertension, renal failure, hypercholesterolemia, diabetes mellitus, peripheral arterial disease, coronary artery disease, hyperhomocysteinemia, and chronic heart failure (Bode-Böger et al. 2007). More strong and convincing evidence of the role of ADMA in the development of cardiovascular diseases comes from prospective clinical studies where higher ADMA levels are a prognostic marker for major cardiovascular events and mortality in populations at high, intermediate or low global vascular risk (Böger et al. 2009). Moreover, in healthy human subjects intravenous infusion of ADMA caused an increase in systemic vascular resistance and mean arterial pressure and reduced heart rate and cardiac output (Kielstein et al. 2004).

In human physiological plasma levels of ADMA range between 0.4 and 0.6 μmol/L but can increase two to threefold in cardiovascular diseases (Bode-Böger et al. 2007). The underlying causes for the elevated plasma ADMA levels seen in cardiovascular diseases might be due to increased synthesis and/or impaired metabolism.

ADMA Synthesis

Arginine methylation of mainly eukaryotic nuclear proteins is an important posttranslational modification that has physiological roles in a number of different cellular processes including RNA binding, protein-protein interaction, signal transduction, transcriptional regulation and DNA repair. It occurs shortly after protein synthesis by protein arginine methyltransferases (PRMTs), a family of enzymes that catalyse the transfer of either one or two methylgroups from S-adenosyl-L-methionine (SAM) to a guanidino nitrogen atom of arginine side chains and S-adenosyl-L-homocysteine (SAH) is a reaction by-product. This post-translational modification and subsequent proteolysis of the methylated proteins releases free methylarginines into the citosol (Nicholson et al. 2009).

Besides ADMA, other forms of methylarginines have been formed in eukaryotes: N-monomethyl-L-arginine (L-NMMA) and symmetrical dimethylarginine (SDMA). L-NMMA is also an endogenous inhibitor of all three NOS isoforms. Because the human plasma levels of ADMA are tenfold greater than those of L-NMMA, ADMA may be regarded as the major endogenous NOS inhibitor and most studies to date have focused on ADMA. SDMA does not affect NOS activity (Vallance et al. 1992). However, we have recently shown that SDMA is able to influence NO synthesis by NOS independent mechanisms since it competes with L-arginine for cellular transport across cationic amino acid transporters (CAT) of system y^+, and thus reduces NO formation by decreasing L-arginine uptake in cultured human endothelial cells (Bode-Böger et al. 2006).

In mammalian cells, PRMTs has been classified into type I (PRMT-1, -3, -4, -6 and -8) and type II (PRMT-5, -7 and -9) enzymes. Both types of PRMT catalyze the formation of L-NMMA as an intermediate. In a second step, type I PRMTs produces ADMA, whereas type II PRMTs catalyzes SDMA. PRMT-1 is the predominant mammalian type I enzyme and it accounts for approximately 85 % of arginine methylation reaction (Nicholson et al. 2009). Therefore, it is also plausible that the modulation of activity and/or expression of PRMT-1 could be a mechanism for altering endogenous ADMA levels. Indeed, it has been demonstrated that the expression of PRMT-1 is increased in coronary heart diseases and in decompensated alcoholic cirrohosis resulting in increased ADMA levels (Nicholson et al. 2009). Moreover, the activity and expression of PRMT-1 are upregulated in response to oxidized lipoprotein, shear stress, lysophosphatidycholine, and angiotensin II, leading to a corresponding increase in ADMA level in human endothelial cells. The activation of PRMT-1 is blocked by antioxidant pyrrollidine dithiocarbamate (PDTC) or diphenyliodonium, a NADPH oxidase inhibitor, suggesting that redox-regulated mechanisms may underlie this effect (Scalera and Bode-Böger 2010). Recently, we have shown that red wines from different growing areas and made from different grapes downregulate PRMT-1 expression by activating sirtuin (SIRT)1 signaling associated with a significant decrease in the ADMA level in human endothelial cells (Scalera et al. 2009b). SIRT1, a NAD+-dependent class III histone deacetylase, is a regulator of proteins and genes involved in the antioxidant response, anti-inflammatory response, anti-apoptotic response, insulin response and gene transcription and has also been shown to play an important role in the cardiovascular system, cell survival and cellular senescence.

ADMA Metabolism

In healthy subjects, total ADMA production can be estimated at 300 µmol per day. A minor part of this amount is eliminated unchanged by renal excretion, but more than 80 % is actively metabolized to L-citrulline and dimethylamine by dimethylarginine dimethylaminohydrolase (DDAH) enzymes (Cooke 2004).

In mammals, there are two distinct isoforms of DDAH (DDAH1 and DDAH2). Both enzymes contain 285 amino acids showing 62 % amino acid identity but are encoded by different genes. The DDAH1 gene maps to chromosome 1p22, whereas DDAH2 gene maps to chromosome 6p21.3 (Palm et al. 2007). Interestingly, the two isoforms have distinct tissue distributions: DDAH1 is mainly expressed in brain, liver and kidney, whereas DDAH2 is most highly expressed in the vascular endothelium, heart, placenta and immune tissues. Thus, there is a correlation in tissue distribution of NOS isoforms and DDAH enzymes. DDAH1 colocalizes with nNOS, while DDAH2 colocalizes with eNOS and iNOS. However, recent studies have shown that both DDAH enzymes are expressed in endothelium but the relative contribution of each DDAH isoform to endothelial ADMA metabolism remains to be determined (Leiper and Nandi 2011).

The colocalization of DDAH enzymes with NOS isoforms supports the hypothesis that

DDAH enzymes by regulating intracellular ADMA levels might provide a mechanism for controlling the local availability of NO and downstream vascular endothelium-dependent responses. The first evidence for this hypothesis came from observations that pharmacological inhibition of DDAH activity leads to increased ADMA concentrations and a reduction in NO-mediated vasodilation in isolated vascular segment, indicating that DDAH activity controls endogenous ADMA concentrations and thus NOS activity. These observations were also demonstrated *in vivo* by using genetic manipulation of DDAH enzymes including viral transfection, knockout, overexpression and small interfering (si)RNAs in mice and rodents (Leiper and Nandi 2011). Knockout or siRNA interference of DDAH1 increases ADMA levels, impairs NO-mediated vascular signaling and leads to a cardiovascular phenotype, i.e., hypertension. In contrast, the overexpression of DDAH1 in transgenic mice reduces circulating ADMA levels and increases NO bioavailability and the mice have a decreased systolic blood pressure. Mice that overexpress DDAH2 similarly display a reduction of ADMA levels and an increase in NO synthesis but no differences in systolic blood pressure under basal conditions. However, mice that overexpress DDAH2 are protected against the hypertensive effects of exogenous ADMA administration.

There is increasing evidence that several different compounds modulate the DDAH activity via transcriptional and/or post-translational mechanisms. The promoter region of DDAH genes contains putative consensus binding sites for multiple transcription factors including signal transducer and activator of transcription, interferon regulatory factor-1, retinoid X receptor-α, peroxisome proliferator-activated receptor (PPAR)-γ, and farnesoid X receptor. Retinoic acid, estrogen, IL-1β, β1-adrenergic receptor antagonist nebivolol and PPAR-γ ligands enhance DDAH expression, whereas DDAH expression is downregulated by tumor necrosis factor-α, oxidized low-density lipoprotein, lipopolysaccharide, high concentration of glucose, coupling factor 6, nicotine and cigarette smoke extract (Palm et al. 2007; Wadham and Mangoni 2009).

The activity of DDAH seems to be exquisitely sensitive to oxidative as well as nitrosative stress because of a critical sulfhydryl group in the active site of the enzyme that contains a cysteine-histidine-glutamate catalytic triad. Inhibition of DDAH by oxidation of the active site cysteine might be considered as a common pathological mechanism by which cardiovascular risk factors cause ADMA accumulation leading to endothelial dysfunction and atherosclerotic events (Palm et al. 2007). Therefore, it has been speculated that pharmacological interventions to reduce oxidative stress could be a useful therapeutic option. Drugs used in the therapy of cardiovascular diseases such as nebivolol, telmisartan and fenofibrate have been reported to enhance DDAH activity in vitro probably by decreasing oxidative stress (Wadham and Mangoni 2009), but each drug has shown contradictory results such as no change or reduce of ADMA levels in clinical studies, although the conflicting results of each drug on ADMA levels could be attributed to the different dose, duration and/or study group used in the studies.

Although increased activity of DDAH is usually desired in cardiovascular diseases, it may have disadvantageous effects when occurring in tumors. Indeed, constitutive overexpression of DDAH1 in C6 rat glioma xenografts resulted in increased tumor growth, increased tumor vascularization and elevated vascular endothelial growth factor (VEGF) secretion. The underlying mechanism is that the increased DDAH activity enhances NO production contributing to tumor angiogenesis by upregulating VEGF-induced neovascularization (Kostourou et al. 2002). Conversely, inhibition of DDAH activity could be a protective mechanism to decrease NO levels in diseases in which overproduction of NO contributes to disease pathology (Leiper and Nandi 2011).

Another minor catabolic pathway that metabolizes ADMA to α–keto-δ-dimethylguanidino valeric acid (DMGV) is mediated by the enzyme alanine-glyoxylate aminotransferase-2 (AGXT2) which is expressed only in kidney and liver. AGXT2 is a 514-amino-acid pyridoxal-phosphate-dependent mitochondrial aminotransferase and the 50-kb gene is located on chromosome 5p13 (Leiper and Nandi 2011). A recent study has demonstrated that overexpression of human AGXT2 decreases ADMA and increases NO production in cultured cells and in mice (Rodionov et al. 2010). Our group has recently developed a liquid chromatography–tandem mass spectrometry method for the quantification of DMGV (Martens-Lobenhoffer et al. 2011).

ADMA in Cellular Senescence and in the Elderly

We investigated the effect of senescence on endogenous level of ADMA by cultured human endothelial cells until terminal growth arrest associated with a senescent phenotype. We found a less than threefold increase in extracellular level of endogenous ADMA accompanied by a twofold increase in the protein expression of ADMA synthesizing enzyme PRMT-1 and a 50 % decrease in the activity and protein expression of ADMA metabolizing enzyme DDAH in senescent cells compared to young cells (Scalera et al. 2009b).

The transport of ADMA is mediated by leucine-insensitive transport system y+, represented by CAT-1 and CAT-2, which moves ADMA between the cytosol and extracellular fluid. Our analysis of ADMA transport system showed a fourfold increase in the mRNA expression of CAT-2B in senescent cells compared to young cells, whereas the expression of CAT-1 remained unaltered (Scalera et al. 2009a).

Our findings are in accordance with human studies showing that plasma levels of ADMA augment with age in healthy individuals. Miyazaki et al. (1999) enrolled 116 asymptomatic subjects (100 male and 16 female; age, 52 ± 1 years) who had no clinical evidence of coronary or peripheral arterial diseases or renal dysfunction and were not taking any medication. The authors demonstrated that plasma levels of ADMA were positively correlated with age. Moreover, Kielstein et al. (2003) showed a more than twofold increase in plasma ADMA concentrations of 24 elderly healthy subjects (13 male and 11 female; age, 69 ± 1 years) compared to 24 young healthy subjects (13 male and 11 female; age, 25 ± 1 years), and this increase was more accentuated in 24 elderly patients with mild to moderate essential hypertension (13 male and 11 female; age, 70 ± 1 years).

In summary, our results suggest that in cellular senescence the increased extracellular levels of endogenous ADMA might be due to enhanced synthesis, decreased metabolism and increased efflux. The relative importance of each component is not yet known, but the aforementioned mechanism might explain the augmented plasma levels of ADMA observed in healthy individuals with age.

Asymmetric Dimethylarginine Accelerated Cellular Senescence

The idea for investigating the effect of exogenous ADMA on cellular senescence came from the findings that the repeated addition of the NO donor S-nitroso-penicillamine prevented age-related downregulation of telomerase activity and thereby inhibiting telomere shortening. Conversely, the repeated addition of the endogenous inhibitor of NOS L-NMMA induces a twofold increase in endothelial senescence by reducing telomerase activity (Vasa et al. 2000). These findings suggest that the synthesis and signaling pathway of NO are associated with the process of cellular senescence.

ADMA Accelerates Cellular Senescence by Inhibited Telomerase Activity and Shortened Telomere Length

In this study, we investigated the link between cellular senescence and endogenous NOS inhibitor ADMA. Human umbilical vein endothelial cells (HUVEC) were cultured in endothelial basal medium (Cell System/Clonetics) supplemented with hydrocortisone (0.5 mg/mL), gentamicin (30 μg/mL), amphotericin B (15 μg/mL), human endothelial growth factor (10 μg/mL), human

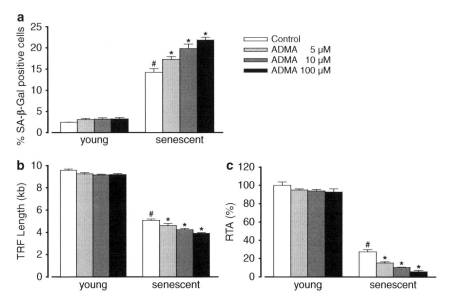

Fig. 1.1 ADMA accelerated the process of cellular senescence (**a**) by increasing the percentage of senescence associated β-galactosidase (SA-β-gal) positive cells, (**b**) by shortening terminal restriction fragment (TRF), und (**c**) by inhibiting the relative telomerase activity (RTA) in a dose-dependent manner. Human endothelial cells were incubated with ADMA (5, 50 and 100 μM) starting from the fourth passage (young) and replaced every 48 h until the tenth passage (senescent). Each point represents the mean ± SEM of results from three experiments. *$P<0.05$ vs. senescent control, #$P<0.01$ vs. young control

fetal growth factor-B (1 μg/mL), VEGF (2 μg/mL), ascorbic acid (75 mg/mL), R^3-IGF-1 (5 μg/mL), heparin (1 mg/mL), and 2 % fetal calf serum until growth arrest (tenth passage). The different concentrations of ADMA (5, 50 and 100 μmol/L, Sigma) were replaced every 48 h starting at the fourth passage. After reaching confluence (between 8 and 9 days), endothelial cells were trypsinized and seeded at a density of 2,500 cells/cm^2 per 75-cm^2 flasks. The endothelial cells were counted in a Neubauer cell chamber (Assistant) and the vitality was determined by means of staining with trypan blue (0.5 %; Sigma) in physiological saline. Viability after trypsinization was usually >95 %.

The most widely used biomarker associated with the senescent phenotype is the SA β-gal activity. Dimri et al. (1995) demonstrated that human fibroblasts undergoing cellular senescence *in vitro* express increased activity of SA β-gal. When applied to human dermal tissue from donors with an age range of 20–90 years, there was a positive correlation between the number of SA β-gal positive cells and the age of donor tissue. Our results showed that senescent cells had a sixfold increase in the percentage of SA β-gal positive cells versus young cells. A long-term treatment with 5, 50 and 100 μmol/L ADMA increased the percentage of SA β-gal positive cells by 21, 39, and 54 %, respectively, compared to control of senescent cells (Fig. 1.1a).

Telomeres are the final region of chromosomes and their main function is to protect the rest of chromosome from degradation and oxidation. The shortening of the length of telomeres is considered one of the most important mechanisms that may lead to the endothelial senescence. In order to examine whether exogenous ADMA-accelerated senescence is associated with the shortening of telomere length, the genomic DNA was isolated, separated and telomere restriction fragment length was measured according to the manufacture's protocol of Telo TAGGG Telomere Length Assay kit (Roche). Our analysis showed that the telomere length was shortened by 50 % in senescent cells compared to

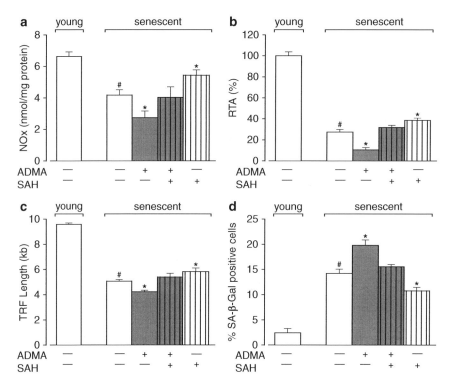

Fig. 1.2 SAH reversed the effect of ADMA on NO synthesis and cellular senescence. Human endothelial cells were incubated with SAH (100 μM) and ADMA (50 μmol/L) starting from the fourth passage (young) and replaced every 48 h until the tenth passage (senescent). (**a**) NO synthesis was measured as nitrate and nitrite metabolites (NOx). (**b**) Relative telomerase activity (RTA). (**c**) Terminal restriction fragment (TRF). (**d**) Percentage of SA-β-gal positive cells. Each point represents the mean ± SEM of results from three experiments. *$P<0.05$ vs. senescent control, #$P<0.01$ vs. young control

young cells, and this shortening was further increased in a dose-dependent manner after the repeated addition of different concentrations of ADMA (Fig. 1.1b).

Because the synthesis of telomeres is associated with enzymatic activity of an RNA protein complex called telomerase, we measured the relative activity of telomerase in HUVEC according to the manufacture's protocol of the TeloTAGGG telomerase polymerase chain reaction ELISA[plus] Kit (Roche). For telomerase activity, 1.5 μg protein was used by polymerase chain reaction. Telomerase activity decreased by 70 % in senescent cells compared to young cells and was further reduced in a dose-dependent manner in the ADMA-treated cells (Fig. 1.1c).

ADMA-Inhibited NO Synthesis in Senescence Is Associated with Enhanced ROS Formation and Downregulated DDAH Activity

In order to provide insight into cellular mechanism of ADMA-accelerated cellular senescence, we determined the NO synthesis as nitrite and nitrate metabolites (NOx) in conditioned media by gas chromatography-mass spectrometry (Scalera et al. 2004). Our analysis showed that the level of NOx decreased by 37 % in senescent cells compared to young cells and reached a 58 % reduction after treatment with exogenous ADMA (Fig. 1.2a).

SAH is an endogenous inhibitor of the ADMA synthesizing enzyme PRMT-1. Böger et al. (2000)

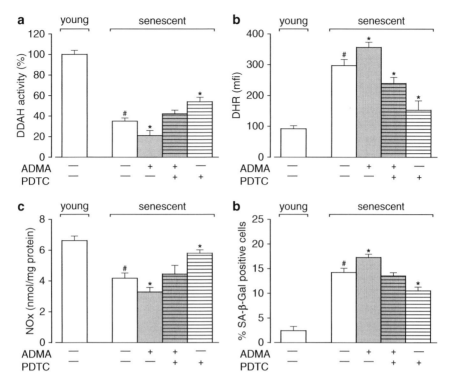

Fig. 1.3 PDTC reversed the effect of ADMA on DDAH activity, ROS formation, NO synthesis and senescence. Human endothelial cells were incubated with PDTC (10 µM) and ADMA (5 µmol/L) starting from the fourth passage (young) and replaced every 48 h until the tenth passage (senescent). (**a**) DDAH activity as percentage of young control. (**b**) Dihydrorhodamin 123 (DHR) as mean fluorescence intensities (mfi). (**c**) Nitrate and nitrite metabolites (NOx). (**d**) Percentage of SA-β-gal positive cells. Each point represents the mean ± SEM of results from three experiments. *$P<0.05$ vs. senescent control, #$P<0.01$ vs. young control

showed that SAH reduced the baseline formation of ADMA by about 40 % in endothelial cells. Therefore, we investigated the effect of SAH on NOx synthesis and senescence by incubation of human endothelial cells with SAH (100 µmol/L) starting from young cells until growth arrest. SAH enhanced NOx levels by 39 % and increased the telomerase activity by 41 % compared with control of senescent cells (Fig. 1.2a, b, respectively). Additionally, the shortening of telomere length was significantly delayed and the percentage of SA β-gal positive cells was significantly decreased (Fig. 1.2c, d, respectively). Interestingly, SAH was able also to reverse the effects of high concentrations of exogenous ADMA on NOx synthesis and senescence (Fig. 1.2).

ADMA is metabolized by DDAH enzymes. To examine whether the activity of DDAH enzymes is impaired in senescence, DDAH activity was determined by assessing the rate of degradation of exogenous ADMA added to the cell lysates (Scalera et al. 2004). We found that DDAH activity decreased 65 % in senescent cells compared to young cells. Surprisingly, ADMA led to further inhibition of DDAH activity (Fig. 1.3a).

Stühlinger et al. (2001) demonstrated that the impairment of DDAH activity caused by incubation of endothelial cells with DL-homocysteine or L-methionine was inversed by the antioxidant PDTC. To test whether PDTC could reversed the impaired DDAH activity in cellular senescence, HUVEC were incubated with PDTC (10 µmol/L) starting of starting from the fourth passage until senescence. We showed that PDTC enhanced DDAH activity by 54 % compared to control of senescent cells and prevented the impairment of DDAH activity by ADMA (Fig. 1.3a).

DDAH activity seems to be particularly susceptible to inhibition by oxidative stress. Because Haendeler et al. (2004) reported a twofold increase in the formation of reactive oxygen species (ROS) in senescent cells compared to young cells, we hypothesized that the impaired DDAH activity is due to increased oxidative stress. Accordingly, we measured the formation of endogenous ROS by dihydrorhodamin 123 (DHR). Endothelial cells were incubated for 20 min at 37 °C with 10 μmol/L DHR. The reaction was stopped by cooling on ice for 1 min and subsequent addition of 500 μl phosphate-buffered saline followed by two washing steps. After final fixation with 1 % paraformaldehyde, the cells were analyzed by flow cytometry (Epics XL-MCL; Coulter). We found that the endogenous levels of ROS were increased about threefold in senescent cells compared to young cells (Fig. 1.3b). ADMA enhanced ROS formation by 20 % compared to control of senescent cells. The antioxidant PDTC reversed the effect of ADMA on endogenous oxidative stress not only, but also led to a 20 % decrease compared to control of senescent cells, whereas PDTC alone reached a 50 % decrease in ROS formation (Fig. 1.3b). Moreover, PDTC-enhanced DDAH activity was associated with a 40 % increase in NO synthesis (Fig. 1.3c) and 26 % decrease in SA β-gal positive cells (Fig. 1.3d) compared to control of senescent cells. In addition, the effect of ADMA on NOx synthesis and SA β-gal positive cells was completely abolished by PDTC (Fig. 1.3c, d, respectively).

Discussion

The results of the present study show that ADMA accelerates the process of cellular senescence by downregulating telomerase activity and increasing the rate of telomere shortening. This effect is associated with a reduction in NO synthesis and an enhancement in ROS formation leading to an impairment of the activity of DDAH, the enzyme that degrades ADMA.

Aging is associated with endothelial dysfunction and an increasing prevalence of cardiovascular diseases. A number of studies have shown the presence of senescent cells in human atherosclerotic plaques obtained from coronary arteries of patients with ischemic heart diseases. The senescent cells are predominately localized on the luminal surface on the plaque and have been identified as endothelial cells, whereas no senescent cells are observed in the internal mammary arteries of the same patients in whom atherosclerotic changes are minimal. Recent studies suggest that the senescence phenotype of endothelial cells can be transformed from healthy to pro-atherosclerotic, implicating a critical role of endothelial cell senescence in the initiation and/or progression of atherosclerosis. Thus, endothelial cell senescence promotes endothelial dysfunction and may contribute to the pathogenesis of age-associated vascular disorders (Minamino and Komuro 2007).

One of the most accepted mechanisms that may contribute to the phenomenon of cellular senescence is the shortening of the telomere length. Several human studies have shown the importance of telomere length in the pathology of age-related cardiovascular diseases. Particularly persuasive in this regard are studies involving individuals with an age-corrected shortage of leukocyte telomeres who have a significant higher risk of developing coronary heart diseases or dying from cardiovascular diseases (Fuster and Andrès 2006). Likewise, the length of telomeres in the endothelium of the abdominal aorta and iliac arteries is inversely correlated with age. Telomere shortening is also more advanced in coronary artery endothelial cells from patients with coronary heart diseases compared with cells from healthy subjects (Minamino and Komuro 2007).

The shortening of telomere length is prevented by the enzyme telomerase. Mouse models lacking telomerase show progressive telomere shortening with each generation, loss of viability when they reach critically short telomeres and are prone to similar cardiovascular disorders to those associated with normal aging in human, such as heart dysfunction and hypertension. In contrast, transgenic mice engineered to express TERT specifically in cardiomyocytes maintain telomerase activity in the adult heart and this correlated with a reduced area of myocardial infarction, a reduced fibrous area, and preservation of systolic function after mechanical and ischemic injury. On the other hand, cardiac-specific TERT expression induces late-onset cardiac hypertrophy (Fuster and Andrès 2006). Further, it could be shown that the introduction of TERT into human endothelial cells can extend their life span and preserve a younger phenotype, underlying the important role of telomerase and telomere stabilization for longevity and functional activity of endothelial cells (Minamino and Komuro 2007).

Emerging data have indicated a possible involvement of NO in the regulation of telomerase activity. Increasing NO bioavailability or eNOS activity delays the process of cellular senescence by activating TERT activity. In contrast, the inhibition of NOS activity accelerates cellular senescence by reduced telomerase activity (Minamino and Komuro 2007).

Results from our study are consistent with these findings. We show that exogenously applied ADMA-inhibited NO synthesis downregulates telomerase activity and accelerates the shortening of telomere length and the onset of cellular senescence. Endothelial cells are capable of synthesizing ADMA by PRMT-1. A major source of methyl groups used for various methylating reactions is SAM, and SAH inhibits the methyltransferase-catalyzed reactions leading to decreased endogenous levels of ADMA (Böger et al. 2000). In our study, SAH-increased NO synthesis delays the process of endothelial senescence by enhanced telomerase activity. We have recently shown that incubation of senescent endothelial cells with SAH for 48 h reduced ADMA concentration by about 40 % compared to control cells (Scalera and Bode-Böger 2010). Therefore, it is tempting to speculate that the effect of SAH on NO synthesis and senescence may be related to the reduced levels of endogenous ADMA.

Continuous cultivation of endothelial cells until senescence results in an increased ROS formation. The increased oxidative stress activates the tyrosine kinase Scr that phosphorylates TERT protein and promotes the transition of the enzyme from nucleus into cytoplasm. This mechanism leads to accelerate the onset of cellular senescence. Incubation with the antioxidant N-acetylcysteine inhibits ROS formation, blocks nuclear transport of TERT and, subsequently, delays the onset of cellular senescence (Haendeler et al. 2004).

In the experiments presented here, we confirm the increase in ROS formation and show a further increase when endothelial cells were incubated with exogenous ADMA. The antioxidant PDTC delays the onset of cellular senescence and prevents the effect of ADMA on senescence by inhibited ROS formation.

A proposed molecular mechanism for ADMA-increased ROS formation is due to uncoupled eNOS that produces large amounts of superoxide rather than NO. Thus, ADMA increase the superoxide formation from uncoupled eNOS by making the transfer of electrons to the heme more favourable via mechanisms involving the modulation of the heme spin-date, altering the electrostatic environment of the heme and/or by altering the structural stability of the active dimer (Scalera and Bode-Böger 2010).

The activity of the enzyme DDAH seems to be particular susceptible to inhibition of oxidative stress because of the presence of sulfhydryl group in the catalytic region of the active site of the enzyme. A wide range of pathological stimuli induce endothelial oxidative stress and consequently reduce DDAH activity in vitro and in vivo (Palm et al. 2007). These observations suggest a role for oxidative stress as a mechanism to control and limit enzymatic activity of DDAH. Results from our study are consistent with these observations. We found that the increased ROS formation is accompanied by a downregulation of DDAH activity in endothelial senescence. Incubation with the antioxidant PDTC reduces ROS formation and consequently upregulates DDAH activity. Moreover, ADMA-increased ROS formation impairs DDAH activity that is prevented by co-incubation with PDTC.

Several studies have demonstrated the central role of DDAH activity in the regulation of ADMA levels and consequently NO synthesis by using DDAH-specific inhibitors and genetic manipulation of DDAH enzymes including viral transfection, knockout, overexpression and siRNAs (Leiper and Nandi 2011). Loss of DDAH activity results in enhanced ADMA level and thereby in reduced NO signaling causing endothelial dysfunction. In contrast, upregulation of DDAH activity reduces ADMA levels and increases NO bioavailability leading to enhanced endothelial functions. Consistent with these findings, we show that PDTC-enhanced DDAH activity increases NO synthesis. Take together, the present data are consistent with the hypothesis that PDTC-reduced oxidative stress enhances DDAH activity leading to increase in NO synthesis and consequently delay the onset of endothelial senescence.

Recent studies have demonstrated that the cellular uptake of exogenous ADMA is significantly reduced in presence of extracellular L-arginine. In this study, we used high concentrations of ADMA (5, 50 and 100 µmol/L), which do not correspond to pathophysiological plasma concentration, but endothelial basal medium contains rather high concentration of L-arginine of approximately 300 µmol/L, which is higher than typical plasma concentration of L-arginine (in the range of 50–100 µmol/L) (Bode-Böger et al. 2007). Therefore, we performed the following experiments in the presence of pathophysiologically relevant concentration of ADMA (2 µmol/L) and physiological level of L-arginine (70 µmol/L). We showed that ADMA also accelerates endothelial senescence and increases oxidative stress. Therapeutically relevant concentrations of L-arginine reverse the effect of ADMA on senescence by increasing NO synthesis and telomerase activity (Scalera et al. 2006). Surprisingly, we found that L-arginine alone displays a Janus face effect on senescence. Instead of delaying senescence, L-arginine alone accelerates the process of senescence by inhibiting NO synthesis, increasing oxidative stress and downregulating telomerase activity. We showed that the translational and posttranslational activation of arginase II is responsible for the paradoxical effect of L-arginine on senescence (Scalera et al. 2009a). We would speculate that elevated extracellular ADMA concentrations might be obviously needed to regulate L-arginine metabolism via arginase in cellular senescence. These results provide a reason to explain the disappointing results of recent randomized placebo controlled study applying L-arginine to patients after myocardial infarction for 6 months (Schulman et al. 2006). L-arginine does not improve vascular stiffness measurements or ejection fraction, but even increased mortality for unknown reasons. In the patient group ADMA levels were not determined. Testing patients for ADMA and L-arginine plasma levels for calculating the L-arginine/ADMA ratio might be an adequate strategy (Bode-Böger et al. 2007).

In contrast to our results presented here, Rentoukas et al. (2012) have recently shown a positive correlation between serum ADMA levels and telomerase activity ($r=0.604$, $p=0.038$) that was measured in circulating peripheral blood mononuclear cells of patients with metabolic syndrome. Unfortunately, the study has serious flaws that limit its guidance for clinicians. First, the number of patients included in the study is very small (29 patients with metabolic syndrome compared to 15 healthy volunteers). Second, the patients and healthy volunteers were not old enough. The mean age of patients was 54 ± 9.9 years (in the range of 31–60 years) and the mean age of healthy volunteers was 56 ± 10 years (in the range of 33–69 years). Finally, the serum ADMA levels were analysed using an ADMA-ELISA kit that overestimates the concentration of ADMA compared to mass spectrometry-based analytical methods. Indeed, the mean ADMA level was 2.07 ± 0.67 µmol/L in patients with metabolic syndrome and 1.15 ± 0.43 µmol/L in healthy volunteers which are two to threefold higher than the plasma ADMA levels measured by mass spectrometry-based analytical methods (Bode-Böger et al. 2007).

It has been shown that in atherosclerotic patients, plasma ADMA levels are positively correlated with risk factors for atherosclerosis. Miyazaki et al. (1999) reported a positively correlation for ADMA with age in individuals with no symptoms of coronary or peripheral artery disease. Celermajer et al. (1994) reported that aging is associated with progressive endothelial dysfunction in normal humans. A question that remains is whether telomere exhaustion is a cause or a consequence of cardiovascular disease and which role ADMA plays in this circle. Given that an accelerated rate of telomere shortening may be expected from the inflammation occurring in atherosclerosis and from the action of several cardiovascular risk factors (e.g., oxidative stress, hypertension, diabetes, smoking), telomere exhaustion may be a surrogate marker of cardiovascular disease. Interestingly, ADMA has been found to be elevated in the same clinical settings. So the question remains, is ADMA the first player? Do people with high plasma levels of ADMA as a cardiovascular risk factor have an accelerated development of endothelial dysfunction on the basis of increased cellular senescence? Our data strengthen the importance of early determination of ADMA and endothelial function in prospective studies as surrogate parameters for the increased risk of future cardiovascular events.

In summary, these data suggest that at concentrations corresponding to plasma levels at pathophysiological conditions, ADMA accelerates cellular senescence and provokes increased oxidative stress. This would provide a mechanism through which elevated plasma ADMA levels observed in the elderly and in the presence of atherosclerotic risk factors might be a hallmark of a hazard of endothelial senescence and atherogenesis. Furthermore, appearance of the senescent phenotype concomitantly with increased oxidative stress within foci of the endothelium could further impair NO synthesis and potentiate endothelial dysfunction, creating a kind of vicious circle. In this view, it could be speculated that therapeutic intervention aimed at reducing ADMA levels may help to attenuate aging-related cardiovascular diseases in an increasing older population, although we have no evidence for this speculation at yet.

References

Bode-Böger SM, Scalera F, Kielstein JT, Martens-Lobenhoffer J, Breithardt G, Fobker M, Reinecke H (2006) Symmetrical dimethylarginine: a new combined parameter for renal function and extent of coronary artery disease. J Am Soc Nephrol 17:1128–1134

Bode-Böger SM, Scalera F, Ignarro LJ (2007) The L-arginine paradox: importance for the L-arginine/asymmetrical dimethylarginine ratio. Pharmacol Ther 114:295–306

Böger RH, Sydow K, Borlak J, Thum T, Lenzen H, Schubert B, Tsikas D, Bode-Böger SM (2000) LDL cholesterol upregulates synthesis of asymmetrical dimethylarginine in human endothelial cells: involvement of S-adenosylmethionine-dependent methyltransferases. Circ Res 21(87):99–105

Böger RH, Maas R, Schulze F, Schwedhelm E (2009) Asymmetric dimethylarginine (ADMA) as a prospective marker of cardiovascular disease and mortality: an update on patient populations with a wide range of cardiovascular risk. Pharmacol Res 60:481–487

Celermajer DS, Sorensen KE, Spiegelhalter DJ, Georgakopoulos D, Robinson J, Deanfield JE (1994) Aging is associated with endothelial dysfunction in healthy men years before the age-related decline in women. J Am Coll Cardiol 24:471–476

Collins K, Mitchell JR (2002) Telomerase in the human organism. Oncogene 21:564–579

Cooke JP (2004) Asymmetrical dimethylarginine: the Uber marker? Circulation 109:1813–1819

Deng Y, Chan SS, Chang S (2008) Telomere dysfunction and tumor suppression: the senescence connection. Nat Rev Cancer 8:450–458

Dimri GP, Lee X, Basile G, Acosta M, Scott G, Roskelley C, Medrano EE, Linskens M, Rubelj I, Pereira-Smith O, Peacocke M, Campisi J (1995) A biomarker that identifies senescent human cells in culture and in aging skin in vivo. Proc Natl Acad Sci U S A 92:9363–9367

Fuster JJ, Andrès V (2006) Telomere biology and cardiovascular disease. Circ Res 99:1167–1180

Haendeler J, Hoffmann J, Diehl JF, Vasa M, Spyridopoulos I, Zeiher AM, Dimmeler S (2004) Antioxidants inhibit nuclear export of telomerase reverse transcriptase and delay replicative senescence of endothelial cells. Circ Res 94:768–775

Harley CB, Futcher AB, Greider CW (1990) Telomeres shorten during ageing of human fibroblasts. Nature 345:458–460

Hayflick L, Moorhead PS (1961) The serial cultivation of human diploid cells strains. Exp Cell Res 25:585–621

Kielstein JT, Bode-Böger SM, Frölich JC, Ritz E, Haller H, Fliser D (2003) Asymmetric dimethylarginine, blood pressure, and renal perfusion in elderly subjects. Circulation 109:172–177

Kielstein JT, Impraim B, Simmel S, Bode-Böger SM, Tsikas D, Frölich JC, Hoeper MM, Haller H, Fliser D (2004) Cardiovascular effects of systemic nitric oxide synthase inhibitor with asymmetric dimethylarginine in human. Circulation 109:172–177

Kostourou V, Robinson SP, Cartwright JE, Whitley GS (2002) Dimethylarginine dimethylaminohydrolase I enhances tumour growth and angiogenesis. Br J Cancer 87:673–680

Lakatta EG, Levy D (2003) Arterial and cardiac aging: major shareholders in cardiovascular disease enterprises. Part I: aging arteries: a "set up" for vascular disease. Circulation 107:139–146

Leiper J, Nandi M (2011) The therapeutic potential of targeting endogenous inhibitors of nitric oxide synthesis. Nat Rev Drug Discov 10:277–291

Martens-Lobenhoffer J, Rodionov RN, Drust A, Bode-Böger SM (2011) Detection and quantification of α-keto-δ-(N(G), N(G)-dimethylguanidino) valeric acid: a metabolite of asymmetric dimethylarginine. Anal Biochem 419:234–240

Minamino T, Komuro I (2007) Vascular cell senescence: contribution to atherosclerosis. Circ Res 100:15–26

Miyazaki H, Matsuoka H, Cooke JP, Usui M, Ueda S, Okuda S, Imaizumi T (1999) Endogenous nitric oxide synthase inhibitor: a novel marker of atherosclerosis. Circulation 99:1141–1146

Nicholson TB, Chen T, Richard S (2009) The physiological and pathophysiological role of PRMT1-mediated protein arginine methylation. Pharmacol Res 60:466–474

Palm F, Onozato ML, Luo Z, Wilcox CS (2007) Dimethylarginine dimethylaminohydrolase (DDAH): expression, regulation, and function in the cardiovascular and renal systems. Am J Physiol Heart Circ Physiol 293:H3227–H3245

Rentoukas E, Tsarouhas K, Kaplanis I, Korou E, Nikolaou M, Marathonitis G, Kokkinou S, Haliassos A, Mamalaki A, Kouretas D, Tsitsimpikou C (2012) Connection between telomerase activity in PBMC and markers of inflammation and endothelial dysfunction in patients with metabolic syndrome. PLoS One 7:e35739

Rodionov RN, Murry DJ, Vaulman SF, Stevens JW, Lentz SR (2010) Human alanine-glyoxylate aminotransferase 2 lowers asymmetric dimethylarginine and protects from inhibition of nitric oxide production. J Biol Chem 285:5385–5391

Scalera F, Bode-Böger SM (2010) Nitric oxide-asymmetric dimethylarginine system in endothelial cell senescence. In: Ignarro LJ (ed) Nitric oxide: biology and pathobiology, vol 2., pp 487–489

Scalera F, Borlak J, Beckmann B, Martens-Lobenhoffer J, Thum T, Täger M, Bode-Böger SM (2004) Endogenous nitric oxide synthesis inhibitor asymmetric dimethyl L-arginine accelerates endothelial cell senescence. Arterioscler Thromb Vasc Biol 24:1816–1822

Scalera F, Martens-Lobenhoffer J, Täger M, Bukowska A, Lendeckel U, Bode-Böger SM (2006) Effect of L-arginine on asymmetric dimethylarginine (ADMA) or homocysteine-accelerated endothelial cell aging. Biochem Biophys Res Commun 345:1075–1082

Scalera F, Closs EI, Flick E, Martens-Lobenhoffer J, Boissel JP, Lendeckel U, Heimburg A, Bode-Böger SM (2009a) Paradoxical effect of L-arginine: acceleration of endothelial senescence. Biochem Biophys Res Commun 386:650–655

Scalera F, Fulge B, Martens-Lobenhoffer J, Heimburg A, Bode-Böger SM (2009b) Red wine decreases asymmetric dimethylarginine via SIRT1 induction in human endothelial cells. Biochem Biophys Res Commun 390:703–709

Schulman SP, Becker LC, Kass DA, Champion HC, Terrin ML, Forman S, Ernst KV, Kelemen MD, Townsend SN, Capriotti A, Hare JM, Gerstenblith G (2006) L-arginine therapy in acute myocardial infarction: the Vascular Interaction with Age in Myocardial Infarction (VINTAGE MI) randomized clinical trial. JAMA 295:58–64

Stühlinger MC, Tsao PS, Her JH, Kimoto M, Balint RF, Cooke JP (2001) Homocysteine impairs the nitric oxide synthase pathway: role of asymmetric dimethylarginine. Circulation 104:2569–2575

Vallance P, Leone A, Calver A, Collier J, Moncada S (1992) Accumulation of an endogenous inhibitor of nitric oxide synthesis in chronic renal failure. Lancet 339:572–575

Vasa M, Breitschopf K, Zeiher AM, Dimmeler S (2000) Nitric oxide activates telomerase and delays endothelial cell senescence. Circ Res 87:540–542

Wadham C, Mangoni AA (2009) Dimethylarginine dimethylaminohydrolase regulation: a novel therapeutic target in cardiovascular disease. Expert Opin Drug Metab Toxicol 5:303–319

Membrane-Derived Extracellular Vesicles from Endothelial Progenitor Cells Activate Angiogenesis

Vincenzo Cantaluppi, Federico Figliolini, Maria Chiara Deregibus, and Giovanni Camussi

Contents

Abstract 17
Introduction 18
Endothelial Progenitor Cells and Neoangiogenesis 18
Extracellular Vesicles (EVs) 19
   Biogenesis of EVs 20
   Pleiotropic Functions of EVs 21
Role of EVs in Tumor Neoangiogenesis 21
EPC-Derived EVs in Tissue Regeneration After IRI 23
References 24

Abstract

The neoformation of blood vessels is a biological process involved in tissue homeostasis and repair as well as in pathologic conditions such as inflammatory diseases and cancer. Endothelial progenitor cells (EPCs) a cell population derived from the bone marrow and circulating in the blood stream, have been shown to take part to these processes. EPCs exert their effects mainly by the release of paracrine factors such as growth factors, cytokines and extracellular vesicles (EVs). EVs are small particles released by different types of activated cells by a membrane sorting process. Recent studies identified EVs as a new mechanism of cell-to-cell communication as they can mediate the exchange of receptors, proteins, bioactive lipids and nucleic acids between cells. EPC-derived EVs were shown to activate an angiogenic program in quiescent endothelial cells through an epigenetic reprogramming due to horizontal transfer of mRNA and microRNA. Whereas in the context of tumor neoangiogenesis this mechanism may be detrimental as it favors tumor vascularization and diffusion, in the context of regenerative medicine, EVs derived from EPCs can be exploited as potential therapeutic option to prevent ischemia-reperfusion injury.

Keywords

Endosomal sorting complex required for transport (ESCRT) • Endothelial progenitor cells and

V. Cantaluppi • F. Figliolini • M.C. Deregibus
G. Camussi (✉)
Molecular Biotechnology Center, Department of Medical Sciences, University of Turin, Turin, Italy
e-mail: giovanni.camussi@unito.it

neoangiogenesis • EPC-derived EVs in tissue regeneration • Extracellular vesicles (EVs) • Multivesicular bodies (MVBs) • Pleiotropic functions • Tumor neoangiogenesis

Introduction

Neoangiogenesis is defined as the process that leads to the formation of new blood vessels in the body. Even though neoangiogenesis is a physiological process essential for tissue homeostasis and repair, the formation of new blood vessels is a common feature of several inflammatory diseases and cancer. Indeed, the neovascularization process is crucial for solid tumor growth and invasion as the vasculature provides metabolic support and access to the circulation, thus favoring metastasis. Tumor angiogenesis does not seem to depend exclusively on normal endothelial cell recruitment from the surrounding vascular network. Recent data suggest alternative strategies for tumor vascularization based on the distinct phenotype of intra-tumor endothelial cells. Indeed, it has been recently shown that tumor-derived endothelial cells are genetically unstable and are different from normal vessel endothelial cells at molecular and functional levels (Bussolati et al. 2011). The enhanced angiogenic properties of tumor endothelial cells have been related to the expression of embryonic genes and to the autocrine production of angiogenic growth factors. Recent studies suggested the potential role of stem cells in tumor angiogenesis and in the phenotypic switch of tumor-derived endothelial cells. Among bone marrow-derived stem cells, endothelial progenitor cells (EPCs) have been shown to localize within sites of vascular injury and within tumors exerting a pro-angiogenic effect mainly due to the release of paracrine factors (Bussolati et al. 2011).

Neo-angiogenesis is a pivotal process also in the regeneration of tissues following ischemia-reperfusion injury (IRI). The regeneration of damaged endothelial cells is of particular interest in the field of cell and organ transplantation where the lack of revascularization is associated with primary or delayed graft function.

Endothelial Progenitor Cells and Neoangiogenesis

In the last years, the role of stem cells in regenerative therapies to heal injured tissues has been considerably expanding. In the adults, stem cells are critical for tissue self-renewal in the hematopoietic system, skin and intestine where an high cell turnover is required. On the other hand, a potential role of stem cells in tissue homeostasis has been described also in organs with a much lower rate of cell turnover such as kidney, lung, skeletal muscle and liver. Stem cells are characterized by their ability to self-renew and to differentiate into a variety of cell types. It has been shown that bone-marrow-derived stem cells have the ability to cross lineage boundaries making components of several tissues.

Recently, an endothelial committed type of progenitor cells, defined as EPCs provoked a great interest in regenerative medicine. EPCs, derived from the hematopoietic stem cell lineage, are produced in the bone marrow and subsequently released into the peripheral blood (Asahara et al. 1997).

EPCs are phenotypically identified by flow cytometry and are characterized by the expression of CD34, CD133 and/or VEGF receptor 2 (KDR). These molecules are shared by other hematopoietic stem/progenitor cells and the presence of hematopoietic contamination of EPCs is debated. Yoder et al. (2007) re-defined EPCs into two different cell populations: ECFCs (endothelial colony-forming cells) and CFU-EC (endothelial cell colony-forming units). ECFCs have endothelial (CD31, CD105, CD144, CD146, vWF, KDR and UEA-1) but not hematopoietic antigens, a robust proliferative potential and the ability to form secondary endothelial colony and vessels *in vivo*. Moreover, ECFCs possess a hierarchy of progenitor stages with different proliferative ability. CFU-ECs derive from hematopoietic stem cells, have both endothelial (CD31, vWF and UEA-1) and hematopoietic antigens (CD45, CD14), but with some myeloid progenitor activity and without ability to form secondary endothelial cell colony or vessels *in vivo*. Thus,

ECFCs are considered "true EPCs" whereas CFU-ECs are considered myeloid progenitors (Yoder et al. 2007). It has been shown that EPCs play a pivotal role in vascular homeostasis and repair, but these cells are also involved in tumor neoangiogenesis. EPCs have the potential to provide ongoing endothelial repair by homing to endothelial injury sites. Indeed, different studies showed that EPCs are recruited to injured vascular sites contributing to the preservation of endothelial integrity. EPC levels are lower in patients with different disease states such as coronary artery disease, diabetes, cerebral stroke and in particular chronic kidney disease. In these diseases, the reduction of EPC number seems due to the down-regulation of pathways responsible for EPC release from the bone marrow by exposure to risk factors rather than to EPC consumption.

Recently, it has been shown that EPCs are recruited in the kidney after IRI and that they are able to favour tissue repair by secretion of pro-angiogenic factors. It has been suggested that EPC paucity and dysfunction are the main mechanisms responsible for accelerated vascular injury in chronic kidney disease (CKD) patients. Moreover, the re-establishment of renal function after transplantation correlated with an increased number and function of circulating EPCs (Li et al. 2010).

EPC repair process is characterized by a series of specific and orchestrated steps: EPCs leave the bone marrow following gradients of growth factors and cytokines that are released into the circulation by injured endothelium and tissues (Schatteman et al. 2007). Once in the circulation, EPCs localize to sites of damage and promote vascular repair. Homing of circulating EPCs to sites of vascular injury is a complex process mainly directed by signalling via key CC- and CXC-chemokines, L-selectin and β-integrins. We previously demonstrated that EPCs exploit a mechanism typical of leukocyte adhesion for their localization at the sites of vascular injury, since L-selectin present on EPC surface can selectively bind to fucosylated oligosaccharide residues which are up-regulated following IRI (Biancone et al. 2004).

In the context of cancer, the recruitment of EPCs may be detrimental as they may favour tumor vascularization. However, the role of EPCs in neo-vessel formation is debated. Whereas in some studies a direct contribution of EPCs to neoformed vessels has been suggested, other studies indicated that EPCs accumulate in the periphery of tumor and favour neoangiogenesis by paracrine mechanisms as they secrete a number of angiogenic factors (Lyden et al. 2001).

Among the soluble angiogenic factors, we recently demonstrated that EPCs released extracellular vesicles (EVs) able to activate an angiogenic program in quiescent endothelial cells by horizontal transfer of mRNA (Deregibus et al. 2007).

Whereas the pro-angiogenic and anti-apoptotic effects of EPC-derived EVs may be detrimental for tumor growth, they may find potential therapeutic applications in ischemic diseases and cell/organ transplantation. In fact, EVs released from EPCs protect the kidney from IRI and improve neoangiogenesis of human islets transplanted in SCID mice through a mechanism related to the epigenetic reprogramming of injured target cells based on the transfer of specific mRNA and microRNA subsets (Cantaluppi et al. 2012a, b).

Extracellular Vesicles (EVs)

Every cell needs to communicate with the environment and with other cells and uses different ways of signaling which can act in close proximity of cells or at long distance. In addition to the well-known cell-to-cell communication based on proteins, cells may relate each other by means of EVs. EVs are extruded from the cell membrane into the extra-cellular space by a universal process that is preserved in prokaryotes and eukaryotes (György et al. 2011). EVs can be isolated from medium of many *in vitro* cell cultures, including tumor cells and from several body fluids such as serum, plasma, urine, amniotic fluid, saliva, milk, bronchoalveolar fluid, etc.

Almost all cells release vesicles and these vesicles differ depending on the cell of origin and on the functional state of the cell from which they are produced (stimulated, apoptotic, tumoral, etc.)

(Simons and Raposo 2009). The classification and the nomenclature of these EVs are at the moment confusing and not well defined. Different names are present in the literature such as microvesicles, microparticles, exosomes, prostasomes, etc. and different protocols of vesicle purification are used. However, all these vesicles share the characteristic to be small and complex structures composed of cytosol surrounded by a lipid bilayer with hydrophilic soluble components resembling that of cell membrane. In addition, they expose many transmembrane proteins able to activate the target cell interacting with cell receptors and allowing recognition and internalization of EVs (Théry et al. 2009). EVs can be defined as carriers of bioactive molecules, since they are able to transfer biological active lipids, functional proteins and nucleic acids (mRNA and microRNA) to a recipient cell modifying its behavior. After their release into the extracellular space, EVs may undergo capture from neighboring cells by means of specific receptor-ligand interactions or enter the circulation and other biological fluids allowing a long distance transport of the bioactive molecules in them enclosed.

Biogenesis of EVs

Most of the studies in this field mainly concentrate on two major classes of EVs: microvesicles and exosomes. The EVs secreted by cells *in vitro* or *in vivo* are a blend of these two types and it is difficult to totally separate them with the actual procedures; therefore many of the studies do not discriminate the two groups of EVs. Microvesicles and exosomes differ not only for the size and density but also for the composition of their membrane, and the mechanism of their formation is not completely known. Microvesicles are a heterogeneous class of vesicles with size ranging from 100 to 1,000 nm and originating by direct blebbing and subsequent detaching from the plasma membrane of cells into the extracellular space in a process under the regulation of flippase, floppase, scramblase, calpain, and gelsolin enzymes involved in the membrane lipid sidedness (Mathivanan et al. 2010). Flippase and floppase generate and maintain membrane phospholipid asymmetry; scramblase, by mixing the lipids between the two layers of the plasma membrane, promotes the collapse of the membrane asymmetry in a mechanism dependent on calcium concentration. In addition, an increased cytosolic calcium concentration, as it happens during cell activation, apoptosis and necrosis, may activate calpain facilitating membrane blebbing and the shedding of microvesicles from the cell membrane. The membrane of microvesicles contains components of membrane lipid rafts such as flotillin-1 and is enriched in cholesterol and sphingolipids with many saturated fatty acids that render the site less fluid than the adjacent membrane, and eligible for vesiculation (Mathivanan et al. 2010). Moreover, microvesicles may show high exposure of phosphatidylserine, and may express tissue factor and cell specific markers.

Exosomes are a homogenous class of vesicles of 30–120 nm in size and exhibit a characteristic protein and lipid composition which allows their identification. Some protein markers such as tumor susceptibility gene 101 (TSG101), Alix, CD9, CD63 and CD81 tetraspanins, and heat shock proteins (HSP60, HSP70, HSP90) are thought to be specific of exosomes, others are frequently present such as annexins and the small GTPases Rabs important for membrane trafficking and fusion; others are specific of the cell from which exosomes derive. The lipid composition of exosomes includes some components of lipid rafts such as cholesterol, ceramide, glycerophospholipids, and sphingolipids (Simons and Raposo 2009).

Exosome biogenesis is mostly unknown, but they are thought to originate from the endosomal compartment by internal budding and formation of intraluminal vesicles (ILVs) inside multivesicular bodies (MVBs). MVBs may have a dual destination: to reach the lysosomes, fuse their membrane with them and release the intraluminal vesicles into lysosomes undergoing degradation, or to reach the cell membrane and undergo release of ILVs into the extra-cellular space, by a process of exocytosis. These exocytic vesicles are named exosomes (Mathivanan et al. 2010).

The sorting and assembling of proteins into ILVs is a not yet defined process involving multiple mechanisms. The finding of Alix, a protein associated with the endosomal sorting complex required for transport (ESCRT) machinery, in many exosomal membrane preparations, suggested the requirement of the ESCRT for the release of exosomes but, whilst the involvement of ESCRT multiprotein complexes in the mechanism of lysosomal degradation is well studied, more ambiguous is the ESCRT function in the process of exosomes biogenesis (Théry et al. 2001). However, a recent study provided evidence for the possibility that the transfer to exosomes from distinct subdomains on the endosomal membrane occurs independently from the ESCRT processing but rather depends on ceramide (Trajkovic et al. 2008).

The EVs derived from EPCs are most probably a mixture of exosomes and microvesicles shed from the cell surface. Transmission electron microscopy studies showed shedding of EVs from EPC surface. The majority of them have a size under 100 nm as judged by electron microscopy and ranged between 60 and 160 nm when evaluated by Nanosight. Molecules expressed on surface of EPC -derived EVs include α4 and β1 integrins, CD34, L-selectin and CD154 (CD40-L) (Cantaluppi et al. 2012a). Conversely, they were negative for HLA class I and class II antigens and for markers of monocytes (CD14) and platelets (P-selectin, CD42b).

Pleiotropic Functions of EVs

After interaction with EVs the behavior of the recipient cell is modified by several mechanisms. EVs may directly stimulate target cells; for instance, the EVs bearing tissue factor may interact with P-selectin expressed on the surface of macrophages, polymorphonuclear neutrophils and platelets and activate these cells (Cocucci et al. 2008). EVs may transfer receptors and ligands between cells as shown for the adhesion molecule CD41 that can be shifted from platelets to endothelial cells (Barry et al. 1998). EVs may transfer biological active proteins as shown for EVs released from endotoxin-activated monocytes that transfer caspase-1 to vascular smooth muscle cells inducing apoptosis (Sarkar et al. 2009).

More recently, EVs were found to transfer genetic information as they contain a selection of mRNA and microRNA in association with ribonucleoproteins involved in the intracellular traffic of RNA (Valadi et al. 2007; Yuan et al. 2009; Collino et al. 2010). The horizontal transfer of mRNA and microRNA induces epigenetic reprogramming of the recipient cells (Ratajczak et al. 2006; Deregibus et al. 2007; Aliotta et al. 2010).

Role of EVs in Tumor Neoangiogenesis

The angiogenic shift is critical for tumor malignancy. Vascularization is needed to provide oxygen and nutrients for tumor growth and to allow its expansion and access to the circulation. For this purpose, tumor cells produce a number of factors that, acting on microenvironment, create favorable conditions for their growth. Among these factors, EVs play a critical role in inducing phenotypic changes in stroma cells including endothelial cells (Fig. 2.1). A number of studies provide evidences that tumor endothelial cell phenotype differs from normal endothelium (Bussolati et al. 2011). EVs released from tumor cells have the potential to induce these phenotype changes, as they have been shown to carry oncogene products, pro-angiogenic factors and nucleic acids involved in modulation of pro-angiogenic pathways. Indeed, tumor EVs have been shown to be involved in the activation of the angiogenic shift (Castellana et al. 2010). It has been shown that the activation of endothelial cells and fibroblasts that creates a favorable environment for tumor growth, can be mediated by EVs released from lung cancer cells (Wysoczynski and Ratajczak 2009). Tumor EVs were also shown to express on their surface matrix metalloproteinases (MMP) and extracellular MMP inducer involved in the extracellular matrix degradation needed for neoangiogenesis (Castellana et al. 2010). Al-Nedawi et al. (2009) demonstrated that glioma-derived EVs contain the oncogenic form of the epidermal growth factor receptor EGFRvIII, that can be horizontally transferred to endothelial cells promoting

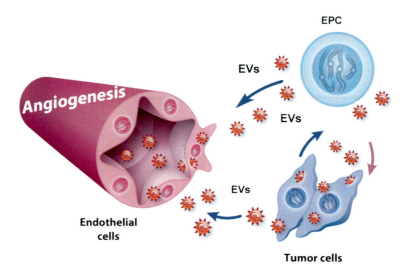

Fig. 2.1 Schematic representation of a bidirectional exchange of genetic information between EPC and tumor cell. The transfer of mRNA and microRNA delivered by EVs may lead to the formation of an environment favourable for tumor development. EPCs recruited within the tumors release EVs that may transfer molecules potentially involved in the stimulation of tumor angiogenesis

activation of EGFRvIII-regulated genes including the vascular endothelial growth factor (VEGF) gene. In turn, EVs derived from activated endothelial cells may propagate the angiogenic signal by transferring the Notch ligand Delta-like 4 (Dll4) to quiescent endothelial cells (Sheldon et al. 2010). Millimaggi et al. (2007) showed that ovarian cancer-derived EVs express CD147/extracellular MMP inducer that activates an angiogenic program in endothelial cells. The critical role of CD147/extracellular MMP inducer shuttled by EVs was shown by using small interfering RNA against CD147 that abrogated the angiogenic activity of EVs. Tumor-derived EVs also convey tetraspanin 8 that favours endothelial cell activation and angiogenesis (Nazarenko et al. 2010). Moreover, cancer-derived EVs may reprogram normal endothelial cells by transferring genetic information from tumor cells to endothelium. The transfer of mRNA and microRNA via EVs has been shown in different experimental settings (Baj-Krzyworzeka et al. 2006; Ratajczak et al. 2006). Skog et al. (2008) showed the transfer of mRNA and microRNA associated to cell migration from glioblastoma to endothelial cells promoting angiogenesis. More recently, it has been shown that EVs shed from tumors contain retro-transposon elements and amplified oncogene sequences (Balaj et al. 2011) and mitochondrial DNA that may be transferred to target cells (Guescini et al. 2010).

We recently demonstrated in the renal clear cell carcinoma that the EVs possessing the angiogenic activity are those released by tumor stem cells (Grange et al. 2011). These EVs, by creating a favorable endothelial environment in the lung, permit metastatic implantation of tumor cells.

In the context of tumor, the role of EPC is debated. Whereas some studies suggest a direct participation of EPC in the formation of new vessels, other studies suggest that they act by a paracrine mechanism (Bussolati et al. 2011). We found that EVs produced by EPC are able to activate quiescent normal endothelial cells as result of horizontal transfer of mRNA related to pro-angiogenic pathways (Fig. 2.1) (Deregibus et al. 2007). The transferred mRNAs are functional as EVs bearing GFP mRNA induce GFP protein synthesis in the recipient cells. In particular, EVs transfer to target endothelial cells mRNAs associated with the PI3K/AKT and eNOS signaling pathways. Protein expression and functional studies showed that PI3K and eNOS are one of the main effectors in this experimental setting.

EPC-Derived EVs in Tissue Regeneration After IRI

Different studies showed that ischemia-reperfusion induces acute and chronic vascular dysfunction characterized by loss of endothelium-dependent vasodilation. IRI generates high levels of free radicals that enhance endothelial dysfunction and may induce cell death (Lefer and Lefer 1993).

IRI is especially harmful in organ transplantation because it represents a key non-immunologic factor that influences early and late loss of graft function. It is well known that endothelial cell impairment and death associated with the accumulation of leukocytes and platelets causing microcirculatory disturbances and graft failure. Endothelial cell death after ischemic preservation is not only necrotic but also apoptotic. Endothelial cells are the main target of harmful mediators during cold ischemia and subsequent warm reperfusion injury. Reperfusion injury after cold storage induces apoptosis by caspase 3, 8, 9 activation. In particular, caspase activation depends primarily on triggering of Bax and Bak, the pro-apoptotic members of Bcl-2 family which are able to start a sequence of events leading to the release of cytochrome C from mitochondria into the cytosol. Caspase-3 is the final executor molecule, responsible for protein cleavage, thereby disabling cellular structure and repair processes (Kuwana and Newmeyer 2003).

The role of EPCs in neovascularization has also been studied in the model of pancreatic islet transplantation. Islet transplantation is a therapeutic option for the treatment of type I diabetes. In this context, bone marrow-derived stem cells have shown to exert a beneficial effect by enhancing neoangiogenesis and by modulating the allo- and autoimmune response. In particular, it has been shown that EPCs are able to specifically localize within the implanted islets and to chimerize with donor vessels. EPCs enhance transplanted islet revascularization with a better graft survival. Moreover, it has been shown that EPCs are recruited to the pancreas in response to islet injury, favouring neovascularization that may facilitate the recovery of injured beta-cells improving islet allograft function (Zampetaki et al. 2008).

We have recently shown that EPCs may exert a paracrine effect on transplanted islets through the release of EVs. EPC-derived EVs are internalized in both endothelial and beta cells enhancing islet vascularisation in a model of xenotransplantation in SCID mice. Of interest, EPC-derived EVs favoured insulin secretion and survival by an anti-apoptotic and pro-angiogenic effect. These effects depend on the horizontal transfer mediated by EVs of specific mRNA and microRNA such as miR-126 and miR-296 (Cantaluppi et al. 2012b).

Another model in which we studied the effects of EPC-derived EVs is kidney IRI. Microvascular injury is a hallmark of kidney subjected to ischemia-reperfusion injury. Indeed, the presence of endothelial dysfunction is associated with an extension phase of renal damage often requiring long term dialysis. This is of particular importance in the setting of kidney transplantation where microvasculature derangement associated with tubular cell injury following ischemia-reperfusion injury are the main causes of the so called delayed graft function (DGF), a clinical syndrome characterized by acute dysfunction of grafted kidneys and defined as the need of dialysis in the first week after transplantation. Several on-going clinical trials and experimental studies showed the protective effect of bone marrow-derived stem cells in kidney regeneration after ischemic injury. EPCs are mobilized from bone marrow following kidney ischemia, form an intra-splenic niche and then migrate to the injured kidney promoting tissue regeneration *via* paracrine mechanisms. We found that EVs derived from EPCs mimic the effect of the cells by preventing acute kidney injury (Fig. 2.2). Indeed, EVs are internalized within endothelial cells of peritubular capillaries and tubular epithelial cells inducing tubular cell proliferation, apoptosis and leukocyte infiltration inhibition, thus leading to functional and morphologic protection. EPC-derived EVs are also able to interfere with the mechanisms of progression toward chronic kidney damage by inhibiting capillary rarefaction, glomerulosclerosis, and tubulointerstitial fibrosis

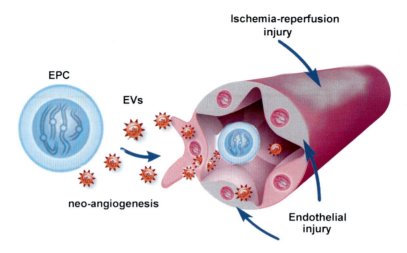

Fig. 2.2 Pro-angiogenic effect of EPC-derived EVs in injured ischemic tissues. EVs may deliver factors that restrain loss of endothelial cells thus limiting injury and that promote endothelial cell proliferation and angiogenesis

(Cantaluppi et al. 2012a). As described for islet transplantation, the renoprotective effect of EVs is due to transfer of mRNA and microRNA from EPCs to injured kidney endothelial and epithelial cells.

In conclusion, EVs emerged as a new mechanism of intercellular exchange of biological information. This may lead to phenotypical and functional changes in the recipient cells. In the context of tumor, EPC-derived EVs may be detrimental as they favor tumor vascularization. At variance, in diseases characterized by endothelial cell injury and loss of vascularization, EPC-derived EVs may find a therapeutic application. Indeed, they limit endothelial injury and stimulate neoangiogenesis.

References

Al-Nedawi K, Meehan B, Kerbel RS, Allison AC, Rak J (2009) Endothelial expression of autocrine VEGF upon the uptake of tumor-derived microvesicles containing oncogenic EGFR. Proc Natl Acad Sci U S A 106:3794–3799

Aliotta JM, Pereira M, Johnson KW, de Paz N, Dooner MS, Puente N, Ayala C, Brilliant K, Berz D, Lee D, Ramratnam B, McMillan PN, Hixson DC, Josic D, Quesenberry PJ (2010) Microvesicle entry into marrow cells mediates tissue-specific changes in mRNA by direct delivery of mRNA and induction of transcription. Exp Hematol 38:233–245

Asahara T, Murohara T, Sullivan A, Silver M, van der Zee R, Li T, Witzenbichler B, Schatteman G, Isner JM (1997) Isolation of putative progenitor endothelial cells for angiogenesis. Science 275:964–967

Baj-Krzyworzeka M, Szatanek R, Weglarczyk K, Baran J, Urbanowicz B, Brański P, Ratajczak MZ, Zembala M (2006) Tumour-derived microvesicles carry several surface determinants and mRNA of tumour cells and transfer some of these determinants to monocytes. Cancer Immunol Immunother 55:808–818

Balaj L, Lessard R, Dai L, Cho YJ, Pomeroy SL, Breakefield XO, Skog J (2011) Tumour microvesicles contain retrotransposon elements and amplified oncogene sequences. Nat Commun 2:180

Barry OP, Praticò D, Savani RC, FitzGerald GA (1998) Modulation of monocyte-endothelial cell interactions by platelet microparticles. J Clin Invest 102:136–144

Biancone L, Cantaluppi V, Duò D, Deregibus MC, Torre C, Camussi G (2004) Role of L-selectin in the vascular homing of peripheral blood-derived endothelial progenitor cells. J Immunol 173:5268–5274

Bussolati B, Grange C, Camussi G (2011) Tumor exploits alternative strategies to achieve vascularization. FASEB J 25:2874–2882

Cantaluppi V, Gatti S, Medica D, Figliolini F, Bruno S, Deregibus MC, Sordi A, Biancone L, Tetta C, Camussi G (2012a) Microvesicles derived from endothelial progenitor cells protect the kidney from ischemia-reperfusion injury by microRNA-dependent reprogramming of resident renal cells. Kidney Int 82:412–427

Cantaluppi V, Biancone L, Figliolini F, Beltramo S, Medica D, Deregibus MC, Galimi F, Romagnoli R, Salizzoni M, Tetta C, Segoloni GP, Camussi G (2012b) Microvesicles derived from endothelial progenitor cells enhance neoangiogenesis of human pancreatic islets. Cell Transplant 21:1305–1320

Castellana D, Toti F, Freyssinet JM (2010) Membrane microvesicles: macromessengers in cancer disease and progression. Thromb Res 125(Suppl 2):S84–S88

Cocucci E, Racchetti G, Meldolesi J (2008) Shedding microvesicles: artefacts no more. Trends Cell Biol 19:43–51

Collino F, Deregibus MC, Bruno S, Sterpone L, Aghemo G, Viltono L, Tetta C, Camussi G (2010) Microvesicles derived from adult human bone marrow and tissue specific mesenchymal stem cells shuttle selected pattern of miRNAs. PLoS One 5:e11803

Deregibus MC, Cantaluppi V, Calogero R, Lo Iacono M, Tetta C, Biancone L, Bruno S, Bussolati B, Camussi G (2007) Endothelial progenitor cell derived microvesicles activate an angiogenic program in endothelial cells by a horizontal transfer of mRNA. Blood 110:2440–2448

Grange C, Tapparo M, Collino F, Vitillo L, Damasco C, Deregibus MC, Tetta C, Bussolati B, Camussi G (2011) Microvesicles released from human renal cancer stem cells stimulate angiogenesis and formation of lung premetastatic niche. Cancer Res 71:5346–5356

Guescini M, Genedani S, Stocchi V, Agnati LF (2010) Astrocytes and Glioblastoma cells release exosomes carrying mtDNA. J Neural Transm 117:1–4

György B, Szabó TG, Pásztói M, Pál Z, Misják P, Aradi B, László V, Pállinger E, Pap E, Kittel A, Nagy G, Falus A, Buzás EI (2011) Membrane vesicles, current state-of-the-art. emerging role of extracellular vesicles. Cell Mol Life Sci 68:2667–2688

Kuwana T, Newmeyer DD (2003) Bcl-2-family proteins and the role of mitochondria in apoptosis. Curr Opin Cell Biol 15:691–699

Lefer DJ, Lefer AM (1993) Nitric oxide homeostasis control as therapy for cardiovascular diseases. Cardiovasc Res 27:2282

Li B, Cohen A, Hudson TE, Motlagh D, Amrani DL, Duffield JS (2010) Mobilized human hematopoietic stem/progenitor cells promote kidney repair after ischemia/reperfusion injury. Circulation 121:2211–2220

Lyden D, Hattori K, Dias S, Costa C, Blaikie P, Butros L, Chadburn A, Heissig B, Marks W, Witte L, Wu Y, Hicklin D, Zhu Z, Hackett NR, Crystal RG, Moore MA, Hajjar KA, Manova K, Benezra R, Rafii S (2001) Impaired recruitment of bone-marrow-derived endothelial and hematopoietic precursor cells blocks tumor angiogenesis and growth. Nat Med 7:1194–1201

Mathivanan S, Ji H, Simpson RJ (2010) Exosomes: extracellular organelles important in intercellular communication. J Proteomics 73:1907–1920

Millimaggi D, Mari M, D'Ascenzo S, Carosa E, Jannini EA, Zucker S, Carta G, Pavan A, Dolo V (2007) Tumor vesicle-associated CD147 modulates the angiogenic capability of endothelial cells. Neoplasia 9:349–357

Nazarenko I, Rana S, Baumann A, McAlear J, Hellwig A, Trendelenburg M, Lochnit G, Preissner KT, Zöller M (2010) Cell surface tetraspanin Tspan8 contributes to molecular pathways of exosome-induced endothelial cell activation. Cancer Res 70:1668–1678

Ratajczak J, Miekus K, Kucia M, Zhang J, Reca R, Dvorak P, Ratajczak MZ (2006) Embryonic stem cell-derived microvesicles reprogram hematopoietic progenitors: evidence for horizontal transfer of mRNA and protein delivery. Leukemia 20:847–856

Sarkar A, Mitra S, Mehta S, Raices R, Wewers MD (2009) Monocyte derived microvesicles deliver a cell death message via encapsulated caspase-1. PLoS One 4:e7140

Schatteman GC, Dunnwald M, Jiao C (2007) Biology of bone marrow-derived endothelial cell precursors. Am J Physiol Heart Circ Physiol 292:H1–H18

Sheldon H, Heikamp E, Turley H, Dragovic R, Thomas P, Oon CE, Leek R, Edelmann M, Kessler B, Sainson RC, Sargent I, Li JL, Harris AL (2010) New mechanism for notch signaling to endothelium at a distance by Delta-like 4 incorporation into exosomes. Blood 116:2385–2394

Simons M, Raposo G (2009) Exosomes–vesicular carriers for intercellular communication. Curr Opin Cell Biol 21:575–581

Skog J, Würdinger T, van Rijn S, Meijer DH, Gainche L, Sena-Esteves M, Curry WT Jr, Carter BS, Krichevsky AM, Breakefield XO (2008) Glioblastoma microvesicles transport RNA and proteins that promote tumour growth and provide diagnostic biomarkers. Nat Cell Biol 10:1470–1476

Théry C, Boussac M, Véron P, Ricciardi-Castagnoli P, Raposo G, Garin J, Amigorena S (2001) Proteomic analysis of dendritic cell-derived exosomes: a secreted subcellular compartment distinct from apoptotic vesicles. J Immunol 166:7309–7318

Théry C, Ostrowski M, Segura E (2009) Membrane vesicles as conveyors of immune responses. Nat Rev Immunol 9:581–593

Trajkovic K, Hsu C, Chiantia S, Rajendran L, Wenzel D, Wieland F, Schwille P, Brügger B, Simons M (2008) Ceramide triggers budding of exosome vesicles into multivesicular endosomes. Science 319:1244–1247

Valadi H, Ekstrom K, Bossios A, Sjostrand M, Lee JJ, Lotvall JO (2007) Exosome-mediated transfer of mRNAs and microRNAs is a novel mechanism of genetic exchange between cells. Nat Cell Biol 9:654–659

Wysoczynski M, Ratajczak MZ (2009) Lung cancer secreted microvesicles: underappreciated modulators of microenvironment in expanding tumors. Int J Cancer 125:1595–1603

Yoder MC, Mead LE, Prater D, Krier TR, Mroueh KN, Li F, Krasich R, Temm CJ, Prchal JT, Ingram DA (2007) Redefining endothelial progenitor cells via clonal analysis and hematopoietic stem/progenitor cell principals. Blood 109:1801–1809

Yuan A, Farber EL, Rapoport AL, Tejada D, Deniskin R, Akhmedov NB, Farber DB (2009) Transfer of microRNAs by embryonic stem cell microvesicles. PLoS One 4:e4722

Zampetaki A, Kirton JP, Xu Q (2008) Vascular repair by endothelial progenitor cells. Cardiovasc Res 78:413–421

Induction of P21-Dependent Senescence: Role of NAE Inhibitor MLN4924

3

Yongfu Pan, Yi Sun, and Lijun Jia

Contents

Abstract	27
Introduction	28
Characteristics of Senescence	28
Senescence in Tumorigenesis and Anti-cancer Therapy	28
Neddylation	29
Neddylation of Cullin-RING E3 Ligase (CRL)	29
NAE Inhibitor MLN4924	30
NAE Inhibitor MLN4924 Inhibits Cullin Neddylation and Inactivates CRL to Suppress the Growth of Cancer Cells	30
MLN4924 Induces Irreversible Cellular Senescence	30
MLN4924-Induced Senescence Is p21-Dependent	30
Discussion	31
References	31

Abstract

Cellular senescence is a state of cell growth-arrest that limits the proliferation of cancer cells, and thus induction of senescence has been regarded as a promising anticancer strategy. Neddylation is a post-translational modification by adding ubiquitin-like molecule NEDD8 to targeted proteins via an enzymatic cascade reaction, and thus regulates the localization, stability and function of neddylated proteins. The most well identified targets of neddylation so far are cullins which function as essential subunits of multiunit Cullin-RING E3 ubiquitin ligases (CRLs). Recently, a specific inhibitor of NEDD8 activating enzyme (NAE), MLN4924, was identified as the first-in-class anticancer agent. We found that MLN4924 induces cellular senescence as a new mechanism of cancer cell growth suppression. Mechanistically, NAE inhibitor MLN4924 blocks cullin neddylation and thus inhibits the activity of CRL. By doing so, MLN4924 induces the accumulation of p21, a well-known CRL substrate and a major mediator of senescence, to trigger cellular senescence, whereas depletion of p21 largely abrogates MLN4924-induced senescence and attenuates the anticancer efficacy of MLN4924. Our study reveals a novel mechanism of MLN4924 action which will facilitate the development of this investigational drug as a novel class of anticancer agent.

Y. Pan (✉) • L. Jia
Cancer Institute, Fudan University Shanghai Cancer Center, Shanghai 200032, China
e-mail: pyf.whu@gmail.com;
jialijun2002@yahoo.com.cn

Y. Sun
Division of Radiation and Cancer Biology, Department of Radiation Oncology, University of Michigan, Ann Arbor, MN 48109, USA
e-mail: sunyi@umich.edu

Keywords

Cellular senescence • Characteristics of senescence • Cullin-RING E3 ligase (CRL) • P21 • NAE inhibitor MLN4924 • Neddylation • Tumorigenesis and anti-cancer therapy

Introduction

Cellular senescence was first defined by Hayflick and colleagues in 1960s as a potential mechanism associated with ageing. In their observation, human fibroblast cells underwent a finite number of population doublings during culture, suggesting there is a replication or biological clock that counts number of times for cellular division and determines the longevity of cells. The postulation of clocking mechanism drives the discovery of telomere shortening in which the end terminus of telomere sequence gradually decreases at the each round of DNA replication. Telomere shortening is now well-known associated with telomerase, a ribonucleoprotein enzyme that extends the 3'end of telomere by adding DNA sequence repeats TTAGGG. Cancer cells possess infinite capacity of replication, which is partially due to relatively higher expression of telomerase in cancer cells than in normal human cells. Likewise, normal human cells could be immortalized by the transfection of the catalytic subunit of human telomerase (Hayflick 2000).

In addition to telomere shortening, diverse cellular stresses, such as DNA damage, oxidative stress and oncogene activation, have been reported to induce senescence. For example, many chemotherapeutics or radiation that induce DNA damage and/or oxidative stress can trigger senescence-like growth arrest (Serrano and Blasco 2001). Furthermore, aberrant activation of oncogenes and signal factors, such as RAS, BRAF, MYC and E2F, can induce cell senescence in vitro and in vivo as well (Braig and Schmitt 2006). Furthermore, epigenetic modification of histone by histone deacetylases (Wagner et al. 2001) and histone methyltransferases (Braig et al. 2005) are also associated with senescence induction.

Characteristics of Senescence

Several biochemical and cellular biomarkers have been developed and used to characterize cellular senescence: (a) Senescent cells express high level of senescence-associated beta-galactosidase (SA-β-gal) which can be detected by classical SA-β-gal staining; (b) senescent cells are in a state of cell growth-arrest with low metabolic activity, displaying an enlarged and flattened shape in morphology (Dimri et al. 1995); (c) DNA damage and DNA damage response (DDR) are usually triggered during the establishment of senescence and required for the maintenance of the phenotype (d'Adda di Fagagna 2008); (d) many senescent cells have senescence associated heterochromatin foci (SAHF) and senescence-associated DNA-damage foci (SDF) (Campisi and d'Adda di Fagagna 2007); and (e) senescent cells may also secret a complex senescence-associated inflammatory cytokines, such as IL-6/IL-8 to communicate their compromised state to the surrounding tissue (Kuilman et al. 2008).

Senescence in Tumorigenesis and Anti-cancer Therapy

Cancer cells acquire infinite capacity of replication, whereas senescence induces cancer cells in a state of grow arrest and replication failure both *in vitro* and *in vivo*. Thus, induction of senescence has been recognized as a new mechanism or strategy for tumor suppression (Campisi 2001). Observations in murine tumorigenesis models and human tumor tissues revealed that senescence mainly occurs at the early stage of tumorigenesis in various tumors of different histotypes, but not in malignant tumors, highlighting a critical role of senescence in prevention of tumorigenesis (Collado and Serrano 2010). Consistently, disruption of senescence pathways by deletion of tumor suppressors, such as p53, p16 and PTEN (Chen et al. 2005) or histone methyltransferase Suv39H1 (Braig et al. 2005) results in malignancy. Moreover, recent studies have convincingly demonstrated that many anticancer agents, such as doxorubicin and retinoid (Roninson 2002), can suppress cancer cell growth by inducing senescence.

Neddylation

Neddylation of Cullin-RING E3 Ligase (CRL)

Neddylation is a type of post-translational modification by adding ubiquitin-like molecule NEDD8 to targeted proteins via an enzyme cascade reaction (Fig. 3.1). NEDD8 (short for neural precursor cell-expressed, developmentally down-regulated protein 8) was first discovered in 1992 and shares 60 % amino acid sequence identity to ubiquitin (Kumar et al. 1992). As a ubiquitin-like protein, NEDD8 conjugates to substrates through a ubiquitin analogous cascade reaction involving E1, E2 and E3. Briefly, NEDD8 is first activated by E1 NEDD8 activating enzyme (NAE), which is composed of UBA3 and NAE1 heterodimer, in an ATP-dependent manner. Activated NEDD8 is then transferred to the E2-conjuagating enzyme UBC12 or UBE2F. Finally, an E3 NEDD8 ligase transfers NEDD8 to substrates (Schulman and Harper 2009).

The process of the conjugation of NEDD8 to cullin is called cullin neddylation, which changes the conformation of Cullin-RING E3 ligase (CRL) complex, leading to its activation (Fig. 3.1) (Girdwood et al. 2011). As the largest cellular ubiquitin ligase family, CRL is responsible for ubiquitination and degradation of a mass of cellular proteins, thus regulating numerous biological processes. Given the fact that neddylation positively regulates CRL and CRL is often

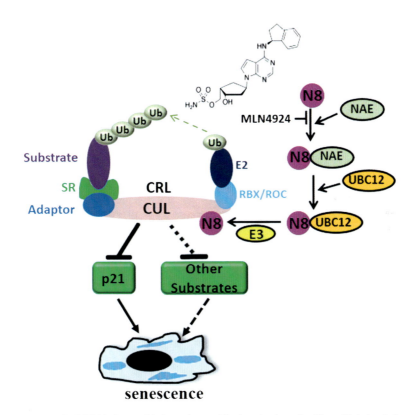

Fig. 3.1 NAE inhibitor MLN4924 induces p21-dependent cellular senescence. As an inhibitor of NAE, MLN4924 binds to NEDD8 and forms MLN4924-NEDD8 adduct that inhibits the activity of NAE, as well as cullin neddylation. The inactivation of cullin neddylation induces the accumulation of substrates of CRL, including p21, which triggers p21-dependent cellular senescence. *N8* NEDD8, *SR* substrate receptor, *CUL* cullin, *CRL* cullin RING E3 ligase

activated in many human cancers (Watson et al. 2011), targeting the neddylation process may serve as a promising strategy for cancer therapy.

NAE Inhibitor MLN4924

NAE Inhibitor MLN4924 Inhibits Cullin Neddylation and Inactivates CRL to Suppress the Growth of Cancer Cells

MLN4924, a specific inhibitor of NAE, was recently discovered via a high-throughput screening (Soucy et al. 2009). In structure, MLN4924 mimics AMP and forms stable MLN4924-NEDD8 adduct, which competes with ATP (Fig. 3.1) (Brownell et al. 2010). By blocking NEDD8 activation by NAE, MLN4924 disrupts the cascade reaction of cullin neddylation, leading to the inactivation of CRL. The inhibition of CRL activity then causes the accumulation of CRL substrates, which trigger a variety of biochemical and biological responses, including DNA re-replication stress and DNA damage response, abnormal cell cycle progression (Soucy et al. 2009), autophagy (Luo et al. 2012; Zhao et al. 2012), as well as apoptosis and senescence (Jia et al. 2011; Lin et al. 2010b), and eventually suppress tumor cell growth both *in vitro* and *in vivo*. Due to its significant anticancer efficacy in preclinical studies, MLN4924 has been advanced into several phase I clinical trials for solid tumors and hematological malignancies.

Previous studies indicated that apoptosis pathway could be activated by MLN4924 in several cancer cell lines (Milhollen et al. 2010; Soucy et al. 2009; Swords et al. 2010). However, the role of apoptosis in MLN4924-induced growth suppression has not been well defined. We recently found that MLN4924 treatment induced robust apoptosis in HCT-116 colon cancer cells, but not in U87 glioblastoma and H1299 lung cancer cells, suggesting that induction of apoptosis is not the only mechanism of MLN4924 action (Jia et al. 2011). Moreover, blockage of apoptosis pathway with a pan-caspase inhibitor z-VAD or siRNA silencing of caspase 3 in treated cells only partially rescued growth suppression in HCT-116, but not in U87 and H1299 cells, further indicating that other mechanisms is involved in the MLN4924-induced growth suppression (Jia et al. 2011).

MLN4924 Induces Irreversible Cellular Senescence

Our further study revealed that HCT-116, U87 and H1299 cancer cells underwent cellular senescence upon exposure to MLN4924 (Fig. 3.1), as evidenced by the following observations: (a) defect in cell cycle progression and growth arrest; (b) enlarged and flattened morphology; (c) positively stained with SA-β-gal; and (d) accumulation of CRL substrates CDT-1 and Orc1 which induces DNA re-replication and DDR. Importantly, MLN4924-induced senescence is irreversible, given the fact that MLN4924 removal after senescence induction failed to restore the proliferation of senescent cells (Jia et al. 2011).

MLN4924-Induced Senescence Is p21-Dependent

The p16/pRB and p53/p21 axes are two major senescence-triggering pathways in response to diverse stresses (Kuilman et al. 2010). To elucidate the key mediators of MLN4924–induced senescence, we determined the activation status of these classical pathways in MLN4924–treated cells. Our results revealed that only p21, a well-characterized substrate of CRL and an effective senescence mediator, but not p53 or p16/pRB, was accumulated in a dose- and time-dependent manner in all three cancer cell lines, suggesting that p21 plays an important role in the induction of senescence (Fig. 3.1). More importantly, genetic deletion of p21 nearly abrogated MLN4924-induced cellular senescence when compared to wild-type p21-expressing cells (Jia et al. 2011). Finally, we showed that the induction of p21-dependent senescence contributed to MLN4924-mediated growth suppression (Jia et al. 2011). Our study convincingly demonstrated that MLN4924-induced senescence is largely p21-dependent.

Discussion

Neddylation pathway is essential for the oncogenic activity of CRL through regulation of cullin neddylation and thus serves as a potential anticancer target (Watson et al. 2011). NAE inhibitor MLN4924 was identified and characterized as a first-in-class anticancer agent, which has been advanced to phase I clinical trials for solid tumors and hematological malignancies, owing to its remarkable anticancer effect and satisfactory drug-tolerance (Soucy et al. 2009). Mechanistically, MLN4924 blocks cullin neddylation to inactivate CRL and results in the accumulation of CRL substrates, which finally triggers multiple cellular responses, including DNA damage response, cell cycle defects, apoptosis and autophagy in a broad panel of cancer cell lines (Lin et al. 2010b; Luo et al. 2012; Soucy et al. 2009; Yang et al. 2012).

Recently we found that MLN4924 induces cellular senescence as a novel anticancer mechanism in multiple cancer cell lines (Jia et al. 2011). We further demonstrated that induction of senescence by MLN4924 is largely dependent on p21, which contributes to MLN4924-induced growth suppression (Jia et al. 2011). Others also found that MLN4924 could induce senescence in HCT116 colon cancer cells (Lin et al. 2010b) and in PC-3 prostate cancer cells in a xenograft tumor model (Lin et al. 2010a).

Induction of senescence as a new mechanism of MLN4924 action provides three useful aspects with regards to future development of MLN4924 as a novel class of anticancer agent (Jia et al. 2011). First, p21 accumulation in tumor cells/tissues could serve as a biomarker to indicate that MLN4924 indeed reaches the target. Second, senescence induced by MLN4924, even at low drug concentrations, is irreversibly, which makes it possible to use low doses of the drug to avoid normal tissue toxicity. Third, MLN4924-induced senescence in cancer cells is a phenomenon independent of p53 status, implying that all human cancers regardless of p53 status can be treated by the drug (Jia et al. 2011). These findings will facilitate the further development of MLN4924 as a promising investigational anticancer agent.

Acknowledgements We thank Millennium Pharmaceutical, Inc. for providing us MLN4924. This work is supported by NCI grants (CA111554 and CA118762) to YS, and National Natural Science Foundation Grant of China (31071204, 81172092), National Basic Research Program of China (973 program, 2012CB910302), Shanghai Pujiang Talent Program (12PJ1400600), the Program for Professor of Special Appointment (Eastern Scholar) at Shanghai Institutions of Higher Learning and Key Project of Shanghai Municipal Health Bureau (2010012) to Lijun Jia.

References

Braig M, Schmitt CA (2006) Oncogene-induced senescence: putting the brakes on tumor development. Cancer Res 66:2881–2884

Braig M, Lee S, Loddenkemper C, Rudolph C, Peters AH, Schlegelberger B, Stein H, Dorken B, Jenuwein T, Schmitt CA (2005) Oncogene-induced senescence as an initial barrier in lymphoma development. Nature 436:660–665

Brownell JE, Sintchak MD, Gavin JM, Liao H, Bruzzese FJ, Bump NJ, Soucy TA, Milhollen MA, Yang X, Burkhardt AL, Ma J, Loke HK, Lingaraj T, Wu D, Hamman KB, Spelman JJ, Cullis CA, Langston SP, Vyskocil S, Sells TB, Mallender WD, Visiers I, Li P, Claiborne CF, Rolfe M, Bolen JB, Dick LR (2010) Substrate-assisted inhibition of ubiquitin-like protein-activating enzymes: the NEDD8 E1 inhibitor MLN4924 forms a NEDD8-AMP mimetic in situ. Mol Cell 37:102–111

Campisi J (2001) Cellular senescence as a tumor-suppressor mechanism. Trends Cell Biol 11:S27–S31

Campisi J, d'Adda di Fagagna F (2007) Cellular senescence: when bad things happen to good cells. Nat Rev Mol Cell Biol 8:729–740

Chen Z, Trotman LC, Shaffer D, Lin HK, Dotan ZA, Niki M, Koutcher JA, Scher HI, Ludwig T, Gerald W, Cordon-Cardo C, Pandolfi PP (2005) Crucial role of p53-dependent cellular senescence in suppression of Pten-deficient tumorigenesis. Nature 436:725–730

Collado M, Serrano M (2010) Senescence in tumours: evidence from mice and humans. Nat Rev Cancer 10:51–57

d'Adda di Fagagna F (2008) Living on a break: cellular senescence as a DNA-damage response. Nat Rev Cancer 8:512–522

Dimri GP, Lee X, Basile G, Acosta M, Scott G, Roskelley C, Medrano EE, Linskens M, Rubelj I, Pereira-Smith O et al (1995) A biomarker that identifies senescent human cells in culture and in aging skin in vivo. Proc Natl Acad Sci U S A 92:9363–9367

Girdwood D, Xirodimas DP, Gordon C (2011) The essential functions of NEDD8 are mediated via distinct surface regions, and not by polyneddylation in Schizosaccharomyces pombe. PLoS One 6:e20089

Hayflick L (2000) The illusion of cell immortality. Br J Cancer 83:841–846

Jia L, Li H, Sun Y (2011) Induction of p21-dependent senescence by an NAE inhibitor, MLN4924, as a mechanism of growth suppression. Neoplasia 13:561–569

Kuilman T, Michaloglou C, Vredeveld LC, Douma S, van Doorn R, Desmet CJ, Aarden LA, Mooi WJ, Peeper DS (2008) Oncogene-induced senescence relayed by an interleukin-dependent inflammatory network. Cell 133:1019–1031

Kuilman T, Michaloglou C, Mooi WJ, Peeper DS (2010) The essence of senescence. Genes Dev 24:2463–2479

Kumar S, Tomooka Y, Noda M (1992) Identification of a set of genes with developmentally down-regulated expression in the mouse brain. Biochem Biophys Res Commun 185:1155–1161

Lin HK, Chen Z, Wang G, Nardella C, Lee SW, Chan CH, Yang WL, Wang J, Egia A, Nakayama KI, Cordon-Cardo C, Teruya-Feldstein J, Pandolfi PP (2010a) Skp2 targeting suppresses tumorigenesis by Arf-p53-independent cellular senescence. Nature 464:374–379

Lin JJ, Milhollen MA, Smith PG, Narayanan U, Dutta A (2010b) NEDD8-targeting drug MLN4924 elicits DNA rereplication by stabilizing Cdt1 in S phase, triggering checkpoint activation, apoptosis, and senescence in cancer cells. Cancer Res 70:10310–10320

Luo Z, Yu G, Lee HW, Li L, Wang L, Yang D, Pan Y, Ding C, Qian J, Wu L, Chu Y, Yi J, Wang X, Sun Y, Jeong LS, Liu J, Jia L (2012) The Nedd8-activating enzyme inhibitor MLN4924 induces autophagy and apoptosis to suppress liver cancer cell growth. Cancer Res 72:3360–3371

Milhollen MA, Traore T, Adams-Duffy J, Thomas MP, Berger AJ, Dang L, Dick LR, Garnsey JJ, Koenig E, Langston SP, Manfredi M, Narayanan U, Rolfe M, Staudt LM, Soucy TA, Yu J, Zhang J, Bolen JB, Smith PG (2010) MLN4924, a NEDD8-activating enzyme inhibitor, is active in diffuse large B-cell lymphoma models: rationale for treatment of NF-{kappa}B-dependent lymphoma. Blood 116:1515–1523

Roninson IB (2002) Tumor senescence as a determinant of drug response in vivo. Drug Resist Updat 5:204–208

Schulman BA, Harper JW (2009) Ubiquitin-like protein activation by E1 enzymes: the apex for downstream signalling pathways. Nat Rev Mol Cell Biol 10:319–331

Serrano M, Blasco MA (2001) Putting the stress on senescence. Curr Opin Cell Biol 13:748–753

Soucy TA, Smith PG, Milhollen MA, Berger AJ, Gavin JM, Adhikari S, Brownell JE, Burke KE, Cardin DP, Critchley S, Cullis CA, Doucette A, Garnsey JJ, Gaulin JL, Gershman RE, Lublinsky AR, McDonald A, Mizutani H, Narayanan U, Olhava EJ, Peluso S, Rezaei M, Sintchak MD, Talreja T, Thomas MP, Traore T, Vyskocil S, Weatherhead GS, Yu J, Zhang J, Dick LR, Claiborne CF, Rolfe M, Bolen JB, Langston SP (2009) An inhibitor of NEDD8-activating enzyme as a new approach to treat cancer. Nature 458:732–736

Swords RT, Kelly KR, Smith PG, Garnsey JJ, Mahalingam D, Medina E, Oberheu K, Padmanabhan S, O'Dwyer M, Nawrocki ST, Giles FJ, Carew JS (2010) Inhibition of NEDD8-activating enzyme: a novel approach for the treatment of acute myeloid leukemia. Blood 115:3796–3800

Wagner M, Brosch G, Zwerschke W, Seto E, Loidl P, Jansen-Durr P (2001) Histone deacetylases in replicative senescence: evidence for a senescence-specific form of HDAC-2. FEBS Lett 499:101–106

Watson IR, Irwin MS, Ohh M (2011) NEDD8 pathways in cancer, Sine Quibus Non. Cancer Cell 19:168–176

Yang D, Tan M, Wang G, Sun Y (2012) The p21-dependent radiosensitization of human breast cancer cells by MLN4924, an investigational inhibitor of NEDD8 activating enzyme. PLoS One 7:e34079

Zhao Y, Xiong X, Jia L, Sun Y (2012) Targeting Cullin-RING ligases by MLN4924 induces autophagy via modulating the HIF1-REDD1-TSC1-mTORC1-DEPTOR axis. Cell Death Dis 3:e386

Regulation of the Novel Senescence Pathway by SKP2 E3 Ligase

Guocan Wang, Yuan Gao, Li Chen, Ying-Jan Wang, and Hui-Kuan Lin

Contents

Abstract	33
Introduction	34
Types of Cellular Senescence	34
Replicative Senescence	34
Oncogene-Induced Senescence	35
Pten-Loss Induced Senescence	36
ARF/P53-Dependent Senescence Pathway	36
SKP2 Regulates an ARF/P53-Indepenent Senescence Pathway	38
Skp2 Is a Member of the F-Box Protein Family	38
Skp2 Targets Protein for Proteolytic and Non-proteolytic Ubiquitination	38
Skp2 Suppresses Cellular Senescence Independent of ARF-p53	39
Strategies for Pro-senescence Therapy	40
Discussion	41
References	42

G. Wang
Department of Genomic Medicine, MD Anderson Cancer Center, University of Texas, Houston, TX, USA
e-mail: GWang6@mdanderson.org

Y. Gao • L. Chen
Department of Molecular and Cellular Oncology, MD Anderson Cancer Center, University of Texas, Houston, TX, USA

Y.-J. Wang
Department of Molecular and Cellular Oncology, MD Anderson Cancer Center, The University of Texas, Houston, TX 77030, USA

Department of Environmental and Occupational Health, National Cheng Kung University, Medical College, Tainan, Taiwan

Abstract

Cellular senescence, a stress response triggered by multiple stimuli, results in a form of irreversible cell cycle arrest that can serve as a critical barrier for cancer development. Various studies have demonstrated the critical role of ARF/p53 pathways in the induction of cellular senescence by activation of oncogenic pathways through overexpression of oncogenes, such as Ras, or by inactivation of tumor suppressor genes, such as PTEN. Recent studies also uncover novel ARF/p53-independent cellular senescence pathways in restricting tumorigenesis. Given that ARF/p53 pathways play an essential role in tumor suppression and are often inactivated in human cancers through deficiency or mutations of ARF or p53, better understanding of these pathways governing the induction of senescence in human cancer will pave the ways for developing effective pro-senescence therapies. Thus, it's important to screen current available drugs that stabilize p53 expression for the ability to

H.-K. Lin (✉)
Department of Molecular and Cellular Oncology, MD Anderson Cancer Center, The University of Texas, Houston, TX 77030, USA

Graduate Institute of Basic Medical Science, China Medical University, Taichung 404, Taiwan

Department of Biotechnology, Asia University, Taichung 404, Taiwan
e-mail: hklin@mdanderson.org

target possibility that these Arf-p53 dependent pathways or by developing novel inhibitors to target the Arf-p53 independent pathways.

Keywords

ARF/P53-dependent senescence pathway • ARF/P53-indepenent senescence pathway • Homologous recombination (HR) • Novel senescence pathway • Oncogene-induced senescence (OIS) • Pro-senescence therapy • Proteolytic and non-proteolytic ubiquitination • PTEN-loss induced senescence • Replicative senescence

Introduction

Cellular senescence, a phenomenon first described by Hayflick and Moorhead (1961) as a state of stable cell cycle arrest, challenges the dogma that normal cells could undergo unlimited proliferation in culture. In addition to such replicative senescence in culture, it is now well-established that premature cellular senescence can be triggered by numerous stress stimuli, including telomere erosion, persistent DNA damage response (DDR), or strong mitogenic signals from activation of oncogenes or loss of tumor suppressor genes (Acosta and Gil 2012; Nardella et al. 2011; Rodier and Campisi 2011). Although senescent cells remain arrested in cell cycle even in the presence of growth factors, these enlarged and flattened senescent cells still have active metabolism, display senescence-associated β-galactosidase (SA-β-gal) activity, and secrete various extracellular matrix and soluble factors known as the senescence-associated secretory phenotype (SASP) or senescence messaging secretome (SMS) (Acosta and Gil 2012; Nardella et al. 2011). Furthermore, senescence-associated heterochromatic foci (SAHF), the facultative heterochromatin domains, are often found in senescent cells as well (Acosta and Gil 2012; Nardella et al. 2011). Thus, cellular senescence is fundamentally different from quiescence or apoptosis.

Importantly, senescence has been implicated to not only act as a physiological barrier against tumor initiation and progression both in vitro and in vivo (Braig et al. 2005; Chen et al. 2005; Herbig et al. 2006; Michaloglou et al. 2005; Xue et al. 2007), but also promote tissue repair and fuel inflammation associated with aging and cancer progression (Rodier and Campisi 2011). Hence, an increasing understanding of the mechanisms that trigger senescence might result in the development of novel cancer therapies that enable better management of human cancers.

Types of Cellular Senescence

Replicative Senescence

Replicative senescence was originally identified as a permanent growth arrest caused by telomere erosion generated by repeated cell passaging. Human telomere ends typically shorten 50–200 base pairs per population doubling (PD). After 50–100 PDs, the telomere ends become critically short and exposed, which will be sensed by cells and trigger DNA damage response (DDR) to detect damaged DNA and subsequently trigger a long-term cell cycle arrest (d'Adda di Fagagna et al. 2003; Herbig et al. 2004; Karlseder et al. 2002; Takai et al. 2003). Mechanistically, these shortened telomere ends recruit molecular markers of DNA double stranded breaks (DSBs), including 53BP1 and Nijmegen breakage syndrome 1 (NBS1), formγ-H2AX-positive senescence-associated DNA damage foci (SDF), and activate a signaling cascade that involves activation of ATM (ataxia telangiectasia mutation), ATR (ATM and Rad3 related), CHK1 (checkpoint kinase 1), and CHK2 (checkpoint kinase 2). This type of replicative senescence triggered by telomere dysfunction is critically dependent on p53-p21 pathways (Fig. 4.2) (Acosta and Gil 2012; Nardella et al. 2011; Rodier and Campisi 2011). In addition, inhibition of DDR leads to a bypass of cellular senescence, further supporting the role of DDR in senescence induction.

Furthermore, replicative senescence has also been observed in vivo. For example, uncapped telomere-induced DDR has been identified in aged animal and cells in vivo, which also contain DNA damage foci with irreparable DSBs (Acosta and

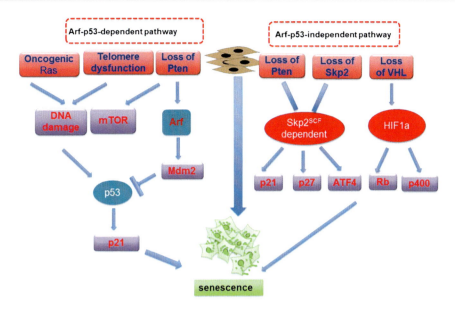

Fig. 4.1 Arf-p53-dependent-independent senescence pathways. Premature cellular senescence can be triggered by overexpression of several oncogenes, such as HRas^{G12V} as well as loss of tumor suppressors, such as PTEN, NF1, VHL, and Arf. Overexpression of Ras or loss of *Pten* leads to Arf/p53 activation and cellular senescence. However, inactivation of *Skp2* in combination of *Pten* or *Arf* leads to an induction of p21, p27 and ATF4 in an Arf/p53 independent manner

Gil 2012; Nardella et al. 2011; Rodier and Campisi 2011), Importantly, genetically engineered mouse models with short telomeres are protected from tumorigenesis through inducing p53-dependent senescence. In addition, both INK4A (also known as p16; a CDKI that operates upstream of RB) and the RB tumor suppressor have important roles in replicative senescence.

Oncogene-Induced Senescence

Oncogene-induced senescence (OIS) was initially described as an in vitro response triggered by cells to prevent oncogenic transformation, suggesting that OIS may serve as a physiological barrier for tumor progression. The initial studies that described such phenomenon used constitutively active form of oncogene Ras (G12Vmutant), whose overexpression resulted in morphological and molecular changes in primary fibroblasts that were indistinguishable from the replicative senescence (Serrano et al. 1997). Subsequent studies demonstrated that overexpression of activated Ras in primary cells lead to hyperproliferation and DNA hyper-replication (Bartkova et al. 2006; Di Micco et al. 2006; Moiseeva et al. 2006; Serrano et al. 1997), which causes activation of an S phase-specific DDR (Di Micco et al. 2006) and triggers senescence response through upregulation of p53 and p16INK4a activity and subsequent cell cycle arrest (Fig. 4.1) (Serrano et al. 1997). Similar to replicative senescence, OIS is not induced in cells that are defective in ATM activity or cells that cannot sense DNA damage or transduce DDR signals to p53 (Di Micco et al. 2006). In addition, multiple Ras effectors also play a role in the senescence response, including Raf/Mek pathway downstream of Ras, and E2F family of transcription factors that are distal effectors of the Ras pathway (Lin et al. 1998; Zhu et al. 1998).

Importantly, a series of studies have demonstrated that the process of senescence is not merely an artifact of in vitro culture shock, but indeed is an important physiological response to activated oncogenes to suppress tumor progression in vivo (Acosta and Gil 2012; Nardella et al. 2011; Rodier and Campisi 2011). Senescence induced by oncogenes such as Kras, Braf, Nras, and Hras has been observed in premalignant lesions. For example, senescent cells can be observed during the early stages of lung and pancreatic tumorigenesis driven by an inducible endogenous Kras^{G12V} oncogene, in lung tumors and melanocytic nevi driven by BrafV600E, in lymphocytes subjected to chemotherapy in Nras^{G12D} transgenic mouse, in the mammary tumors driven by high Hras expression and in prostatic intraepithelial neoplasia in Akt1 transgenic mouse. However, OIS is not consistently observed in lung lesions of the KRas^{G12D} knock-in mice. Interestingly, hyperproliferation is observed in the mammary gland when Hras^{G12V} oncogene is expressed at low levels, suggesting that different HRas expression levels can affect cellular phenotypes, and may have important implications for the mechanism of induction of cellular senescence. Thus, deepening our understanding of the molecular basis and tissue specificity of OIS will not only provide novel insights into how cellular senescence is regulated but also aid in developing therapeutic strategies and assessments for cancer treatments.

Pten-Loss Induced Senescence

Senescence can be elicited not only by oncogene activation, but also by loss of tumor suppressor, such as phosphatase and tensin homolog (PTEN) (Chen et al. 2005) and von Hippel–Lindau (VHL) (Young et al. 2008) (Fig. 4.1). PTEN is among the most frequently lost or mutated tumor suppressor gene that opposes the PI3K/AKT pathway through dephosphorylating phosphoinositide-3, 4, 5-triphosphate. Acute inactivation of *Pten* or overexpression of active *Akt1* in mouse embryonic fibroblasts (MEFs) induces a profound senescence response (Alimonti et al. 2010; Chen et al. 2005). However, Pten-inactivation-induced senescence (PICS) displays some unique features that are different from OIS. PICS can occur in cells that are treated with aphidicolin, which blocks S phase entry and prevents DNA replication (Chen et al. 2005), whereas senescence is abrogated in HRAS^{G12V}-induced OIS (Acosta and Gil 2012; Nardella et al. 2011; Rodier and Campisi 2011). In addition, the formation of senescence-associated DNA damage foci is not observed in PICS, and inhibition of ATM activity has no effect on PICS induction (Alimonti et al. 2010). Moreover, p53 upregulation mainly resulting from mTOR activation and mTOR-mediated translation (Alimonti et al. 2010) is distinct from the phosphorylation-mediated activation of p53 in OIS (Di Micco et al. 2006). Furthermore, the genetic inactivation of ARF at the Cdkn2a locus does not significantly affect p53 levels in PICS or prevent PICS induction in prostate tumorigenesis (Chen et al. 2009), but results in a marked reduction of p53 in OIS (Acosta and Gil 2012; Nardella et al. 2011; Rodier and Campisi 2011). Recently, a novel phosphatase-independent function of nuclear PTEN has been shown to promote the upregulation of INK4A through ETS2, thereby contributing to PICS (Song et al. 2011).

Importantly, the relevance of PICS in vivo is supported by the prostate-specific inactivation of *Pten* in mouse model (Chen et al. 2005). Furthermore, concomitant inactivation of *Trp53* with *Pten* in the mouse prostate leads to a bypass of senescence, illustrating the critical role of p53 in PICS and the importance of PICS in preventing rapid progression of pre-malignant lesions to aggressive cancer (Chen et al. 2005; Nardella et al. 2011). Recently, *Smad4* inactivation in mouse prostate results in a bypass of PICS, whereas p53 levels as well as functions are not dramatically affected in the *Pten/Smad4* compound mutant mice (Ding et al. 2011). These findings suggest that there may be different pathways to bypass PICS, such as inactivation of p53 or Smad4. Although it is clear that PICS plays a crucial role in blocking tumor progression in prostate tumorigenesis, the relevance of PICS in restricting other tumor development is not clear.

ARF/P53-Dependent Senescence Pathway

The tumor suppressor p53 is key player in maintaining genomic integrity in response to various stress stimuli through binding to specific response elements in DNA and modulates the expression of genes that play a major roles in cell cycle arrest, apoptosis, maintenance of genetic integrity, inhibition of angiogenesis as well as cellular senescence (Vogelstein and Kinzler 2004; Whibley et al. 2009).

DDR is induced in replicative senescence, OIS, and DNA-damaging agent treatments through activation of checkpoint kinases ATM and ATR, which results in p53 phosphorylation, stabilization, and subsequent transcriptional activation of its target genes, including *CDKN1A* (encoding the protein p21) (Acosta and Gil 2012; Nardella et al. 2011; Rodier and Campisi 2011) (Fig. 4.1). In addition, p53 is also essential for PICS (Chen et al. 2005), even though DDR is not required in this process (Alimonti et al. 2010; Nardella et al. 2011). On the contrary, silencing or inactivation of *p53* by viral onco-proteins, such as SV40 large T antigen or HPV-16 E6 protein, suppresses senescence response. The relevance of p53 in senescence also has been validated in vivo in multiple studies. Genetic inactivation of p53 bypasses senescence and promotes tumor progression in BRaf^{V600E}-induced lung tumors, HRas^{G12V}-induced mammary tumors, and Pten-loss induced prostate tumors (Acosta and Gil 2012; Nardella et al. 2011; Rodier and Campisi 2011). In addition, endogenous p53 reactivation in established murine sarcomas and liver carcinomas results in tumor regression due to cellular senescence (Acosta and Gil 2012; Nardella et al. 2011; Rodier and Campisi 2011). Interestingly, apoptosis is induced by p53 reactivation in the context of lymphomas, suggesting the tissue specific function of p53 in regulating apoptosis and senescence.

Tumor suppressor ARF, an upstream regulator of p53, has been extensively studied for its role in cellular senescence. The best-known function of ARF is to prevent MDM2-mediated p53 ubiquitination and degradation, which results in p53 induction and activation. Similar to p53, *Arf* deficiency in MEFs suppresses replicative senescence and leads to cell immortalization, whereas induction of *Arf* expression induces p53-dependent senescence in fibroblasts (Fig. 4.1) (Acosta and Gil 2012; Nardella et al. 2011; Rodier and Campisi 2011). In addition, ARF expression is strongly induced in response to oncogene overexpression, such as Ras and E2F, resulting in p53-dependent senescence (Acosta and Gil 2012; Nardella et al. 2011; Rodier and Campisi 2011). Interestingly, it has been shown that tumor suppressors ARF and INK4A (p16), encoded by the *CDKN2A* locus, are derepressed and induced in the presence of senescence stimuli (Acosta and Gil 2012; Nardella et al. 2011; Rodier and Campisi 2011). However, ARF is dispensable for the DDR- induced OIS (Acosta and Gil 2012; Nardella et al. 2011; Rodier and Campisi 2011). Interestingly, ARF regulates senescence response in MEFs under the condition of PICS in a p53-independent manner (Fig. 4.1) (Alimonti et al. 2010; Chen et al. 2009). In addition, the induction of p53 in PICS is independent of ARF, but dependent of mTOR-mediated translation both in vitro in MEFs and in vivo in mouse prostate epithelium (Fig. 4.1) (Alimonti et al. 2010; Chen et al. 2009). Specifically, *Pten* loss in prostate epithelial cells drives p53-dependent senescence, but concomitant inactivation of *Arf* and *Pten* surprisingly does not suppress p53 activation and cellular senescence. However, *Arf* deficiency in MEFs abolishes cellular senescence upon complete *Pten* inactivation (Chen et al. 2009), indicating that p53-triggered senescence is regulated variably in different cell types.

p21, a p53 target gene, is highly expressed in senescent cells and induces premature senescence when it is ectopically expressed in cells, even in the absence of p53, supporting its role as an important downstream effector for p53-mediated cellular senescence (Fig. 4.1) (Wang et al. 1999). However, inactivation of *p21* expression in human fibroblasts bypasses senescence (Brown et al. 1997), but not in MEFs, suggesting the role of p21 as a downstream effector p53 in cellular senescence varies among species. Given the complexity of the p53 transcriptional network, it remains unclear whether other p53 target genes are also involved in p53-dependent senescence pathways.

SKP2 Regulates an ARF/P53-Indepenent Senescence Pathway

Skp2 Is a Member of the F-Box Protein Family

Ubiquitination, a process by which ubiquitin is covalently attached to the lysine residue of a target protein for proteasomal degradation, is mediated by three enzymes, E1 (ubiquitin-activating enzyme), E2 (ubiquitin-conjugating enzymes (UBCs)) and E3s (ubiquitin ligases) (Fig. 4.2a). This process is reversible as ubiquitin covalently attached to a target protein can be removed by deubiquitinating enzymes (DUBs). Thus, the level and activity of a target protein is critically dependent on the balance between the activities of ubiquitinating and deubiquitinating enzymes.

Skp2, a critical component of Skp2^{SCF} ubiquitin E3 ligase, was originally identified as an interacting partner of Skp1 and Cyclin A/Cdk2/Cks1 and was found to be overexpressed in many cancer cell lines (Frescas and Pagano 2008). Skp2 is a member of the F-Box protein family, which constitutes one of the four subunits of the SCF ubiquitin protein ligase complex (SKP1-cullin-F-box), whose function is to facilitate the transfer of ubiquitin from E2 ubiquitin conjugating enzymes to protein substrates and promote their degradation in the proteasome (Fig. 4.2b) (Frescas and Pagano 2008). All F-box proteins contain an F-box motif consisting of about 45–50 amino acids. Both genetic and biochemical analyses have demonstrated that the SCF E3 ligase targets a variety of important proteins for ubiquitin-dependent proteasomal degradation (Frescas and Pagano 2008). Interestingly, it seems that phosphorylation of the substrates on either serine or threonine is required for SCF-mediated protein degradation (Frescas and Pagano 2008).

Skp2 Targets Protein for Proteolytic and Non-proteolytic Ubiquitination

The function of Skp2 in targeting proteins, such as p27, p21, Foxo1, for ubiquitination and subsequent proteasomal degradation has been extensive studied (Fig. 4.2c). As one of the best-characterized SCF complexes, multiple Skp2

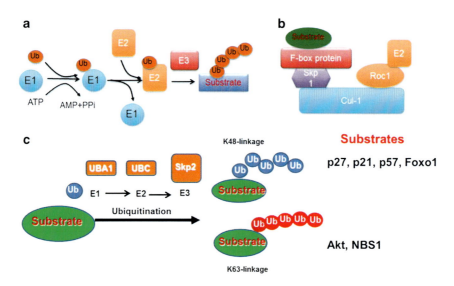

Fig. 4.2 Skp2 belongs to the F-Box protein family. (**a**). A simplified scheme for the ubiquitin proteasome system. Posttranslational modification of protein with ubiquitin is mediated by three enzymes: the E1 activating enzyme, the E2 conjugated enzyme, and the E3 ligase. (**b**). SCF E3 ligase complex is composed of Cul-1, Skp1, Roc1, and an F-Box protein, such as Skp2. (**c**). Skp2 can target protein for K48-linked polyubiquitination as well as K63-linked polyubiquitination

substrates, which are involved in multiple cellular processes such as cell cycle and apoptosis, have been identified (Frescas and Pagano 2008). Among the various Skp2 substrates, cell cycle inhibitor p27, a well-known tumor suppressor, has been demonstrated to be a critical substrate of Skp2 by genetically engineered mouse model (Frescas and Pagano 2008). Furthermore, Skp2 is not only overexpressed in human cancers from various origins but also is strongly associated with poor prognosis, accompanied by a decrease in p27 expression (Frescas and Pagano 2008), suggesting that Skp2 may play a critical role in tumorigenesis in human cancers. In addition to p27, many additional substrates of Skp2^{SCF} have been identified, such as p21, p57, Myc and Foxo1 (Frescas and Pagano 2008). However, the relevance of these proteins to Skp2 function remains to be determined.

In addition to the well-established role of Skp2 in proteasomal degradation pathway, our recent work suggests that Skp2 also targets proteins for non-proteolytic K63 ubiquitination (Fig. 4.1c) (Chan et al. 2012; Wu et al. 2012). We demonstrated that Skp2 is responsible for ErbB-receptor-mediated Akt ubiquitination and membrane recruitment in response to EGF, in turn promoting aerobic glycolysis and tumorigenesis. Importantly, Skp2 deficiency results in a profound decrease in Akt activity and Glut1 expression, leading to suppressing mammary tumor development in the *MMTV-Neu* mice. Moreover, we further demonstrated that Skp2 overexpression in human breast cancers not only serve as a prognosis marker for poor survival in Her2-positive patients, but also as a potential therapeutic targets as Skp2 silencing renders Her2-overexpressing tumors more sensitive to Herceptin treatment. We also identified a role for Skp2 in the regulation of DNA damage response through inducing K63-ubiquitination of NBS1. Specifically, K63-ubiquitination of NBS1 promotes its interaction with ATM and subsequent ATM recruitment to the DNA foci for activation (Wu et al. 2012). As a result, inactivation of Skp2 results in a defect in homologous recombination (HR) repair and an increase in IR sensitivity. These findings collectively suggest that Skp2-mediatetd K63-ubiquitination may play a broader role in the regulation of signal transduction cascades.

Skp2 Suppresses Cellular Senescence Independent of ARF-p53

Although the ARF/p53 pathway has been demonstrated to play a critical role in cellular senescence response, recent reports suggest that cellular senescence responses can also be induced in an ARF/p53 independent manner (Fig. 4.1). *VHL* (von Hippel-Lindau), a tumor suppressor that is frequently mutated in many human cancers (Kaelin 2007), promotes the ubiquitination of oncogenic hypoxia-inducible factor (HIF) and its subsequent degradation. Interestingly, acute inactivation of *VHL* causes cellular senescence *in vitro* and *in vivo*, which critically depends on Rb activation and p400 reduction, but not HIF accumulation or p53 activation (Young et al. 2008). In addition, oncogenic Ras-induced OIS in human mammary epithelial cells (HMEC) with finite lifespan depends on p16/RB, but not p53 (Cipriano et al. 2011). Furthermore, anti-cancer compounds, such as resveratrol and doxorubicin, trigger p53-independent premature senescence in human squamous cell carcinoma (SCC) cells through mTOR-dependent inhibition of SIRT1 and subsequent increase in oxidative stress-mediated DNA damage (Back et al. 2011).

Several lines of evidence suggest that Skp2 may also play a role in senescence response. The overexpression of the human T lymphotropic virus type 1 (HTLV-1) Tax protein as well as VHL inactivation (Young et al. 2008) trigger Skp2 down regulation during cellular senescence. Our recent work provides strong genetic evidence that Skp2 is a critical player in cellular senescence response (Fig. 4.2) (Lin et al. 2010). Using *Skp2*-deficient MEFs, we demonstrated that inactivation of *Skp2* did not trigger cellular senescence response on its own, although stabilization of Skp2 substrates p21 and p27 was observed (Lin et al. 2010). Surprisingly, *Skp2* deficiency profoundly induced senescence response in the transformed cells, such as MEFs with overexpression of oncoproteins Ras and E1A or $Pten^{+/-}$ MEFs (Lin et al. 2010). Given that E1A overexpression enables Ras to overcome the ARF/p53-depdendent cellular senescence response, senescence response due to the inactivation of Skp2 may be independent of ARF/p53. In addition, these findings suggest that inactivation of Skp2 may suppress tumorigenesis in the presence of potent oncogenic stimuli. We have tested this using two cancer prone mouse models, $Pten^{+/-}$ mice, which develop spontaneous cancers, most notably in the lymph node and adrenal glands (Di Cristofano et al. 1998; Lin et al. 2010), and $p19Arf^{-/-}$ mice, which mainly develop sarcoma and lymphoma (Kamijo et al. 1999). Consistent with our findings in vitro, compared to the $Pten^{+/-}$ mice, we observed a significant reduction in the adrenal tumor formation and lymphadenopathy in the Pten/Skp2 compound mutant mice ($Pten^{+/-}$; $Skp2^{-/-}$), suggesting Skp2 inactivation strongly protects the tumor-prone mice from tumorigenesis (Lin et al. 2010). Moreover, tumorigenesis induced by prostate-specific inactivation of tumor suppressor Pten or induced by inactivation of tumor suppressor p19Arf was also strongly suppressed. Such tumor suppression may be attributed to the strong senescence response induced by Skp2 deficiency, as we found that senescent cells were significantly increased in the pre-malignant tissues from the Pten and Skp2 compound mutant mice such as lymph nodes and prostate, whereas no senescent cells were observed in the lymph nodes and only very few in prostates from the Pten mutant mice. In addition, there was a significant decrease in cell proliferation as determined by Ki67 staining in the pre-malignant tissues from the Pten and Skp2 compound mutant mice as compared to the Pten mutant mice. Importantly, we demonstrated that the senescence response triggered by Skp2 deficiency in these tumor prone mouse models (Pten inactivation or Arf loss) does not result in p53 activation or DNA damage response (DDR), but instead leads to a significant increase in the expression of cell cycle inhibitors p27, p21 and endoplasmic reticulum stress protein ATF4. Importantly, knockdown of p27, p21 and ATF4 all partially rescue the senescence response resulting from Skp2 inactivation, suggesting that they may synergistically contribute to the Skp2 regulated senescence response (Lin et al. 2010).

Strategies for Pro-senescence Therapy

Since cellular senescence has emerged as an important physiological barrier for suppressing tumor progression, there is an increasing interest in the development of novel pro-senescence therapy for human cancers. Given that the tumor suppressor p53 is an important effector for senescence, targeting the p53 pathway may be a potential approach to induce senescence for treating human cancers. However, since p53 is one of the most frequently mutated or loss tumor suppressors in human cancers, the status of p53 in tumors is a critical factor that determines the pro-senescence approaches used in cancer treatment. Extensive research on the biology of p53 has led to the discovery of several small molecules that can promote and restore p53 activity (Nardella et al. 2011). Among these, nutlins, found to inhibit the interaction of MDM2 with p53, was shown to induce senescence in cancer cells. In addition, in a preclinical model of pro-senescence therapy based on PICS, nutlins were able to strongly enhance the p53 response (Alimonti et al. 2010; Nardella et al. 2011). Moreover, intensive efforts have been made to develop methods to restore the functions of the mutant p53, as loss of p53 function through mutation of its DNA-binding domain (DBD) occurs in many tumors, resulting in alteration in the folding of the wild-type p53 protein and its normal transcriptional program. Recently, it has been demonstrated that novel drugs, such as PRIMA-1MET (also known as APR-246) and ellipticine, can interact with mutant p53 to induce conformational change of its structure, resulting in reversion from mutant p53 to wild type p53 activity and reactivation of some p53 transcriptional targets, such as CDKN1A.

Another potential target for pro-senescence therapy is oncoprotein Myc. Numerous studies have shown that Myc is overexpressed in many human cancers due to retroviral transduction, retroviral promoter or enhancer insertion, chromosomal translocation, and gene amplification. In addition, studies using Myc transgenic mice revealed that Myc overexpression can promote cancers. Moreover, several mouse tumor models that are driven by MYC have recently been demonstrated that inactivation of Myc in established tumors results in tumor regression, and cellular senescence has been identified as a specific mechanism to suppress the development of various tumor types, including lymphoma, osteosarcoma and hepatocellular carcinoma. These findings suggest that targeting Myc is a potential approach for cancer treatments. Although it remains difficult to target transcription factors directly, Myc can be inactivated by an artificial dimerization partner known as Omomyc, resulting in tumor regression along with apoptosis and senescence responses (Soucek et al. 2008). However, such established tumors should have competent p53 and INK4A to remain sensitive to the transforming growth factor-β signaling pathway in order to elicit senescence induction upon Myc inactivation.

Given that many tumors present with monoallelic loss of *PTEN*, especially at early onset, targeting of tumor suppressor PTEN to elicit senescence may be one possible therapeutic approach. In support of this notion, it has recently been shown that the transient use of PTEN-targeting drugs, such as VO-OHpic, can temporarily and selectively inactivate PTEN activity in the tumor cells expressing one allele of Pten, which induces a senescence response through a signaling short circuit that is driven by hyperactivation of a PI3K–AKT–mTOR–p53 signaling pathway (Alimonti et al. 2010; Nardella et al. 2011), whereas only a marginal increase in the activation of the AKT–mTOR signaling pathway due to a transient decrease in PTEN activity in wild-type cells that express PTEN at normal levels (Alimonti et al. 2010; Nardella et al. 2011).

Since p53 is the commonly mutated gene in human cancers, any approach for pro-senescence strategy relying on intact p53 cannot be applied to tumors with *p53* loss or inactivation. We have demonstrated that inactivation of *SKP2* sensitizes tumor cells to senescence induction through the accumulation of p27, p21 and ATF4, even in the absence of an intact p53 pathway, suggesting that targeting *p53*-independent cellular senescence pathways may be the key for ensuring success of human cancer treatments. MLN4924, a NEDD8 activating enzyme inhibitor that inhibits Skp2 SCF complex activity and is currently in phase I clinical trials, has been shown to trigger p53-independent senescence and repress prostate tumors with p53 inactivation (Lin et al. 2010). Thus, it will be useful to develop Skp2-specific small molecule inhibitors, as Skp2 inhibition will not only result in cell cycle arrest, but also senescence. It will be important to test the efficacy of Skp2 inhibitors alone or in combination with other therapies in vivo, which may be beneficial for the treatment of human cancers. Another potential target for induction of p53-independent cellular senescence in cancer therapy is Cdk2, as its inactivation also elicits p53-independent cellular senescence. In fact, several small molecule inhibitors of Cdk2 have already been developed and among them, two were able to induce senescence in Myc-overexpressed leukemia cancer cells with ablated p53 function (Hydbring et al. 2010). However, future studies are necessary to determine the efficacy of such Cdk2 inhibitors to suppress tumor growth in vivo. Moreover, a recent study provides genetic and pharmacological evidence demonstrating that ablation of Cdk4 provokes senescence responses to attenuate K-Ras-driven lung adenocarcinoma (Puyol et al. 2010).

Discussion

Although cellular senescence was originally identified as response induced by cell culture stress, it is now well-recognized that senescence response serve as a natural and intrinsic physiological response to prevent caner progression by eliminating damaged or arrested cells (Acosta and Gil 2012; Nardella et al. 2011; Rodier and Campisi 2011). Even though the regulatory mechanisms remain obscure, the complex signaling network underlying senescence response may offer many opportunities for novel therapeutic discovery for cancer treatments.

Since many conventional chemotherapeutic agents and radiotherapy regimens are able to induce apoptosis and also trigger senescence-like response, the relative contribution of senescence to the therapeutic activities of these cancer treatments still remains to be determined. In order to establish a drug discovery program to target the senescence pathway, several important questions still remain to be addressed. First, in order to determine the relative contribution of senescence to activities of current standard of care treatment for cancer, we need to establish the "gold standard" marker to reliably and efficiently detect and measure senescence response *in vitro* and *in vivo*. Second, pro-senescence therapy can be regarded as a form of targeted therapy, which will require intensive investigation to determine which cancer types are more sensitive to pro-senescence therapy. In addition, the possible feedback mechanisms exist in various targeted therapy should be studied. Third, considering the importance of cancer initiating cell in tumor recurrence and metastasis, it will be intriguing to see whether pro-senescence therapy can have an effect on such cell populations. Fourth, we need to further understand how the senescence response operates with regard to its tumor suppression and tumor promotion processes, as senescence response may result in some tumor-promoting effects, such as SASP acquired by the senescent cells to trigger the malignant phenotype of neighboring cells (Krtolica et al. 2001). Fifth, we are still lacking robust compounds to induce senescence response.

Nevertheless, the advances in identifying p53-dependent and p53-independent senescence pathways not only shed new light on cellular senescence programs, but also pave the ways for developing potent senescence-inducing agents for "pro-senescence" therapy in the clinical settings.

Acknowledgements This work is supported in part by National Institutes of Health grants (R01CA136787-01A2 and R01CA149321-01), MD Anderson Trust Scholar Fund, a grant from Cancer Prevention Research Institute of Texas and by a New Investigator Award from the Department of Defense (PC081292) to H.K. Lin.

References

Acosta JC, Gil J (2012) Senescence: a new weapon for cancer therapy. Trends Cell Biol 22:211–219

Alimonti A, Nardella C, Chen Z et al (2010) A novel type of cellular senescence that can be enhanced in mouse models and human tumor xenografts to suppress prostate tumorigenesis. J Clin Invest 120:681–693

Back JH, Rezvani HR, Zhu Y et al (2011) Cancer cell survival following DNA damage-mediated premature senescence is regulated by mammalian target of rapamycin (mTOR)-dependent inhibition of sirtuin 1. J Biol Chem 286:19100–19108

Bartkova J, Rezaei N, Liontos M et al (2006) Oncogene-induced senescence is part of the tumorigenesis barrier imposed by DNA damage checkpoints. Nature 444:633–637

Braig M, Lee S, Loddenkemper C et al (2005) Oncogene-induced senescence as an initial barrier in lymphoma development. Nature 436:660–665

Brown JP, Wei W, Sedivy JM (1997) Bypass of senescence after disruption of p21CIP1/WAF1 gene in normal diploid human fibroblasts. Science 277:831–834

Chan CH, Li CF, Yang WL et al (2012) The Skp2-SCF E3 ligase regulates Akt ubiquitination, glycolysis, herceptin sensitivity, and tumorigenesis. Cell 149:1098–1111

Chen Z, Trotman LC, Shaffer D et al (2005) Crucial role of p53-dependent cellular senescence in suppression of Pten-deficient tumorigenesis. Nature 436:725–730

Chen Z, Carracedo A, Lin HK et al (2009) Differential p53-independent outcomes of p19(Arf) loss in oncogenesis. Sci Signal 2:ra44

Cipriano R, Kan CE, Graham J et al (2011) TGF-beta signaling engages an ATM-CHK2-p53-independent RAS-induced senescence and prevents malignant transformation in human mammary epithelial cells. Proc Natl Acad Sci U S A 108:8668–8673

d'Adda di Fagagna F, Reaper PM, Clay-Farrace L et al (2003) A DNA damage checkpoint response in telomere-initiated senescence. Nature 426:194–198

Di Cristofano A, Pesce B, Cordon-Cardo C et al (1998) Pten is essential for embryonic development and tumour suppression. Nat Genet 19:348–355

Di Micco R, Fumagalli M, Cicalese A et al (2006) Oncogene-induced senescence is a DNA damage response triggered by DNA hyper-replication. Nature 444:638–642

Ding Z, Wu CJ, Chu GC et al (2011) SMAD4-dependent barrier constrains prostate cancer growth and metastatic progression. Nature 470:269–273

Frescas D, Pagano M (2008) Deregulated proteolysis by the F-box proteins SKP2 and beta-TrCP: tipping the scales of cancer. Nat Rev Cancer 8:438–449

Hayflick L, Moorhead PS (1961) The serial cultivation of human diploid cell strains. Exp Cell Res 25(3):585–621

Herbig U, Jobling WA, Chen BP et al (2004) Telomere shortening triggers senescence of human cells through a pathway involving ATM, p53, and p21(CIP1), but not p16(INK4a). Mol Cell 14:501–513

Herbig U, Ferreira M, Condel L et al (2006) Cellular senescence in aging primates. Science 311:1257

Hydbring P, Bahram F, Su Y et al (2010) Phosphorylation by Cdk2 is required for Myc to repress Ras-induced senescence in cotransformation. Proc Natl Acad Sci U S A 107:58–63

Kaelin WG (2007) Von Hippel-Lindau disease. Annu Rev Pathol 2:145–173

Kamijo T, Bodner S, van de Kamp E et al (1999) Tumor spectrum in ARF-deficient mice. Cancer Res 59:2217–2222

Karlseder J, Smogorzewska A, de Lange T (2002) Senescence induced by altered telomere state, not telomere loss. Science 295:2446–2449

Krtolica A, Parrinello S, Lockett S et al (2001) Senescent fibroblasts promote epithelial cell growth and tumorigenesis: a link between cancer and aging. Proc Natl Acad Sci U S A 98:12072–12077

Lin AW, Barradas M, Stone JC et al (1998) Premature senescence involving p53 and p16 is activated in response to constitutive MEK/MAPK mitogenic signaling. Genes Dev 12:3008–3019

Lin HK, Chen Z, Wang G et al (2010) Skp2 targeting suppresses tumorigenesis by Arf-p53-independent cellular senescence. Nature 464:374–379

Michaloglou C, Vredeveld LC, Soengas MS et al (2005) BRAFE600-associated senescence-like cell cycle arrest of human naevi. Nature 436:720–724

Moiseeva O, Mallette FA, Mukhopadhyay UK et al (2006) DNA damage signaling and p53-dependent senescence after prolonged beta-interferon stimulation. Mol Biol Cell 17:1583–1592

Nardella C, Clohessy JG, Alimonti A et al (2011) Pro-senescence therapy for cancer treatment. Nat Rev Cancer 11:503–511

Puyol M, Martin A, Dubus P et al (2010) A synthetic lethal interaction between K-Ras oncogenes and Cdk4 unveils a therapeutic strategy for non-small cell lung carcinoma. Cancer Cell 18:63–73

Rodier F, Campisi J (2011) Four faces of cellular senescence. J Cell Biol 192:547–556

Serrano M, Lin AW, McCurrach ME et al (1997) Oncogenic ras provokes premature cell senescence associated with accumulation of p53 and p16INK4a. Cell 88:593–602

Song MS, Carracedo A, Salmena L et al (2011) Nuclear PTEN regulates the APC-CDH1 tumor-suppressive complex in a phosphatase-independent manner. Cell 144:187–199

Soucek L, Whitfield J, Martins CP et al (2008) Modelling Myc inhibition as a cancer therapy. Nature 455:679–683

Takai H, Smogorzewska A, de Lange T (2003) DNA damage foci at dysfunctional telomeres. Curr Biol 13:1549–1556

Vogelstein B, Kinzler KW (2004) Cancer genes and the pathways they control. Nat Med 10:789–799

Wang Y, Blandino G, Givol D (1999) Induced p21waf expression in H1299 cell line promotes cell senescence and protects against cytotoxic effect of radiation and doxorubicin. Oncogene 18:2643–2649

Whibley C, Pharoah PD, Hollstein M (2009) p53 polymorphisms: cancer implications. Nat Rev Cancer 9:95–107

Wu J, Zhang X, Zhang L et al (2012) Skp2 E3 ligase integrates ATM activation and homologous recombination repair by ubiquitinating NBS1. Mol Cell 46:351–361

Xue W, Zender L, Miething CD et al (2007) Senescence and tumour clearance is triggered by p53 restoration in murine liver carcinomas. Nature 445:656–660

Young AP, Schlisio S, Minamishima YA et al (2008) VHL loss actuates a HIF-independent senescence programme mediated by Rb and p400. Nat Cell Biol 10:361–369

Zhu J, Woods D, McMahon M, Bishop JM (1998) Senescence of human fibroblasts induced by oncogenic Raf. Genes Dev 12:2997–3007

Oncogene-Induced Senescence: Role of Mitochondrial Dysfunction

Olga Moiseeva and Gerardo Ferbeyre

Contents

Abstract 45

Senescence 46

Mitochondria 46
Energy Metabolism in Normal Cells 47
Energy Metabolism in Cancer Cells 47
Energy Metabolism During Senescence 47

Role of P53 in Regulation of Mitochondrial
Biogenesis 49

Discussion 49

References 51

Abstract

Mitochondrial biogenesis is activated in response to different signals and environmental stimuli. Recent data demonstrate an important role for mitochondrial biogenesis in the development of cellular senescence. Normal cellular response to increased energy demand after oncogene stimulation includes the induction of mitochondrial respiration. However, strong and prolonged activation of oxidative phosphorylation leads to oxidative stress, nuclear and mitochondrial DNA damage, mitochondrial dysfunction and senescence. Mutations in tumor suppressors in cooperation with activated oncogenes trigger the shift from mitochondrial respiration to aerobic glycolysis supplying the cells with ATP and metabolic substrates. It has been shown that the reverse shift has antiproliferative effect on cancer cells. As in normal proliferating cells mtDNA concentration is proportional to the oxidative capacity of the cell, the antitumor therapy can be based on the stimulation of mtDNA replication. It is expected that such therapy will stimulate mitochondrial respiration and additional disruption of oxidative phosphorylation will produce more free radicals, DNA damage response and senescence.

Keywords

Cancer cells • Energy metabolism • Mitochondrial dysfunction • Mitochondrial single-stranded DNA-binding protein (mtSSB)

O. Moiseeva (✉) • G. Ferbeyre
Departmente de Biochimie, Universite de Montreal,
C.P. 6128, Succ. Centre-Ville, Montreal, QC
H3C 3J7, Canada
e-mail: olga.moiseeva@umontreal.com

- Oxidative phosphorylation (OXPHOS)
- Senescence-associated heterochromatin foci (SAHF) • Senescence-associated secretory phenotype (SASP) • Tumor development

Senescence

In 1961 Leonard Hayflick discovered that primary cells divide only a finite number of times in culture (Hayflick and Moorhead 1961). This phenomenon of permanent proliferative arrest is known as a replicative senescence. Later it was shown that primary cells undergo a permanent cell cycle arrest in response to oncogene overexpression and other stress stimuli. A key property of senescent cells is that they are incapable of triggering the expression of genes required for proliferation in response to growth factors (Seshadri and Campisi 1990).

Senescence is characterized by several biomarkers, among which senescence-associated β-galactosidase activity (SA-β-gal) is the most characteristic one. In most cases senescence is also accompanied by increased production of free radical species (ROS), DNA-damage response, formation of senescence-associated heterochromatin foci (SAHF), production and release of a variety of pro-inflammatory cytokines and chemokines, so called senescence-associated secretory phenotype (SASP).

It has been proposed that different mechanisms cooperate to promote and maintain cellular senescence. One of them is a DNA-damage response, which is caused by ROS (Lee et al. 1999), excessive replication and short telomeres (Di Micco et al. 2006). DNA damage response is mediated by ATM and ATR through their target p53. Another tumor suppressive pathway involved in the development of senescence is p16-RB. This pathway appears to be required for SAHF formation, which represents DAPI-stained condensed chromatin. Functionally SAHF have been implicated in transcriptional silencing of proliferation-promoting genes. The role of senescence-associated cytokines in cellular senescence is not clear but they may also stimulate ROS production (Acosta et al. 2008).

Replicative senescence is a well-characterized process caused by progressive attrition of the telomeres, chromosome ends. Shot telomeres are recognized as DNA breaks that trigger DNA-damage response leading to cell cycle arrest in G1.

Some environmental factors, for example, oxidative stress, ionizing radiation, numerous chemical compounds trigger stress-induced premature senescence (SIPS) (Brack et al. 2000).

As it has been mentioned above, activated oncogenes, genes encoding proteins that in activated form are involved in transformation of immortalized cells, also trigger senescence, called oncogene-induced senescence (OIS). Serrano et al. (1997) discovered that expression of oncogenic *ras* induced senescence via activation of p53 and p16 pathways in primary fibroblasts. It has been proposed that OIS provides the barriers for cellular transformation *in vitro* and *in vivo* (Sarkisian et al. 2007).

Mitochondria

Mitochondria are intracellular organelles with their own genome. There are multiple copies of mitochondrial DNA (mtDNA) per cell and the amount of mtDNA varies according to the cellular metabolism. Within the mitochondria mtDNA is packaged into nucleoprotein complex called nucleoid. In human, mtDNA encodes 13 mRNAs of essential oxidative phosphorylation (OXPHOS) members as well as the mitochondrial rRNAs and tRNAs (Attardi and Schatz 1988). Mitochondria are the major source of ROS in the cell. Mitochondrial ROS is a by-product of oxidative phosphorylation. Among mitochondrial complex I and complex III, the first one is the major superoxide producer. In normally working mitochondria superoxide production is rather low and it is rapidly converted into hydrogen peroxide by manganese superoxide dismutase (MnSOD). Hydrogen peroxide in the mitochondrial matrix is converted into water by glutathione peroxidases or diffuse from mitochondrial matrix into cytoplasm where catalase converts it into water. Superoxide, produced by complex I, releases inside the mitochondrial matrix and causes mtDNA damage. As mtDNA lacks protective histones and have less efficient DNA repair systems than the nucleus, mutation rate of mitochondrial genome is tenfold higher than nuclear genome (Polyak et al. 1998). mtDNA mutations may cause malfunction of the respiratory chain which in turn leads to increased generation of ROS and further DNA damage.

Energy Metabolism in Normal Cells

The concentration of ATP, ADP, and AMP reflect the cellular energy state and can be expressed as adenylate energy charge. The energy charge is similar for the most cells, while the ATP level varies among tissues and species. According to Atkinson (1968), cells stabilize their energy charge. Decrease in ATP concentration during ATP-consuming processes and concomitant increase of AMP should lead to AMP degradation or to stimulation of ATP production for energy charge stabilization (Ataullakhanov and Vitvitsky 2002). It has been demonstrated that normal cells stimulate OXPHOS activity and ATP production on ATP demand (Nagino et al. 1989). OXPHOS is under a tight control by the amount of mtDNA in cell. In normal proliferating cells mtDNA concentration is proportional to the oxidative capacity of the cell (Rocher et al. 2008). The mitochondrial D-loop contains the sites that initiate replication and transcription, fitting the model that all transcription enhancers also enhance replication and vice versa.

Energy Metabolism in Cancer Cells

Tumor development is a complex process that not only involves the activation of oncogenes and repression of tumor suppressors, but also changes in mitochondrial biogenesis. In 1924 Otto Warburg discovered that cancer cells have an activated glycolysis even in presence of oxygen, called aerobic glycolysis (Warburg et al. 1924). This switch from OXPHOS towards aerobic glycolysis in cancer cells is also known as a Warburg effect. The phenomenon of Warburg effect remains unclear. Cancer cells can use aerobic glycolysis to produce substrates to synthesize proteins, nucleic acids, and fatty acids. To satisfy increasing demand in ATP rapidly dividing cells facilitate glucose transport and transcription of a number of genes that encode the proteins of glycolytic pathway. The well-established shift from OXPHOS to glycolysis in cancer cells may indicate that mitochondrial metabolism is somehow defective. However, there are no convincing data showing that mitochondrial dysfunction is a cause of Warburg effect or that the OXPHOS is irreversibly impaired in cancer cells. Some somatic mutations were found in tumors, but mitochondrial function was largely not affected (Polyak et al. 1998).

Some mechanisms have been proposed to induce aerobic glycolysis. One of them is the stimulation of glycolysis by oncogenes. Involvement of oncogenic signaling in development of Warburg effect has been demonstrated during RAS and AKT transformation. Oncogenic activation of the AKT pathway results in increased glucose uptake, enhanced glucose phosphorylation by hexokinase that facilitates the glycolytic rate. Glycolytic activation inhibits OXPHOS, for example, by competition for NADH (Greiner et al. 1994). Hypoxic cancer cells environment can also stimulate glycolysis through activation of HIF-1. However, as it has been demonstrated Warburg effect is not universal for cancers and probably mostly it is the consequence of hypoxia in solid tumors (Guppy et al. 2002). Some types of cancer cells use enhanced glycolysis coupled with increased OXPHOS (Ramanathan et al. 2005), while transformed human mesenchymal stem cells depend mostly on OXPHOS (Funes et al. 2007). Thus, mitochondrial biogenesis in cancer cells show a variety depending on cell type and genetics.

Energy Metabolism During Senescence

As a senescence is a barrier for transformation, mitochondrial response to activated oncogenes during senescence could help us to understand what mitochondrial alterations prevent tumorigenesis. It has been shown that the mitochondrial ROS production was a major cause of replicative senescence. The reduction of mitochondrial ROS inhibited telomere shortening and DNA-damage response (Passos et al. 2007). Nearly all stresses that induce senescence are thought to increase intracellular ROS (Fridovich 1998). However, it is not clear why mitochondrial ROS are increased during senescence. Our data demonstrate that OIS induced by oncogenic RAS is also accompanied by ROS production and DNA damage (Moiseeva et al. 2009). Activated RAS triggers premature senescence after initial burst of proliferation (Serrano et al. 1997). During this proliferation phase, there is an increase in mitochondrial biogenesis in RAS-expressing cells to cope with the increased energy demand stimulated by RAS. Facilitated OXPHOS leads to ROS production and mtDNA damage that further increases ROS level in the cells. Finally, mitochondrial malfunction leads to decrease of mitochondrial membrane potential, ATP shortage, activation of AMPK, high mtDNA level and senescence (Moiseeva et al. 2009; Hutter et al. 2004). The level of RAS expression correlates with the strength of the senescence response. High level of RAS leads to induction of the senescence program, whereas low level of RAS leads to cellular hyperproliferation. RAS-induced senescence can be bypassed by expressing E7 and small hairpin RNA against p53 (shp53) in primary human fibroblasts. Interestingly, E7 and depletion of p53 prevent induction of mtDNA replication (Fig. 5.1) and ROS production (Moiseeva et al. 2009) as well as coexpression of E1A, MYC and RAS, resulting in fibroblasts transformation (Fig. 5.1). It is not clear how these rapidly proliferating cells satisfy the increased energy demand, but it is not due to the normal response increasing mtDNA level. However, E7shp53RAS-expressing fibroblasts produce ATP through OXPHOS, since an inhibition of ATP synthesis by oligomycin restored the senescence phenotype (Fig. 5.2). These results indicate that mitochondrial dysfunction is the major factor inducing OIS and that the bypass of senescence involves a correction of its energy limitations.

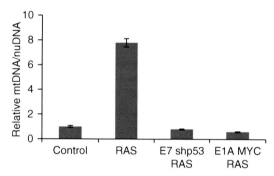

Fig. 5.1 mtDNA level in human primary fibroblasts expressing different combination of indicated proteins

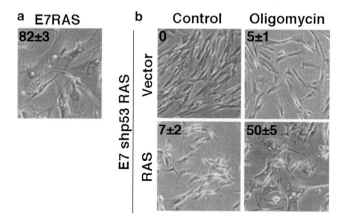

Fig. 5.2 Oligomycin induces SA-β-gal staining in human primary fibroblasts expressing E7 shp53 RAS. (**a**) Positive control (senescent fibroblasts after 8 days of E7 RAS infection), (**b**) E7 shp53 RAS expressing fibroblasts with and without oligomycin treatment. Cells were treated with 7 μg/ml of oligomycin for 6 days

Role of P53 in Regulation of Mitochondrial Biogenesis

It is well established that cellular senescence is induced and maintained through Rb and p53 pathways. However, in most cases p53 depletion is sufficient to bypass senescence phenotype indicating the key role of p53 in senescence development. Furthermore, p53 plays a key role in cancer and loss of p53 results in Warburg effect. P53 is a predominantly nuclear protein, lacking a mitochondrial targeting sequence. After stress-induced stabilization p53 translocates to the mitochondria via HMD2 promoting monoubiquitylation (Marchenko et al. 2007). P53 can be implicated in mitochondrial metabolism at two levels, at the level of transcription in the nucleus and in mitochondria as p53 was found in mitochondrial matrix. In the nucleus, p53 regulates the transcription of p53R2 subunit of ribonucleotide reductase. In the mitochondria, p53 physically interacts with mitochondrial Pol γ and mtDNA, and facilitates the mtDNA replication in response to mtDNA damage induced by ROS (Achanta et al. 2005). Depletion of p53 is accompanied by reduction in mtDNA level, TFAM and p53R2 (Lebedeva et al. 2009). Mitochondrial single-stranded DNA-binding protein (mtSSB) is also involved in mtDNA replication (Clayton 1982). P53 physically interacts with mtSSB and mitochondrial transcription factor A (TFAM) (Yoshida et al. 2003), which is essential for the transcription and replication of mtDNA, since fragments of RNA transcripts from mtDNA are required for priming mtDNA replication. Also p53 stimulates OXPHOS by inducing the expression of SCO2, the protein of cytochrome c oxidase complex (Matoba et al. 2006) and inhibits glycolysis by expressing TP53-induced glycolysis and apoptosis regulator (TIGAR) (Bensaad et al. 2006).

Discussion

The mitochondria are known to play a central role in such important cellular processes as energy metabolism, apoptosis, ROS production and calcium signaling. Recent data demonstrate the involvement of mitochondria in development of cellular senescence. Numerous reports elucidate dramatic changes in mitochondrial biogenesis during tumorigenesis and aging. Furthermore, the cellular response is different depending on the nature of stimulus. For instance, normal cells stimulate OXPHOS in response to energy demand. At the same time cancer cells in many cases use aerobic glycolysis to compensate their energy consumption.

Concerning to the Mitochondrial Free Radicals Theory of Aging, proposed by Denham Harman, aging is the result of the accumulation of oxidative damage caused by free radicals of mitochondrial origin generated as by-products during normal metabolism (Harman 1956). Free radicals damage mitochondrial DNA mtDNA affecting mitochondrial functions and ATP production. To compensate the reduction of OXPHOS activity aging cells stimulate mtDNA replication since mtDNA transcription can be regulated by mtDNA copy number. Indeed, the mtDNA copy number was found to increase with age, but the amount of mitochondrial protein did not change (Barrientos et al. 1997; Pesce et al. 2001). It means that some mtDNA is not transcribed or mtDNA is transcribed at lower level. *In vitro* data also demonstrate that cells in replicative senescence have increased mtDNA content (Passos et al. 2007; Hutter et al. 2004; Lee et al. 2002). Our results with RAS-induced senescence show that stimulation of mtDNA replication starts as soon as 2 days after infection when fibroblasts increase their proliferation rate and energy demand. However, during that time the level of ROS production is increased, leading to mitochondrial malfunction and ATP depletion, which in turn further stimulates mtDNA replication (Moiseeva et al. 2009). Therefore, OIS is also characterized by increased mtDNA content. We suggest that normal cells stimulate OXPHOS in response to increased energy demand but prolonged strong OXPHOS activation is a cause of excessive ROS production, mitochondrial dysfunction and senescence. Similarly, mitochondrial dysfunction caused by expression of shRNA against Rieske iron sulfur protein (RISP) leads to induction of mtDNA replication, ROS production and senescence (Moiseeva et al. 2009).

Interestingly, cancer cells have the same or less mtDNA compare to normal cells (Simonnet et al. 2002; Wallace 2005). Moreover, oncocytic or oxyphilic tumors, which have a four to sevenfold increase in mtDNA, are usually benign tumors, rarely inducing metastasis (Tallini 1998). Cancer cells have mutations, which change the energy metabolism from OXPHOS to aerobic glycolysis. Initially it has been suggested that Warburg effect in cancer cells is induced by impaired OXPHOS. However, there is no evidence of whether mtDNA mutations themselves contribute to the development of the aerobic glycolysis. In many cancer cell types OXPHOS capacity is normal and changes in energy metabolism can be explained as a switch from OXPHOS to glycolysis. Immediately after H-RasV12/E1A transformation of MEFs there is a sharp increase in oxygen consumption, ROS production and cell death. Then there is smooth transition towards higher glycolytic rates later during tumorigenesis. Authors suggest that the OXPHOS induction in early transformed cells may be a general concept that precedes the Warburg effect and tumorigenic development (deGroof et al. 2009). Another group have also demonstrated that at the beginning of tumorigenesis BJ fibroblasts had increase in mitochondrial mass while the most transformed cells didn't have elevated mtDNA content (Ramanathan et al. 2005). Importantly, cellular senescence occurs in premalignant human lesions and mDNA levels may help to identify those lesions. The study of the different stages of tumorigenesis revealed that DNA damage response activates in the early dysplastic stage and it is lost progressively during cancer progression (Bartkova et al. 2005). These results demonstrate the correlation between increased mtDNA content and OXPHOS from one side and DNA damage and senescence from another in premalignant lesions. It means that cancer cells escape senescence by preventing OXPHOS activation in response to ATP demand. We suggest that malignant cells can prevent OXPHOS stimulation by several mechanisms. Cells can increase the efficiency of mitochondrial respiration, for instance, by inducing the expression of mitochondrial chaperons. Then, cells can stabilize their energy charge by decreasing the adenylate pool through AMP degradation. Some oncogenes trigger the shift from OXPHOS to aerobic glycolysis.

Recent data support the idea that the opposite shift is also possible. What is important that the reverse switch to OXPHOS correlate with cancer growth retardation *in vitro* and *in vivo*. Rossignol et al. (2004) showed that the substrate availability can promote the changes in mitochondria biogenesis. In galactose/glutamine medium HeLa cells have decreased proliferating rate and increased mitochondrial respiration. Interestingly, mtDNA was not increased, but cells grown in galactose had mitochondria with more cristae.

Other reports describe the shift from glycolysis to OXPHOS in cancer cells with stable knockdown of lactate dehydrogenase A (Fantin et al. 2006; Le et al. 2010). Lactate dehydrogenase is an enzyme that converts pyruvate to lactate. Accumulated pyruvate stimulates the OXPHOS leading to diminution of tumorigenicity and compromising cell growth under hypoxia.

Oncogene Bcr-Abl has also been implicated to play a role in glycolysis, and inhibition of Bcr-Abl by Imatinib (STI571) seems to reverse the Warburg effect by switching glucose metabolism from glycolysis to mitochondrial oxidative phosphorylation (Gottschalk et al. 2004). Activation of OXPHOS was accompanied by cell growth inhibition of chronic myelogenous leukemia cells.

Another report of switching from aerobic glycolysis to OXPHOS came from Bonnet et al. (2007). It has been shown that dichloroacetate (DCA), well-known inhibitor of pyruvate dehydrogenase kinase (PDK), shifted metabolism of three cancer cell lines from glycolysis to glucose oxidation. DCA inhibited proliferation, induced apoptosis and diminished tumor growth *in vivo*.

The data support the idea that aerobic glycolysis is reversible and its prevention leads to cell growth inhibition or apoptosis. However, the shift from aerobic glycolysis to OXPHOS does not induce mtDNA replication, indicating that an additional stimulation of mtDNA replication might be required to trigger ROS production and cellular senescence in cancer cells.

References

Achanta G, Sasaki R, Feng L, Carew JS, Lu W, Pelicano H, Keating MJ, Huang P (2005) Novel role of p53 in maintaining mitochondrial genetic stability through interaction with DNA Pol gamma. EMBO J 24:3482–3492

Acosta JC, O'Loghlen A, Banito A, Guijarro MV, Augert A, Raguz S, Fumagalli M, Da Costa M, Brown C, Popov N, Takatsu Y, Melamed J, d'Adda di Fagagna F, Bernard D, Hernando E, Gil J (2008) Chemokine signaling via the CXCR2 receptor reinforces senescence. Cell 133:1006–1018

Ataullakhanov FI, Vitvitsky VM (2002) What determines the intracellular ATP concentration. Biosci Rep 22:501–511

Atkinson DE (1968) The energy charge of the adenylate pool as a regulatory parameter. Interaction with feedback modifiers. Biochemistry 7:4030–4034

Attardi G, Schatz G (1988) Biogenesis of mitochondria. Annu Rev Cell Biol 4:289–333

Barrientos A, Casademont J, Cardellach F, Ardite E, Estivill X, Urbano-Marquez A, Fernandez-Checa JC, Nunes V (1997) Qualitative and quantitative changes in skeletal muscle mtDNA and expression of mitochondrial-encoded genes in the human aging process. Biochem Mol Med 62:165–171

Bartkova J, Horejsi Z, Koed K, Kramer A, Tort F, Zieger K, Guldberg P, Sehested M, Nesland JM, Lukas C, Orntoft T, Lukas J, Bartek J (2005) DNA damage response as a candidate anti-cancer barrier in early human tumorigenesis. Nature 434:864–870

Bensaad K, Tsuruta A, Selak MA, Vidal MN, Nakano K, Bartrons R, Gottlieb E, Vousden KH (2006) TIGAR, a p53-inducible regulator of glycolysis and apoptosis. Cell 126:107–120

Bonnet S, Archer SL, Allalunis-Turner J, Haromy A, Beaulieu C, Thompson R, Lee CT, Lopaschuk GD, Puttagunta L, Bonnet S, Harry G, Hashimoto K, Porter CJ, Andrade MA, Thebaud B, Michelakis ED (2007) A mitochondria-K+ channel axis is suppressed in cancer and its normalization promotes apoptosis and inhibits cancer growth. Cancer Cell 11:37–51

Brack C, Lithgow G, Osiewacz H, Toussaint O (2000) EMBO WORKSHOP REPORT: Molecular and cellular gerontology Serpiano, Switzerland, September 18–22, 1999. EMBO J 19:1929–1934

Clayton DA (1982) Replication of animal mitochondrial DNA. Cell 28:693–705

deGroof AJC, teLindert MM, van Dommelen MMT, Wu M, Willemse M, Smift AL, Winer M, Oerlemans F, Pluk H, Fransen JAM, Wieringa B (2009) Increased OXPHOS activity precedes rise in glycolytic rate in H-RasV12/E1A transformed fibroblasts that develop a Warburg phenotype. Mol Cancer 8:54–67

Di Micco R, Fumagalli M, Cicalese A, Piccinin S, Gasparini P, Luise C, Schurra C, Garre' M, Nuciforo PG, Bensimon A, Maestro R, Pelicci PG, d'Adda di Fagagna F (2006) Oncogene-induced senescence is a DNA-damage response triggered by DNA hyper-replication. Nature 444:638–642

Fantin VR, St-Pierre J, Leder P (2006) Attenuation of LDH-A expression uncovers a link between glycolysis, mitochondrial physiology, and tumor maintenance. Cancer Cell 9:425–434

Fridovich I (1998) Oxygen toxicity: a radical explanation. J Exp Biol 201:1203–1209

Funes JM, Quintero M, Henderson S, Martinez D, Qureshi U, Westwood C, Clements MO, Bourboulia D, Pedley RB, Moncada S, Boshoff C (2007) Transformation of human mesenchymal stem cells increases their dependency on oxidative phosphorylation for energy production. Proc Natl Acad Sci U S A 104:6223–6228

Gottschalk S, Anderson N, Hainz C, Eckhardt SG, Serkova NJ (2004) Imatinib (STI571)-mediated changes in glucose metabolism in human leukemia BCR-ABL-positive cells. Clin Cancer Res 10:6661–6668

Greiner EF, Guppy M, Brand K (1994) Glucose is essential for proliferation and the glycolytic enzyme induction that provokes a transition to glycolytic energy production. J Biol Chem 269:31484–31490

Guppy M, Leedman P, Zu X, Russel V (2002) Contribution by different fuels and metabolic pathways to the total ATP turnover of proliferating MCF-7 breast cancer cells. Biochem J 364:309–315

Harman D (1956) A theory based on free radical and radical chemistry. J Gerontol 11:298–300

Hayflick L, Moorhead PS (1961) The serial cultivation of human diploid cell strains. Exp Cell Res 25:585–621

Hutter E, Renner K, Pfister G, Stockl P, Jansen-Durr P, Gnaiger E (2004) Senescence-associated changes in respiration and oxidative phosphorylation in primary human fibroblasts. Biochem J 380:919–928

Le A, Cooper CR, Gouw AM, Dinavahi R, Maitra A, Deck LM, Royer RE, Jagt DLV, Semenza GL, Dang CV (2010) Inhibition of lactate dehydrogenase A induces oxidative stress and inhibits tumor progression. Proc Natl Acad Sci U S A 107:2037–2042

Lebedeva MA, Eaton JS, Shadel GS (2009) Loss of p53 causes mitochondrial DNA depletion and altered mitochondrial reactive oxygen species homeostasis. Biochim Biophys Acta 1787:328–334

Lee AC, Fenster BE, Ito H, Takeda K, Bae NS, Hirai T, Yu ZX, Ferrans VJ, Howard BH, Finkel T (1999) Ras proteins induce senescence by altering the intracellular levels of reactive oxygen species. J Biol Chem 274:7936–7940

Lee HC, Yin PH, Chi CW, Wei YH (2002) Increase in mitochondrial mass in human fibroblasts under oxidative stress and during replicative cell senescence. J Biomed Sci 9:517–526

Marchenko ND, Wolff S, Erster S, Becker K, Moll UM (2007) Monoubiquitylation promotes mitochondrial p53 translocation. EMBO J 26:923–934

Matoba S, Kang JG, Patino WD, Wragg A, Boehm M, Gavrilova O, Hurley PJ, Bunz F, Hwang PM (2006) P53 regulates mitochondrial respiration. Science 312:1650–1653

Moiseeva O, Bourdeau V, Roux A, Deschenes-Simard X, Ferbeyre G (2009) Mitochondrial dysfunction contributes to oncogene-induced senescence. Mol Cell Biol 29:4495–4507

Nagino M, Tanaka M, Nishikimi M, Nimura Y, Kubota H, Kanai M, Kato T, Ozawa T (1989) Stimulated rat liver mitochondrial biogenesis after partial hepatectomy. Cancer Res 49:4913–4918

Passos JF, Saretzki G, Ahmed S, Nelson G, Richter T, Peters H, Wappler I, Birket MJ, Harold G, Schaeuble K, Birch-Machin MA, Kirkwood TBL, von Zglinicki T (2007) Mitochondrial dysfunction accounts for the stochastic heterogeneity in telomere-dependent senescence. PLoS Biol 5:1138–1151

Pesce V, Cormio A, Fracasso F, Vecchiet J, Felzani G, Lezza AM, Cantatore P, Gadaleta MN (2001) Age-related mitochondrial genotypic and phenotypic alterations in human skeletal muscle. Free Radic Biol Med 30:1223–1233

Polyak K, Li Y, Zhu H, Lengauer C, Willson JKV, Markowitz SD, Trush MA, Kinzler KW, Vogelstein B (1998) Somatic mutations of the mitochondrial genome in human colorectal tumors. Nat Genet 20:291–293

Ramanathan A, Wang C, Schreiber SL (2005) Perturbation profiling of a cell-line model of tumorigenesis by using metabolic measurements. Proc Natl Acad Sci U S A 102:5992–5997

Rocher C, Taanman J-W, Pierron D, Faustin B, Benard G, Rossignol R, Malgat M, Pedespan L, Letellier T (2008) Influence of mitochondrial DNA level on cellular energy metabolism: implications for mitochondrial diseases. J Bioenerg Biomembr 40:59–67

Rossignol R, Gilkerson R, Aggeler R, Yamagata K, Remington SJ, Capaldi RA (2004) Energy substrate modulates mitochondrial structure and oxidative capacity in cancer cells. Cancer Res 64:985–993

Sarkisian CJ, Keister BA, Stairs DB, Boxer RB, Moody SE, Chodosh LA (2007) Dose-dependent oncogene-induced senescence in vivo and its evasion during mammary tumorigenesis. Nat Cell Biol 9:493–505

Serrano M, Lin AW, McCurrach ME, Beach D, Lowe SW (1997) Oncogenic ras provokes premature cell senescence associated with accumulation of p53 and p16INK4a. Cell 88:593–602

Seshadri T, Campisi J (1990) Repression of c-fos transcription and an altered genetic program in senescent human fibroblasts. Science 247:205–209

Simonnet H, Alazard N, Pfeiffer K, Gallou C, Beroud C, Demont J, Bouvier R, Schagger H, Godinot C (2002) Low mitochondrial respiratory chain content correlates with tumor aggressiveness in renal cell carcinoma. Carcinogenesis 23:759–768

Tallini G (1998) Oncocytic tumors. Virchows Arch 433:5–12

Wallace DC (2005) A mitochondrial paradigm of metabolic and degenerative diseases, aging, and cancer: a dawn for evolutionary medicine. Annu Rev Genet 39:359–407

Warburg O, Posener K, Negelein E (1924) The metabolism of tumors. Biochem Z 152:319–344

Yoshida Y, Izumi H, Torigoe T, Ishiguchi H, Itoh H, Kang D, Kohno K (2003) P53 physically interacts with mitochondrial transcription factor A and differentially regulates binding to damaged DNA. Cancer Res 63:3729–3734

Interleukin-6 Induces Premature Senescence Involving Signal Transducer and Activator of Transcription 3 and Insulin-Like Growth Factor-Binding Protein 5

Hirotada Kojima, Hiroyuki Kunimoto, Toshiaki Inoue, and Koichi Nakajima

Contents

Abstract	53
Introduction	54
Roles of Cytokines in Premature Senescence	54
IL-6, Soluble IL-6 Receptor, gp130 and Signaling Molecules	55
Induction of Premature Senescence of Human Fibroblasts by IL-6/Soluble IL-6R	56
Senescent Human Fibroblasts Express IL-6 and IL-6Rα Chains and Constitutively Activated STAT3	56
IL-6/sIL-6R Causes Premature Senescence of Primary Human Fibroblasts in ROS/DDR/p53-Dependent Manner	56
STAT3-IGFBP5 Axis Is Critical in IL-6-Dependent Senescence-Inducing Circuit	57
Discussion	58
References	59

Abstract

Normal cells undergo senescence in response to telomere erosion, various stresses causing DNA damage, and certain cytokines. One such cytokine, interleukin-6 (IL-6), a multifunctional cytokine, can act on multiple lineages of cells together with soluble IL-6 receptor to induce cell proliferation, differentiation, and even promotion of tumorigenesis. We studied the molecular mechanisms by which IL-6 and soluble IL-6R (sIL-6R) cause premature senescence using primary human TIG3 fibroblasts. Stimulation of TIG3 cells with IL-6/sIL-6R sequentially caused generation of reactive oxygen species (ROS) as early as day 1, followed by DNA damage, p53 accumulation, and finally senescence on days 8–10. Signal transducer and activator of transcription 3 (STAT3) was required for the early and late events leading to senescence, including the early-phase increase of ROS and senescence-associated secretary phenotype (SASP) occurring 4–5 days after IL-6/sIL-6R stimulation. Interestingly, the STAT3 function was indirect, and insulin-like growth factor-binding protein 5 (IGFBP5) secreted into the supernatants was identified as the STAT3-downstream molecule responsible for the IL-6/STAT3-induced ROS generation and premature senescence. IGFBP5 was consistently expressed from the initial phase through the entire senescence process, the profile being quite different from that of SASP. Thus,

---

H. Kojima • H. Kunimoto • K. Nakajima (✉)
Department of Immunology, Graduate School of Medicine, Osaka City University, 1-4-3, Asahi-machi, Abeno-ku, Osaka 545-8585, Japan
e-mail: knakajima@med.osaka-cu.ac.jp

T. Inoue
Division of Human Genome Science, Department of Molecular and Cellular Biology, School of Life Sciences, Faculty of Medicine, Tottori University, Yonago, Japan

IL-6/sIL-6R forms a senescence-inducing circuit involving the STAT3–IGFBP5 axis as a key triggering and reinforcing component.

Keywords

Cytokines • GP130 and signaling molecules • Human umbilical vein endothelial cells (HUVECs) • Insulin-like growth factor-binding protein 5 (IGFBP5) • Interleukin-6 (IL-6) • Reactive oxygen species (ROS) • Senescence-associated secretary phenotype (SASP) • Signal transducer and activator of transcription 3 (STAT3) • Soluble IL-6R (sIL-6R)

Introduction

Normal human cells undergo senescence in response to various insults. Cellular senescence is a state characterized by the inability of cells to proliferate despite the presence of nutrients and mitogens, but with maintaining cell viability and metabolic activity (Kuilman et al. 2010; Rodier and Campisi 2011; Sikora et al. 2011). Cellular senescence is becoming recognized as an important program to escape from unregulated proliferation, tumorigenesis *in vivo*, in addition to the cell-death program (Kuilman et al. 2010; Rodier and Campisi 2011; Sikora et al. 2011).

Cells show several types of cellular senescence, including replicative senescence, oncogene-induced senescence (OIS), and stress-induced senescence, including exposure to cytokines (Kuilman et al. 2010; Sikora et al. 2011). Replicative senescence was first described by Hayflick and Moorhead in 1961, who demonstrated that after a finite number of cell divisions, normal human fibroblasts cultured in conditions that drive their continuous replication eventually reach an arrested state in which they do not respond to mitogenic stimuli despite being metabolically active. This type of senescence is now known to result from the DNA damage response (DDR) due to telomere shortening below a certain threshold (d'Adda di Fagagna et al. 2003).

Induction of senescence is also the frequent outcome of oncogenic mutations in normal cells (Collado and Serrano 2010). A number of actively mutated oncogenes, including Ras, Raf, MEK, and c-Myc, or inactivated negative regulatory molecules, including PTEN, have been shown to induce premature senescence (Collado and Serrano 2010). Stress-induced premature senescence of normal cells results from response to stressors such as oxidative stress, ionizing/non-ionizing radiation, and DNA damage reagents. Oncogene-induced senescence and stress-induced senescence are independent of telomere shortening (Kuilman et al. 2010; Sikora et al. 2011).

Roles of Cytokines in Premature Senescence

Another important insult causing premature senescence in certain cells is exposure to certain cytokines, including IFNα, IFNγ, TGFβ, and the IGFBP family members IGFBP3, IGFBP5, and IGFBP7 (Debacq-Chainiaux et al. 2008; Frippiat et al. 2001; Kim et al. 2007, 2009; Moiseeva et al. 2006; Wajapeyee et al. 2008). Interestingly, recent studies have begun to show significant transcriptome changes in cells undergoing senescence, which result in the secretion of a large number of factors including cytokines, chemokines, and their receptors (Acosta et al. 2008; Coppé et al. 2008). This phenomenon is called the senescence-associated secretory phenotype (SASP) (Coppé et al. 2008; Rodier and Campisi 2011) or senescence-messaging secretome (SMS) (Kuilman and Peeper 2009). These SASP or SMS factors may cause growth arrest or reinforce premature senescence in a cell-autonomous fashion, recruit immune cells for clearance of senescent cells, contribute to tumorigenesis, modulate tissue repair, and possibly contribute to aging (Kuilman et al. 2010; Rodier and Campisi 2011). Kuilman et al. (2008) showed that activated BRAF causes the production of IL-6 and IL-6Rα chains in primary human fibroblasts, and that such IL-6 with IL-6Rα chains are required for the oncogene-induced senescence in an autocrine fashion. In a different type of cells, Acosta et al. (2008) showed that CXCR2 and its ligands produced by senescent cells are required for maintaining the senescence process.

IL-6 has been implicated in the pathogenesis of a variety of diseases often associated with inflammation and aging, including rheumatoid arthritis and tumorigenesis (Ershler and Keller 2000). Considering that both IL-6 and soluble IL-6Rα (sIL-6Rα) chains are produced at sufficient concentrations in various tissues in infection, inflammation, tumor infiltration, and aging and that the signal transducing receptor subunit of IL-6R, gp130, is expressed ubiquitously in most cells, it is possible that the IL-6 and sIL-6Rα produced by surrounding cells might affect the progress of senescence in certain cells that do not have the IL-6Rα chains.

IL-6, Soluble IL-6 Receptor, gp130 and Signaling Molecules (Fig. 6.1)

IL-6 is a multifunctional cytokine that regulates cell proliferation, survival, and differentiation and enhances cellular function in multiple lineages of cells (Hirano et al. 1997). IL-6 acts on cells by binding to the specific IL-6 receptor (IL-6R) composed of an IL-6Rα chain (also known as IL-6Rα, IL-6R, or CD126) and a signal-generating receptor β chain gp130 (also known as CD130 or IL-6Rβ). The gp130 receptor molecule is a common component of multiple receptors for the IL-6 family of cytokines, such as IL-6, oncostatin M, leukemia inhibitory factor, ciliary neurotrophic factor, IL-11, and other factors (Hirano et al. 1997). IL-6Rα, which interacts with IL-6, is mainly expressed in hepatocytes and hematopoietic cells, such as T-cells, monocytes/macrophages, activated B-cells, and neutrophils. In contrast, a wide range of cells expresses gp130.

Interestingly, soluble isoforms of the IL-6Rα chains are generated by alternative splicing or by limited proteolysis of the a-disintegrin-and-metalloproteinase (ADAM) gene family members ADAM10 and ADAM17. The soluble IL-6R binds IL-6 and complexes with gp130 to form an IL-6–IL-6R complex for IL-6 signaling. Because inflamed, infected, tumor infiltrated, and aging tissues contain both IL-6 and sIL-6R that are produced or shed by myeloid cells, most of the cells in such tissues with gp130 can be activated by IL-6 and sIL-6R. Upon interaction of IL-6 with IL-6R complex, gp130 molecules become dimerized, and associated JAK tyrosine kinases are activated due to the close interaction. Then, activated JAK kinases phosphorylate the cytoplasmic domain of gp130 at the specific tyrosine residues, which are required for activation of downstream signal transduction pathways. Src-homology 2 domain-containing protein tyrosine phosphatase 2 (SHP2) is recruited to the phosphorylated YSTV motif and phosphorylated and activated by JAK kinases, which in turn activate

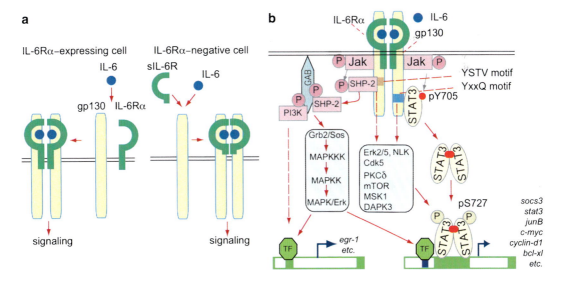

Fig. 6.1 IL-6 receptor system and its signal transduction pathways. (**a**) Interleukin-6 classic and trans-signaling. (**b**) Major signaling pathways of IL-6R

both the Ras-Raf-ERK1/2 pathway and the PI3K-AKT pathway though the Gab family (Hirano et al. 1997). Signal transducer and activator of transcription 3 and some STAT1 are recruited to the phosphorylated YXXQ motifs and phosphorylated by JAK kinases at the critical tyrosine residues, Y705 for STAT3 and Y701 for STAT1 (Hirano et al. 1997). The activated STAT3 and STAT1 dimerize with each other, making STAT3 or STAT1 homodimers and STAT3/STAT1 heterodimers. These activated and dimerized STAT molecules enter the nucleus and bind to the specific DNA sequences in the regulatory regions of their target genes. These three major pathways coordinately determine the IL-6 function depending on the cellular context through regulation of a set of genes. Among them, the STAT3 pathway often plays a major role in multiple IL-6 functions. The IL-6 receptor system, its signal transduction pathways, and their major target genes are depicted in Fig. 6.1.

Induction of Premature Senescence of Human Fibroblasts by IL-6/Soluble IL-6R

Senescent Human Fibroblasts Express IL-6 and IL-6Rα Chains and Constitutively Activated STAT3 (Fig. 6.2)

To understand the role of pro-inflammatory cytokines, especially IL-6, in the regulation of cellular senescence, we have chosen primary human fibroblast TIG3 cells. These cells show a senescent phenotype, as exemplified by growth arrest, flattened shape, and senescence-associated β-galactosidase (SA-β-Gal) activity, after ~55 population doublings (PD). At this stage, the older TIG3 cells show constitutive expression of the mRNAs for IL-6 and IL-6 receptor α chains, while younger TIG3 cells at PD33 do not express either. Remarkably, STAT3 is also constitutively activated in older TIG3 cells without exogenous IL-6 stimulation. This observation prompted us to examine whether IL-6 plus soluble IL-6R (IL-6/sIL-6R) induces premature senescence in a young primary human TIG3 fibroblast. We found that this was the case. By using this senescence induction system, we characterized the course of senescence progression initiated by IL-6/sIL-6R. We will describe here how IL-6/sIL-6Rα chains initiate and reinforce the progress of premature senescence in TIG3 cells through a senescence-inducing circuit involving the STAT3-IGFBP5 axis, which we recently reported (Kojima et al. 2012).

IL-6/sIL-6R Causes Premature Senescence of Primary Human Fibroblasts in ROS/DDR/p53-Dependent Manner

When IL-6 and soluble IL-6R were administered at 50 ng/ml each to cultures of young TIG3 cells at PD33, the cells showed the phenotypes of senescence, with growth arrest and SA-β gal activity, at around 8 days (Fig. 6.2b). To

a

	young TIG3	old TIG3
IL-6	−	+
IL-6Rα	−	+
gp130	+	+
activated STAT3	−	+
SA-β-gal	−	+

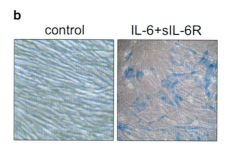

b

Fig. 6.2 IL-6 plus sIL-6R causes senescence in TIG3 cells. (**a**) Summary of expression levels of IL-6, IL-6Ra, and IL-6-related phenotype of young and old human fibroblasts. (**b**) TIG3 cells were treated with IL-6/sIL-6R or vehicle for 8 days and then subjected to SA-β-Gal assay

understand the sequential events during the 8 days leading to final senescence, we detected changes in the events often observed in the progress to senescence. The p53 levels were first examined. The levels of p53 protein declined on day 2 and gradually increased thereafter. The importance of p53 was tested by knocking down p53 using shRNA against p53 mRNA. With an almost complete loss of p53, TIG3 fibroblasts showed no sign of senescence and proliferated well in the presence of IL-6/sIL-6R, indicating that p53 is essential for the IL-6/sIL-6R-induced premature senescence.

What are the events preceding increases in the p53 level? ROS generation and the subsequent DNA damage response (DDR) have been reported to be involved in p53-dependent premature cellular senescence in many cases (Di Micco et al. 2006). The ROS levels, detected by the activation of H_2DCFDA fluorescence in response to ROS, increased as early as day 1 and reached higher levels thereafter. To test the role of ROS, we used the antioxidant N-acetyl cysteine (NAC) to counteract the generated ROS. Pretreatment of TIG3 fibroblasts with NAC before IL-6/sIL-6R stimulation inhibited the ROS-dependent DNA damage response and premature senescence. Thus, IL-6/sIL-6R induces the premature senescence of fibroblasts in a ROS/DDR/p53-dependent manner. Because we detected increases in both $p16^{INK4a}$ and $p15^{INK4b}$, it is likely that both the p53- and Rb-mediated pathways are involved in the IL-6/sIL-6R-induced premature senescence.

STAT3-IGFBP5 Axis Is Critical in IL-6-Dependent Senescence-Inducing Circuit (Fig. 6.3)

As explained in the previous section regarding the IL-6 receptor system and its signaling pathways, IL-6 activates three major pathways derived from phosphorylated gp130 at specific tyrosine residues through JAK tyrosine kinases. These include the STAT3 pathway, which is derived from the pYXXQ motif, and the SHP2-Ras-Raf-ERK1/2 pathway and the Gab-PI3K-Akt pathway, both of which are derived from the pYSTV motif (Hirano et al. 1997). We tested the role of STAT3 using two types of dominant-negative STAT3s, STAT3F and

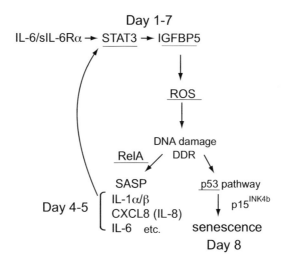

Fig. 6.3 A model for the senescence-inducing circuit involving the IL-6-STAT3-IGFBP5 axis. IGFBP5 produced in a STAT3-dependent manner causes the initial ROS generation, and subsequent DDR and SASP (expression of IL-1α, IL-1β, IL-6 and CXCL8/IL-8), and then prolonged expression of IGFBP5 in conjunction with components of SASP drive the circuit generating more ROS and severe DNA damage, leading to p53-dependent premature senescence. Inhibition of any constituent, STAT3, IGFBP5, ROS, SASP or p53, attenuates the IL-6/sIL-6R-induced premature senescence. Molecules that are underlined are indispensable components

STAT3D (Nakajima et al. 1996). TIG3 cells expressing either HA-tagged STAT3F or HA-tagged STAT3D showed little ROS increase at the early phase and very little increase in SA-β-galactosidase activity on day 8 in response to IL-6/sIL-6R, indicating that STAT3 activity was essential for the IL-6/sIL-6R-induced early events leading to premature senescence.

Another important feature associated with senescence is expression of a variety of cytokines and their receptors, including IL-1α, IL-1β, CXCL8 (IL-8), and IL-6 (Acosta et al. 2008; Kuilman et al. 2008). This phenomenon is termed the senescence-associated secretory phenotype (SASP) (Coppé et al. 2008). In IL-6/sIL-6R-induced premature senescence, the mRNA expressions of IL-1α, IL-1β, IL-8 and IL-6 molecules dramatically increased on days 4 and 5 after ROS generation and the DNA-damage response. This SASP was strongly repressed in the TIG3 fibroblasts expressing dominant-negative STAT3. Considering the late increases of mRNAs in SASP and the sequential events of ROS generation, DDR and SASP, the STAT3 effect on SASP should be indirect. Therefore, we searched for molecule(s) mediating the STAT3 action that is responsible for subsequent events leading to the final premature senescence.

We assumed that TIG3 cells secrete such molecule(s) in the early phase of response to IL-6/sIL-6R because no change of culture medium enhanced the IL-6/sIL-6R-induced premature senescence. To avoid the effect of exogenous IL-6/sIL-6R, we used TIG3 cells expressing a chimeric receptor composed of the G-CSF receptor extracellular domain, the gp130 transmembrane and a truncated cytoplasmic region, with 133 intracellular amino acid residues containing both the YSTV-motif and a YXXQ motif, YRHQ (Abe et al. 2001). We also used dominant-negative STAT3F to determine if the activity is STAT3-dependent. The culture supernatants from TIG3 cells stimulated with granulocyte colony-stimulating factor for the chimeric receptors for 2 days showed the senescence-inducing activity. This activity was dependent on the STAT3 activity. After searching for secreted factors with senescence-inducing activity, including insulin-like growth factor binding protein 3 (IGFBP3), IGFBP5 and IGFBP7, we finally identified IGFBP5 as the major STAT3-dependent senescence-inducing protein in the supernatants.

The characteristics of IGFBP5 in the IL-6/sIL-6R-induced premature senescence were as follows. IGFBP5 mRNA was induced from an early phase in a STAT3-dependent manner and expressed throughout the course, with a peak at day 5 and sustained levels on days 6 and 7. IGFBP5 itself had senescence-inducing activity in TIG3 cells with increased ROS generation from earlier days. An IGFBP5 knockdown reduced the senescence-inducing activity in supernatants, and an IGFBP5 knockdown inhibited the IL-6/sIL-6R-induced ROS increase and premature senescence. All of these results were consistent with the notion that IGFBP5, which is produced in a STAT3-dependent manner, is a major triggering molecule for IL-6/gp130-induced premature senescence in TIG3 cells, and that this STAT3-IGFBP5 axis establishes a circuit to reinforce premature senescence.

Discussion

We have described how IL-6/sIL-6R induces premature senescence in human primary fibroblasts and presented a model for the progression of premature senescence through a circuit initiated by IL-6/sIL-6R based on our recent work. This circuit is composed of IL-6, STAT3, IGFBP-5, ROS, DDR, SASPs, and p53. Inhibition of any of these components in the circuit blocked progression to the subsequent process leading to premature senescence, indicating that each component was essential for this circuit.

RelA, a component of NFκB, was also important for this circuit. A RelA knockdown in TIG3 cells suppressed the IL-6/sIL-6R-induced expression of mRNAs for IL-1α, IL-1β, IL-8 and IL-6, which occurred on days 4 and 5 as a consequence of DDR, and inhibited the IL-6/sIL-6R-induced premature senescence (Kojima and Nakajima, unpublished data). Therefore, it is likely that IL-6 expressed as one of the SASPs, is involved in reinforcing the senescence-inducing circuit by activating the STAT3–IGFBP5 axis. Other cytokines produced as SASPs may be involved in this circuit.

STAT3 plays multiple roles depending on the context or condition of cells (Hirano et al. 1997). In some conditions, STAT3 is involved in growth arrest and differentiation, and even in apoptosis. In other conditions, STAT3 is involved in proliferation and cell survival. Persistently activated STAT3 in some conditions can mediate tumorigenesis by protecting cells from apoptotic stimuli and by promoting cell-cycle progression in a variety of cancers or leukemias (Yu et al. 2009). We showed here that STAT3 is essential for the process leading to premature senescence by initiating the STAT3–IGFBP5 axis-dependent senescence-inducing circuit. It is, therefore, important to understand the mechanisms by which STAT3, in combination with other cellular conditions, regulates its target genes essential for various outcomes.

IGFBP5 has been shown to mediate various biological functions by virtue of IGF-dependent or -independent mechanisms (Beattie et al. 2006). We mention only two reports related to these topics. Chapman et al. (1999) reported that STAT3-dependent IGFBP5 production is involved in mammary gland cell apoptosis in the process of mammary gland involution, probably by inhibiting the function of IGF1 through IGFBP5-binding to IGF1. Kim et al. (2007) reported that IGFBP5 expression is upregulated during replicative senescence in human umbilical vein endothelial cells (HUVECs), and that IGFBP5 induces cellular senescence in HUVECs via a p53-dependent pathway; they also demonstrated the increased IGFBP5 proteins in fibrotic stromal matrix of aging-associated atherosclerotic lesion of arteries.

Several studies have shown that the levels of IL-6 and sIL-6R increase with aging and aging-associated diseases (Daynes et al. 1993; Ershler and Keller 2000). Based on our work, it is possible that IL-6/sIL-6R in the environment contributes to the tissue aging process, and that the production of IL-6 and sIL-6R is not merely an outcome of aging, but may cause or enhance senescence in some conditions. What really determines the role of IL-6/STAT3 in processes ranging from cell proliferation, cell survival, differentiation and apoptosis to cellular senescence remains to be addressed.

Acknowledgements This work was supported in part by the Ministry of Education, Culture, Sports, Science, and Technology of Japan; the Japan Health Foundation and the Kampou Science Foundation. The authors gratefully acknowledge their appreciation to members of the Nakajima laboratory for helpful discussions, to Junko Kanbara for excellent technical assistance.

References

Abe K, Hirai M, Mizuno K, Higashi N, Sekimoto T, Miki T, Hirano T, Nakajima K (2001) The YXXQ motif in gp130 is crucial for STAT3 phosphorylation at Ser727 through an H7-sensitive kinase pathway. Oncogene 20:3464–3474

Acosta JC, O'Loghlen A, Banito A, Guijarro MV, Augert A, Raguz S, Fumagalli M, Da Costa M, Brown C, Popov N et al (2008) Chemokine signaling via the CXCR2 receptor reinforces senescence. Cell 133:1006–1018

Beattie J, Allan GJ, Lochrie JD, Flint DJ (2006) Insulin-like growth factor-binding protein-5 (IGFBP-5): a critical member of the IGF axis. Biochem J 395:1–19

Chapman RS, Lourenco PC, Tonner E, Flint DJ, Selbert S, Takeda K, Akira S, Clarke AR, Watson CJ (1999) Suppression of epithelial apoptosis and delayed mammary gland involution in mice with a conditional knockout of Stat3. Genes Dev 13:2604–2616

Collado M, Serrano M (2010) Senescence in tumours: evidence from mice and humans. Nat Rev Cancer 10:51–57

Coppé JP, Patil CK, Rodier F, Sun Y, Muñoz DP, Goldstein J, Nelson PS, Desprez PY, Campisi J (2008) Senescence-associated secretory phenotypes reveal cell-nonautonomous functions of oncogenic RAS and the p53 tumor suppressor. PLoS Biol 6:2853–2868

d'Adda di Fagagna F, Reaper PM, Clay-Farrace L, Fiegler H, Carr P, Von Zglinicki T, Saretzki G, Carter NP, Jackson SP (2003) A DNA damage checkpoint response in telomere-initiated senescence. Nature 426:194–198

Daynes RA, Araneo BA, Ershler WB, Maloney C, Li GZ, Ryu SY (1993) Altered regulation of IL-6 production with normal aging. Possible linkage to the age-associated decline in dehydroepiandrosterone and its sulfated derivative. J Immunol 150:5219–5230

Debacq-Chainiaux F, Pascal T, Boilan E, Bastin C, Bauwens E, Toussaint O (2008) Screening of senescence-associated genes with specific DNA array reveals the role of IGFBP-3 in premature senescence of human diploid fibroblasts. Free Radic Biol Med 44:1817–1832

Di Micco R, Fumagalli M, Cicalese A, Piccinin S, Gasparini P, Luise C, Schurra C, Garre' M, Nuciforo PG, Bensimon A et al (2006) Oncogene-induced senescence is a DNA damage response triggered by DNA hyper-replication. Nature 444:638–642

Ershler WB, Keller ET (2000) Age-associated increased interleukin-6 gene expression, late-life diseases, and frailty. Annu Rev Med 51:245–270

Frippiat C, Chen QM, Zdanov S, Magalhaes JP, Remacle J, Toussaint O (2001) Subcytotoxic H2O2 stress triggers a release of transforming growth factor-beta 1, which induces biomarkers of cellular senescence of human diploid fibroblasts. J Biol Chem 276:2531–2537

Hayflick L, Moorhead PS (1961) The serial cultivation of human diploid cell strains. Exp Cell Res 25:585–621

Hirano T, Nakajima K, Hibi M (1997) Signaling mechanisms through gp130: a model of the cytokine system. Cytokine Growth Factor Rev 8:241–252

Kim KS, Seu YB, Baek SH, Kim MJ, Kim KJ, Kim JH, Kim JR (2007) Induction of cellular senescence by insulin-like growth factor binding protein-5 through a p53-dependent mechanism. Mol Biol Cell 18:4543–4552

Kim KS, Kang KW, Seu YB, Baek SH, Kim JR (2009) Interferon-gamma induces cellular senescence through p53-dependent DNA damage signaling in human endothelial cells. Mech Ageing Dev 130:179–188

Kojima H, Kunimoto H, Inoue T, Nakajima K (2012) The STAT3-IGFBP5 axis is critical for IL-6/gp130-induced premature senescence in human fibroblasts. Cell Cycle 11:730–739

Kuilman T, Peeper DS (2009) Senescence-messaging secretome: SMS-ing cellular stress. Nat Rev Cancer 9:81–94

Kuilman T, Michaloglou C, Vredeveld LC, Douma S, van Doorn R, Desmet CJ, Aarden LA, Mooi WJ, Peeper DS (2008) Oncogene-induced senescence relayed by an interleukin-dependent inflammatory network. Cell 133:1019–1031

Kuilman T, Michaloglou C, Mooi WJ, Peeper DS (2010) The essence of senescence. Genes Dev 24:2463–2479

Moiseeva O, Mallette FA, Mukhopadhyay UK, Moores A, Ferbeyre G (2006) DNA damage signaling and p53-dependent senescence after prolonged beta-interferon stimulation. Mol Biol Cell 17:1583–1592

Nakajima K, Yamanaka Y, Nakae K, Kojima H, Ichiba M, Kiuchi N, Kitaoka T, Fukada T, Hibi M, Hirano T (1996) A central role for Stat3 in IL-6-induced regulation of growth and differentiation in M1 leukemia cells. EMBO J 15:3651–3658

Rodier F, Campisi J (2011) Four faces of cellular senescence. J Cell Biol 192:547–556

Sikora E, Arendt T, Bennett M, Narita M (2011) Impact of cellular senescence signature on ageing research. Ageing Res Rev 10:146–152

Wajapeyee N, Serra RW, Zhu X, Mahalingam M, Green MR (2008) Oncogenic BRAF induces senescence and apoptosis through pathways mediated by the secreted protein IGFBP7. Cell 132:363–374

Yu H, Pardoll D, Jove R (2009) STATs in cancer inflammation and immunity: a leading role for STAT3. Nat Rev Cancer 9:798–809

A Role for the Nuclear Lamina Shape in Cell Senescence and Aging

7

Christiaan H. Righolt and Vered Raz

Contents

Abstract	61
Introduction	62
The Shape of the Cell Nucleus Changes During Biological Processes	62
Quantification of the Nuclear Lamina	63
Detection of the Nuclear Lamina	63
Biophysical Properties of the Nuclear Lamina in Aging Cells	65
Cell Classification Based on Features of the Nuclear Lamina	66
Discussion	66
References	68

C.H. Righolt
Department of Imaging Science and Technology,
TU Delft, Delft, The Netherlands

Manitoba Institute of Cell Biology, University of Manitoba, Winnipeg, Canada
e-mail: righoltc@cc.umanitoba.ca

V. Raz (✉)
Department of Human Genetics, Leiden University Medical Center, Leiden, The Netherlands
e-mail: v.raz@lumc.nl

Abstract

Proteins of the nuclear lamina define the nuclear shape and determine nuclear integrity. Lamin proteins bind to intranuclear proteins and the chromatin, and play a regulatory role in a wide spectrum of nuclear functions such as chromatin organization, epigenetic transcription, replication and repair. In addition, the nuclear lamina is physically connected to the outer nuclear lamina and, therefore, to structural elements off the cytoskeleton and the plasma membrane. Structural changes in the nucleus and at the plasma membrane are, therefore, possibly connected. Deformation of the nuclear lamina shape will affect chromatin architecture by binding of the nuclear lamina to the chromatin. This subsequently results in changes in nuclear and cellular functions. The shape of the nuclear lamina dramatically changes during aging and in senescence. Here we will describe how changes in the nuclear lamina can be measured and discuss implications on nuclear function.

Keywords

Cell classification • Curvature • Fluorescence intensity • Fluorescent recovery after photobleaching (FRAP) • Human mesenchymal stem cells (hMSCs) • Hutchinson Gilford Progeria Syndrome (HGPS) • Inner nuclear membrane (INM) • Nuclear envelope (NE) • Nuclear lamina

Introduction

The Shape of the Cell Nucleus Changes During Biological Processes

The nuclear envelope (NE) forms a boundary between the nucleus and the cytoplasm in eukaryotic cells, creating a separation between nuclear and cytoplasmic activities. Selective and regulated physical connections between these two cell compartments are administrated via the nuclear pore protein complex. The nuclear pores are embedded within the outer nuclear membrane (ONM) and the inner nuclear membrane (INM). The inner and outer nuclear membranes are bilipid structures supported by a network of proteins, which provide the structure its regulatory role. The INM is underlined within the nucleus by the dynamic fibrous network of the nuclear lamina.

The nuclear lamina is predominantly composed of lamin proteins, which are dynamically anchored to the INM via post-translational modifications. Lamin proteins are expressed in all eukaryote organisms and vertebrates and are divided into A- and B-types. These proteins play a central role in the integrity of the nuclear lamina. In human, both *lamin A*- and *lamin B*- encoding genes are required for viability, whereas in mice *lamin A* knockdown is viable and *lamin B* knockdown is lethal. Silencing of *lamin B* causes dramatic changes in the lamin A meshwork, whereas removal of *lamin A-types* has no effect on structural organization of lamin LB filaments (Shimi et al. 2008). This indicates a crucial and indispensable role for LB1 in maintaining the structural integrity of the nuclear lamina.

Lamin proteins are targeted into the INM via a farnesyl hydrophobic group. The farnesyl group is attached to the C-terminus of the molecule in a post-translational modification forming the mature molecule (Parnaik 2008). B-type lamins remain farnesylated, because their maturation is terminated at this step. Post-translational modification of lamin A at its C-terminus includes additional steps that promote a reversible anchorage to the INM. A dynamic anchorage of lamin A to the nuclear lamina has implications in the regulatory role of the nuclear lamina. Moreover, the expression of A-type lamins is developmentally regulated, whereas the majority of B-type lamins are ubiquitously expressed. Within an organism the composition of the nuclear lamina varies between cell types and biological conditions, such as differentiation, division and senescence.

Mutations in the lamin A gene in humans cause a large number of heritable dominant diseases with a broad spectrum of symptoms- collectively called laminopathies. Mutations in lamin A serve as a helpful tool in the quest for regulatory and structural roles of the nuclear lamina in nuclear and cellular functions. A point mutation in the last intron-exon boundary in the *lamin A* gene causes Hutchinson Gilford Progeria Syndrome (HGPS). The gene product of this point mutation, named progerin, activates a cryptic splice site that results in a deletion of the C-terminus. The mobility of progeria protein is considerably slower compared to the mobility of mature lamin A protein. As a result, progerin irreversibly associates with the lamina filaments, whereas the wild-type protein association with the nuclear lamina is dynamic and reversible (Goldman et al. 2004).

Progerin accumulation in the nuclear lamina causes premature cell senescence, associated with genome-wide transcriptional changes and telomere shortening, the hallmark of cell senescence (Scaffidi and Misteli 2006; Cao et al. 2011). HGPS patients are characterized by premature aging. While normally an aging-associated tissue degeneration initiates only after midlife, in HGPS aging symptoms starts already in 2–3 year old babies. In germ cells, progerin is a result of a *de novo* mutation, and the prevalence of HGPS is approximately only one in eight million (Gordon et al. 2011). Despite the low prevalence, progerin is used as a molecular model unraveling the role of lamin A in nuclear function, development and aging. Recent studies demonstrated that progerin expression is found during physiological aging (McClintock et al. 2007; Scaffidi and Misteli 2006). The mechanism that activates progerin expression during aging is not fully understood.

The nuclear envelop, including the nuclear lamina, is reversibly demolished during cell division in mammalian cells. During the prophase of mitosis the structure is disintegrated and after cytokinesis – when the two daughter cells are formed – the nuclear lamina is re-formed from existing lamina proteins.

The nuclear envelope is also degraded during cell death. Degradation of the nuclear lamina during programmed cell death is, however, accompanied by cleavage of lamin proteins. Changes in the structure of the nuclear lamina in human mesenchymal stem cells (hMSCs) that are programmed to apoptosis precede DNA fragmentation, the hallmark of cell apoptosis. Cleavage of lamins by caspase-3 follows the structural deformation of the nuclear lamina structure (Raz et al. 2006). This suggests that structural deformation of the nuclear lamina precedes irreversible degradation of lamins during apoptosis. Since apoptosis is normally activated during aging and contributes to aging-associated tissue degeneration, structural changes in the nuclear lamina exhibited in senescent cells were also investigated.

When propagated *in vitro*, hMSCs enter premature growth arrest within only 6–7 passages (Abdallah and Kassem 2008). Premature growth arrest, which is associated with a set of molecular markers, characterizes senescent cells (Hayflick 1965). Cells from aged donors often exhibit a slow growth rate and enter growth arrest faster than cells from young individuals (Bonab et al. 2006).

Senescence of hMSCs is accompanied by spatial changes in the structure of the nuclear lamina (Raz et al. 2008). Structural changes include bending of the nuclear envelop and intruding of the nuclear lamina into the nuclear space creating intranuclear structures (Fig. 7.1a). Those structural changes are detected already after 6–7 passages, preceding the expression of molecular markers, which is observed only after 8–9 passages (Righolt et al. 2011b). An example of spatial changes in the shape of the nuclear lamina is shown in Fig. 7.1a. Structural changes in the shape of the nuclear lamina are also exhibited in hMSCs from young and old donors (Fig. 7.1). The shape of the nuclear lamina is also distorted in cells that express mutations in lamin A and in cells from patients that carry mutations in lamin A. Moreover, structural changes of the nuclear lamina are also observed in aging *C. elegans* animals (Haithcock et al. 2005). Together this indicates that structural changes of the nuclear lamina are highly associated with aging and cell senescence. A quantitative description of the nuclear lamina structure in cells could provide an unbiased and robust description of the status of this structure in cells.

Quantification of the Nuclear Lamina

Detection of the Nuclear Lamina

The nuclear lamina is a three-dimensional structure, architectural changes should, therefore, be investigated in living cells. This can be achieved by expression of lamin proteins fused to fluorescent proteins (Broers et al. 1999). High overexpression of lamin proteins leads, however, to structural deformation. Only cells with low expression level should, therefore, be considered. The fluorescent signal is visualized by fluorescence microscopy. A three-dimensional image is then obtained from Z-stacks acquired from living cells by a confocal microscope. This analysis suggests that spatial changes in the nuclear lamina of senescent hMSCs are initiated in the vertical direction (z-direction) (Righolt et al. 2011b). To determine quantitative changes in the structure, the pixels associated with the nuclear lamina have to be automatically identified and analyzed, in order to provide an unbiased description of these changes. The field of image-processing provides the tools to create such an automated method.

The first step for developing an automated analysis of the nuclear shape is the automatic recognition of the nuclear lamina structure from images of nuclei expressing lamin-A GFP. It is important to note that lamin A (and lamin B) proteins are found in both the nucleoplasm as soluble proteins, or bound to the nuclear lamina structure. In this procedure only lamina proteins that are associated with the nuclear lamina are

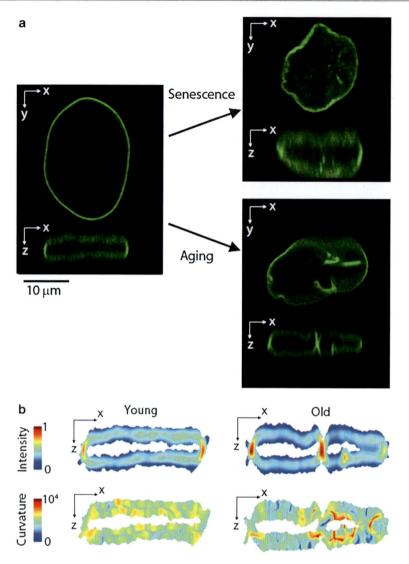

Fig. 7.1 (a) Cross sections of confocal images of the nuclear lamina in human mesenchymal stem cells. Images were taken from living cells expressing lamina A fused to GFP. Lamin A-GFP transgene was expressed in cells by a lentivirus expression system. Images show spatial changes in the structure of the nuclear lamina in cross sections in a lateral (x,y)-plane and in a (x,z)-plane. *Left panel* show a typical nucleus of an early passage from a 35 year old donor. *Upper right panel* shows a nucleus from the same donor but at a late passage number. This nucleus shows changes in the structure of the nuclear lamina which are typical to senescent cells. *Lower right panel* shows a typical nucleus of an early passage from a donor 81 year old. A slice through the cell is shown in a lateral (x,y)-plane and in a (x,z)-plane. (b) The changes in the shape of the nuclear lamina are associated with redistribution of lamin A at the nuclear lamina. From an "even distribution" at the opposite sides of the structure in young cells (*left upper image*), the lamin-A proteins show a heterogeneous distribution along the nuclear lamina (*right upper image*). High protein accumulation is demonstrated with *red* and low with *dark blue*, the color scale is linear. The curvature of the lamina locally increases along the structure (*lower right panel*) compared to the young cell (*lower left panel*). High curvature is indicated with *red* and low values with *dark blue* on a logarithmic scale. These are the values for the same cross sections as shown in a

included in the analysis. Several studies reported the recognition of the shape of the cell nucleus in two-dimensions (Lammerding et al. 2006; Vermolen et al. 2008). Those analyses only account for the edge of the nucleus and would not provide a complete architectural description. For example, spatial changes including intranuclear structures are excluded from these analyses. More recently, an automatic method that detects the nuclear lamina three-dimensional structure has been developed (Righolt et al. 2011a). Once pixels of the nuclear lamina are segmented, quantitative analysis of the lamina architecture can be performed.

Biophysical Properties of the Nuclear Lamina in Aging Cells

In young and healthy hMSCs the nuclear shape, and thus the bending of the nuclear lamina, in senescent cells can be described by curvature measures. Folding of the nuclear lamina would then be associated with high curvature values. In addition, fluorescence intensity is a measure for protein concentration, high intensity values of lamin-A, therefore, indicate protein accumulation Images of lamin A-GFP in young cells indicate that high intensity values are exhibited mainly in the "short edges" of the nuclear surface (Fig. 7.1a). In aging or senescent cells areas of protein accumulation are distributed throughout the lamina structure. Analysis of the nuclear lamina in young and senescent cells indicates that the main changes are associated with distribution of intensity and curvature along the structure (examples are depicted in Fig. 7.1b). Folded and intranuclear structures often contain high intensity values and are often surrounded by high curvature values (Fig. 7.1b). Architectural changes of the nuclear lamina can, therefore, be objectively described by an increase in the intensity distribution and curvature values during cell senescence.

How to quantify the structure of the nuclear lamina? A simple way to quantify morphological differences between two objects is to compare the length-to-width ratio of these objects. This measure is not, however, suited to describe spatial changes of the nuclear lamina that are associated with senescence and aging. Aged cells may not necessarily differ much from young cells in terms of its length-to-width ratio. Spatial changes could be locally restrained and should, therefore, be described by measures that describe local changes, which include bending of the structure and change in protein distribution within the nuclear lamina structure Fig. 7.1b.

Mathematically, the curvature is defined as the reciprocal radius of the circle that follows the contour at that point. When the contour is relatively flat, this circle will be large and the curvature will be small. When the contour bends sharply, this circle will be small and the curvature will be high. The more bended the nuclear lamina is, the higher the curvature value will be. To describe folding of the nuclear lamina, curvatures values from the entire three-dimensional structure are averaged, and the curvature of an individual cell is described with a single average value. The curvature value is then normalized to the size of the cell to compare between cells. When comparing curvature values between young and aged cells, bending of the nuclear lamina are effectively quantified (Righolt et al. 2011a).

The second biophysical feature describes redistribution of lamin-A proteins in the nuclear lamina structure Fig. 7.1b. Fluorescent intensity can, however, only be used as a relative quantity. This means intensity values should only be compared between samples after proper normalization. Here the values are normalized to 0 at the background level and 1 at the top intensity level. The interpretation of the average normalized intensity is counter-intuitive. When the maximum absolute intensity increases, the average normalized intensity will decrease. The following example will illustrate this: Situation one: Suppose the average fluorescent intensity of the lamin proteins is f above the background and the maximum intensity that occurs naturally is $2f$ above the background, then these numbers will become ½ and 1 respectively for the normalized intensity, i.e., the average normalized intensity is 0.5. Situation two: Suppose the proteins are redistributed such that the maximum intensity is

$3f$ above the background, while the average still is f above the background, then the average normalized intensity will be 0.33. This shows how the average normalized intensity decreases while the maximum absolute intensity increases. In addition to the mean intensity, the skewness of intensity also significantly changes between aging and young cells (Righolt et al. 2011b). Together this suggests that changes occur in protein distribution of lamin A during aging and senescence and that this protein accumulates. To retrieve spatial information of the curvature and intensity distribution over the entire three-dimensional structure, both the curvature and the normalized intensity values are averaged over the nuclear lamina and are represented with a single average number per cell (Righolt et al. 2011a).

A confined accumulation of proteins is often a result of a decrease in protein mobility. Protein mobility can be measured by the fluorescent recovery after photobleaching (FRAP) procedure. A detailed description of this procedure can be found in (Reits and Neefjes 2001). In brief: Cells expressing a GFP-fused protein are imaged with a confocal microscope. A region of interest is photo-bleached with a laser beam and the fluorescence recovery in this region is measured during time. The fluorescence recovery is a measure of protein mobility. This procedure was applied to demonstrate differences in protein mobility between lamin A- and lamin B- types. Consistent with the regulatory role of lamin A, lamin A is, in general, more mobile compared to lamin B. In addition, mutations in *lamin A* that cause laminopathies are less mobile, compared to the wild type protein (Dechat et al. 2010).

FRAP of lamin A-GFP at the nuclear lamina revealed that lamin A mobility at the nuclear envelop was lower at regions with high protein accumulation than for regions with a low fluorescence intensity (Righolt et al. 2011b). Since the high fluorescence regions were mainly localized at bended regions and intranuclear structures. This suggests that folding of the nuclear lamina is associated with local accumulation of lamin A.

Cell Classification Based on Features of the Nuclear Lamina

So far we have discussed how features of the nuclear lamina can help to understand structural changes in a single cell. Biophysical features of the nuclear lamina can be represented with a single number per cell for every feature as discussed before. Those features can, therefore, be used to describe the status of cell population when measured from cell populations. This procedure can be applied to robustly test the correlation between changes in lamina shape during aging or senescence. With statistical classification tools, e.g. linear regression, it is possible to determine whether cell populations can be classified based on the nuclear lamina structure. Since all values are normalized, it is also possible to compare between different cell populations and cell types (Righolt et al. 2011b). The result of this robust analysis demonstrated that young and aged cell populations can be discriminated based on mean intensity and curvature values of the nuclear lamina (Fig. 7.2).

Overall, this analysis shows that the mean intensity of lamin A decreased from the young to the old and senescent cells. This indicates changes in lamin A distribution in the nuclear lamina during aging and senescence. Changes in the average curvature between young cells and senescent and old cells are smaller compared to the intensity (Fig. 7.3). The average curvature increases, however, significantly more in apoptotic cells compared to young or senescent cells (Righolt et al. 2011b; Fig. 7.3). Since cell apoptosis is increased in a population of senescent cells, this suggests that a major increase in folding of the nuclear lamina is secondary to increase in local protein accumulation.

Discussion

The nuclear lamina affects a broad spectrum of nuclear activities including chromatin architecture, epigenetic chromatin modifications, transcriptional activities, via binding of transcription

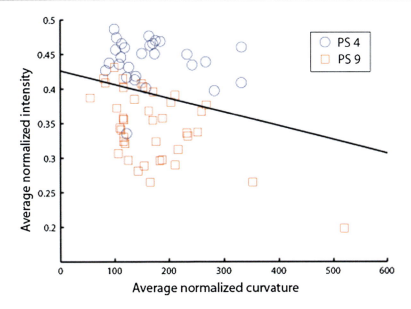

Fig. 7.2 An example of cell classification based on measured features of the nuclear lamina is shown. For living human mesenchymal stem cells at passage 4 (PS 4; *blue circles*) and passage 9 (PS 9; *red squares*) the average normalized curvature and intensity is plotted in the scatter plot. The two groups can be classified (separated) by a linear classifier – the *black line* in the scatter plot – with an error rate of 13 % (see Righolt et al. 2011a for technical details). This shows that these features can discriminate between the younger and older cells

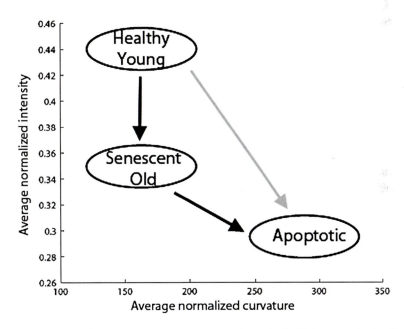

Fig. 7.3 Measurements of the features of the nuclear lamina from a combination of cell populations show that redistribution of lamin-A proteins in old and senescent cells is associated with a decrease in the normalized intensity. In apoptosis the shape of the nuclear lamina changes more dramatically. This is associated with an increase in the normalized curvature and decrease in normalized intensity. Cells can also go into apoptosis directly from their healthy state. Apoptotic cell can also be found in senescent cell cultures (Adapted from Righolt et al. (2011b))

factors (Peric-Hupkes and van Steensel 2010), DNA damage response Gonzalez-Suarez et al. (2009), and telomere length, the hallmark of cell senescence in certain cell types (Cao et al. 2011). The structure of the nuclear lamina dramatically changes in disease and during aging. It is unclear whether the structure of the lamina directly affects nuclear activity. The measured biophysical features of the nuclear lamina give rise to some hypotheses about a structural-functional role of the nuclear lamina. Based on lamina quantification studies the following sequence of events can be suggested: First, protein mobility decreases, leading to confined lamin A accumulation along the structure. This process is analogous to the entrapment of progerin proteins in the nuclear lamina. Second, the structure bends possibly due to local mechanical pressure. The bending process might be started after the proteins accumulate beyond some critical range. At that stage the structural pressure might become too large to maintain the smooth ellipsoidal shape and the nuclear lamina would deform to a more irregular shape.

It is not clear how lamin A mobility reduces during aging and senescence. Protein accumulation and aggregation is among the hallmarks of cellular aging. Protein turnover of nuclear proteins is regulated by the ubiquitin proteasome system, whose activity declines during aging. In addition it is not fully understood what the implications of lamin A accumulation at the nuclear lamina are. In regions with accumulation of lamin A, accumulation and aggregation of telomeres were found (Raz et al. 2008). Aggregation of heterochromatin into foci also marks senescent cells. Lamin A binds heterochromatic regions, but it is yet to be determined whether aging-associated heterochromatic foci are association with reduced mobility of lamin A. In HGPS, the progerin protein has reduced mobility and accumulates at the nuclear envelope. In cells expressing progerin mutant global changes in chromatin organization are associated with enhanced interactions with a specific subset of genes, which possibly affect transcription levels. However, it is unclear how transcription levels are regulated by chromatin binding to lamin A (Kubben et al. 2012). Here we described evidence that structural changes of the nuclear lamina are associated and induced during cell senescence and aging. Future studies should further investigate the functional consequences of structural changes of the nuclear lamina on nuclear function and nuclear architecture. In addition, the implications of changes in the shape of the nuclear lamina on cytoplasmic functions and the plasma membrane, both cellular compartments play a central role in aging and senescence are also required future studies.

References

Abdallah BM, Kassem M (2008) Human mesenchymal stem cells: from basic biology to clinical applications. Gene Ther 15:109–116

Bonab MM, Alimoghaddam K, Talebian F, Ghaffari SH, Ghavamzadeh A, Nikbin B (2006) Aging of mesenchymal stem cell in vitro. BMC Cell Biol 7:14

Broers JL, Machiels BM, van Eys GJ, Kuijpers HJ, Manders EM, van Driel R, Ramaekers FC (1999) Dynamics of the nuclear lamina as monitored by GFP-tagged A-type lamins. J Cell Sci 112:3463–3475

Cao K, Blair CD, Faddah DA, Kieckhaefer JE, Olive M, Erdos MR, Nabel EG, Collins FS (2011) Progerin and telomere dysfunction collaborate to trigger cellular senescence in normal human fibroblasts. J Clin Invest 121:2833–2844

Dechat T, Adam SA, Taimen P, Shimi T, Goldman RD (2010) Nuclear lamins. Cold Spring Harb Perspect Biol 2:a000547

Goldman RD, Shumaker DK, Erdos MR, Eriksson M, Goldman AE, Gordon LB, Gruenbaum Y, Khuon S, Mendez M, Varga R, Collins FS (2004) Accumulation of mutant lamin A causes progressive changes in nuclear architecture in Hutchinson-Gilford progeria syndrome. Proc Natl Acad Sci U S A 101:8963–8968

Gonzalez-Suarez I, Redwood AB, Perkins SM, Vermolen B, Lichtensztejin D, Grotsky DA, Morgado-Palacin L, Gapud E, Sleckman BP, Sullivan T, Sage J, Stewart CL, Mai S, Gonzalo S (2009) Novel roles for A-type lamins in telomere biology and the DNA damage response pathway. EMBO J 28:2414–2427

Gordon LB, Brown WT, Collins FS (2011) Hutchinson-Gilford Progeria syndrome. In: Pagon RA, Adam MP, Bird TD et al (eds) GeneReviews. University of Washington, Seattle, 1993–2013

Haithcock E, Dayani Y, Neufeld E, Zahand AJ, Feinstein N, Mattout A, Gruenbaum Y, Liu J (2005) Age-related changes of nuclear architecture in Caenorhabditis elegans. Proc Natl Acad Sci U S A 102:16690–16695

Hayflick L (1965) The limited in vitro lifetime of human diploid cell strains. Exp Cell Res 37:614–636

Kubben N, Adriaens M, Meuleman W, Voncken JW, van Steensel B, Misteli T (2012) Mapping of lamin A- and progerin-interacting genome regions. Chromosoma 121(5):447–464 [Epub ahead of print]

Lammerding J, Fong LG, Ji JY, Reue K, Stewart CL, Young SG, Lee RT (2006) Lamins A and C but not lamin B1 regulate nuclear mechanics. J Biol Chem 281:25768–25780

McClintock D, Ratner D, Lokuge M, Owens DM, Gordon LB, Collins FS, Djabali K (2007) The mutant form of lamin A that causes Hutchinson-Gilford progeria is a biomarker of cellular aging in human skin. PLoS One 2:e1269

Parnaik VK (2008) Role of nuclear lamins in nuclear organization, cellular signaling, and inherited diseases. Int Rev Cell Mol Biol 266:157–206

Peric-Hupkes D, van Steensel B (2010) Role of the nuclear lamina in genome organization and gene expression. Cold Spring Harb Symp Quant Biol 75:517–524

Raz V, Carlotti F, Vermolen BJ, Van der Poel E, Sloos WC, Knaan-Shanzer S, De Vries AA, Hoeben RC, Young IT, Tanke HJ, Garini Y, Dirks RW (2006) Changes in lamina structure are followed by spatial reorganization of heterochromatic regions in caspase-8-activated human mesenchymal stem cells. J Cell Sci 119:4247–4256

Raz V, Vermolen BJ, Garini Y, Onderwater JJ, Mommaas-Kienhuis MA, Koster AJ, Young IT, Tanke H, Dirks RW (2008) The nuclear lamina promotes telomere aggregation and centromere peripheral localization during senescence of human mesenchymal stem cells. J Cell Sci 121:4018–4028

Reits EA, Neefjes JJ (2001) From fixed to FRAP: measuring protein mobility and activity in living cells. Nat Cell Biol 3:E145–E147

Righolt CH, Raz V, Vermolen BJ, Dirks RW, Tanke H, Young IT (2011a) Molecular image analysis: quantitative description and classification of the nuclear lamina in human mesenchymal stem cells. Int J Mol Imaging 2011:723283

Righolt CH, van't Hoff ML, Vermolen BJ, Young IT, Raz V (2011b) Robust nuclear lamina-based cell classification of aging and senescent cells. Aging 3:1192–1201

Scaffidi P, Misteli T (2006) Lamin A-dependent misregulation of adult stem cells associated with accelerated ageing. Nat Cell Biol 10:452–459

Shimi T, Pfleghaar K, Kojima S, Pack CG, Solovei I, Goldman AE, Adam SA, Shumaker DK, Kinjo M, Cremer T, Goldman RD (2008) The A- and B-type nuclear lamin networks: microdomains involved in chromatin organization and transcription. Genes Dev 22:3409–3421

Vermolen BJ, Garini Y, Young IT, Dirks RW, Raz V (2008) Segmentation and analysis of the three-dimensional redistribution of nuclear components in human mesenchymal stem cells. Cytometry A 73:816–824

Upregulation of Alpha-2-Macroglobulin in Replicative Senescence

8

Li Wei Ma, Guo Dong Li, and Tan Jun Tong

Contents

Abstract	71
Introduction	72
Alpha-2-Macroglobulin	72
Changes of α2M Level During Replicative Senescence	73
Alpha-2-Macroglobulin Level as Senescence Biomarker	74
Regulation of Alpha-2-Macroglobulin in Replicative Senescence	74
Functional Promoter Region Determination	74
Positive Element Identification	75
Discussion	77
References	79

L.W. Ma • G.D. Li • T.J. Tong (✉)
Research Center on Aging, Department of Biochemistry and Molecular Biology, Peking University Health Science Center, 38 Xueyuan Road, Beijing 100191, People's Republic of China
e-mail: tztong@bjmu.edu.cn

Abstract

Replicative senescence is a well-established model system for studying the molecular basis of aging. Using high-density complementary deoxyribonucleic acid (cDNA) arrays, we identified alpha-2-macroglobulin (α2M) as a differentially expressed gene in replicative senescence. Our study demonstrated a positive linear-correlation of the mRNA level of α2M with cumulative population doublings (PDL) of human fibroblasts. The levels of α2M increased in senescence cells, but not in quiescent state of cultured fibroblasts, and remained stable in immortal HeLa cells. Moreover, the mRNA level of α2M in leucocytes showed significant difference between newborn and old human body. These results indicate that the up-regulated expression of α2M could be a universal age-related phenotype and the mRNA level of α2M may be used as a biomarker of aging *in vitro* and *in vivo*. To further explore the mechanism responsible for the up-regulation of α2M in senescent, We identified a novel transcriptional regulatory element, the α2M transcription enhancement element (ATEE), within the α2M promoter. This element differentially activates α2M expression in senescent *versus* young fibroblasts. Electrophoretic mobility shift assays revealed abundant complexes in senescent cell nuclear extracts compared with young cell nuclear extracts. The DNase I footprint revealed the protein-binding core sequence through which the protein binds the ATEE. Mutation within ATEE selectively abolished α2M promoter activity in senescent (but not young) cells. These results indicated the ATEE, as a positive transcription regulatory element, contributes to the up-regulation of α2M during replicative senescence.

Keywords

Alpha-2-macroglobulin (α2M) • Electrophoretic mobility shift assays (EMSA) • Functional promoter region determination • α2M transcription enhancement element (ATEE) • Positive element identification • Replicative senescence • Reporter gene assay • Trans-acting factors

Introduction

Cellular senescence is a well-established model system for studying the molecular basis of aging. Normal human fibroblasts do not divide indefinitely. This property, termed the finite replicative lifespan of cells, leads to a permanent and irreversible arrest of cell division by a process termed replicative or cellular senescence. To understand the mechanisms that regulate cellular senescence, many genetic factors and environmental factors that affect the processes of cell senescence have been studied. Besides accumulation of cell generations, cellular senescence occurs as a consequence of exposure to subcytotoxic conditions, stimulation with cytokines, overexpression of proto-oncogenes and tumor suppressor genes, and deficiency of JunD (Toussaint et al. 2002). However, a reliable biomarker for cellular age and further study of the gene expression profile of aging are needed to elucidate the mechanism of cellular senescence.

Although no consensus exists as to what constitutes an acceptable biomarker of cellular aging, the suggested criteria generally include that it (1) is highly reproducible and reflects biological age; (2) shows significant age-related changes within a relatively brief period of time; (3) is easily measured and the sample for measurement is small; (4) is stable in immortal cells. Markers characterizing the transition from young replicating to old growth-arrested cultured cells in different experimental systems were reported to be an increase in neutral β-galactosidase activity (Dimri et al. 1995), telomere shortening (Allsopp et al. 1992), over-expression of collagenase (Sottile et al. 1989), two-chain cathepsin B (DiPaolo et al. 1992) and Alzheimer disease β-amyloid precursor protein (APP) (Alder et al. 1991), increasing resistance of DNA to damage by H_2O_2 (Caldini et al. 1998), reduced inducibility of c-fos in response to serum stimulation (Seshadri and Campisi 1990), as well as reduced expression of heat shock protein 70 in response to stress (Choi et al. 1995) and EPC-1 in response to serum deprivation (Tresini et al. 1999). The most widely used biomarkers of cellular senescence are neutral β-galactosidase (SA-β-gal) activity and telomere shortening. Recent reports state that SA-β-gal activity is variable in different culture conditions have made it unreliable as a biomarker of cellular senescence (Severino et al. 2000). Telomere shortening as an aging biomarker is also challenged by the existence of telomere-independent senescence (Chen et al. 2001).

To identify a reliable biomarker for cellular age and further study the gene expression of aging, we profiled the gene expression difference between aged and young cultured human embryonic lung fibroblasts (2BS) by high-density complementary deoxyribonucleic acid (cDNA) arrays (Ma et al. 2004). Among the differentially expressed genes, alpha-2-macroglobulin (α2M) was selected for further study. Alpha-2-macroglobulin is a well-characterized proteinase inhibitor with high abundance in human plasma. According to the cDNA array data, α2M was highly expressed both in aged and young fibroblasts, so quantification of its mRNA will less be interfered by 'noise'. Moreover, the significant difference in its expression level between aged and young 2BS cells makes α2M an ideal candidate as senescence biomarker.

Alpha-2-Macroglobulin

Alpha-2-macroglobulin (α2M) is the major representative of a plasma protein family named as α-macroglobulins (αMs), which includes complement components C3 and C4 as well as pregnancy zone protein (PZP). The α2M gene is a single-copy gene in the human genome and is located on chromosome l2pl2-13. As a 718 kDa homotetrameric glycoprotein, α2M is one of the most abundant proteins in human plasma with a concentration of 2–4 mg/ml (Scottrup-Jensen et al. 1989). Alpha-2-macroglobulin is well characterized as an extra-cellular pan-proteinase inhibitor and as a carrier of specific growth factors, including transforming growth factor-β (TGF-β) and nerve growth factor β (NGF-β) (Crookston et al. 1994). In fact, the cleavage of a peptide bond at the bait region of the α2M molecule by a protease leads to a conformational change, which is referred to as activated α2M (α2M*). TGF-β1 preferentially binds to α2M*, whereas TGF-β2 binds both α2M and α2M* with an affinity higher than that of TGF-β1 (Liu et al. 2001). Additionally, many lines of biological evidence suggested the contribution of α2M to the etiology of Alzheimer's disease. For example, α2M mediates the clearance and degeneration of Alzheimer β-protein by making a complex with serine protease or enhancing internalization into the cells through low-density lipoprotein related protein (Qiu et al. 1999). Moreover, it was observed that increased expression of α2M is related to the benign prostatic hyperplasia and plays an important role in regulating the benign and malignant prostatic growth (Lin et al. 2005).

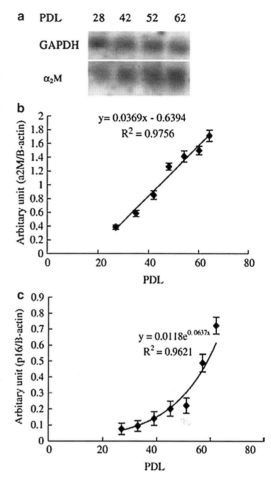

Fig. 8.1 The mRNA level of α_2M during replicative senescence. (**a**) Northern blotting analysis of the time course of changes in gene expression of α2M in the aging process of cultured 2BS cells. GAPDH was used as a normalization control for northern blot. (**b**) Relationship of the gene expression of α2M and cumulative PDL. (**c**) Relationship of the gene expression of p16^{INK4a} and cumulative PDL (RNA preparations from three different experiments)

Changes of α2M Level During Replicative Senescence

Compared with young 2BS cells (PD28), the expression of α2M in aged cells (PD64) was upregulated more than eight times as demonstrated by cDNA array. To further confirm the correlation between α2M expression and senescent extent, northern blot was performed with RNA samples from four representative 2BS PDLs, including PD28, PD42, PD52 and PD62. The northern hybridization result showed that the gene expression of α2M increased gradually during the aging of cultured fibroblasts (Fig. 8.1a). In contrast to northern hybridization, RT-PCR is easier to perform and more sensitive, and is a more suitable method for detecting an aging biomarker. Therefore, the gene expression of α2M was measured by RT-PCR using RNA preparations from 2BS cells of PD27, PD35, PD42, PD48, PD54, PD60 and PD64 (Fig. 8.1b). Mean mRNA level of α2M was then plotted versus cumulative PDL, and regression analysis showed a linear increase in mRNA level of α2M with cumulative PDL as modeled by the equation (Y=0.0369X−0.6394, R^2=0.9756). Moreover, the senescence dependent up-regulation of α2M expression is not limited to the cultured fibroblasts. The peripheral blood leucocytes from old human donors demonstrated a significant increase compared to the new-born babies group. Another important phenomenon we need to point out is that overexpressing or inhibiting α2M almost does not influence the aging process of 2BS cells, and expression of senescence associated genes such as p16 and p21 are also not affected (Sun et al. 2005).

Although unable to fully substitute replicative senescence, stress-induced premature senescent cell is also extensively used in aging research. The expression of α2M is detected in 2BS cells treated with sub-lethal H_2O_2, which is most acceptable model of stress-induced premature senescence (Toussaint et al. 2002). Similarly, there is a remarkable increase of α2M expression in H_2O_2 induced premature senescent cells. As contrast, α2M expression is not affected in quiescent status of 2BS cells achieved through serum deprivation. Finally, α2M expression in HeLa cells is much lower and stable, even after more than ten passages. The constant expression level of α2M in tumor cells complies well with the criterion mentioned above for cell aging biomarkers.

Alpha-2-Macroglobulin Level as Senescence Biomarker

Except α2M, $p16^{INK4A}$ and $p21^{CIP1}$ are also the target molecules worthy of our special attention. $p16^{INK4A}$ (also known as p16 and MTS1), is generally recognized as a tumor suppressor and senescence associated gene. It has been previously reported by our research group that p16 over-expression in human fibroblasts could induce permanent and irreversible growth arrest while down-regulating p16 level led to a prolonged life span (Duan et al. 2001). As for α2M measuring, the same RNA preparation and statistical methods were also used to determine the gene expression of p16 (Ma et al. 2004). Mean mRNA level of p16 increased in 2BS cells with cumulative PDL (Fig. 8.1b), and the increase can be modeled as an exponential equation ($Y = 0.0118e^{0.0637X}$, $R^2 = 0.9621$). However, there is no statistical significance between neighboring groups below PD54. In addition, $p21^{CIP1}$ is another crucial gene participating in the regulation of cell senescence, but no affirmative statistical correlation between p21 expression and cell senescence has yet been established. Therefore, neither p16 nor p21 is ideal biomarker for cellular senescence evaluation.

Considering the criteria for biomarkers of cellular aging, α2M, as a single gene, is one of the most suitable candidates. The increase of α2M expression in 2BS cells is dependent on the cell senescent extent and demonstrates a linear correlation. Sub-lethal H_2O_2 treatment could increase the expression level of α2M whereas the α2M level is not altered in serum deprivation induced cell growth arrest. Alpha-2-macroglobulin keeps a week but stable level in long term passage of tumor cells. The expression level of α2M is easy to detect and quantify. Finally, the varying pattern of α2M expression is reproducible in samples extracted from clinical human donors.

Aging is such a complicated process that the development and regulation of which could not be achieved by a single molecule. Although one or two specific genes might be obtained and used to measure the cellular senescence, a comprehensive marker, such as a regression equation based on several variables, seems more reasonable and accurate. Of course, the variables or parameters for the comprehensive marker should not be confined to the molecular level; all senescence related morphological, cytochemical or other indicators might be involved. In fact, an ideal comprehensive marker not only could provide a quantitative measuring of senescent extent but also could forecast the maximal lifespan of specific cell cultures or even model organisms.

Regulation of Alpha-2-Macroglobulin in Replicative Senescence

Functional Promoter Region Determination

To investigate the molecular mechanism underlying α2M up-regulation in senescence, we determined the transcription activation region within

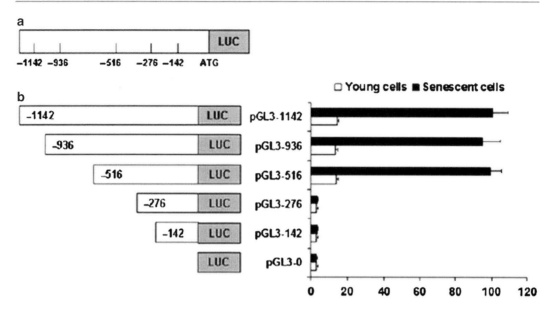

Fig. 8.2 Deletion analysis of the α2M promoter activities in young and senescent 2BS cells. (**a**) Serial deletions of 1.1-kb region of the α2M gene promoter were inserted to the upstream of the LUC report gene of pGL3-basic vector. The numbers represent the positions relative to the first ATG of α2M ORF. (**b**) The constructs of the serial deletions of human α2M promoter were transfected into young and senescent cells. LUC activity was measured according to the description in materials and methods; the promoter-less vector pGL3-basic (pGL3-0) was used as a control. The *left panel* is a schematic representation of the constructs used in transcription activity assays. The *right panel* displays the relative promoter activity of the different constructs. The promoter relative activity was calculated as the ratio of −1,142 bp promoter activity in senescent cells which was arbitrarily set at 100 %. All the results were normalized to pRL-CMV activity and represent the mean ± SD for three independent experiments

the α2M promoter required for the over-expression of α2M in senescent fibroblasts by deletion analysis. DNA fragments from different region of the α2M promoter were inserted into the upstream of the LUC reporter gene in the report plasmids with the correct 5′ to 3′ orientation. These plasmids were transiently transfected in both young and senescent 2BS cells together with a CMV LUC construct. Their relative efficiency in promoting the expression of the LUC gene was normalized to the CMV LUC activity. The mean result of independent transfection experiments is shown in Fig. 8.2, together with a schematic representation of the constructs used. We observed very low expression of LUC in young and senescent cells transfected by pGL3-142 and pGL3-276. The expression of LUC, under the control of the −516, −916 and −1,142 bp of the α2M promoter, increased ~7–8 folds in senescent cells relative to that in young cells. Deletion of the fragment of −516 to −276 bp caused a marked decrease of LUC expression (Fig. 8.2). These results suggested that the region from −516 to −276 bp contains the important elements with a strong regulatory influence on the over-expression of α2M in senescent cells.

Positive Element Identification

To further define the *cis*-elements involved in the transcriptional regulation of the α2M gene, we carried out DNase I footprinting assays to identification of the functional elements within the α2M promoter. The assays were performed on the region of the α2M promoter from −516 to −276 bp with nuclear extracts from young and senescent cells (Fig. 8.3). The two DNase I-protected regions were found and designated as region A and B. The region A was from −426 to −409 bp with the sequence of 5′-GCCTGCCCAGTGTTGCTT-3′. The region B was from −390 to −381 bp with the sequence of 5′-CCAGCTATT-3′. The region A was found to be protected from DNase I digestion

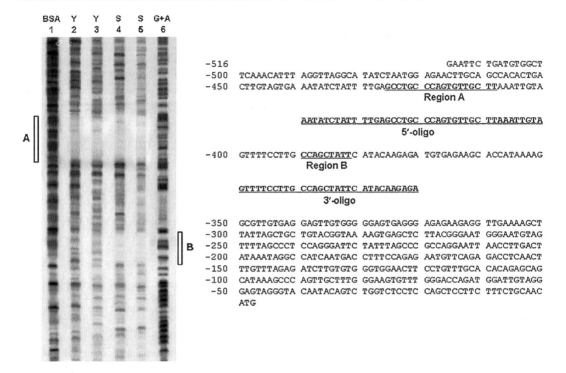

Fig. 8.3 Identification of the functional elements within the α2M promoter by DNase I footprinting analysis of the region from −516 to −276 bp of the human α2M promoter. The fragment was 3′ end labeled (*coding strand*) and incubated with nuclear extracts of young (Y) and senescent (S) cells. Complexes were digested with DNase I in the presence of 40 μg bovine serum albumin (BSA) (*lane 1*, control), 80 μg (*lane 2*) and 120 μg (*lane 3*) nuclear extracts of young cells, 80 μg (*lane 4*) and 120 μg (*lane 5*) nuclear extracts of senescent cells. Aliquots of the probes were chemically treated as for Maxam–Gilbert sequencing of G + A (*lane 6*) residues to give markers. Positions of protected regions are indicated as A and B. The region A spans −426 to −409 with the sequence of 5′-GCCTGCCCAGTGTTGCTT-3′ and the region B spans −390 to −381 with the sequence of 5′-CCAGCTATT-3′. The *right panel* show the partial sequence of α2M promoter

with extracts from both young and senescent 2BS cells, while the region B was only detected by the extracts from the senescent 2BS cells (Fig. 8.3). This suggested that there were some specific transcriptional factors expressed from senescent cells binding to region B.

To identify the trans-acting factors interacting with the region A and B in DNase I footprinting, we further performed electrophoretic mobility shift assays (EMSA) with young and senescent fibroblasts extracts. Oligonucleotides from −436 to −376 bp of the α2M promoter were end-labeled with γ-^{32}p and used as probes. Two double-stranded synthetic oligo-nucleotide probes (the 5′- oligo, −440 to −401 bp covering the footprinted A, and 3′- oligo, −400 to −371 bp covering the footprinted region B) were used for competition. We detected five retarded bands (Li et al. 2011). Bands a, b, c, and e could be competed by the 5′-oligo oligonucleotides, while band d could only be detected and competed by the 3′-oligo oligonucleotides in the presence of extract from senescent 2BS cells (Li et al. 2011). The results further suggest that a protein may interact with the 3′-oligo oligonucleotides which contain the footprinted region B. This result is consistent with that of DNA footprinting.

To determine the molecular weight of putative trans-acting factors binding to the region from −436 to −376 bp of the α2M promoter, we further performed Southwestern blotting with young and senescent fibroblast extracts. The probes were the same as that for EMSA extracts. In young cells we detected an 18 kDa and a 13 kDa binding proteins, while a 61 kDa and a 41 kDa binding protein were observed in senescent cells (Li et al. 2011). The 61, 18 and 13 kDa proteins could be competed by 5′-oligo oligonucleotides (Li et al. 2011). The intensity of 41 kDa binding protein declined greatly when competed by 3′- oligo oligonucleotides, which corresponds to the band d from EMSA, suggesting a protein interacted with 3′-oligo oligonucleotide containing the footprinted region B (Li et al. 2011).

We performed the reporter gene assay to confirm the functional significance of region A and B on the promoter activity of the α2M gene. The nucleotides of −436 to −400 bp (defined as "A"-MUT) containing the footprinted region A and −400 to −377 bp (defined as "B"-MUT) containing the footprinted region B from −516 bp of the α2M promoter were deleted. Transient transfections were performed to assess the effect of these mutations on transcription activation activity of the −516 to −276 bp fragments in both young and senescent 2BS cells. The pGL3-"A"-MUT with the deletion of the −436 to −400 bp regions showed there was no significant effect on the expression of the reporter gene in both young and senescent cells (Li et al. 2011). The result suggested that region A may not affect promoter activity of the α2M gene in both young and senescent 2BS cells. The pGL3-"B"-MUT with the deletion of the region of −400 to −377 bp decreased the expression of the reporter gene by 93 % in senescent cells, but no significant change observed in young cells (Li et al. 2011), suggesting region B may play an important role in α2M promoter activity.

Discussion

Replicative senescence of human diploid fibroblasts has been used as a well-established model system for studying aging. To find a new biomarker candidate of aging, we have surveyed the gene expression alterations in senescent fibroblasts by cDNA arrays. We found that the gene expression level of α2M was up-regulated in senescent fibroblasts, and confirmed this finding by northern hybridization and RT-PCR (Ma et al. 2004). Alpha-2-macroglobulin protein fragment was reported to be accumulated in senescent fibroblasts, and down-regulated in immortal cells, but the protein fragment was considered to be taken up from the culture medium and fragmented by the cells (Kondo et al. 2001). The relatively high amount of α2M protein in bovine serum makes it complicated to detect the protein expression of α2M in cultured cells. The gene expression level of α2M can easily be measured by semi-quantitative RT-PCR with a small sample and it is a more suitable biomarker candidate of aging then well-known senescence-associated gene such as p16^{INK4a}. Furthermore, our study showed that mRNA level of α2M has a positive linear-correlation with cumulative PDL (Fig. 8.1b). Additional assays revealed that the mRNA level of α2M was up-regulated in premature senescence induced by sublethal H_2O_2 but unchanged in quiescent cultured fibroblasts. These results suggested that up-regulated gene expression of α2M can be used to mark the progress of cellular senescence in fibroblasts like telomere shortening, rather than act as a biomarker of growth arrest like SA-β-gal. Alpha-2-macroglobulin protein level in plasma of healthy subjects has been reported to correlate with age (Wager et al. 1982). The comparison of α2M mRNA levels from leucocytes between new-born and old human body showed significant difference, while the gene expression level of α2M in immortal HeLa cells was very low and stable (Ma et al. 2004). These results indicate that up-regulated expression of α2M could be a universal age-related phenotype. Further experiments are necessary to identify if α2M protein or mRNA level can be a biomarker of aging in other types of cells in vitro or in vivo.

It has been previously reported by our research group that over-expression of p16^{INK4a} in human fibroblasts induced permanent and irreversible growth arrest at a PDL below the Hayflick limit, and fibroblasts harboring an antisense construct to p16^{INK4a} mRNA had a prolonged life span (Duan et al. 2001). Our results in this study show that gene expression of p16^{INK4a} at passage was exponentially correlated with cumulative PDL (Fig. 8.1c), in keeping with previous reports on p16^{INK4a}, and provides further evidence that up-regulation of p16^{INK4a} mRNA level is one of the causes of cellular senescence. The linear relationship between gene expression level of α2M and cumulative PDL suggests that up-regulated expression of α2M may play a different role from that of p16^{INK4a}. Alpha-2-macroglobulin is the major binding protein for TGF-β1 and TGF-β2. Alpha-2-macroglobulin and TGF-β1 were reported to be induced in both replicative senescence and H$_2$O$_2$-induced senescence of human fibroblasts (Frippiat et al. 2001). It was reported by Frippiat et al. (2001) that TGF-β1 is responsible for the induction of biomarkers of replicative senescence. The up-regulation of α2M may be caused by the progression of cellular aging to bind and neutralize the activity of TGF-β1 or other proteases up-regulated in senescent cells. Alpha-2-macroglobulin* has been demonstrated to increase DNA synthesis and cell division in macrophages (Misra and Pizzo 2002).

We next studied the transcription regulatory mechanism of α2M in senescence and characterized a novel enhancer, ATEE, located in the region of −390 to −381 bp (Region B) upstream of the first ATG of the ORF of α2M (Li et al. 2011). This enhancer might contribute strongly to the over-expression of α2M in senescent 2BS cells. The sequence analysis of this region didn't identified consensus sequence of *cis*-acting transcription factors that have been reported. The original transfection experiments showed that the −516 bp α2M promoter activity increased by ~7–8 folds in the senescent fibroblast compared with young cells, and a potential element located in the region of −516 to −276 bp may regulate the α2M expression. Two protected sites in the promoter region were subsequently identified using the DNase I footprinting assay. The region B (−390 to −381 bp) was only detected in the senescent cells (Li et al. 2011). In order to define the function of region B, EMSA and the site-directed deletion assays were performed. The results of the EMSA showed that only the gel mobility-retarded DNA-protein complex d was in competition with the unlabeled 3′-oligo oligonucleotides. Furthermore, the complex d was only present in nuclear extracts from the senescent 2BS cells (Li et al. 2011). This result suggests that a protein might be bound to this region in senescent cells. In the deletion mutation experiment, we found that region B was a positive regulation element related to the increased expression of α2M in senescent cells. In senescent cells, deletion of this region showed a 93 % decrease of the α2M promoter activity. In contrast, the promoter activity only decreased by 22.4 % in young cells (Li et al. 2011). Totally, these results suggested that ATEE, acting as a transcription enhancer, might be responsible for the increased expression of α2M during senescence.

In addition, the molecular weight of the proteins binding to the fragment (−436 to −376 bp, encompassing region A and B) of the α2M promoter was examined by the Southwestern blotting. The results showed that the intensity of a 41 kDa protein which was only present in the senescent 2BS cells decreased significantly by use of a 100-fold molar unlabeled 3′-oligo oligonucleotide as competitor (Li et al. 2011). The 41 kDa protein might act as a transcriptional activator by binding to ATEE, which needs to be confirmed in the future.

The other footprinted region (region A) was detected in both young and senescent 2BS cells. It also didn't contain consensus sequence of *cis*-acting transcription factors that have been reported. The EMSA demonstrated that the gel mobility-retarded DNA-protein a, b, c, and e were related to region A because they were competed for the 5′-oligonucleotide encompassing region A. Consistent with the results of the DNase I footprinting assay and the EMSA, the Southwestern blotting analysis showed that the 61, 18, and 13 kDa proteins could also be competed by the 5′-oligo oligonucleotides (Li et al. 2011). However, the deletion of this region showed no significant change in promoter activity in both young and senescent cells (Li et al. 2011), suggesting that the proteins bound to region A might not contribute to the differential expression of α2M during senescence.

Together, our data indicate that a novel transcription enhancer, ATEE, contributes to the differential expression of α2M during cellular senescence. ATEE is different from the interleukin-6 response elements of α2M reported previously (Hocke et al. 1992; Lerner et al. 2003; Yoo et al. 2001), which suggests a different signal transduction pathway in regulating α2M expression. Use of the ATEE to search transcription factor databases revealed no homologies to known human transcription factor binding sites, suggesting that the protein(s) binding to the ATEE are novel. Experiments to identify the putative protein by a one-hybrid assay are under way and should reveal important insights regarding its role(s) in cellular aging. Further study on the regulation of $α_2M$ will contribute greatly to understand both the mechanism of replicative senescence and its role in the etiology of Alzheimer's disease.

Acknowledgements This work was supported by grants 2013CB530801 from National Basic Research of China and grants 31000609,81170319 from National Natural Science Foundation of China.

References

Alder MJ, Coronel C, Shelton E, Seegmiller JE, Dewji NN (1991) Increased gene expression of Alzheimer disease beta-amyloid precursor protein in senescent cultured fibroblasts. Proc Natl Acad Sci U S A 88:16–20

Allsopp RC, Vaziri H, Patterson C, Goldstein S, Younglai EV, Futcher AB, Greider CW, Harley CB (1992) Telomere length predicts replicative capacity of human fibroblasts. Proc Natl Acad Sci U S A 89:10114–10118

Caldini R, Chevanne M, Mocali A, Tombaccini D, Paoletti F (1998) Premature induction of aging in sublethally H2O2-treated young MRC5 fibroblasts correlates with increased glutathione peroxidase levels and resistance to DNA breakage. Mech Ageing Dev 105:137–150

Chen QM, Prowse KR, Tu VC, Purdom S, Linskens MH (2001) Uncoupling the senescent phenotype from telomere shortening in hydrogen peroxide-treated fibroblasts. Exp Cell Res 265:294–303

Choi AMK, Pignolo RJ, Rhys CMJ, Cristifalo VJ, Holbrook NJ (1995) Alterations in the molecular response to DNA damage during cellular aging of cultured fibroblasts: reduced Ap-1activation and collagenase gene expression. J Cell Physiol 164:65–74

Crookston KP, Webb DJ, Wolf BB, Gonias SLJ (1994) Classification of alpha 2-macroglobulin-cytokine interactions based on affinity of noncovalent association in solution under apparent equilibrium conditions. J Biol Chem 269:1533–1540

Dimri GP, Lee X, Basile G, Acosta M, Scott G, Roskelley C, Medrano EE, Linskens M, Rubelj I, Pereira-Smith O (1995) A biomarker that identifies senescent human cells in culture and in aging skin in vivo. Proc Natl Acad Sci U S A 92:9363–9367

DiPaolo BR, Pignolo RJ, Cristifalo VJ (1992) Overexpression of two-chain form of cathepsin B in senescent WI-38 cells. Exp Cell Res 201:500–505

Duan JM, Zhang ZY, Tong TJ (2001) Senescence delay of human diploid fibroblast induced by anti-sense p16INK4a expression. J Biol Chem 276:48325–48331

Frippiat C, Chen QM, Zdanov S, Magalhaes JP, Remacle J, Toussaint O (2001) Subcytotoxic H2O2 stress triggers a release of transforming growth factor-beta 1, which induces biomarkers of cellular senescence of human diploid fibroblasts. J Biol Chem 276:2531–2537

Hocke GM, Barry D, Fey GH (1992) Synergistic action of interleukin-6 and glucocorticoids is mediated by the interleukin-6 response element of the rat alpha 2 macroglobulin gene. Mol Cell Biol 12:2282–2294

Kondo T, Sakaguchi M, Namba M (2001) Two-dimensional gel electrophoretic studies on the cellular aging: accumulation of alpha-2-macroglobulin in human fibroblasts with aging. Exp Gerontol 36:487–495

Lerner L, Henriksen MA, Zhang X, Darnell JJ (2003) STAT3-dependent enhanceosome assembly and disassembly: synergy with GR for full transcriptional increase of the alpha 2-macroglobulin gene. Genes Dev 17:2564–2577

Li RZ, Ma LW, Huang Y, Zhang ZY, Tong TJ (2011) Characterization of a novel positive transcription regulatory element that differentially regulates the alpha-2-macroglobulin gene in replicative senescence. Biogerontology 12:517–525

Lin VK, Wang SY, Boetticher NC, Vazquez DV, Saboorian H, McConnell JD, Roehrborn CG (2005) Alpha(2) macroglobulin, a PSA binding protein, is expressed in human prostate stroma. Prostate 63:299–308

Liu Q, Ling TY, Shieh HS, Johnson FE, Huang SS (2001) Identification of the high affinity binding site in transforming growth factor-beta involved in complex formation with alpha 2-macroglobulin. Implications regarding the molecular mechanisms of complex formation between alpha 2-macroglobulin and growth factors, cytokines, and hormones. J Biol Chem 276:46212–46218

Ma H, Li RZ, Zhang ZY, Tong TJ (2004) mRNA level of alpha-2-macroglobulin as an aging biomarker of human fibroblasts in culture. Exp Gerontol 39:415–421

Misra UK, Pizzo SV (2002) Regulation of cytosolic phospholipase A2 activity in macrophages stimulated with receptor-recognized forms of alpha 2-macroglobulin: role in mitogenesis and cell proliferation. J Biol Chem 277:4069–4078

Qiu Z, Strickland DK, Hyman BT, Rebeck GW (1999) Alpha 2-macroglobulin enhances the clearance of endogenous soluble beta-amyloid peptide via low-density lipoprotein receptor-related protein in cortical neurons. J Neurochem 73:1393–1398

Scottrup-Jensen L, Sand O, Kristensen L, Fey GH (1989) The alpha-macroglobulin bait region. Sequence diversity and localization of cleavage sites for proteinases in five mammalian alpha-macroglobulins. J Biol Chem 264:15781–15789

Seshadri T, Campisi J (1990) Repression of c-fos transcription and an altered genetic program in senescent human fibroblasts. Science 247:205–209

Severino J, Allen RG, Balin S, Balin A, Cristifalo VJ (2000) Is â-galactosidase staining a maker of senescence in vitro and in vivo. Exp Cell Res 257:162–171

Sottile J, Mann DM, Diemer V, Millis AJ (1989) Regulation of collagenase and collagenase mRNA production in early- and late-passage human diploid fibroblasts. J Cell Physiol 138:281–290

Sun Y, Li RZ, Zhang ZY, Tong TJ (2005) Functions of á-2-macroglobulin in human diploid fibroblasts during aging. Chin J Geriatr 24:858–861

Toussaint O, Remacle J, Dierick JF, Pascal T, Frippiat C, Zdanov S, Magalhaes JP, Royer V, Chainiaux F (2002) From the Hayflick mosaic to the mosaics of ageing. Role of stress-induced premature senescence in human ageing. Int J Biochem Cell Biol 34:1415–1429

Tresini M, Pignolo RJ, Allen RG, Cristifalo VJ (1999) Effects of donor age on the expression of a marker of replicative senescence (EPC-1) in human dermal fibroblasts. J Cell Physiol 179:11–17

Wager V, Wagnerova M, Zvara K (1982) Age correlation of alpha-2-macroglobulin levels in healthy subjects. Physiol Bohemoslov 31:359–364

Yoo JY, Wang W, Desiderio S, Nathans D (2001) Synergistic activity of STAT3 and c-Jun at a specific array of DNA elements in the alpha 2-macroglobulin promoter. J Biol Chem 276:26421–26429

Elevation of Ceramide in Senescence: Role of Sphingolipid Metabolism

Mark E. Venable

Contents

Abstract	81
Introduction	82
Sphingolipids in Biology	82
Sphingolipids in Cancer	82
Ceramide in Biology	83
Ceramide Mechanisms of Action	84
Ceramide and Senescence	85
Mechanisms of Ceramide Elevation	86
Implications	86
Future Directions	87
References	87

M.E. Venable (✉)
Biology Department, Appalachian State University,
572 Rivers St, Boone, NC 28608, USA
e-mail: venableme@appstate.edu

Abstract

Sphingolipids play a wide variety of roles in biological systems. In cancer, the sphingolipids ceramide and sphingosine-1-phosphate play offsetting roles in balancing cell proliferation with senescence and survival with death. By tipping the balance toward more ceramide and less sphingosine-1-phosphate cells are less prone to survive and proliferate. Cellular senescence is thought to have evolved as a mechanism to limit cancer growth. Senescent cells produce higher levels of ceramide and this appears to contribute to this phenotype. Ceramide is at the hub of metabolism of a variety of sphingolipids. Ceramide levels are determined through the balance of a large group of enzymes. In senescence, several of these are modulated in a coordinated fashion such that ceramide levels are kept elevated compared to their proliferative cell counterparts. Ceramide can work through a variety of targets including PKC, PP2A, PP1, phospholipase D, kinase suppressor of Ras and cathepsin D. By blocking the action of the phospholipase D and PKC ceramide can, in part, maintain these cells in a postproliferative state. There are a variety of pharmaceutical tools available to modulate sphingolipid levels in cells through the enzymes that metabolize them. These have proven very promising for potential cancer drugs.

Keywords

Ceramide-activated protein kinase • Ceramides • Dihydroceramide • Glucosylceramide synthesis • Glycosphingolipids and sphingomyelin • Human diploid fibroblasts (HDF) • Human umbilical vein endothelial cells (HUVECs) • Lipidomics profiles • Protein phosphatases • Sphingolipid metabolism

Introduction

Sphingolipids in Biology

Sphingolipids have long been known to play vital structural roles in cell membranes. Sphingomyelin plays an important role in the structure of the plasma membrane and glucosphingolipids are key components of the glycocalyx. Glycosphingolipids and sphingomyelin play a key role in lipid raft structure and function affecting many aspects of signaling. Many roles for sphingolipids have also been found in intra- and intercellular signaling. Glycosphingolipids play key roles in cell:cell recognition. Thirty years ago these compounds were recognized as playing roles in cellular interactions that are key to cellular differentiation and proliferation (Hakomori 1981).

Ceramides have been found to play important roles in cell cycle arrest, differentiation, senescence, apoptosis and stress responses. Sphingosine mediates apoptosis and cell cycle arrest. Sphingosine-1-phosphate is a potent mitogen and involved in inflammation, migration, angiogenesis and cell survival. Other sphingolipids play key roles in cell signaling as well. Defects in sphingolipid metabolism have been shown to have devastating impacts for cells and patients.

Sphingolipids in Cancer

Recognition of sphingolipids as key mediators in cancer development, survival and chemoresistance is quickly emerging. Researchers have found decreases in ceramide concentration in cancers that are consistent with cancer growth and survival. Observations have also been made of increases in sphingosine-1-phosphate in a variety of tumors consistent with mitosis and cell survival as well as inflammation. Recent reviews bring together many instances of how sphingolipids and some of the enzymes that metabolize them are altered in various types of cancer (Oskouian and Saba 2010; Furuya et al. 2011).

Cell viability and proliferation can be determined, in part, by ceramide concentration. Cancer cells often find ways to lower their ceramide concentration to allow for survival and proliferation. This area is thoroughly reviewed by Ryland et al. (2011). Ceramide can be metabolized to glycolipids, sphingomyelin, ceramide-1-phosphate or sphingosine or can be made from these lipids or dihydroceramide. The concentration of ceramide is determined by the concentrations the relevant enzymes and their activation state.

Glucosylceramide synthesis is upregulated in many cancer cells and catalyzes the effective removal of ceramide that would otherwise induce apoptosis in damaged cells hence leading to increased cell survival (Liu et al. 1999). Other changes in glycosphingolipids, including the sulfides, are associated with cancer transformation, metastatic potential and treatment outcomes (Morichika et al. 1996; Barth et al. 2011). Other superproliferative cells have been found to lower their ceramide concentration using ceramidase. Often this is linked to its conversion to sphingosine-1-phosphate (S1P) that is discussed below (Hannun and Obeid 2008). The location of ceramide within the cell plays a key in cell response. Increasing the transport of ceramide using CERT away from its molecular target(s) and/or to be metabolized can also allow cancer cells to circumvent the apoptosis or cell cycle arrest that ceramide imposes (Scheffer et al. 2011).

The study of sphingolipids in cancer treatment has also proven fruitful. Fernandez et al. (2010) and Ryland et al. (2011) describe how, based on our current knowledge of sphingolipid action, sphingolipid targets are showing promise as useful in cancer therapies and that markers of ceramide metabolism are useful for determining the potential effectiveness of chemotherapy (Fernandez et al. 2010; Ryland et al. 2011).

Neutralization of ceramide via phosphorylation or glycosylation in malignant cells has been linked to multidrug chemoresistance (Barth et al. 2011). Ceramide has been shown to play a role in transducing the effects of certain chemopreventive agents such as curcumin and resveratrol (Oskouian and Saba 2010). Ceramide's mechanisms of action are described below.

The story is more complex, however, and studies using lipidomics techniques show that there are differences between the various molecular species of ceramide (Hannun and Obeid 2011; Ryland et al. 2011) in how they function and are regulated. Indeed, in some cases certain species of ceramide are actually increased in some cancers and may promote growth or simply not inhibit growth while other mediators promote proliferation. It appears that the nearly 30 known distinct enzymes that metabolize ceramide are specific to certain cellular compartments and certain species of sphingolipid and are differentially regulated (Hannun and Obeid 2011; Ryland et al. 2011). Molecular species specificity for certain of enzymes and signaling targets accounts for much of what was earlier seen as discrepancies in outcomes of modulating sphingolipid levels.

Sphingosine-1-phosphate on the other hand acts to stimulate mitosis through the action of a G-protein-coupled receptor. These receptors are in the same gene family (Edg, LPA_{1-3} and $S1P_{1-5}$) that also includes the receptors for lysophosphatidic acid that have been found to have strong linkages to cancer development (Im 2010). S1P plays a role in the regulation of survival, proliferation and migration of tumor cells (Ryland et al. 2011). Higher levels of S1P are linked to cell survival and poor survival rate in some cancer patients. Neutralization of S1P has been shown to improve cell death. Alterations in S1P receptors are also associated with the cancer phenotype. Sphingosine-1-phosphate can oppose apoptosis through the downregulation of Bax and the activation of Akt (Oskouian and Saba 2010).

The related compound ceramide-1-phosphate has been shown to have the potential to promote cancer growth by causing a blockage of apoptosis and DNA replication (Levi et al. 2010). In cancers there tends to be a decrease in cell cycle arrest- and apoptosis-promoting ceramide and an increase in growth promoting and apoptosis inhibiting sphingosine-1-phosphate. Changes in sphingolipid concentrations have the potential to set up an environment in which cells can transition into a malignant state.

Through the pharmacologic modulation of sphingolipid levels, apoptosis may be initiated in cancer cells that are otherwise resistant (Oskouian and Saba 2010; Burns and Luberto 2011; Canals et al. 2011; Delgado et al. 2012). Strategies largely aim to increase ceramide or to block ceramide removal or S1P synthesis. Other drugs include ceramide mimetics and S1P antagonists. In several cases it has been found that existing chemotherapy agents possess certain of these activities (Oskouian and Saba 2010).

Ceramide in Biology

In early studies ceramides were found to play important roles in cell cycle arrest, differentiation, senescence and apoptosis (Hannun and Obeid 2008). These outcomes are related in that they lead to an outcome that opposes proliferation and cancer development. Later, ceramides were found to play a role in the response to certain types of cell stress (Hannun and Obeid 2008; Nikolova-Karakashian and Rozenova 2010; Nagai et al. 2011). While a large body of work has shown ceramide to play important roles in cell cycle arrest, differentiation, senescence and stress response these areas still require significant work to clarify how these outcomes are differentially controlled.

One of the difficulties in determining the roles that sphingolipid mediators play in a given disease state is that these compounds are readily interconvertible through the action of a variety of enzymes that can be selective for specific molecular species and can be compartmentalized such that the predominating product may differ between cellular compartments at the same time (Hannun and Obeid 2011). In addition, some of these signaling molecules are also metabolic intermediates that will be present at detectable levels regardless of the status of the cell.

It has been known for some time that ceramide and dihydroceramide have very different bioactivity. Dihydroceramide is considered to be primarily a metabolic intermediate although recent evidence suggests that it also plays a signaling role (Ryland et al. 2011). The role of ceramide as a lipid autacoid has been firmly established. Exogenous ceramides with different fatty acids have been used to stimulate biological effects. Short chain ceramides are effectively delivered using an amphipathic solvent vehicle such as ethanol or dimethylsulfoxide. Some significant differences between C_2- C_6 and C_8 ceramides have been reported but will not be reviewed here. Delivery of longer chain ceramides into cells has been shown to be possible using dodecane. When short chain ceramides are delivered the fatty acid may then be replaced in the cell by a long chain fatty acid and its subsequent metabolism may also determine cell fate as described above (Chapman et al. 2010).

The isoforms of the enzymes that synthesize ceramides show a high degree of chain length specificity in developing the mixture found in cells (Grösch et al. 2011). The profile of ceramides in a cell at a given time can be regulated by the expression and activation of the enzymes. How the various ceramides are moved through the cell is to some extent regulated by the specificity of the CERT proteins. Chain length of endogenous ceramides is now recognized to have a dramatic impact on signaling. Ceramide targets show a considerable degree of specificity and therefore the physiological outcome is dependent on which species are produced where.

It is important to recognize that differential dosing of ceramide can lead to different outcomes. Lower doses of ceramide induce, depending on the cells, cell-cycle arrest or senescence while still higher concentrations initiate apoptosis. Lipid mediators like ceramide partition to and function in the two dimensional membrane. When exogenous ceramides are delivered to cells what matters is how much mediator is used relative to the amount of cell membrane. Hence, when a certain population of cells is treated with 500 nM ceramide in 10 ml medium then the dose is effectively double that if the volume is 5 ml. This effect can be buffered with albumin in the medium that reduces the effective dose by preventing or slowing the partitioning of ceramide into the membrane. When measuring changes in endogenous sphingolipid levels, these are normalized to levels of other lipids such as total lipid phosphorous.

Ceramide Mechanisms of Action

Molecular targets of ceramide have been elucidated in a variety of ways and with difficulty due to a variety of nonspecific interactions being possible. Following the discovery that sphingosine could regulate PKC (Canals et al. 2011) the door was opened to the possibility of sphingolipids specifically modulating the action of other proteins. It is now known that PKCζ is directly inhibited by ceramide, having impacts on RNA splicing (Pettus et al. 2002).

Phosphatidylcholine phospholipase D isoforms have been implicated in intracellular vesicle trafficking, endocytosis, exocytosis, actin cytoskeleton dynamics, cell proliferation, differentiation, migration and survival and has been linked to cancer development (Peng and Frohman 2012). Working with human diploid fibroblasts we found synergistic activation of PLD by PKCα and the GTP binding protein ARF in mitotic signaling (Venable et al. 1996). In that system, ceramide inhibited the ability of PKCα to activate PLD although basal activation by ARF was not affected. This helped explain how growth factor activation of senescent cells failed to activate cell replication.

Certain protein phosphatases have now been found to be activated by ceramide (Perry et al. 2012). The original ceramide-activated protein phosphatase was identified as PP2A. Later, PP1A was found to be a ceramide target and now PP2C has been shown to be activated by ceramide. This pathway is active in the induction of apoptosis, growth arrest and inflammation.

A ceramide-activated protein kinase was identified as the kinase suppressor of Ras (KSR) (Pettus et al. 2002). Ceramide has also been shown to interact directly with Raf. Cathepsin D has been shown to play an important role in apoptosis. Ceramide generated by acid sphingomyelinase has been shown to play a key role in its regulation (Pettus et al. 2002). Ceramide and S-1-P have been shown to play a key role in apoptosis through Bcl-2 (Oskouian and Saba 2010). Ceramide also acts to dephosphorylate and inactivate Akt (Oskouian and Saba 2010). Evidence for other targets has been proposed and likely others will be determined in the future. The mediators described above have clear implications for cell outcomes related to cancer.

Ceramide and Senescence

While studying the regulation of diacylglycerols induced by mitogens in human diploid fibroblasts (HDF) at different population doubling levels we noticed that ceramide was present at substantially higher levels in senescent cells than low-passage cells. This finding was possible because the diacylglycerol kinase method for quantifying diacylglycerol concentrations also measures ceramide. This was fascinating because of the interplay these classes of mediators show in many different systems. After further study we and others determined that ceramide levels remain low until a significant fraction of the cells in the population have attained senescence (Venable et al. 1995; Meacci et al. 1996). Later when studying senescence in human umbilical vein endothelial cells (HUVECs) we made the same observation that ceramide concentration increases with the onset of replicative senescence. Assessments in other cell types remains to be studied.

We knew that this was simply a correlation and did not show causation; therefore, we treated low-passage cells with ceramide to determine whether the hallmarks of senescence were induced. We found that in HDF and later in HUVECs that ceramide treatment (cell-permeable C_6-ceramide) at levels reflecting endogenous levels in senescent cells shut down DNA replication, AP-1 activation, Rb phosphorylation and induced the senescence-associated beta-galactosidase. Hence, ceramide was all that was needed to mimic cellular senescence under these standard conditions. It is interesting to note that HUVECs are exquisitely sensitive to ceramide and an increase of only twofold is needed to induce senescence. In general the dose-response curves for effects of ceramide are quite steep when compared with most other mediators and changes in ceramide of two to fivefold often spans from no to maximum effect.

Interestingly, Modrak et al. (2009) show that treatment of pancreatic cancer cells with gemcitabine induced a form of premature senescence in a large percentage of the cells. This protected the cells from death. Addition of sphingomyelin, as source for synthesizing more ceramide, reversed this protection. The same result was seen when exogenous C_8-ceramide was given. A model was proposed in which the chemotherapy agent provides protection by inducing enough ceramide to induce senescence but not apoptosis and that by treating cells in any of a variety of ways to increase ceramide levels that cancer cells are sensitized and chemotherapy leads to improved outcomes. This is supported by more recent work by Giovannetti et al. (2010) using a ceramide analog.

Since ceramide is a mediator of several types of stress it may follow that this could lead to characteristics of senescence. Telomere shortening is the basic process that leads to replicative senescence. One mechanism by which this could occur is by ceramide leading to telomere shortening. Ceramide was shown to effectively inhibit telomerase (Saddoughi et al. 2011) through the activation of Myc degradation as down regulation of the hTERT promotor (Wooten-Blanks et al. 2007). There have been numerous links made between stress and telomere shortening (Calado and Young 2009) and the elevation of ceramide leading to telomere shortening may provide a mechanism for this to occur.

Stress of various types has been shown to cause a type of premature senescence (Muller 2009). In a model of glycated collagen-induced premature senescence ceramide is proposed to play a role in autophagy as a mechanism of actuating senescence (Patschan et al. 2008). Senescence-inducing stressors will generally lead to some form of genotoxic insults and DNA damage response (Rodier and Campisi 2011). One of the characteristics of this type of senescence is that they secrete multiple factors that can set up other cells for replication. These include growth factors and cytokines (Rodier and Campisi 2011).

Mechanisms of Ceramide Elevation

Sphingolipid metabolism is quite complex and key for this discussion are those enzymes that influence ceramide levels. These are thoroughly described in the excellent review by Merrill (2011). The first level of regulation is in those enzymes that directly produce or consume ceramide. Ceramides may be formed *de novo* via sphingosine and dihydroceramide. This multistep pathway has been shown to be regulatory in determining the concentration of endogenous ceramide in certain situations such as cell stress (Nikolova-Karakashian and Rozenova 2010). In our research in senescent cells we found no change in overall *de novo* synthesis (Venable et al. 2006).

Ceramide may also be generated by the salvage pathway (sphingomyelinase) from sphingomyelin. Under stress conditions an acid sphingomyelinase is activated and ceramide is generated (Oskouian and Saba 2010). Specific stimulation through TNFα leads to the activation of a neutral sphingomyelinase. In senescent cells we found a slight but significant increase in acid sphingomyelinase activity but a large increase in neutral sphingomyelinase. We took this to mean that the neutral enzyme was more significant in the senescent-associated ceramide increases. Breakdown of glycosphingolipid also leads to ceramide and in senescent fibroblasts we found a small but significant increase in that pathway.

Conversely, ceramide concentrations may be increased by reducing the utilization of ceramide for other products (Venable et al. 2006). Ceramide may be hydrolyzed to sphingosine and fatty acid by a variety of ceramidases. In our studies we analyzed the alkaline and acidic forms. Whereas there was no change in the alkaline activity the acidic activity was reduced by a factor of 4. Ceramide may be glycosylated to form any of a number of compounds. Flux through the glucosylceramide synthase pathway was blunted in senescent fibroblasts but unchanged in endothelial cells. Enzyme activities reflected this result. When we analyzed the sphingomyelin synthesis activity we found flux through the pathway in intact cells slightly depressed but no measureable change in sphingomyelin synthase activity. Taken as a whole the results showed that there is a general shift in metabolism leading to increased ceramide levels in senescent cells. From these types of studies one cannot conclude if it is indeed the combined effect of all these changes or if one particular enzyme were responsible. That will require an in depth molecular approach.

Implications

Clearly, alterations in sphingolipid metabolism can lead to dramatic differences in cellular outcome. As lipidomics pushes forward, we continue to find changes in levels of classes and species of sphingolipids. The changes in the relevant enzymes' expression and activity that lead to these changes will require more research before the overall framework becomes clear. The impacts of these changes on the cell are complex. The outcome for the cell likewise has important implications for patients in that there is too much or too little cell survival or proliferation. Sphingolipid biology clearly is somewhat more complex in that the signaling molecules are themselves also metabolic intermediates and that relatively subtle changes in their concentrations can have dramatic and even opposed outcomes for the cell. There is great interest in developing new drugs that modulate sphingolipids for cancer treatment and the results obtained thus far are very promising for effective therapies in that arena as described above.

The implications of having some senescent cells in tissues that oppose growth while others promote growth are perplexing. With this seeming paradox, the path toward a therapy is less clear, although the model of senescent cell progression may help to address this (Rodier and Campisi 2011). Much work remains to determine the extent to which ceramide plays a role in different forms of senescence. With findings showing some cancers having elevated ceramide levels suggests the possibility that this dichotomy exists in diverse populations of senescent cells as well. It would be interesting to determine if there is a link between elevated ceramide and cancer promoting senescent cells. It needs to be determined whether there is a clear difference between these two types/stages of senescent cells in their sphingolipid content or metabolism. If it is determined that modulating sphingolipids content in both functional types toward more S1P or toward ceramide can lead to improved prognosis then an approach similar to that being used for the cancer cells might be used. Otherwise, a more targeted approach will be needed.

Future Directions

One area that requires better understanding is how small changes in ceramide can have such important differences in cellular outcome, particularly senescence or apoptosis. The careful work of elucidating how the concentration of ceramide can be so closely controlled needs to be done. Determination of binding characteristics of ceramide molecular species and other sphingolipids to their respective target proteins will lead to better models of how modulation of the activities of the proteins controlling ceramide profiles affect cellular outcomes.

Lipidomics profiles of different cell types that have become senescent due to reaching their Hayflick limit and those that are in a stress-induced premature senescent state will help determine how these states compare. Ceramide target proteins need to be further analyzed in replicatively senescent cells to better understand how ceramide helps maintain the senescent phenotype. In many cases findings made in the lipid field remain isolated, largely because many molecular biologists know little about lipids and do not know what to do with what they have learned. Hence, models need to be developed that integrate what has been learned in the lipidomics and lipid signaling realms with models of protein signaling pathways in senescence and the other areas impacted by the enigmatic group of molecules called sphingolipids.

References

Barth BM, Cabot MC, Kester M (2011) Ceramide-based therapeutics for the treatment of cancer. Anti Cancer Agents Med Chem 11:911–919

Burns TA, Luberto C (2011) Sphingolipid metabolism and leukemia: a potential for novel therapeutic approaches. Anti Cancer Agents Med Chem 11:863–881

Calado RT, Young NS (2009) Telomere diseases. N Engl J Med 361:2353–2365

Canals D, Perry DM, Jenkins RW, Hannun YA (2011) Drug targeting of sphingolipid metabolism: sphingomyelinases and ceramidases. Br J Pharmacol 163:694–712

Chapman JV, Gouazé-Andersson V, Messner MC, Flowers M, Karimi R, Kester M, Barth BM, Liu X, Liu YY, Giuliano AE (2010) Metabolism of short-chain ceramide by human cancer cells—implications for therapeutic approaches. Biochem Pharmacol 80:308–315

Delgado A, Fabriàs G, Bedia C, Casas J, Abad J (2012) Sphingolipid modulation: a strategy for cancer therapy. Anti Cancer Agents Med Chem 12:285–302

Fernandez LE, Gabri MR, Guthmann MD, Gomez RE, Gold S, Fainboim L, Gomez DE, Alonso DF (2010) Ngcgm3 ganglioside: a privileged target for cancer vaccines. Clin Dev Immunol 2010;814397–814404

Furuya H, Shimizu Y, Kawamori T (2011) Sphingolipids in cancer. Cancer Metastasis Rev 30:567–576

Giovannetti E, Leon L, Bertini S, Macchia M, Minutolo F, Funel N, Alecci C, Giancola F, Danesi R, Peters G (2010) Study of apoptosis induction and deoxycytidine kinase/cytidine deaminase modulation in the synergistic interaction of a novel ceramide analog and gemcitabine in pancreatic cancer cells. Nucleosides Nucleotides Nucleic Acids 29:419–426

Grösch S, Schiffmann S, Geisslinger G (2011) Chain length-specific properties of ceramides. Prog Lipid Res 51:50–62

Hakomori S (1981) Glycosphingolipids in cellular interaction, differentiation, and oncogenesis. Annu Rev Biochem 50:733–764

Hannun YA, Obeid LM (2008) Principles of bioactive lipid signalling: lessons from sphingolipids. Nat Rev Mol Cell Biol 9:139–150

Hannun YA, Obeid LM (2011) Many ceramides. J Biol Chem 286:27855–27862

Im DS (2010) Pharmacological tools for lysophospholipid gpcrs: development of agonists and antagonists for lpa and S1P receptors. Acta Pharmacologica Sinica 31:1213–1222

Levi M, Meijler MM, Gómez-Muñoz A, Zor T (2010) Distinct receptor-mediated activities in macrophages for natural ceramide-1-phosphate (C1P) and for phospho-ceramide analogue-1 (pCera-1). Mol Cell Endocrinol 314:248–255

Liu Y-Y, Han T-Y, Giuliano AE, Ichikawa S, Hirabayashi Y, Cabot MC (1999) Glycosylation of ceramide potentiates cellular resistance to tumor necrosis factor-α-induced apoptosis. Exp Cell Res 252:464–470

Meacci E, Vasta V, Neri S, Farnararo M, Bruni P (1996) Activation of phospholipase d in human fibroblasts by ceramide and sphingosine: evaluation of their modulatory role in bradykinin stimulation of phospholipase D. Biochem Biophys Res Commun 225:392–399

Merrill AH Jr (2011) Sphingolipid and glycosphingolipid metabolic pathways in the era of sphingolipidomics. Chem Rev 111:6387

Modrak DE, Leon E, Goldenberg DM, Gold DV (2009) Ceramide regulates gemcitabine-induced senescence and apoptosis in human pancreatic cancer cell lines. Mol Cancer Res 7:890–896

Morichika H, Hamanaka Y, Tai T, Ishizuka I (1996) Sulfatides as a predictive factor of lymph node metastasis in patients with colorectal adenocarcinoma. Cancer 78:43–47

Muller M (2009) Cellular senescence: molecular mechanisms, in vivo significance, and redox considerations. Antioxid Redox Signal 11:59–98

Nagai K, Takahashi N, Moue T, Niimura Y (2011) Alteration of fatty acid molecular species in ceramide and glucosylceramide under heat stress and expression of sphingolipid-related genes. Adv Bio Chem 1:35–48

Nikolova-Karakashian MN, Rozenova KA (2010) Ceramide in stress response. In: Chalfant C, Del Poeta M (eds) Sphingolipids as signaling and regulatory molecules. Springer, New York, pp 86–108

Oskouian B, Saba JD (2010) Cancer treatment strategies targeting sphingolipid metabolism. In: Chalfant C, Del Poeta M (eds) Sphingolipids as signaling and regulatory molecules. Springer, New York, pp 185–205

Patschan S, Chen J, Polotskaia A, Mendelev N, Cheng J, Patschan D, Goligorsky MS (2008) Lipid mediators of autophagy in stress-induced premature senescence of endothelial cells. Am J Physiol Heart Circ Physiol 294:H1119–H1129

Peng X, Frohman MA (2012) Mammalian phospholipase D physiological and pathological roles. Acta Physiologica 204:219–226

Perry DM, Kitatani K, Roddy P, El-Osta M, Hannun YA (2012) Identification and characterization of PP2C activation by ceramide. J Lipid Res 53(8):1513–1521. doi:10.1194/jlr.M025395

Pettus BJ, Chalfant CE, Hannun YA (2002) Ceramide in apoptosis: an overview and current perspectives. Biochim Biophys Acta Mol Cell Biol Lipids 1585:114–125

Rodier F, Campisi J (2011) Four faces of cellular senescence. J Cell Biol 192:547–556

Ryland LK, Fox TE, Liu X, Loughran TP, Kester M (2011) Dysregulation of sphingolipid metabolism in cancer. Cancer Biol Ther 11:138–149

Saddoughi SA, Garrett-Mayer E, Chaudhary U, O'Brien PE, Afrin LB, Day TA, Gillespie MB, Sharma AK, Wilhoit CS, Bostick R (2011) Results of a phase II trial of gemcitabine plus doxorubicin in patients with recurrent head and neck cancers: serum c18-ceramide as a novel biomarker for monitoring response. Clin Cancer Res 17:6097–6105

Scheffer L, Rao Raghavendra P, Ma J, K Acharya J (2011) Ceramide transfer protein and cancer. Anti Cancer Agents Med Chem 11:904–910

Venable ME, Lee JY, Smyth MJ, Bielawska A, Obeid LM (1995) Role of ceramide in cellular senescence. J Biol Chem 270:30701–30708

Venable ME, Bielawska A, Obeid LM (1996) Ceramide inhibits phospholipase D in a cell-free system. J Biol Chem 271:24800–24805

Venable ME, Webb-Froehlich LM, Sloan EF, Thomley JE (2006) Shift in sphingolipid metabolism leads to an accumulation of ceramide in senescence. Mech Ageing Dev 127:473–480

Wooten-Blanks LG, Song P, Senkal CE, Ogretmen B (2007) Mechanisms of ceramide-mediated repression of the human telomerase reverse transcriptase promoter via deacetylation of SP3 by histone deacetylase 1. FASEB J 21:3386–3397

Molecular Signals Underlying Hair Follicle Morphogenesis and Cutaneous Regeneration

10

Xusheng Wang and Yaojiong Wu

Contents

Abstract	89
Introduction	90
Signals for Hair Follicle Morphogenesis	90
The Induction and Morphogenesis of Hair Follicles	90
Hair Follicle Patterning	94
The Quiescence and Activation of Hair Follicle Stem Cells	95
Stem Cells and Niche Cells for Cutaneous Regeneration	95
Hair Follicle Stem Cells Regenerate Epidermal Appendages	95
DP Cells Induce Hair Follicle Neogenesis	97
Mesenchymal Stem cells in Cutaneous Repair and Regeneration	97
Discussion	99
References	99

X. Wang (✉) • Y. Wu
Life Science Division, Tsinghua University Graduate
School at Shenzhun, L406A, Tsinghua Campus,
Shenzhen 518055, China
e-mail: wu.yaojiong@tsinghua.edu.cn

Abstract

Hair follicle morphogenesis is initiated during embryo development, and undergoes regularly cycling process after birth. As a functional mini-organ, hair follicle morphogenesis undergoes in an environment with dynamic and alternating changes of numerous molecular signals. Studies in the past decades of genetically modified mouse models have lead to tremendous progresses in elucidating the molecule signals regulating hair follicle development. In the adult mouse skin, hair follicles serve as a reservoir of cutaneous stem cells. Stem cells derived from the bulge and dermal papilla of the hair follicle are capable of forming *de novo* epidermal appendages. But signals guiding these stem cells to regenerate cutaneous structures have been poorly understood. On the other hand, studies in the past decade indicate that born marrow derived mesenchymal stem cells exert a great potential in improving wound healing. Advances in our understanding of the molecular regulation of hair follicle morphogenesis and regeneration may help develop novel therapeutic strategies to facilitate improved wound healing.

Keywords

Connective tissue growth factor (CTGF) • Dermal papilla (DP) cells • Edar signaling • Embryonic day (ED) • Hair follicle morphogenesis • Keratinocyte growth factor (KGF) and epidermal growth factor (EGF) signals •

Label-retaining cells (LRCs) • Mesenchymal stem cells • Platelet-derived growth factor-A (PDGF-A) • Stem and niche cells • Transforming growth factor-β (TGF–β)

Introduction

The development of embryonic hair follicles starts with the presence of placodes in the early epidermis, which is followed by the formation of dermal-condensation (DC). Thereafter, the placode extends downward into the dermis and forms a primary hair follicle structure. For all events in embryonic hair follicle development, hair follicle initiation and patterning are central, which are under tight guidance of dynamic signals. Wnt/β-catenin signaling and ectodysplasin (Eda)-A1/nuclear factor kappa-B (NF-κB) pathway have been identified as essential components in controlling hair follicle formation. Wnt signaling has been considered as the earliest and the predominant factor for hair follicle development. Meanwhile, defects in Eda-A1/NF-κB pathway result in hypohidrotic (anhidrotic) ectodermal dysplasia (HED). Moreover, bone morphogenetic protein (BMP), Sonic Hedgehog (Shh), Dickkopf-1/4 (Dkk-1/4), noggin, and adenomatosis polyposis coli down-regulated 1 (APCDD1) are also involved in hair follicle morphogenesis. Wnt and Dkk-1, together with Edar and BMP, act as activator/inhibitor pairs to determine the patterning of hair follicle via a reaction-diffusion (RD) mechanism. The mature hair follicle persistently goes through three phases: growth (anagen), regression (catagen), and rest (telogen), which are associated with corresponding changes in functional status of hair follicle stem cells (Hardy 1992).

Acute cutaneous wounds normally heal in a very orderly and efficient manner characterized by four distinct, but overlapping phases: hemostasis, inflammation, proliferation and remodeling. In postpartum humans and mammals, injury to the skin heals not by the regeneration of the tissue to the pre-injured form but by the formation of scar tissue, despite of the existence of cutaneous stem cells. Cutaneous regeneration marked by regeneration of functional appendage organs, such as hair follicles, sweat glands and sebaceous glands, and less fibrosis in the wounded tissue. Evidence suggests that poor cutaneous regeneration is largely due to the lack of appropriate inductive signals for cutaneous stem cells. Therefore, knowledge about molecular signals in hair follicle morphogenesis and cutaneous stem cell activation may help develop novel therapeutic strategies to enhance cutaneous regeneration and improve wound healing (Wang et al. 2012).

Signals for Hair Follicle Morphogenesis

The Induction and Morphogenesis of Hair Follicles

Approximately at mouse embryonic day (ED) 14.5, the placode is formed as the earliest structure of a hair follicle. It is formed as the consequence of focal epithelial proliferation induced by signals arising from the underlying mesenchyme, which is called the "first dermal signal" (Hardy 1992) (Fig. 10.1 i, ii). The composition of the "first dermal signal" has not been fully elucidated. On the contrary, numerous activators and inhibitors for hair follicle induction have been identified. Some factors that exert significant effects on hair follicle morphogenesis, such as Wnt, BMP, Shh, Dkk-1/4, noggin, APCDD1 and Edar, have been studied extensively.

Wnt/β-catenin signaling is essential for the initiation of the placode. Conditional ablation of β-catenin in mouse skin epithelium prevents hair placode formation in the embryo, and ectopic Dkk1 (a Wnt inhibitor) expression results in "nothing happens without Wnt" in terms of epithelial appendage formation—no hair follicles, no teeth, and no sebaceous glands. EdaA1/EdaR/NF-κB signaling is also needed for placode formation of guard and zig-zag hairs. EdaA1/EdaR/NF-κB signaling in the placode induction is probably via the effect of NF-κB for cell proliferation and survival. While Wnt activities are detected as early as ED 13.5, Eda/Edar/NF-κB

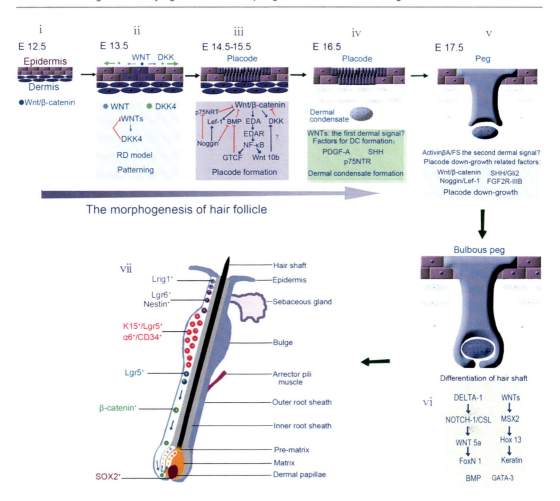

Fig. 10.1 The morphogenesis and cycling of hair follicle. (i) The initial signal directing hair follicle formation arises in the mesenchyme in the dermis. (ii) The DKK4 as a direct target of Wnt/β-catenin, in return inhibits the Wnt signal, and DKK4 is smaller and diffuses faster than Wnt. Both molecules are involved in Reaction-Diffusion (RD) model, a proposed mechanism for hair follicle patterning. (iii) Placodes (follicular epithelium) are induced as a result of Wnt activity in the epidermis. (iv) Signals from the epithelium induce the clustering of mesenchymal cells to form a dermal condensate. (v) The placode proliferates and grows downward into the dermis in response to signals (termed the second dermal signal) from the dermal condensate. (vi) The dermal condensate becomes enveloped by follicular epithelial cells to form the dermal papilla. (vii) The bulbous peg differentiated into outer root sheath, inner root sheath and hair shaft and distinct hair follicle stem cells located in different areas of the hair follicle and exert various functions, and the DP cell is SOX2 positive. Lgr, Leucine-rich repeat-containing GPCR; K15, keratin 15; Blimp1, B-lymphocyte-induced maturation protein 1; Lrig1, Leucine-rich repeats and immunoglobulin-like domains protein 1. (viii) Mature hair follicles engaged in regular Anagen-Catagen-Tolegen cycling, and several factors influence the transition (Adapted from Wang et al. 2012)

activities are not observed in the skin until ED 14.5. Additional data reveal that focal Wnt/β-catenin signaling occurs independent of Eda-A1/Edar/NF-κB signaling, and β-catenin is required for the activation of Eda-A1/Edar/NF-κB signaling in the developing skin epithelium (Zhang et al. 2009). Interestingly, a recent study showed that sustained epithelial β-catenin activity resulted in precocious and excessive induction of hair follicles even in the absence of Eda/Edar/NF-κB signaling. These studies suggest that Wnt/β-catenin is upstream of Eda and plays a dominant role during hair follicle induction (Narhi et al. 2008).

Many different factors are expressed in the placode once it is formed, including fibroblast growth factors (FGF), TGF-β2, Delta1, Noggin, Follistatin (FS), Gremlin, MSX1 and MSX2, and all these factors promote hair follicle development. Conversely, the BMPs including BMP2, BMP4 and BMP7, act as inhibitors of follicle formation, which are expressed in the mesenchyme or placode in response to Edar and Wnt/β-cantenin. The infusing BMPs interact with Edar and Wnt/β-cantenin signaling to regulate the formation, size and pattern of hair follicles.

The placode emits epidermal signals to the mesenchyme and stimulate the later to condensate. In addition, the second dermal signals from the condensate instruct the placodal epithelial cells to proliferate and invaginate into the dermal condensate. The epithelial-mesenchymal signaling interaction is crucial for hair follicle formation. Wnt signaling likely acts as the epidermal signal to induce the dermal condensate. Evidence supporting this conclusion is that the Wnt-responsive TOPGAL reporter gene is expressed in the dermal condensate, and the dermal condensate fails to develop in the absence of epithelial β-catenin. Platelet-derived growth factor-A (PDGF-A) is also required for normal hair follicle development, since mice lacking PDGF-A are characterized with small dermal papillae, dermal sheath abnormalities and thin hair. Additionally, PDGF-A is expressed in the placode, whereas its receptor is expressed in the dermal condensate. Thus, Wnt and PDGF-A molecules are strong candidate components of the first epithelial signal that induce the formation of dermal condensate. Immunological studies or in situ hybridization show that fibronectin, β1 integrin and Notch-1 are expressed in chick skin dermal condensations. Previous reports have implicated important roles for integrins in dermal condensate formation. It is hypothesized that integrins exert effects in the migration of dermal fibroblasts to the appendage via engagement on fibronectin. Notch signaling has also been suggested to play a role in bud and hair follicle formation, and ligand/Notch interactions induce the Notch intracellular domain (NICD) to be released, which could interact with integrins and act as a transcription factor. It is speculated that the interaction between NICD and integrins switch integrins to a high affinity state, hence stabilize the dermal condensation. p75 neurotrophin receptor (p75NTR) is expressed in dermal condensates in developing hair follicles, and loss of function mutations in p75NTR gene affects the rate of hair follicle morphogenesis, suggesting that neurotrophins may be involved at early steps of follicle formation. Syndecan-1 is another factor that is specifically expressed in dermal condensates during murine hair follicle morphogenesis, but further analysis shows that syndecan-1 is not required for follicle initiation and development (Richardson et al. 2009a) (Fig. 10.1 iii).

The condensate sends signals to nascent follicle keratinocytes to stimulate their proliferation and down-growth into the underlying dermis, and to form more mature follicles. The signal arising from dermal condensates is termed as the second dermal signal; its components have not been fully elucidated. Hair germs comprising epidermal placodes and associated dermal condensates were detected in both wild-type and Shh−/− embryos, but progression through subsequent stages of follicle development was blocked in mutant skin. Thus the "second dermal signal" regulation of proliferation and down-growth of the follicular epithelium is likely to be activated by Shh. Further studies suggest that Shh is regulated by Wnt/β-catenin. The down-stream factors that response to Shh are Gli1/2 and Pct1(full spell), since the level of Gli1/2 mRNA was markedly reduced in both the epithelial and mesenchymal components of Shh−/− primary hair germs, while the expression of Ptc1 was reduced in Shh mutant hair germs. In addition, hepatocyte growth factor/scatter factor might also play a role in the "second dermal signal" between the follicular mesenchyme and epithelium (Fig. 10.1 iv, v). Once the primary hair follicle structure (follicle peg) is induced, it subsequently differentiates into the outer root sheath, inner root sheath and hair shaft, thus forming the bulbous peg. Hair follicle differentiation is regulated by multiple signaling pathways. Among these signals, Homeobox C13 (Hoxc13) has been shown to be essential for proper hair shaft differentiation, and Hoxc13 regulate the activity of Foxn 1 for the terminal differentiation of hair (figure vi). Heparanase 1 is suggested as a key participant of inner root sheath differentiation program. Its expression was primarily located in the inner root sheath of human hair follicles, and restricted to anagen phase. Furthermore, inhibition of heparanase in cultured hair follicles induced a catagen-like process. Dlx3 is another crucial transcriptional regulator of hair formation and regeneration. The defect of Dlx3 in the epidermis results in complete alopecia due to failure of the hair shaft and inner root sheath formation. Significantly, Dlx3 is positioned downstream of Wnt signaling and upstream of other transcription factors that regulates hair follicle differentiation, such as Hoxc13 and Gata3 (Hwang et al. 2008).

Wnt signal is modulated by other proteins. First, several proteins have been detected to interplay with components of canonical Wnt/β-catenin signal pathway. β-catenin/LEF1/TCF transcriptional complex, for example, activates numerous target genes. Dkk-1/4, which acts as an antagonist for Wnt signaling, exerts its inhibitory function by direct binding to low-density lipoprotein receptor-related proteins 5 and 6 (LRP5/6) co-receptors, resulting in their endocytosis and eventual degradation. Interestingly, Dkk-4 is induced by canonical Wnt signaling. The Dkk-4 promoter interacts directly with LEF-1/β-catenin complex, and hence Dkk-1/4 are proposed to act in a negative feedback loop to attenuate canonical Wnt signaling (Gregory et al. 2003). Unexpectedly, Dkk-4 is also identified as the direct target of Eda/Edar/NF-κB signaling, suggesting that Edar signaling is an indirect antagonist of canonical Wnt signaling.

Several other signals appear to be involved in hair follicle morphogenesis besides the Wnt (Schmidt-Ullrich and Paus 2005). As described earlier, Edar signaling is also essential for the formation of hair follicles. Edar signaling is mediated by extracellular ligand Eda and its receptor Edar, which is a member of the growing tumor necrosis factor (TNF) super-family. Mutation of one of the two genes causes identical ectodermal dysplasia phenotypes in mice and humans (Mikkola et al. 1999). In humans, the mutation characterized with a pathophysiological phenotype includes sparse head hair, abnormal teeth and defects of sweat glands. The ectodysplasin family of ligands includes two closely related isoforms and trimeric type II membrane proteins, Eda-A1 and Eda-A2, have been shown to bind to and activate two different receptors, Edar and X-linked Eda-A2 receptor (Xedar), respectively. The activated Edar or Xedar then leads to activation of NF-κB transcription factors, which are transported to the nucleus and manipulate the activity of corresponding target genes. Further studies suggested that NF-κB is essential and sufficient to induce Shh expression, which activates Gli2 in the skin and mediates mitogenic signals of Shh by transcriptional activation of cyclin D1 and cyclin D2, resulting in the proliferation of target cells. Shh is not required for initiating hair follicle development, but is essential for controlling placode ingrowth and morphogenesis of the hair follicle. These findings support the speculation that Shh mediates the NF-κB transmitted Eda A1/EdaR signaling to control post-initiation hair placode ingrowth.

Troy is a member of the TNF receptor super family and is homologous with Edar in its ligand-binding domain. Lymphotoxin-alpha (LT-α) has been identified as a functional ligand of Troy. Further analyses reveal that Troy signaling acts redundantly with Edar signaling in regulating the initiation of hair follicle development. Intriguingly, Lgr4 (Leucine-rich repeat containing G-protein coupled receptor 4), one of the genes identified as novel G protein-coupled receptors (GPCRs), has been postulated as a novel gene class in regulating the development of hair follicles. Lgr4/K5 knockout (KO) mice showed sparse head hair and focal alopecia behind their ears. This abnormal phenotype was similar to those observed in tabby (Eda) and downless (Edar) mice. Further studies suggest that there might be interactions between the BMP signal and Lgr4 signal in the regulation of placode formation (Mohri et al. 2008).

BMPs play pivotal roles in the regulation of hair follicle development. BMPs compose a large subgroup within the transforming growth factor-β (TGF-β) gene superfamily. Receptors binding the TGF-β ligands consist of two subfamilies, type I and type II receptors. There are three type-I receptors (BMP type-I A receptor, BMP type I B receptor and activin A type I receptor, ACVRI) and three type II receptors (BMPRII, activin type IIA receptor (ActRIIA) and ActRIIB). Upon ligand binding to the receptor, Smads in cytoplasm are phosphorylated and translocate to the nucleus, where they act as transcription factors and participate in the regulation of target gene expression. In mice, inhibition of BMP signaling by overexpressing the BMP antagonist Noggin results in a marked increase in size of anagen hair follicle and the replacement of zig-zag hairs by awl-like hairs. By contrast, constitutive deletion of Noggin is characterized by the lack of the induction of secondary hair follicles. In chicks, delivery of BMPs inhibits local feather formation. In addition, several inhibitors of BMPs, expressed in developing skin, including Noggin, Smad7, and Sostdc1/Ectodin/WISE are themselves transcriptional targets of BMPs, likely acting as feedback inhibitors of the signaling pathway. These studies suggest that BMPs function as dominant negative regulators for hair follicle formation (Schmidt-Ullrich and Paus 2005).

Hair Follicle Patterning

In mice hair follicles initiate in different waves and space in a regular pattern. Many studies attempt to reveal the mechanisms underlying this phenomenon. The RD (reaction-diffusion) model has been proposed as a mechanism for hair follicle patterning. The system is via positive and negative feedback regulation of an activator/inhibitor pair, where the inhibitors are thought to diffuse and decay faster as compared to the activators. Since canonical Wnt signaling is essential and sufficient for the induction of hair and feather follicles, and appears to be the most upstream regulator of hair follicle morphogenesis (Gat et al. 1998), Wnt signaling represents an appealing candidate for the primary signal that manipulates follicle distribution. Dkk-1/4 exerts their inhibitory property on Wnt signaling by binding to the LRP5/LRP6 component of the Wnt receptor complex. Expression of the Dkk-1 is directly controlled by secreted Wnts. Dkk-4 is also regulated by canonical Wnt signaling, as five LEF/TCF consensus binding motifs are found within 700 base pairs upstream of the transcriptional start site of DKK-4. In addition, Dkks should defuse faster than Wnts since Wnt proteins are about 20–60 % larger than Dkks. Hence, Wnts and their inhibitors Dkk-1/4 satisfy the criteria of the RD model. Further studies using combined experimental and computational modeling approaches, support Wnt and its inhibitor Dkk as the primary determinants of murine hair follicle spacing (Fig. 10.1 ii).

Hair follicle pattern formation is accomplished with additional and more complicated molecular interaction besides the Wnt-Dkk RD model. The interplay between Edar and BMPs is also responsible for the distribution of hair follicle. Edar signaling is essential for hair follicle placode formation, and rapid positive-feedback Edar signaling is coupled with the induction of BMPs, which in turn inhibits Edar signaling. Connective tissue growth factor (CTGF), with up-regulated expression in follicle, is another target of Edar signaling. CTGF binds to and inhibits BMPs in a manner analogous to that of Noggin (Abreu et al. 2002). As a result, BMPs can only exert their inhibitory activity at a distance from follicles. Consequently, the hair follicle placodes is formed with a certain interval between each other, and organized in a relatively regular pattern. In conclusion, Edar–BMPs interactions (activation–inhibition) direct hair follicle patterning by modulating signal receptivity, rather than restricting the localization of an inducing ligand. Intriguingly, a recent study found that Wnt/BMP pair is coopted to mediate interactions among follicles in the population besides in individual hair follicles, thus likely coordinating the global regeneration pattern of cycling hair follicles (Plikus et al. 2011).

The determination of interfollicular fate is equally crucial for the patterning of the epidermis. BMP signaling exhibits inhibition for the follicle or bud formation in the adjacent interfollicular or interbud tissue, but whether it promotes interfollicular or interbud fate and patterning is unclear. Keratinocyte growth factor (KGF) and epidermal growth factor (EGF) signals have been identified as candidates for blocking hair follicle induction and promoting interfollicular epidermal fate in developing mouse skin, which are speculated as promoting cell growth and proliferation in the interfollicular epidermis. EGF signaling is mediated by EGFR and its associated ligands including TGF-α, heparin-binding EGF-like growth factor (HB-EGF), amphiregulin (AR), betacellulin (BTC), epiregulin (EPR), and epigen (EPGN). KGF signaling is transmitted via KGF and its receptor FGFR2 (IIIb). Immunofluorescence analysis shows that EGFR and FGFR2 (IIIb) expression is downregulated in hair follicle placodes. Constitutive EGF and KGF signaling can inhibit hair follicle morphogenesis. Similarly, EGF- and KGF-treated skin hair follicles are characterized with the absence of placodes and dermal condensates, coupled with promoted interfollicular epidermal phenotype. Thus KGF and EGF signals promote epidermal fate at the expense of reduction in hair follicle development (Richardson et al. 2009b). Therefore, balanced signals for follicular fate and interfollicular fate may be crucial for developing hair follicles in normal density.

The Quiescence and Activation of Hair Follicle Stem Cells

Hair follicle stem cells located in the follicle bulge are regulated by the surrounding microenvironment, or niche (Fig. 10.1 vii). Quiescent adult stem cells reside in specialized niches where they become activated to proliferate and differentiate during tissue homeostasis and injury. Cotsarelis and colleagues were the first to identify label-retaining cells (LRCs) in the bulge region. Then Fuchs and colleagues developed an elegant method to isolate live LRCs by generating a doxycycline-controlled histone2B–green fluorescent protein (H2B-GFP) mouse line. The majority of these H2B-GFP LRCs expressed the stem cell marker CD34. Sorted and cultured CD34+ bulge cells from a keratin 14-GFP transgenic mouse were transplanted and shown to generate the entire hair follicle. These studies demonstrate that LRCs (here identified as CD34$^+$ and K14$^+$) in the

hair follicle represent functional stem cells (Li and Clevers 2010). The activation of hair follicle stem cells is cyclic, in which periodic β-catenin activity has been involved. Indeed, in the hair follicle bulge region, the inhibition of Wnt signaling by Wnt inhibitors such as dickkopf3 (Dkk3), secreted frizzled-related protein 1 (Sfrp1), and disabled homolog 2 (Dab2) is likely to promote stem cell quiescence. In adult mouse, regeneration occurs in waves within the follicle population, implying coordination among adjacent follicles and the extrafollicular environment. It has been reported by Chuong and colleagues that periodic expression of BMP2 and BMP4 in the dermis also regulates this process. Further studies indicate that a follicle progresses through cycling stages by continuous integration of inputs from intrinsic follicular and extrinsic environmental signals based on universal patterning principles. Signaling from the Wnt/BMP activator/inhibitor pair is integrated to mediate interactions among follicles in the population. Recently, nuclear factor of activated T-cells, cytoplasmic 1 (NFATc1) has been shown to be preferentially expressed by hair follicle stem cells in their niche, where its expression is activated by BMP signaling upstream and it acts downstream to transcriptionally repress CDK4 and maintain stem cell quiescence. As stem cells become activated during hair growth, NFATc1 is downregulated, relieving CDK4 repression and activating proliferation. When calcineurin/NFATc1 signaling is suppressed, stem cells are activated prematurely, resulting in precocious follicular growth. These results suggest a functional role for calcium-NFATc1-CDK4 circuitry in governing stem cell quiescence (Horsley et al. 2008).

Stem Cells and Niche Cells for Cutaneous Regeneration

Hair Follicle Stem Cells Regenerate Epidermal Appendages

The regeneration of organs is a common and widespread adaptive capability among metazoan creatures. Many larval and adult animals are able to regenerate large sections of their body plan after amputation, and this usually restores the structures that were removed by the operation. One well known research model is that adult urodeles can regenerate their limbs by local formation of a mesenchymal growth zone or blastema. Unfortunately, the regeneration potential tends to discount in great extent for mammalians and humans. Adult skin consists of a keratinized stratified epidermis and an underlying thick layer of collagen-rich dermal connective tissue providing support and nourishment. Hairs and glands, appendages of the skin, are derived from the epidermis but project deep into the dermal layer. When an adult human skin tissue is wounded deep into the level of hair bulbs in the dermis that causes a complete loss of hair follicles, the repairing epithelium does not regenerate hair follicles and other epidermal appendage structures (Martin 1997). Previous studies have indicated that the absence of hair follicle regeneration in the healing wound is due to the lack of proper inductive signals from the underlying wound dermis. Notably, some recent studies suggest that limited hair follicle regeneration may occur in small wounds in mice and Wnt signal seems to be involved (Ito et al. 2007).

Stem cells in the hair follicle bulge do not normally contribute cells to the epidermis, while after epidermal injury, cells from the bulge are recruited into the epidermis and migrate in a linear manner toward the center of the wound, and forming a marked radial pattern (Fig. 10.2). Meanwhile, although the bulge-derived cells acquire an epidermal phenotype, most are eliminated from the epidermis over several weeks, suggesting that bulge stem cells respond rapidly to epidermal wounding by generating short-lived 'transient amplifying' cells responsible for acute wound repair (Ito et al. 2005). Further studies suggest that follicular cells may undergo reprogramming to become repopulating epidermal progenitors following injury (Levy et al. 2007).

Stem cells with various surface markers have been identified. In mid and late anagen, nestin, a neural stem cell marker, is located to cells in the upper outer-root sheath as well as in the bulge area while not in the hair matrix bulb. Some studies suggest that stem cells in the budge capable of regenerating hair follicles may also express nestin. Keratin-15(Krt-15, also known as K15) is highly expressed in the bulge (Fig. 10.1 vii), but lower levels of expression are present in the basal layers of the lower follicle outer-root sheath (ORS) and the epidermis. K15 labeled bulge cell analysis *in vivo* indicating the K15 positive bulge cell contribute to the maintenance of the three major epithelial cell type of the cutaneous epithelium, meanwhile the contribution of K15 positive bulge cell to the generation of new hair follicle and hair is observed at the onset of anagen during the normal hair follicle cycling. $Lgr6^+$ cells, which reside in a previously uncharacterized region directly above the follicle bulge, appear to contribute to hair neogenesis (Snippert et al. 2010). CD34 and α6-integrin have been considered as markers for murine hair follicle stem cells. When budge-derived stem cells positive expressing surface CD34 and α6-integrin were engrafted with newborn dermal cells *de novo* hair follicles formed (Blanpain et al. 2004). Leucine-rich repeats and immunoglobulin-like domains protein (Lrig)1 expression is detected at the junction between the infundibulum and the sebaceous gland (SG) (Fig. 10.1 vii). Lrig1-expressing cells can give rise to all of the adult epidermal lineages in skin reconstitution assays. Lgr5, a marker of intestinal stem cells, is expressed in actively cycling cells in the lower outer root sheath of anagen hair follicles. $Lgr5^{high}$ cells isolated from telogen mouse skin were able to form large colonies with high efficiency. In a skin reconstitution assay, the $Lgr5^{high}$ cells exhibited much superior capacity in regenerating hair follicles than $CD34^+$ cells. Further study suggests that $Lgr5^+$ cells provide a constant flux of stem cells downward toward the dermal papilla during the growth phase of the hair follicle, and these cells seem to exert a elevated β-catenin expression (Jaks et al. 2008). Taken together, recent studies have identified various markers for epidermal stem cells. However, it is unclear whether they represent different types of stem cells or stem cells in different differentiation stages. Future skin reconstitution assays with defined cell types will help identify essential stem cells for hair follicle neogenesis, and thus hair follicle regeneration will become a more controllable procedure.

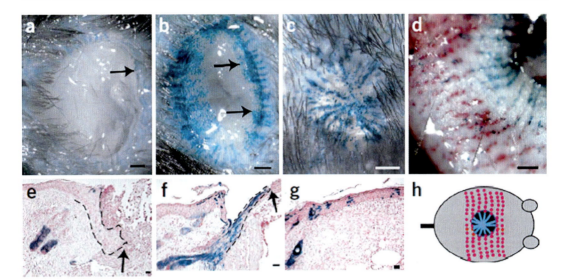

Fig. 10.2 Lineage analysis of bulge cells after wounding. (**a**, **e**) Gross appearance (**a**) and histology (**e**) of wound at 2 days after a full thickness excision of 4 mm. (**b**, **f**) At 5 days after wounding, *blue streaks* suggested that clones of cells derived from the bulge move centripetally. (**c**, **g**) At 8 days after wounding, the epidermis had completely reepithelialized. Bulge-derived cells were present in much of the wound area. (**d**) Blue cells emanated from hair follicle openings (*stained red*) at 4 days after wounding. (**h**) Diagram depicting perilesional follicles (*red dots*) with bulge-derived cells (*blue streaks*) traveling from peripheral follicles to the center of the wound. *Arrows* indicate leading edge of reepithelialization. *Broken lines* delineate epidermis. Scale bars in (**e**–**g**), 25 mm; in (**a**–**d**), 500 mm (Ito et al. 2005)

DP Cells Induce Hair Follicle Neogenesis

Dermal papilla (DP) cells, which are specialized cells derived from the mesenchyme, have been known to provide unique and critical signals for hair follicle development (Reynolds and Jahoda 1992). When engrafted with embryonic mouse epidermis into the skin, adult DP cells are capable of inducing hair follicle formation (Qiao et al. 2008). Signals responsible for the inductive property of DP cells have not been fully elucidated. A previous study indicated that DP cells selectively expressed Wnt-5a, but not Shh (Kishimoto et al. 2000), suggesting a role of Wnt signal in DP cells-mediated follicle neogenesis. Indeed, forced expression of Wnt-5a in the deeper wound induced glandular structures of the epidermis which projected deep into the newly formed dermis, and some structures were similar to hair follicles. Consistent with that found in embryonic hair follicle development, Wnt signal appears to be central in hair follicle neogenesis in the regeneration of adult skin. However, Wnt signal alone seems not to be sufficient for perfect appendage regeneration; and recent studies suggest that BMP signals are involved. Loss of BMPR1A in DP cells resulted in the loss of their signature characteristics *in vitro* and their hair follicle inductive property *in vivo*. These results also imply complex epithelial–mesenchymal crosstalks in appendage regeneration.

Mesenchymal Stem cells in Cutaneous Repair and Regeneration

A growing body of evidence suggests that the bone marrow (BM) derived cells may play an important role in cutaneous repair/regeneration. BM-derived cells have been found in the epidermis, hair follicles, sebaceous glands and dermis of the adult skin. Our recent studies and others suggest that BM-derived mesenchymal stem cells (BM-MSC) may be the BM cells responsible for cutaneous regeneration (Wu et al. 2010). When grafted in a matrix gel into excisional wounds, glandular appendage-like structures formed by

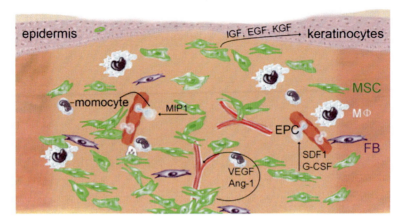

Fig. 10.3 Paracrine effect of Bone marrow-derived mesenchymal stem cells (BM-MSCs) in wound healing. BM-MSCs (*green*) in a cutaneous wound release growth factors such as IGF-1, EGF, and KGF to promote the proliferation of keratinocytes, release proangiogenic cytokines such as VEGF-a and Ang-1 to enhance angiogenesis, release chemokines such as MIP-1 to recruit monocytes into the wound, and release cytokines such as SDF-1 and G-CSF to recruit EPCs into the wound. Abbreviations: *ang-1* angiopoietin-1, *EGF* epidermal growth factor, *EPC* endothelial progenitor cell, *FB* fibroblast, *G-CSF* granulocyte colony-stimulating factor, *IGF* insulin-like growth factor, *KGF* keratinocyte growth factor, *MIP-1* macrophage inflammatory protein-1, *MSC* mesenchymal stem cell, *SDF1* Stromal cell-derived factor-1, *VEGF* vascular endothelial growth factor

cells derived from transplanted BM-MSC were seen, which resembled developing sweat or sebaceous glands; however, these structures were short-lived and disappeared when wounds completely closed. Consistent with our findings, epithelial cells derived from endogenous BM cells were found to appear in the wound transiently during wound healing. Taken together, these results suggest that BM-MSC may not provide long-term self-renewal stem cells for keratinocytes and appendages, but may form precursors of endogenous permanent cutaneous structures. Previous studies suggest certain molecules exclusively released by BM-derived cells in the skin are necessary for cutaneous regeneration, and BM-MSC are capable of releasing some of these molecules (Wu et al. 2010). For example, BM transplant treatment of adult collagen XVII (Col17)-deficient mice induced donor-derived Col17 expression associated with the recovery of hemidesmosomal structures, and MSC have the potential to produce Col17. In a recent work, we found that wounds receiving BM-MSC had significantly increased numbers of appendages formed by host cells. BM-MSCs in the dermis appear to induce keratinocytes to migrate downward into the dermis and form glandular structures while subcutaneous preadipocytes or dermal fibroblasts do not. In a previous work, we found that BM-MSC secret a large variety of growth factors, cytokines and extracellular matrix molecules, which are distinctively different in expression levels from those released by dermal fibroblasts. Some of the highly expressed cytokines have been known to be involved in tissue regeneration of other organs. These results suggest that BM-MSC may release molecules required for cutaneous regeneration. It has been known that fetal skin is capable of regeneration, and the event is associated with a lack of inflammation. This is in co-incidence with higher levels of circulating MSC in the fetal blood stream. Recent studies indicate that paracrine factors from BM-MSC are more involved with cell growth and regeneration while paracrine factors from dermal fibroblasts are associated with inflammation (Chen et al. 2008). Now many studies have indicated that BM-MSC suppress the proliferation of T-lymphocytes and other inflammatory cells. These results imply that fetal skin may contain higher levels of MSC and lower levels of fibroblasts than adult skin. Thus MSCs may promote cutaneous regeneration by contributing critical niche molecules (Fig. 10.3).

Discussion

The embryonic hair follicle development and adult hair follicle regeneration are highly related processes; both rely on populations of stem cells to undergo a highly coordinated and stepwise program of differentiation to produce the completed structure. Convincing evidence indicates that cutaneous stem cells exist in the adult skin, which are capable of forming *de novo* hair follicles and other appendage structures. Recent studies suggest that the absence of regeneration to the wounded skin may be not due to malfunction of cutaneous stem cells, but due to the lack of proper inductive signals for the stem cells. Because of the presence of diverse gene knockout mouse models in the past decades, tremendous progresses have been made in understanding of signals that regulate hair follicle morphogenesis. However, much less has been understood in signals for hair follicle neogenesis and cutaneous regeneration. Recent studies suggest that certain signals critical for hair follicle morphogenesis may also play important roles in *de novo* hair follicle regeneration. With a better understanding of cutaneous stem cells and their niches, novel molecular therapies for cutaneous regeneration will be developed.

Acknowledgements This work was supported by grants (No. 30971496, U1032003) from Natural Science Foundation of China to Y Wu.

References

Abreu JG, Ketpura NI, Reversade B, De Robertis EM (2002) Connective-tissue growth factor (CTGF) modulates cell signalling by BMP and TGF-beta. Nat Cell Biol 8:599–604

Blanpain C, Lowry WE, Geoghegan A, Polak L, Fuchs E (2004) Self-renewal, multipotency, and the existence of two cell populations within an epithelial stem cell niche. Cell 5:635–648

Chen L, Tredget EE, Wu PY, Wu Y (2008) Paracrine factors of mesenchymal stem cells recruit macrophages and endothelial lineage cells and enhance wound healing. PLoS One 4:e1886

Gat U, DasGupta R, Degenstein L, Fuchs E (1998) De Novo hair follicle morphogenesis and hair tumors in mice expressing a truncated beta-catenin in skin. Cell 5:605–614

Gregory CA, Singh H, Perry AS, Prockop DJ (2003) The Wnt signaling inhibitor dickkopf-1 is required for reentry into the cell cycle of human adult stem cells from bone marrow. J Biol Chem 30:28067–28078

Hardy MH (1992) The secret life of the hair follicle. Trends Genet 2:55–61

Horsley V, Aliprantis AO, Polak L, Glimcher LH, Fuchs E (2008) NFATc1 balances quiescence and proliferation of skin stem cells. Cell 2:299–310

Hwang J, Mehrani T, Millar SE, Morasso MI (2008) Dlx3 is a crucial regulator of hair follicle differentiation and cycling. Development 18:3149–3159

Ito M, Liu Y, Yang Z, Nguyen J, Liang F, Morris RJ, Cotsarelis G (2005) Stem cells in the hair follicle bulge contribute to wound repair but not to homeostasis of the epidermis. Nat Med 12:1351–1354

Ito M, Yang Z, Andl T, Cui C, Kim N, Millar SE, Cotsarelis G (2007) Wnt-dependent de novo hair follicle regeneration in adult mouse skin after wounding. Nature 7142:316–320

Jaks V, Barker N, Kasper M, van Es JH, Snippert HJ, Clevers H, Toftgard R (2008) Lgr5 marks cycling, yet long-lived, hair follicle stem cells. Nat Genet 11:1291–1299

Kishimoto J, Burgeson RE, Morgan BA (2000) Wnt signaling maintains the hair-inducing activity of the dermal papilla. Genes Dev 10:1181–1185

Levy V, Lindon C, Zheng Y, Harfe BD, Morgan BA (2007) Epidermal stem cells arise from the hair follicle after wounding. FASEB J 7:1358–1366

Li L, Clevers H (2010) Coexistence of quiescent and active adult stem cells in mammals. Science 5965:542–545

Martin P (1997) Wound healing—aiming for perfect skin regeneration. Science 5309:75–81

Mikkola ML, Pispa J, Pekkanen M, Paulin L, Nieminen P, Kere J, Thesleff I (1999) Ectodysplasin, a protein required for epithelial morphogenesis, is a novel TNF homologue and promotes cell-matrix adhesion. Mech Dev 2:133–146

Mohri Y, Kato S, Umezawa A, Okuyama R, Nishimori K (2008) Impaired hair placode formation with reduced expression of hair follicle-related genes in mice lacking Lgr4. Dev Dyn 8:2235–2242

Narhi K, Jarvinen E, Birchmeier W, Taketo MM, Mikkola ML, Thesleff I (2008) Sustained epithelial beta-catenin activity induces precocious hair development but disrupts hair follicle down-growth and hair shaft formation. Development 6:1019–1028

Plikus MV, Baker RE, Chen CC, Fare C, de la Cruz D, Andl T, Maini PK, Millar SE, Widelitz R, Chuong CM (2011) Self-organizing and stochastic behaviors during the regeneration of hair stem cells. Science 6029:586–589

Qiao J, Philips E, Teumer J (2008) A graft model for hair development. Exp Dermatol 6:512–518

Reynolds AJ, Jahoda CA (1992) Cultured dermal papilla cells induce follicle formation and hair growth by transdifferentiation of an adult epidermis. Development 2:587–593

Richardson GD, Bazzi H, Fantauzzo KA, Waters JM, Crawford H, Hynd P, Christiano AM, Jahoda CA (2009a) KGF and EGF signalling block hair follicle induction and promote interfollicular epidermal fate in developing mouse skin. Development 13:2153–2164

Richardson GD, Fantauzzo KA, Bazzi H, Maatta A, Jahoda CA (2009b) Dynamic expression of Syndecan-1 during hair follicle morphogenesis. Gene Expr Patterns 6:454–460

Schmidt-Ullrich R, Paus R (2005) Molecular principles of hair follicle induction and morphogenesis. Bioessays 3:247–261

Snippert HJ, Haegebarth A, Kasper M, Jaks V, van Es JH, Barker N, van de Wetering M, van den Born M, Begthel H, Vries RG, Stange DE, Toftgard R, Clevers H (2010) Lgr6 marks stem cells in the hair follicle that generate all cell lineages of the skin. Science 5971:1385–1389

Wang X, Tredget EE, Wu Y (2012) Dynamic signals for hair follicle development and regeneration. Stem Cells Dev 1:7–18

Wu Y, Zhao RC, Tredget EE (2010) Concise review: bone marrow-derived stem/progenitor cells in cutaneous repair and regeneration. Stem Cells 5:905–915

Zhang Y, Tomann P, Andl T, Gallant NM, Huelsken J, Jerchow B, Birchmeier W, Paus R, Piccolo S, Mikkola ML, Morrisey EE, Overbeek PA, Scheidereit C, Millar SE, Schmidt-Ullrich R (2009) Reciprocal requirements for EDA/EDAR/NF-kappaB and Wnt/beta-catenin signaling pathways in hair follicle induction. Dev Cell 1:49–61

Role of Chromatin-Remodeling Factor Jun Dimerization Protein 2 (JDP2) in Cellular Senescence

Kazunari K. Yokoyama and Kung-Kai Kuo

Contents

Abstract	101
Introduction	102
Transcription Factor Jun Dimerization Protein 2 (JDP2)	103
Senescence-Associated Heterochromatin Foci SAHF and Histone Chaperones	103
ATP-Dependent Chromatin-Remodeling Complexes	104
Relationship Between SAHF and the pRb and p53 Pathways	105
PRC1 and PRC2 Silencing	105
JDP2-Mediated Regulation of the *INK4a/ARF* Locus	106
Discussion	108
References	108

K.K. Yokoyama (✉)
Graduate Institute of Medicine, Kaohsiung Medical University, 100 Shih-Chuan 1st Road, San Ming District, 807 Kaohsiung, Taiwan
e-mail: kazu@kmu.edu.tw

K.-K. Kuo
Division of Hepatobiliopancreatic Surgery, Department of Surgery, Kaohsiung Medical University Hospital, 100 Tzyou 1st Road, San Ming District, 807 Kaohsiung, Taiwan
e-mail: kazu@kmu.edu.tw

Abstract

Senescence is an irreversible event that occurs during the G1 transition of the cell cycle and is elicited by replicative exhaustion or in response to stresses such as DNA damage, chemotherapeutic drugs, or aberrant expression of oncogenes. Senescence-associated heterochromatin foci, repress the expression of proliferation-promoting genes, thereby contributing to senescence-associated proliferation arrest. This arrest is regulated by chromatin-remodeling factors and two tumor suppressor proteins, p53 and Rb, whose expression is controlled by the *CDKN2A* locus. Jun dimerization protein 2 (JDP2)-deficient mouse embryonic fibroblasts are resistant to replicative senescence. Oxygen induces the expression of JDP2, which then inhibits the recruitment of polycomb repressive complexes (PRCs) 1 and 2 to the promoter of the gene encoding $p16^{Ink4a}$ and $p19^{Arf}$, resulting in inhibition of the methylation of histone H3K27. These results suggest that the PRC controlled by JDP2 plays an important role in replicative senescence. Here we review the possible contribution of the histone chaperone JDP2 to induction of replicative senescence.

Keywords

ATP-dependent chromatin-remodeling complexes • Brahma (BRM) • Cellular senescence • Chromatin-remodeling factor • INK4a/ARF locus • Jun dimerization protein 2 (JDP2) • Polycomb repressive complexes (PRCs) • pRb and p53 pathways • Senescence-associated heterochromatin foci (SAHF)

Introduction

Replicative senescence, as originally defined by Hayflick and Moorhead (1961), can be triggered by excessive cell divisions, thereby shortening telomeres and generating double-strand breaks (Wright and Shay 2001). This senescence response is essentially a DNA-damage response and can be induced by ionizing radiation, activated oncogenes, or replicative stress (Campisi 2005; Campisi and d'Adda di Fagagna 2007). Senescence is also linked to a variety of morphological changes, such as an enlarged cell shape, decreased protein synthesis and degradation, resistance to apoptosis, increased activity of some lysosomal enzymes including senescence-associated β-galactosidase (Itahana et al. 2003), and the appearance of γ-H2AX, a marker of DNA damage (Herbig and Sedivy 2006).

Senescence-associated heterochromatin foci (SAHF) have been reported to play a role in the silencing of proliferation-promoting genes and thereby contribute to the proliferation arrest (Narita et al. 2003). The chromatin is much more compact in cells exhibiting SAHF than in normal-interphase growing cells. In 4′-6 diamino-2-phenylindole-stained senescent human cells, SAHF appears as 30–50 bright, punctate DNA foci. Chromatin from cells with SAHF is more resistant to nuclease digestion than chromatin from growing cells (Narita et al. 2003). Individual SAHF in a senescent cell arises from the condensation of an individual chromosome.

Chromosome regions in constitutive heterochromatin, such as the pericentromeric and telomeric region, are located at the periphery of SAHF (Funayama et al. 2006). SAHF contains several common markers of heterochromatin, including hyperacetylated histones, methylated H3K9, and bound HP1, but SAHF does not contain H3S10P, H2BS14P, or H3S28P. SAHF is also depleted of the linker histone H1 and is enriched in macroH2A and HMGA protein (Costanzi and Pehrson 2001; Narita et al. 2006). Cellular senescence is an important barrier to the development of cancer in mammals because it arrests proliferation and prevents neoplastic progression of cells harboring oncogenic lesions (Campisi 2005); however, accumulation of dysfunctional senescent cells contributes to mammalian aging (Campisi and d'Adda di Fagagna 2007).

At present, little is known about the relationship between chromatin remodeling and cellular senescence. It is well known that the Rb and p53 tumor suppressor pathways are among the master regulators of senescence. Inactivation of these two pathways abolishes senescence in humans and mice (Wright and Shay 2001). The expression of Rb and p53 is regulated by two distinct proteins, $p16^{Ink4a}$ and $p14^{Arf}$ ($p19^{Arf}$ in mouse), respectively, both of which are encoded by the *CDKN2A* locus (Fig. 11.1). Although human cells lacking pRB and p53 circumvent senescence, most of these cells ultimately cease proliferating because of telomere shortening (Reddel 1998). The p53 pathway exerts its effects through activation of downstream target genes, including the cell cycle inhibitor $p21^{Cip1}$, whose expression is increased in senescence cells. Thus, the p53 pathway comprises at least three proteins; p53, $p14^{Arf}$, and Hdm2 (Sherr and McCormick 2002). By contrast, the Rb pathway is thought to comprise at least four proteins; $p16^{Ink4a}$, cyclin D1, cdk4, and pRb (Nevins 2001). By inhibiting cyclin D/cdk4 kinases, $p16^{Ink4a}$ activates pRb. The pRb pathway inhibits cell proliferation through numerous downstream effectors. For example, pRb inhibits the E2F family of transcription factors, whose target genes are necessary for progression through S-phase (Nevins 2001).

Jun dimerization protein (JDP2) functions as a transcription factor and a histone chaperone (Aronheim et al. 1997; Jin et al. 2001, 2006). JDP2 plays a critical role in cell growth, cell differentiation, and senescence (Wang et al. 2011). The chromatin-remodeling factors that regulate the association and dissociation of polycomb repressive complexes (PRCs), which are controlled by JDP2, might be important players in the senescence program. In this review,

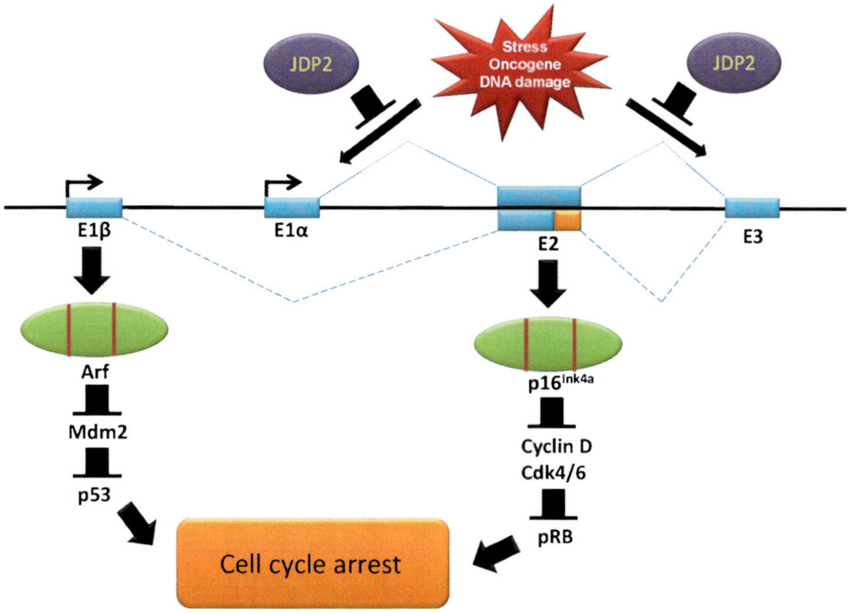

Fig. 11.1 Model of $p19^{Arf}$ and $p16^{Ink4a}$ proteins generated from CDKN2A locus. Signaling from stress, oncogenic activation, and DNA damage activates the transcription from exon 1α and exon 1α of $p19^{Arf}$ and $p16^{Ink4a}$ genes, respectively. As the result of splicing, $p19^{Arf}$ and $p16^{Ink4a}$ share common exon2 and exon 3 but are translated from alternative reading frames that encoded different amino acid sequences. $p19^{Arf}$ and $p16^{Ink4a}$ activate the Rb and p53 exons, respectively

we describe the properties of the histone chaperone JDP2, which interacts with core histones and can modify the structure of chromatin during transcription and the process of replicative senescence (Nakade et al. 2009).

Transcription Factor Jun Dimerization Protein 2 (JDP2)

JDP2 is a protein in the AP-1 family and binds to the TPA-responsive element or cAMP-responsive element on DNA (Aronheim et al. 1997; Jin et al. 2001). JDP2 forms heterodimers with c-Jun, ATF-2, CAAT-/enhancer binding protein γ and the progesterone receptor, and represses the AP-1-mediated activation of transcription (Wang et al. 2011). JDP2 modulates the expression of cyclins D1 and A2, and p21, which have opposing effects on cell cycle progression (Pan et al. 2010). The induction of JDP2 expression is observed during the differentiation of F9 cells, muscle cells, and osteoclasts. JDP2 might provide a threshold for exit from the cell cycle and commitment to differentiation (Pan et al. 2010). We reported previously that JDP2 represses the transactivation mediated by p300 (Jin et al. 2006) and inhibits histone acetylation induced by p300, CREB-binding protein (CBP), and p300/CBP-associated factor. The overexpression of JDP2 appears to repress the retinoic acid-induced acetylation of lysines 8 and 16 of histone H4 and some amino-terminal lysine residues of histone H3. Thus, JDP2 is a histone chaperone that binds directly to core histones and facilitates the assembly of nucleosomes *in vitro*, and inhibits histone acetyltransferase (Jin et al. 2006) and histone methyltransferase activities (Nakade et al. 2009).

Senescence-Associated Heterochromatin Foci SAHF and Histone Chaperones

Senescence-associated exit of cell cycle appears to be linked to the process of chromosome condensation during formation of SAHF. Two chaperones, HIRA and ASF1a, drive chromosome condensation during SAHF assembly in human cells (Zhang et al. 2005). Because HIRA-containing chaperone complexes such as ASF1a deposit the histone variant H3.3 preferentially over the canonical histone H3.1 in a DNA replication-independent manner (Tagami et al. 2004), it is unlikely that deposition of histone H3.3 is linked to transcriptional activation (Ahmad and Henikoff 2002). Instead, deposition of histone H3.3 might be associated with any type of major remodeling of chromatin, perhaps as a way to "reset" histone modification. During SAHF formation, HIRA/ASF1a might use histone H3.1 as a substrate. Consistent with this idea, ASF1a interacts sequentially with histones H3.1 and H3.3, and inactivation of HIRA in mouse embryonic stem cells affects the nuclear mobility of both histones H3.3 and H3.1 (Tagami et al. 2004). After HIRA/ASF1a-driven chromosome condensation, histone H3 is methylated at lysine 9, creating a binding site for HP1, and the histone variant macroH2A is incorporated. The factors responsible for the incorporation of macroH2A are unknown but are likely to include an ATP-dependent remodeling factor such as a histone exchange factor (Bruno et al. 2003). Formation of SAHF appears to depend on the dynamic relocalization of several nuclear components such as histone chaperone HIRA (Zhang et al. 2005). HIRA is known to translocate into a specific sub-nuclear organelle, the promyelocytic leukemia (PML) nuclear body. Human cells contain 20–30 PMLs and many other nuclear bodies, which are typically 0.1–1 µM in diameter and are enriched in PML protein, and many other nuclear regulatory proteins (Borden 2002).

ATP-Dependent Chromatin-Remodeling Complexes

ATP-dependent chromatin-remodeling complexes open up chromatin by inducing nucleosome sliding or eviction, and by mediating chromatin looping. These complexes are widely conserved and can be subdivided into different families on the basis of the sequence and structure of the ATPase subunit.

The TrxG member Brahma (BRM) is a bromodomain-containing protein in *Drosophila Melanogaster* and is homologous to yeast SWI/SNF (Switch/sucrose nonfermentable) and mammalian BRM and BRG1 (also known as SMARCA4); it was identified in screens for suppressors of polycomb complex group (PcG)-mediated homeotic transformations. SWI/SNF complexes are variable and nonredundant, and regulate the chromatin structure of many genes implicated in the cell cycle, cell signaling, cell proliferation and chromosome segregation (Ho and Grabtree 2010).

Human SWI/SNF2 complexes (hSWI/SNF or BRM/BRG1-associated factor complex) contain either BRM or BRG1, the two SW12/SNF2p homologues present in human cells. Human SNF5 gene (hSNF5) induction leads to G1 cell cycle arrest and cellular senescence. This hSNF5-induced cellular senescence is dependent on the $p16^{Ink4a}$/pRb pathway, but not on the $p14^{Arf}$/p53 pathway (Oruetxebarria et al. 2004). Of note, human hBrm and Brg1 are two candidate tumor suppressor genes that have been implicated in repression of E2F target genes and induction of pRb protein. Brg1 is involved in cell growth arrest, apoptosis, and senescence, and both Rb and p53 are indispensable for Brg1-induced senescence (Napolitano et al. 2007). Because E2F target genes are incorporated into SAHF and the formation of SAHF requires intact pRB pathways (Narita et al. 2003), hBrm and/or Brg1 might be involved in SAHF assembly, perhaps through deposition of macroH2A. Ishikawa and coworkers proposed that HMGA proteins displaced histone H1 from SAHF (Funayama et al. 2006). Lowe and coworkers showed that, in contrast to HP1, HMGA1 was required for formation of SAHF, as judged by chromosome condensation (Narita et al. 2006). The displacement of histone H1 and deposition of HMGA proteins are likely to be early events in the formation of SAHF, and the mechanistic relationships require defining.

Relationship Between SAHF and the pRb and p53 Pathways

pRb and p53 appear to be required at the specific downstream points in the formation of SAHF. Significantly, formation of SAHF driven by ectopically expressed p16^{Ink4a}, an activator of the pRb pathway, also requires ASF1a. This observation suggests that the HIRA/ASF1a and pRb pathways act in parallel to form SAHF. Consistent with this idea, the pRb tumor suppressor protein and the HIRA and ASF1a histone chaperones are well-established regulators of chromatin structure. Through an interaction with the E2F transcription factor, pRb protein facilitates formation of heterochromatin at E2F target genes. One way to explain the cooperation with the pRb and HIRA/ASF1a pathways is that pRb initiates heterochromatin formation at the promoter of E2F target genes and this heterochromatin acts as a nucleation site for HIRA/ASF1a-mediated large scale chromosome condensation. Consistent with this idea, E2F target genes such as cyclin A are incorporated into SAHF, and pRb colocalizes partially with SAHF (Narita et al. 2003). Alternatively, pRb and HIRA might act in concert with a single protein, such as DNAJA2, which plays a key role in SAHF formation (Ye et al. 2007). The observation that activation of the HIRA/ASF1a/SAHF assembly pathway through HIRA's localization to PML bodies is independent of pRb and p53 (Ye et al. 2007) suggests the exciting possibility that some early events in the senescence program are independent of these two presumed master regulators of the senescence program. However, we do not know the exact relationship between SAHF and the pRb and p53 pathways.

PRC1 and PRC2 Silencing

The PcG genes encode a family of conserved epigenetic regulators that were discovered as repressors of the genes that control body segmentation in *Drosophila*. In mammals, PcG proteins regulate genes involved in development, differentiation, and survival through epigenetic mechanisms (Schuettengruber et al. 2011; Fig. 11.2).

The PRC2 protein complex includes four proteins; Ezh2, Suz12, embryonic ectoderm development (Eed) and retinoblastoma-binding protein p48 (RBAP48). The catalytic subunit of this complex is Ezh2, a methyltransferase that methylates H3K27 through its SET domain encoded catalytic site (Sawarkar and Paro 2010). Suz12, Eed and RBAP48 are noncatalytic subunits of this complex. Suz12 and Eed are required for optimal Ezh2 histone methyltransferase activity. The interactions between Suz12, Eed, and Ezh2 result in a significant increase in Ezh2 catalytic activity, indicating that the full complex is required for optical trimethylated histone H3 lysine K27 (H3K27me3) formation. The PRC1 protein complex includes a core of four proteins; Ring1B, Bmi-1, PH1, and CBX (Simon and Kingston 2009). The catalytic subunit of this complex, Ring1B, ubiquitinates H2AK119 and is optimally active in association with Bmi-1. An important role of the CBX protein is its association with H3K27me3 to anchor the PRC1 complex to chromatin. PRC1 has been proposed to ubiquitinate H2AK119 as part of the process leading to a closed chromatin state. Modification of chromatin by PRC2 and PRC1 is a coordinated sequential process. The first step in the polycomb complex-mediated modification is performed by PRC2. The Ezh2 protein of PRC2 catalyzes trimethylation of lysine 27 of histone H3. In the next step, the CBX protein of PRC1 binds H3K27me3 to another PRC1 complex at this site and the PRC1 Ring1B protein catalyzes ubiquitination of histone H2A at lysine K119. These events lead to chromatin compaction-mediated suppression of transcription (Beisel and Paro 2011). Although the mechanism underlying this transcriptional repression is not well understood, chromatin compaction might block transcription factor-DNA interaction or inhibit transcription elongation. Bmi-1-deficient fibroblasts exhibit slower proliferation and faster senescence through the regulated expression of the *Ink4a/Arf* locus. This locus encodes two tumor suppressor genes, p16^{Ink4a} and p14^{Arf} (p19^{Arf} in mouse), and

Polycomb Repressive Complexes (PRC) in human

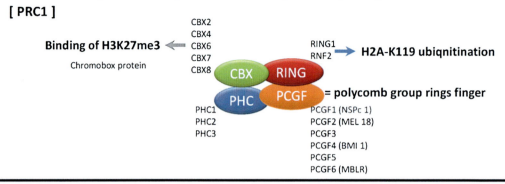

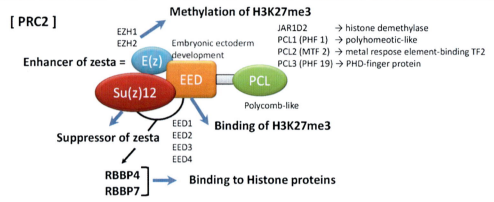

Fig. 11.2 Schematic representation of subunits and functions of polycomb repressive complexes PRC1 and PRC2 involved. The components of PRC1 and PRC2 complexes and their functions are listed. *CBX* Chromobox protein, *EED* Embryonic ectoderm development, *ESC* Extra sex comb, *E(Z)* Enhancer of zesta, *EZH* Enhancer of zesta homologue, *MTF2* Metal response element-binding transcription factor 2, *PC* Polycomb, *PCGF* Polycomb group ring finger, *PH* Polyhomeotic, *PHC* Polyhomeotic-like, *PHF* PHD-finger protein, *RING* Ring finger protein, *RNF2* RING finger protein 2, *SCE* Sex coms extra, *SU(Z)* Suppressor of zesta

controls the cell cycle progression. $p16^{Ink4a}$ inhibits the kinase activity of the cyclin D-dependent kinases, cdk4 and cdk6, and prevents their association with cyclins. This ultimately leads to reduced pRB phosphorylation and reduced cell cycle progression through G1. The $p16^{Ink4a}$ level is normally reduced during cell cycle progression. $p14^{Arf}$ promotes cell cycle arrest by interacting with MDM2 and by reducing its ability to target p53 for degradation. This also ultimately leads to cell cycle arrest (Lowe and Sherr 2003). In the presence of Bmi-1, the level of $p16^{Ink4a}$ and $p14^{Arf}$ level are reduced, thereby permitting the cells to proliferate. Thus, there is an inverse relationship between $p16^{Ink4a}$ level and expression of Bmi-1 (Cordisco et al. 2010).

JDP2-Mediated Regulation of the *INK4a/ARF* Locus

We have shown that the expression of JDP2 regulates the senescence-competent genes, $p16^{Ink4a}$ and $p19^{Arf}$, in response to accumulating oxidative stresses. We analyzed the oxygen-dependent proliferation of mouse embryonic fibroblasts (MEFs) from $Jdp2^{-/-}$ mice in the presence of an environmental (20 %) or low (3 %) oxygen level (Nakade et al. 2009). The $Jdp2^{-/-}$ MEFs continued to divide, even after 6 weeks, whereas the wild-type (WT) MEFs nearly stopped proliferating and entered senescence when exposed to environmental oxygen level. These results demonstrate that MEFs lacking JDP2 can escape from the irreversible

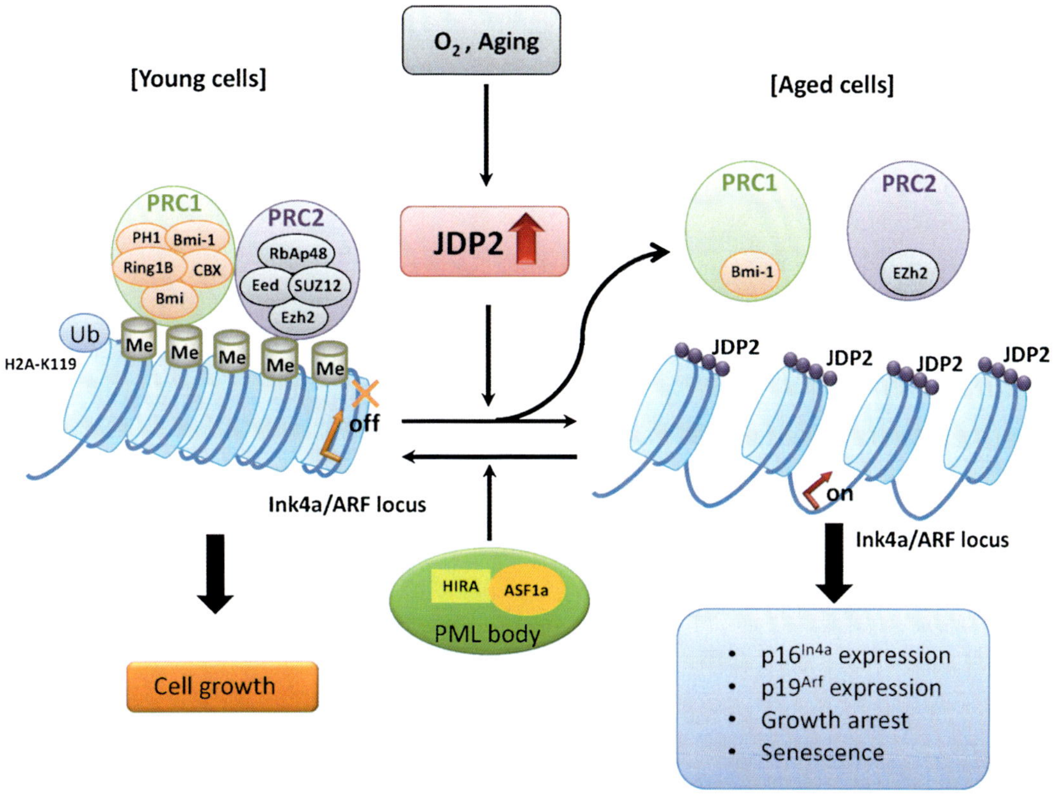

Fig. 11.3 Schematic representation of the epigenetic control of the expression of the genes encoded $p19^{Arf}$ and $p16^{Ink4a}$ by JDP2. The exposure of young MEFs to oxygen stress or aging stress leads to the accumulation of JDP2. JDP2 binds to histones and inhibits the methylation of H3K27me3 at $p19^{Arf}$ and $p16^{Ink4a}$ locus. As a result, polycomb complex PRC1 and PRC2 fail to form the stable repressive complexes and are released from this locus. The consequent expression of $p19^{Arf}$ and $p16^{Ink4a}$ in the aged cells leads to growth arrest and senescence stage. This hypothesis is derived from our results (Nakade et al. 2009)

growth arrest in the presence of an environmental oxygen level. A ChIP assay demonstrated that methylation of H3K27 at the $p16^{Ink4a}/p19^{Arf}$ locus was greater in $Jdp2^{-/-}$ MEFs than in WT MEFs, and that the binding of PRC1 and PRC2 to the $p16^{Ink4a}$ and $p19^{Arf}$ promoters is more efficient in $Jdp2^{-/-}$ MEFs than in WT MEFs. These observations suggest that, in the absence of JDP2, H3K27 is methylated by PRC2 and the $p16^{Ink4a}/p19^{Arf}$ locus is silenced by PRC1. By contrast, the increased expression of JDP2 helps to release PRC1 and PRC2 from the $p16^{Ink4a}/p19^{Arf}$ locus, thereby reducing H3K27 methylation. Thus, we think that JDP2 is an important factor in the regulation of cellular senescence. The loss of JDP2 allows MEFs to escape senescence and, conversely, the overexpression of JDP2 induces cell cycle arrest (Fig. 11.3). The absence of JDP2 reduces the expression of both $p16^{Ink4a}$ and Arf, which inhibit cell cycle progression. Identification of a gene-specific DNA-binding protein, namely JDP2, as a nucleosome assembly factor suggests that H3.3 might be deposited directly at specific locations by site-specific DNA-binding proteins that also have histone chaperone activity. It will be interesting to examine the binding preference of JDP2 for histone H3.1 compared with H3.3.

It is an open question whether the various mechanisms involved are context or gene dependent, or whether they act generally and in a combinatorial manner. In mammals, several DNA-binding proteins have also been implicated in the targeting of PcG proteins. What role in this process could be attributed to any of the other possible targeting mechanisms described above? Considering the importance of epigenetic silencing mechanisms in diverse cellular functions such as gene control, genome stability, and chromosome segregation, it is not surprising that cells have evolved multiple pathways for assembling PcG-dependent repressive chromatin domains and heterochromatin. Alternatively, some pathways could act separately on a specific set of target genes, whereas others could act in a combinational manner. In recent years, a diverse set of ncRNAs has been identified, suggesting the possibility that ncRNAs initiate PcG silencing. It remains unknown how the RNA molecules can mediate the recruitment of PcG proteins to specific sites and which genomic sites are targeted by ncRNA. Further studies are required to dissect the various mechanisms underlying the targeting of epigenetic silencing targeting mechanisms and their interdependencies.

Discussion

JDP2 partially binds to the core histone or nucleosome in a DNA-sequence-specific manner or histone subset-specific manner, and histone acetyltransferase or histone methyltransferase might not have access to the nucleosome *in vitro*. A ChIP assay demonstrated that JDP2 inhibits at least the acetylation of histones H4K8 and H4K16, although we have not identified other precise residues of histone H3 acetylation (Jin et al. 2006). Moreover, JDP2 associates with histone H3K27 and blocks the methylation of histones (Nakade et al. 2009). Thus, we assume that the interaction between JDP2 and the nucleosome is specific to the DNA sequence or histone modification, and thus only certain restricted sets of histones might be associated with JDP2 *in vivo*. Addressing these precise functions in the context of epigenesis will help us understand the regulation of senescence, differentiation, and viral infection in a broader context.

Understanding how JDP2 promotes aging and senescence, whether by inducing cellular senescence or by decreasing the frequency of cell cycle entry, is important. How the oxygen level and aging induce JDP2 gene expression is also interesting.

Further understanding of the *in vivo* regulation of the $p16^{Ink4a}/p19^{Arf}/p15^{Ink4b}$ locus by oxidative stress is needed. The factors and processes involved in reprogramming evoke cellular senescence. Cellular senescence is thought to contribute to tissue aging, raising the question of whether SAHF contribute directly to aging. In support of this idea, markers of increased heterochromatin, including activation of the HIRA/ARF1a pathway, have been reported in the skin of aging primates (Herbig and Sedivy 2006). Thus, to understand the role of heterochromatin in tissue aging in humans, it is necessary to understand the gene network and environmental conditions that influence the rate of normal human aging.

We propose that the role of JDP2 in the induction of cellular senescence involves the accumulation of oxidative stress and/or other environmental stimuli during aging, which upregulates JDP2 expression in primary untransformed cells. Increased protein level of JDP2 helps to remove PRC1 and PRC2, which are responsible for the methylation of histone H3, from the $p16^{Ink4a}/Arf$ locus, leading to increased $p16^{Ink4a}$ and Arf expression and entry into senescence. Our findings provide new insights into the molecular mechanisms, by which senescence is induced in the context of the epigenetic regulation of the $p16^{Ink4a}/Arf$ locus.

Acknowledgments This work was supported by grants from the NSC (NSC-102-2320-B-037-047-MY3; NSC-102-2314-B-038-004-MY2), the NHRI (NHRI-EX101-10109BI, NHRI-102AI-PDCO-03010201) of Taiwan and KMU (KMER006; KMU-EM-99-3; KMV-ER006, to KKY).

References

Ahmad K, Henikoff S (2002) The histone variant H3.3 marks active chromatin by replication-independent nucleosome assembly. Mol Cell 9:1191–1200

Aronheim A, Zaudi E, Hennemann H, Elledge SJ, Karin M (1997) Isolation of an AP-1 repressor by a novel method for detecting protein-protein interactions. Mol Cell Biol 17:3094–3102

Beisel C, Paro R (2011) Silencing chromatin: comparing modes and mechanisms. Nat Rev Genet 12:123–135

Borden KL (2002) Pondering the promyolocytic leukemia protein (PML) puzzles: possible functions for OML nuclear bodies. Mol Cell Biol 22:5259–5269

Bruno M, Flaus A, Stockdale C, Rencurel C, Ferreira H, Owen-Hughes T (2003) Histone H2A/H2B dimer exchange by ATP-dependent chromatin remodeling activities. Mol Cell 12:1599–1606

Campisi J (2005) Senescent cell, tumor suppression, and organismal aging: good citizens, bad neighbors. Cell 120:513–522

Campisi J, d'Adda di Fagagna F (2007) Cellular senescence: when bad things happen to good cells. Nat Rev Cell Biol 8:729–740

Cordisco S, Maurelli R, Bondanza S, Stefanini M, Zambruno G, Guerra L, Dellambra E (2010) Bmi-1 reduction plays a key role on physiological and premature aging of primary human keratinocytes. J Invest Dermatol 130:1048–1062

Costanzi C, Pehrson JR (2001) MACROH2A2, a new member of the MACROH2A core histone family. J Biol Chem 276:21776–21784

Funayama R, Saito M, Tanobe H, Ishikawa F (2006) Loss of linker histone H1 in cellular senescence. J Cell Biol 175:869–880

Hayflick L, Moorhead PS (1961) The serial cultivation of human diploid cell strains. Exp Cell Res 25:585–621

Herbig U, Sedivy JM (2006) Regulation of growth arrest in senescence: telomere damage is not the end of the story. Mech Aging Dev 127:16–24

Ho L, Grabtree GR (2010) Chromatin remodeling during development. Nature 465:474–484

Itahana K, Zou Y, Itahana Y, Martinez JL, Beausejour C, Jacob JJ, Van Lohuizen M, Band V, Campisi J, Dimri GP (2003) Control of the replicative life span of human fibroblasts by p16 and the polycomb protein Bmi-1. Mol Cell Biol 23:389–401

Jin C, Ugai H, Song J, Murata T, Nili F, Sun K, Horikoshi M, Yokoyama K (2001) Identification of mouse Jun dimerization protein 2 as a novel repressor of ATF-2. FEBS Lett 489:34–41

Jin C, Kato K, Chimura T, Yamasaki T, Nakade K, Murata T, Li H, Pan J, Zhao M, Sun K, Chiu R, Ito T, Nagata K, Horikishi M, Yokoyama K (2006) Regulation of histone acetylation and nucleosome assembly by transcription factor JDP2. Nat Struct Mol Biol 13:331–338

Lowe SW, Sherr CJ (2003) Tumor suppression by Ink4a-Arf: progress and puzzles. Curr Opin Genet Dev 13:77–83

Nakade K, Pan J, Yamasaki T, Murata T, Wasylyk B, Yokoyama K (2009) JDP2 (Jun dimerization protein-2)-deficient mouse embryonic fibroblasts are resistant to replicative senescence. J Biol Chem 284:10808–10817

Napolitano MA, Cipollaro M, Cascino A, Melone MAB, Giordano A, Galderisl U (2007) Brg1 chromatin remodeling factor is involved in cell growth arrest, apoptosis and senescence of rat mesenchymal stem cells. J Cell Sci 120:2904–2911

Narita M, Nunez S, Hear E, Lin AW, Hearn SA, Spector DL, Hannon GJ, Lowe SW (2003) Rb-mediated heterochromatin formation and silencing of E2F target genes during cellular senescence. Cell 113:703–716

Narita M, Narita M, Krizhanovsky V, Nuñez S, Chicas A, Hearn SA, Myers MP, Lowe SW (2006) A novel role for high-mobility group a proteins in cellular senescence and heterochromatin formation. Cell 126:503–514

Nevins JR (2001) The Rb/E2F pathway and cancer. Hum Mol Genet 10:699–703

Oruetxebarria I, Venturini F, Kekarainen T, Houweling A, Zuijderduijin JMP, Mphd-Sarip A, Vries RG, Hoeben RC, Verrijzer CP (2004) $p16^{INK4a}$ is required for hSNF5 chromatin remodeler-induced cellular senescence in malignant Rhabdoid tumor cells. J Biol Chem 279:3807–3816

Pan J, Nakade K, Huang Y-C, Zhu Z-W, Masuzaki S, Hasegawa H, Murata T, Yoshiki A, Yamaguchi N, Lee C-H, Yang W-C, Tsai E-M, Obata Y, Yokoyama K (2010) Suppression of cell-cycle progression by Jun dimerization protein-2 (JDP2) involves downregulation of cycling-A2. Oncogene 29:6245–6256

Reddel RR (1998) Genes involved in the control of cellular proliferation potential. Ann N Y Acad Sci 854:8–19

Sawarkar R, Paro R (2010) Interpretation of developmental signaling at chromatin: the polycomb perspective. Dev Cell 19:651–661

Schuettengruber B, Martinez A-M, Iovino N, Cavalli G (2011) Trithorax group proteins: switching genes on and keeping them active. Nat Rev Mol Cell Biol 12:799–814

Sherr CJ, McCormick F (2002) The RB and p53 pathways in cancer. Cancer Cell 2:103–112

Simon JA, Kingston RE (2009) Mechanisms of polycomb gene silencing; known and unknown. Nat Rev Mol Biol 10:697–708

Tagami H, Ray-Gallet D, Almouzni G, Nakatani Y (2004) Histone H3.1 and H3.3 complexes mediate nucleosome assembly pathways dependent or independent of DNA synthesis. Cell 116:51–61

Wang S-W, Lee J-K, Ku C-C, Chiou S-S, Lin C-L, Ho M-F, Wu D-C, Yokoyama K (2011) Jun dimerization protein 2 in oxygen restriction; control of senescence. Curr Pharm Des 17:2278–2289

Wright WE, Shay JW (2001) Cellular senescence as a tumor-protection mechanism: the essential role of counting. Curr Opin Genet Dev 11:98–103

Ye X, Zerlanko B, Zhgang R, Somaiah N, Lipinski N, Adams PD (2007) Definition of pRB- and p53-dependent and independent steps in HIRA/ASF1a-mediated formation of senescence-associated heterochromatin foci (SAHF). Mol Cell Biol 27:2454–2465

Zhang R, Poustovoitov MV, Ye X, Santos HA, Chen W, Daganzo SM, Erzberger SM, Serebriiskii IG, Canutescu AA, Dunbrack RL, Pehrson JR, Berger JM, Kaufman PD, Adams PD (2005) Formation of MacroH2A-containing senescence heterochromatin foci and senescence driven by ASF1a and HIRA. Cell 8:19–30

Induction of Cellular Senescence: Role of Mitogen-Activated Protein Kinase-Interacting Kinase 1

12

Samira Ziaei and Naoko Shimada

Contents

Abstract	111
Introduction	112
Cellular Senescence	112
Cellular Senescence and Cancer	112
Cellular Senescence and Aging	112
Stress-Activated MAPKS in Cellular Senescence	112
The p38 Pathway in Replicative Senescence	113
Heterogeneous Nuclear Ribonucleoproteins	113
P38 MAP Kinase Pathway and Regulation of HNRNP A1 Protein	114
MNK1 Regulation of the Subcellular Distribution of HNRNP A1	115
The Roles of MNK1 in Cellular Senescence	117
References	117

S. Ziaei (✉) • N. Shimada
Department of Biology, The City College of New York,
160 Convent Ave, New York, NY 10031, USA
e-mail: ziaeisamira@gmail.com

Abstract

Heterogeneous nuclear ribonucleoprotein (hnRNP) family members are the most abundant components of messenger ribonucleoprotein complexes (mRNPs) and play regulatory roles in a variety of biogenesis of mRNA. hnRNP A1 is a member of the hnRNP A/B subfamily, is highly abundant, and is involved in pre-mRNA and mRNA metabolism such as alternative splicing, mRNA export, splice site selection, mRNA turnover, and translation. Recent studies have shown that stress stimuli such as osmotic shock or UVC irradiation induce cytoplasmic accumulation of hnRNP A1. The cytoplasmic accumulation is concomitant with an increase in its phosphorylation and that requires p38 MAPK. We have previously demonstrated that hnRNP A1 protein shows diminished expression level and altered subcellular distribution in senescent HS74 fibroblasts. In this study, we observed that phosphorylated hnRNP A1 protein levels decreased as a result of MNK1 inhibition and that reciprocal binding occurs between hnRNP A1 and MNK1. These data implicate MNK1 as the kinase in the p38 MAPK pathway that activates hnRNP A1 in IMR-90 fibroblasts. Furthermore, we demonstrate that inhibition of MNK1 activity modulates the phosphorylation and subcellular distribution of hnRNP A1 protein. These results suggest a role for MNK1 in the regulation of hnRNP A1 in senescent cells. This is the first report, to our knowledge, that shows a link between MNK1 and cellular senescence.

Keywords

Cellular senescence • Heterogeneous nuclear ribonucleoproteins (hnRNP) • Mitogen-activated protein kinase-interacting kinase 1 (MNK1) • Mitogen-activated protein kinases (MAPKs) • Osmotic shock/UVC irradiation • Phosphorylation response • P38 map kinase pathway • Replicative senescence • Tumor necrosis factor (TNF)

Cellular Senescence

Cellular senescence was first described by Hayflick and Moorhead (1961) as a state of irreversible cell cycle arrest seen in cultured normal human fibroblasts after a limited number of cell divisions. This phenomenon, often called the Hayflick limit or replicative senescence, is not specific to human fibroblasts but is seen in many types of cells. Senescent cells exhibit several characteristic features: irreversible growth arrest, flattened cellular morphology, enlarged cytoplasm and nuclei, and senescence-associated heterochromatic foci (Dimri et al. 1995). Positive staining for senescence-associated β-galactosidase (SA-β-gal) activity is another well-known feature of senescent cells, although SA-β-gal -positive cells have also been observed in stress-induced conditions, unrelated to the state of senescence (Severino et al. 2000). Whereas quiescent cells exhibit transient and reversible growth arrest, senescent cells are arrested in the G1/G0 phase of the cell cycle and do not enter the S phase, even when they are treated with strong mitogenic stimuli such as growth factors.

Cellular Senescence and Cancer

Cancer is widely known to be a disease that is triggered by the accumulation of multiple genomic mutations in cells that suffer mainly from various stressors inducing DNA damage. Apoptosis is one of the tumor-suppressive mechanisms that eliminate such damaged cells. Accumulating evidence suggests that cellular senescence also plays a critical role in tumor suppression. Mice deficient in p53, a major tumor suppressor with diverse functions, including induction of cellular senescence, have been shown to be susceptible to spontaneous tumorigenesis (Reed 1999). Recently, it has been demonstrated that tumorigenesis upon inactivation of another tumor suppressor, PTEN in the prostate, is suppressed through p53-dependent cellular senescence. In this model, combined inactivation of PTEN and p53 elicits early onset of invasive prostate cancer, indicating the critical role of cellular senescence in tumor suppression (Donehower et al. 1992). Moreover, it has been shown that telomere dysfunction-induced cellular senescence suppresses tumor progression by a p53-dependent mechanism (Chen et al. 2005). These findings strongly suggest that cellular senescence, as well as apoptosis, contributes to tumor suppression.

Cellular Senescence and Aging

In addition to the beneficial role of cellular senescence in tumor suppression, possible harmful roles of cellular senescence have also been proposed. It has been reported that age-dependent accumulation of senescent cells exhibiting an increased expression of the INK4 family protein $p16^{INK4a}$ and positive SA-β-gal staining is observed in islet β-cells and forebrain progenitor cells (Collado et al. 2007). Moreover, accumulated senescent cells often secrete several cytokines and growth factors, which may promote transformation of neighboring cells (Molofsky et al. 2006). Cellular senescence may thus facilitate the development of late-onset cancer and trigger age-related diseases.

Stress-Activated MAPKS in Cellular Senescence

The MAPK cascades are extensively studied signaling pathways that respond to stimuli from outside the cell and are evolutionarily well conserved in cells from yeasts to humans. Each cascade is typically composed of three hierarchical protein kinases: mitogen-activated protein kinases (MAPKs), MAPK kinases (MAPKKs), and MAPKK kinases (MAPKKKs) (Widmann et al. 1999). MAPKKKs phosphorylate and activate MAPKKs, and activated MAPKKs in turn phosphorylate and activate MAPKs (Widmann et al. 1999). In mammals, there are MAPK cascades that converge on extracellular signal regulated (ERKs), c-Jun N-terminal kinases (JNKs), and p38 MAPKs (p38s) (Maruyama et al. 2009). Among these cascades, the ERK cascade is thought to be activated mainly by growth factors and induces cell proliferation, survival, and differentiation. On the other hand, JNKs and p38s were originally identified as kinases activated in response to a diverse array of cellular stressors such as pro-inflammatory cytokines, UV irradiation, ROS, heat and osmotic shock, and DNA damage, and are therefore called stress-activated MAPKs (Maruyama et al. 2009). The JNK and p38 pathways have been shown to be critically involved in the control of cellular stress responses such as cell death and survival (Widmann et al. 1999).

The p38 Pathway in Replicative Senescence

The p38 pathway has also been shown to be involved in replicative senescence. In human fibroblasts undergoing replicative senescence, p38 was found to be activated, and expression of hTERT appeared to abrogate such p38 activation, suggesting that p38 is activated through telomere shortening (Iwasa et al. 2003). In rabbit articular chondrocytes, inhibition of p38 activity by SB203580 or by expression of dominant negative MKK6 stimulated proliferation and partially delayed the onset of senescence (Kang et al. 2005). These findings suggest that activation of the p38 pathway is involved in replicative senescence induced by telomere shortening.

Constitutive activation of p38 by expression of an active mutant of MKK3 or MKK6 induced premature senescence in primary human fibroblasts (Iwasa et al. 2003). Furthermore, depletion or inactivation of Wip1 phosphatase, a p53-regulated gene product suppressing p38 activity (Takekawa et al. 2000), activated the p38 pathway, and caused cells to enter premature senescence (Iwasa et al. 2003). These findings demonstrate that the p38 pathway plays a crucial role in inducing cellular senescence. Overexpression of a constitutively active mutant of MEK1, a MAPKK for ERKs, has been shown to activate the MKK3/6-p38 pathway in human fibroblasts, resulting in premature senescence (Maruyama et al. 2009). Moreover, inhibition of MEK activity by its inhibitor U0126 has been found to abolish the ability of oncogenic Ras to activate MKK3 and MKK6 (Maruyama et al. 2009). These findings suggest that activation of the p38 pathway in Ras-induced premature senescence is dependent on activation of the Raf1-MEK-ERK pathway. Recently, MINK, an Ste20 family kinase, has been proposed as an activator of p38 in Ras-induced cell cycle arrest (Nicke et al. 2005). MINK is activated via an unknown mechanism following Ras-induced ERK activation and ROS generation, and appears to require ASK1/MAP3K5 and Tpl-2/MAP3K8 as MAPKKKs for p38 activation in this setting (Nicke et al. 2005).

Heterogeneous Nuclear Ribonucleoproteins

Heterogeneous nuclear ribonucleoproteins (hnRNP) make up a significant subclass of known RNP. They were first described as a group of 6 chromatin associated RNA binding proteins which bound nascent polymerase II transcripts (Dreyfuss et al. 1993). They assembled on nascent RNA polymerase II transcripts to form hnRNP complexes, where they co-localize with small nuclear (sn) RNPs. In fact, as nascent hnRNAs emerge from the transcriptional machinery, and during their entire nuclear life, they are associated with a set of proteins collectively termed hnRNPs that are not stable components of the nuclear RNP complexes such as snRNPs (Dreyfuss et al. 1993). These proteins bind to specific sequences on pre-mRNA which are important for pre-mRNA processing such as the

5' and 3' splice sites and the polypyrimidine stretch (Nakielny and Dreyfuss 1997).

All hnRNP A/B family proteins shuttle dynamically between the nucleus and cytoplasm (Dreyfuss et al. 1993). hnRNP A1 protein and its splice variants comprise the most common gene products in the hnRNP A/B family. It has been shown that activation of the p38 MKK$_{3/6}$/p38 stress-signaling pathway in mammalian cells results in both hyperphosphorylation and cytoplasmic accumulation of hnRNP A1 and affects alternative splicing regulation (van der Houven van Oordt et al. 2000). Allemand and colleagues mapped the stress-induced phosphorylation sites in hnRNP A1 to a stretch of serines located adjacent to the M9 motif, which mediates bidirectional transport of hnRNP A1 (Allemand et al. 2005). This phosphorylation event abrogates interactions between hnRNP A1 and its import receptor, transportin, resulting in its cytoplasmic accumulation.

It has been shown that osmotic shock or UVC irradiation induce cytoplasmic accumulation of hnRNP A1 (van der Houven van Oordt et al. 2000). The cytoplasmic accumulation is concomitant with an increase in its phosphorylation and requires p38 MAPK (Guil et al. 2006). We have previously demonstrated that hnRNP A1 protein levels show diminished expression and altered subcellular distribution in senescent HS74 fibroblasts (Hubbard et al. 1995; Zhu et al. 2002). These findings raise the possibilities that there is a relationship between hnRNP A1 and p38 MAPK proteins and suggest that hnRNP A1 may play a significant role in cellular senescence under the control of p38 MAPK pathway. However, the precise molecular mechanisms by which this pathway might regulate hnRNP A1 have yet to be identified.

We have previously demonstrated that hnRNP A1 and p38 MAPK interact *in vivo* and that the p38 MAPK pathway regulates the expression level and subcellular distribution of hnRNP A1 (Shimada et al. 2009). Inhibition of p38 MAPK increased the level of hnRNP A1 protein expression in young and G0-arrested IMR-90 cells suggesting a p38 MAPK-dependent regulation. Our findings of decreased protein expression and increased phosphorylation of hnRNP A1 during senescence indicate that the p38 MAPK pathway might regulate the stability of hnRNP A1 protein via phosphorylation. We have also shown that the phosphorylation level of hnRNP A1 was elevated in senescent cells. Recent studies have shown that stress stimuli such as osmotic shock or UVC irradiation induce cytoplasmic accumulation of hnRNP A1 (van der Houven van Oordt et al. 2000; Guil et al. 2006). The cytoplasmic accumulation is concomitant with an increase in its phosphorylation and that requires p38 MAPK (van der Houven van Oordt et al. 2000). We have previously demonstrated that hnRNP A1 protein shows diminished expression level and altered subcellular distribution in senescent HS74 fibroblasts (Hubbard et al. 1995; Zhu et al. 2002). Furthermore, we demonstrated that hnRNP A1 forms a complex with phosho-p38 MAPK *in vivo* and its expression level and subcellular distribution are regulated by the p38 MAPK pathway in IM-90 human fibroblasts, derived from fetal lung (Shimada et al. 2009). These findings raise the possibilities that there is a relationship between hnRNP A1 and p38 MAPK proteins and suggest that hnRNP A1 plays a significant role in cellular senescence under the control of p38 MAPK. However, the precise molecular mechanisms by which the p38 MAPK pathway might regulate hnRNP A1 have yet to be identified.

In the experimental goals outlined in this work, we investigate the molecular mechanisms responsible for the regulation of hnRNP A1 downstream of p38 MAPK. Furthermore, evidence for involvement of MNK1 as a putative modulator of hnRNP A1 protein levels will be examined. We hypothesize that MNK1 regulates the phosphorylation and the subcellular distribution of hnRNP A1 and that MNK1 may play a role in the induction of senescence.

P38 MAP Kinase Pathway and Regulation of HNRNP A1 Protein

Because hnRNP A1 was observed to be phosphorylated by the p38 MAPK pathway (Allemand et al. 2005), we investigated whether there is a physical interaction between phosho-hnRNP A1 and MNK1 by co-immunoprecipitation assays. Young IMR-90 cells were treated in the presence or absence of varying concentrations of CGP 57380, a specific inhibitor of MNK1 kinase activity, for 3 days. The cell lysates were then immunoprecipitated with the hnRNP A1 specific monoclonal antibody, 4B10. Subsequently, membranes were probed with a phospho-serine antibody. A 34 kDa band was detected, representing phosho-hnRNP A1. The phosphorylation level of hnRNP A1 decreased with increasing concentrations of CGP 57380, indicating that MNK1 kinase activity is required for phosphorylation of hnRNPA1 *in vivo* (Ziaei et al. 2012). Immunoprecipitation of hnRNP A1 by 4B10 antibody was confirmed by reprobing these membranes with antibodies specific for hnRNP A1.

We next determined whether or not there is a physical interaction between hnRNP A1 and MNK1 using co-immunoprecipitation studies in the presence or absence of CGP 57380. Cell lysates were immunoprecipitated with 4B10 antibody and then the membranes were probed with MNK1 antibody. MNK1 co-immunoprecipitated with hnRNP A1 as indicated by a 50 kDa band in the complex. We next conducted the reciprocal analysis to ascertain whether hnRNP A1 was co-immunoprecipitated using a MNK1 specific antibody. A faint 34 kDa band was detected in the complex, representing the presence of hnRNP A1 (Ziaei et al. 2012). These results indicate that MNK1 forms a complex with hnRNP A1 *in vivo*.

We have shown previously that the phosphorylation level of hnRNP A1 was elevated in senescent cells (Shimada et al. 2009). This may be responsible for a role for hnRNP A1 in establishing the senescent morphology of cells. We hypothesize that the phosphorylation of hnRNP A1 by the p38 MAPK pathway is critical for its subcellular distribution.

In regards to the stoichiometry of phosphorylated hnRNP A1 protein, we have previously observed that the majority of A1 protein is phosphorylated relative to total hnRNP A1 protein levels in senescent cells compared to young cells (Shimada et al. 2009). This data suggests that the phosphorylation status of hnRNP A1 affects the protein's subcellular localization and cytoplasmic accumulation in senescent cells (Shimada et al. 2009).

MNK1 Regulation of the Subcellular Distribution of HNRNP A1

It has been reported that mitogen-activated protein kinase-interacting kinase 1 (MNK1) might possess a MAPK-binding domain that allows it to bind to ERK and p38 MAPK followed by phosphorylation through these two kinases, but not JNK (Waskiewicz et al. 1997). MNK1 phosphorylation of eukaryotic initiation factor-4E (eIF4E), which is a translation initiation factor that binds to the 5′ cap structure of eukaryotic cytosolic mRNAs, has been well studied (Pyronnet 2000). Recently, AU-rich element (ARE) binding proteins, including hnRNP A1, have been identified as MNK1 substrates *in vitro* (Coulthard et al. 2009).

We have previously shown that there is diminished expression and altered subcellular distribution of hnRNP A1 protein in senescent human diploid fibroblast cells, which terminated their cell growth and reached cell cycle arrest (Hubbard et al. 1995; Zhu et al. 2002). We have recently demonstrated that hnRNP A1 and p38 MAPK interact *in vivo* and that the p38 MAPK pathway regulates the expression level and subcellular distribution of hnRNP A1 (Shimada et al. 2009). However, the molecular mechanisms responsible for the regulation of hnRNP A1 downstream of p38 MAPK in human diploid fibroblasts have yet to be identified.

We have previously demonstrated that activation of p38 MAPK is required for the cytoplasmic accumulation of hnRNP A1 in senescent cells (Shimada et al. 2009). It has been previously shown that MNK1/2 mediated stress-induced phosphorylation and cytoplasmic accumulation of hnRNP A1 in HeLa cells (Guil et al. 2006). We found that MNK formed a complex with hnRNP A1 and modulated its phosphorylation in young cells. These findings raised the possibility that MNK1 might potentially regulate the subcellular distribution of hnRNP A1 downstream of p38 MAPK in senescent cells.

To asses this, we treated young and senescent IMR-90 cells with CGP 57380 or an equivalent amount of DMSO, then determined the localization pattern of hnRNP A1 by immunocytochemistry. CGP 57380 is a selective inhibitor of MNK1 ($IC_{50} = 2.2$ μM) with no inhibitory activity against p38 MAPK, JNK1, ERK1/2, PKC, or Src-like kinases (Knauf et al. 2001). Control young cells displayed a predominant nuclear localization of the protein while control senescent cells showed predominantly cytoplasmic accumulation (Ziaei et al. 2012), as previously reported by us (Zhu et al. 2002). Young CGP 57380 treated cells did not show any alteration in their nucleo-specific localization pattern compared to control cells (Ziaei et al. 2012). In contrast, CGP 57380 treatment inhibited the cytoplasmic accumulation of hnRNP A1 in senescent cells, as the majority of the protein was localized in the nucleus (Ziaei et al. 2012). These results indicate that the accumulation of hnRNP A1 in the cytoplasm of senescent cells requires the kinase activity of MNK1. We subsequently determined the localization pattern of MNK1 before and after inhibition with CGP 57380. CGP 57380 inhibition did not induce any change in MNK1 subcellular distribution in young and senescent cells, indicating that the kinase activity of MNK1 is not required for its subcellular distribution. The localization patterns were similar to that of hnRNP A1, suggesting that MNK1 and hnRNP A1 may physically interact with each other in both young and senescent cells, further supporting our co-immunoprecipitation results. Another substrate for p38 MAPK is MAPKAPK2, which is known to phosphorylate hnRNP A0 and has structural similarity to other hnRNPs (Rousseau et al. 2002). However, siRNA inhibition of MAPKAPK2 did not result in changes in either location or protein levels of hnRNP.

The phosphorylation response to stressors has been shown to increase with age. It has been demonstrated that growth factors increase locomotor activity in both Parkinson's and aging models and increase dopamine bioavailability and ser31 tyrosine hydroxylase phosphorylation in the substantia nigra (Salvatore et al. 2009). Furthermore, blockage of the p38 MAPK signaling pathway has been shown to promote skin aging (172). These results indicate that protein phosphorylation as response to stress stimuli is upregulated as a function of age. Therefore, further research is warranted to elucidate the mechanisms by which signaling pathways regulate gene expression during aging.

We previously showed age-dependent changes in the expression level and subcellular distribution of hnRNP A1 (Shimada et al. 2009). However, the molecular mechanisms that regulate these changes have not been elucidated. Additionally, we have demonstrated that the p38 MAPK pathway regulates hnRNP A1 protein levels. The p38 MAPK pathway has been shown to play an essential role in induction of senescence and our studies suggest that there is an interplay between this pathway and hnRNP A1. Previous studies have demonstrated that hnRNP A1 are elevated in tumor cells and human cancers (He et al. 2010). Considering the fact that cellular senescence acts as a tumor suppression mechanism *in vivo*, hnRNP A1 might be a putative target for cancer therapy.

p38 MAPK plays a causative role in cellular senescence. We have previously shown that p38 MAPK is required for the subcellular distribution of heterogeneous nuclear ribonucleoprotein A1 (hnRNP A1) in senescent cells and regulates the protein level in young cells (Shimada et al. 2009). In this study, we investigated molecular mechanisms responsible for the regulation of hnRNP A1 downstream of p38 MAPK during senescence. We found that MNK1 modulates the subcellular distribution of hnRNP A1. This result suggests that p38 MAPK regulates hnRNP A1 localization in MNK1-dependent manner. We suspect that phosphorylation of hnRNP A1 by MNK1 is critical for its subcellular distribution. This hypothesis is supported by the findings of others that MNK1 phosphorylates three serine residues of hnRNP A1 at its C-terminal peptide, F-peptide (Buxadé et al. 2005). Phosphorylation of the F-peptide results in hnRNP A1 cytoplasmic accumulation by reducing its interaction with the nuclear receptor, transportin (Allemand et al. 2005).

The Roles of MNK1 in Cellular Senescence

MNKs have been reported to promote protein synthesis, cell cycle progression, and proliferation in tumor cells through phosphorylation of its substrate, eIF4E (Bianchini et al. 2008), which is a central regulator for cap-dependent translation initiation. In contrast, we found an age-dependent increase in MNK1 and phospho-MNK1 expression and eIF4E phosphorylation in senescent cells and that MNK1 activity was required for the phosphorylation of eIF4E in the cells (Ziaei et al. 2012). Senescence is delayed and lifespan is extended in *Caenorhabditis elegans* mutants that are defective in eIF4E (Syntichaki et al. 2007). The MNK1-eIF4E pathway in senescent cells appears to have an opposite regulating role in protein synthesis compared to its role in tumor cells. It is has been shown that protein synthesis is elevated in carcinogenesis, whereas the rate of general protein synthesis is diminished during aging (Tavernarakis 2007). Therefore, we suggest a putative role for MNK1 in senescence. Possible mechanisms that can explain the age-dependent effects by MNK1 is that the increase in the phosphorylation of eIF4E by MNK1might decrease cap-dependent translation followed by a decline in *de novo* synthesis of the proteins required for longevity. It would be interesting to assess whether the rate of protein synthesis during senescence is affected by inhibition of MNK1 activity. However, the role of MNK has not been examined in senescent mice.

Our findings suggest that the MNK1-induced changes in localization of hnRNP A1 contribute to the genetic features of senescence through altered regulation of its target mRNAs. In general, genes required for stress and immune/inflammatory responses are up-regulated during senescence (Bishop et al. 2010). Some of these genes may be potentially regulated by MNK1/hnRNP A1 pathway. Support for this notion comes from the fact that MNK1 induces the synthesis of tumor necrosis factor (TNF) α via hnRNP A1 phosphorylation in Jurkat cells (Jo et al. 2008). Activation of MNK1 by stress stimuli causes phosphorylation and cytoplasmic accumulation of hnRNP A1 in stress granules (99) whereas stress responsive mRNAs are translated. Further analysis of RNA metabolism regulated by the MNK1-hnRNP A1 pathway could reveal potential molecular mechanisms, that establish the senescent phenotype. To address this issue, we inhibited MNK1 activity with CGP 57380 in young and pre-senescent (late passage) IMR90 fibroblasts to determine a potential role for MNK1 in cell growth. We found that CGP 57380 treatment inhibited the growth of both young and late passage cells (Ziaei et al. 2012). These results indicate that MNK1 has pleiotropic effects. We measured senescence-associated β-galactosidase (a biomarker for replicative senescence) in MNK1 inhibited cells and did not observe an induction of senescence (Ziaei et al. 2012). Therefore, further investigation is warranted to examine eIF4E or some other unknown downstream target of MNK1 and their contributions to senescence.

This is the first report, to our knowledge, that shows a link between MNK1 phosphorylation and hnRNP A1 during cellular senescence. Our findings imply that the MNK pathway could provide putative novel target proteins to treat age-related disorders, including Alzheimer's disease, heart disease, cataract, diabetes, and cancer. The continued study of the role of MNK pathways in hnRNP A1 functionality will extend our knowledge of signaling pathways beyond that of tumorigenesis and inflammatory responses. Further studies are warranted to decipher the precise molecular mechanisms by which MNK1 regulates cellular senescence through hnRNP A1.

References

Allemand E, Guil S, Myers M, Moscat J, Cáceres JF, Krainer AR (2005) Regulation of heterogeneous nuclear ribonucleoprotein A1 transport by phosphorylation in cells stressed by osmotic shock. Proc Natl Acad Sci U S A 102:3605–3610

Bianchini A, Loiarro M, Bielli P, Busà R, Paronetto MP, Loreni F, Geremia R, Sette C (2008) Phosphorylation of eIF4E by MNKs supports protein synthesis, cell cycle progression and proliferation in prostate cancer cells. Carcinogenesis 29:2279–2288

Bishop NA, Lu T, Yankner BA (2010) Neural mechanisms of ageing and cognitive decline. Nature 464:529–535

Buxadé M, Parra JL, Rousseau S, Shpiro N, Marquez R, Morrice N, Bain J, Espel E, Proud CG (2005) The Mnks are novel components in the control of TNF alpha biosynthesis and phosphorylate and regulate hnRNP A1. Immunity 23:177–189

Chen Z, Trotman LC, Shaffer D, Lin HK, Dotan ZA, Niki M, Koutcher JA, Scher HI, Ludwig T, Gerald W, Cordon-Cardo C, Pandolfi PP (2005) Crucial role of p53-dependent cellular senescence in suppression of Pten-deficient tumorigenesis. Nature 436:725–730

Collado M, Blasco MA, Serrano M (2007) Cellular senescence in cancer and aging. Cell 130:223–233

Coulthard LR, White DE, Jones DL, McDermott MF, Burchill SA (2009) p38(MAPK): stress responses from molecular mechanisms to therapeutics. Trends Mol Med 15:369–379

Dimri GP, Lee X, Basile G, Acosta M, Scott G, Roskelley C, Medrano EE, Linskens M, Rubelj I, Pereira-Smith O, Peacocke M, Campisi J (1995) A biomarker that identifies senescent human cells in culture and in aging skin in vivo. Proc Natl Acad Sci U S A 92:9363–9367

Donehower LA, Harvey M, Slagle BL, McArthur MJ, Montgomery CA Jr, Butel JS, Bradley A (1992) Mice deficient for p53 are developmentally normal but susceptible to spontaneous tumours. Nature 356:215–221

Dreyfuss G, Matunis MJ, Pinol-Roma S, Burd CG (1993) hnRNP proteins and the biogenesis of mRNA. Annu Rev Biochem 62:289–321

Guil S, Long JC, Cáceres JF (2006) hnRNP A1 relocalization to the stress granules reflects a role in the stress response. Mol Cell Biol 26:5744–5758

Hayflick L, Moorhead PS (1961) The serial cultivation of human diploid cell strains. Exp Cell Res 25:585–621

He ZY, Wen H, Shi CB, Wang J (2010) Up-regulation of hnRNP A1, Ezrin, tubulin β-2C and Annexin A1 in sentinel lymph nodes of colorectal cancer. World J Gastroenterol 16:4670–4676

Hubbard K, Dhanaraj SN, Sethi KA, Rhodes J, Wilusz J, Small MB, Ozer HL (1995) Alteration of DNA and RNA binding activity of human telomere binding proteins occurs during cellular senescence. Exp Cell Res 218:241–247

Iwasa H, Han J, Ishikawa F (2003) Mitogen-activated protein kinase p38 defines the common senescence-signaling pathway. Genes Cells 8:131–144

Jo OD, Martin J, Bernath A, Masri J, Lichtenstein A, Gera J (2008) Heterogeneous nuclear ribonucleoprotein A1 regulates cyclin D1 and c-myc internal ribosome entry site function through AKT signaling. J Biol Chem 283:23274–23287

Kang S, Jung M, Kim CW, Shin DY (2005) Inactivation of p38 kinase delays the onset of senescence in rabbit articular chondrocytes. Mech Ageing Dev 126:591–597

Knauf U, Tschopp C, Gram H (2001) Negative regulation of protein translation by mitogen-activated protein kinase-interacting kinases 1 and 2. Mol Cell Biol 21:5500–5511

Maruyama J, Naguro I, Takeda K, Ichijo H (2009) Stress-activated MAP kinase cascades in cellular senescence. Curr Med Chem 16:1229–1235

Molofsky AV, Slutsky SG, Joseph NM, He S, Pardal R, Krishnamurthy J, Sharpless NE, Morrison SJ (2006) Increasing p16INK4a expression decreases forebrain progenitors and neurogenesis during ageing. Nature 443:448–452

Nakielny S, Dreyfuss G (1997) Nuclear export of proteins and RNAs. Curr Opin Cell Biol 9:420–429

Nicke B, Bastien J, Khanna SJ, Warne PH, Cowling V, Cook SJ, Peters G, Delpuech O, Schulze A, Berns K, Mullenders J, Beijersbergen RL, Bernards R, Ganesan TS, Downward J, Hancock DC (2005) Involvement of MINK, a Ste20 family kinase, in Ras oncogene-induced growth arrest in human ovarian surface epithelial cells. Mol Cell 20:673–685

Pyronnet S (2000) Phosphorylation of the cap-binding protein eIF4E by the MAPK-activated protein kinase Mnk1. Biochem Pharmacol 60:1237–1243

Reed JC (1999) Mechanisms of apoptosis avoidance in cancer. Curr Opin Oncol 11:68–75

Salvatore MF, Pruett BS, Spann SL, Dempsey C (2009) Aging reveals a role for nigral tyrosine hydroxylase ser31 phosphorylation in locomotor activity generation. PLoS One 4:e8466

Severino J, Allen RG, Balin S, Balin A, Cristofalo VJ (2000) Is beta-galactosidase staining a marker of senescence in vitro and in vivo? Exp Cell Res 257:162–171

Shimada N, Rios I, Moran H, Sayers B, Hubbard K (2009) p38 MAP kinase-dependent regulation of the expression level and subcellular distribution of heterogeneous nuclear ribonucleoprotein A1 and its involvement in cellular senescence in normal human fibroblasts. RNA Biol 6:293–304

Syntichaki P, Troulinaki K, Tavernarakis N (2007) eIF4E function in somatic cells modulates ageing in Caenorhabditis elegans. Nature 445:922–926

Takekawa M, Adachi M, Nakahata A, Nakayama I, Itoh F, Tsukuda H, Taya Y, Imai K (2000) p53-inducible wip1 phosphatase mediates a negative feedback regulation of p38 MAPK-p53 signaling in response to UV radiation. EMBO J 19:6517–6526

Tavernarakis N (2007) Protein synthesis and aging: eIF4E and the soma vs. germline distinction. Cell Cycle 6:1168–1171

van der Houven van Oordt W, Diaz-Meco MT, Lozano J, Krainer AR, Moscat J, Caceres JF (2000) The MKK(3/6)-p38-signaling cascade alters the subcellular distribution of hnRNP A1 and modulates alternative splicing regulation. J Cell Biol 149:307–316

Waskiewicz AJ, Flynn A, Proud CG, Cooper JA (1997) Mitogen-activated protein kinases activate the serine/threonine kinases Mnk1 and Mnk2. EMBO J 16:1909–1920

Widmann C, Gibson S, Jarpe MB, Johnson GL (1999) Mitogen-activated protein kinase: conservation of a three-kinase module from yeast to human. Physiol Rev 79:143–180

Zhu D, Xu G, Ghandhi S, Hubbard K (2002) Modulation of the expression of p16INK4a and p14ARF by hnRNP A1 and A2 RNA binding proteins: implications for cellular senescence. J Cell Physiol 93:19–25

Ziaei S, Shimada N, Kucharavy H, Hubbard K (2012) MNK1 expression increases during cellular senescence and modulates the subcellular localization of hnRNP A1. Exp Cell Res 318:500–508

Mechanisms of Premature Cell Senescence

13

Julien Maizel, Jun Chen, and Michael S. Goligorsky

Contents

Abstract	121
Introduction	122
Replicative and Stress-Induced Premature Cell Senescence (SIPS)	122
Molecular Signatures of Premature Cell Senescence	122
Lysosomal Dysfunction Heralds Sirtuin Depletion and SIPS	124
Relevance of Sirtuins to Cancer	124
Senescence Induces a Sessile Cell Phenotype and Encapsulation of These Cells by ECM: Role of Collagen XVIII and Endostatin	126
Future Perspectives of Employing SIPS for Cancer Therapies	127
References	127

J. Maizel • J. Chen
Department of Medicine, Renal Research Institute,
New York Medical College, Valhalla, NY 10595, USA

M.S. Goligorsky (✉)
Departments of Medicine, Pharmacology
and Physiology, Renal Research Institute, New York
Medical College, Valhalla, NY 10595, USA
e-mail: michael_goligorsky@nymc.edu

Abstract

Cell senescence and malignant transformation are intricately and reciprocally related. The search is on for the ways to induce cancer cell senescence; this explains the growing interest in mechanisms inducing premature cell senescence. Despite many similarities between replicative senescence and stress-induced premature senescence (SIPS), the major distinction between the two lies in the fact that the replicative senescence is almost exclusively associated with the attrition of telomeres, whereas the SIPS is not. In the following pages we shall outline some molecular signatures of SIPS (including Wnt pathway, telomere-independent mechanism driven by a member of polycomb group histone methyltransferase, caveolin-1, sirtuins, and microRNA pathways, among others). Further on we shall describe the role of lysosomal dysfunction in the development of SIPS and sirtuin depletion via released cathepsins. The relevance of sirtuins in many cancers is briefly outlined. We summarize the chapter by suggesting several pathways that could be utilized for induction of SIPS in cancer therapy.

Keywords

Cancer therapies • Caveolin-1 • Lysosomal dysfunction • Mechanisms of premature cell senescence • Replicative and stress-induced premature cell senescence (SIPS) • Sirtuins • Wnt pathway

Introduction

Cell senescence and malignant transformation are intricately and reciprocally related as poignantly illustrated by the fact that senescent cells are histochemically detectable in premalignant, but not in malignant tumors. Therefore, one of the questions posed by cancer biology is whether by restoring the ability to undergo cell senescence or by pharmacologically enforcing it on a transformed cell could slow down tumor growth? Uncontrolled proliferation and halted differentiation of cancer cells result in the mass effect, i.e., tumor growth. The diversity of pathways responsible for the uncontrolled proliferation of cancer cells is beyond the scope of this precis. Suffice it to state that the current thinking links uncontrolled proliferation with the ability of cancer cells to evade molecular pathways inducing cell cycle arrest and thus escape senescence and death. Despite persistently high levels of p19, p53 and p21, senescence programs are halted in cancer cells due to overexpression of Id1, which induces refractoriness to cell cycle arrest mediated by p21 (Swarbrick et al. 2008) and, in some cancers, upregulation of sirtuins and miR-22, as discussed below. Therefore, the search is on for the ways to induce cancer cell senescence and the hope is that this strategy may reap therapeutic benefits. Hence, understanding the pro-senescence pathways is essential for this search.

Replicative and Stress-Induced Premature Cell Senescence (SIPS)

Studies by Hayflick and Moorhead (1961) demonstrated that the number of cell cycles is limited and when a cell reaches this limit, it seizes to divide or perform differentiated functions and enters the arrested state during which it remains metabolically active and secrets an array of products collectively known as "senescence-associated secretory products" (SASP) that affect the neighboring cells. This scenario describing the exhaustion of cell cycling is referred to as "replicative senescence". Subsequent studies revealed the main mechanism of replicative senescence as being the attrition of telomeres that leads to genomic instability. This mechanism is most prevalent in aging. Aging leads to a gradual loss of normally functioning cell mass and the specialized cell functions.

Chronic diseases, similar to aging, cause a progressive substitution of proper tissue architecture with senescent, cell cycle-arrested, actively metabolizing and energy consuming cells, which become functionally deficient. In contrast to aging, however, this is due to the stress-induced premature senescence (SIPS), one of the causes of the developing functional deficit. Despite many similarities between replicative senescence and SIPS, as documented by the DNA screen detecting parallels in modified genes, including genes involved in the regulation of cell proliferation, defense, DNA damage, morphogenesis, extracellular matrix, and prostaglandin synthesis (Debacq-Chainiaux et al. 2008), the major distinction between the two lies in the fact that the replicative senescence is almost exclusively associated with the reduction of telomerase activity and attrition of telomeres, whereas the SIPS does not require these events, thus conferring potential reversibility onto this process.

Molecular Signatures of Premature Cell Senescence

Molecular mechanisms of senescence in general and SIPS in particular have been extensively reviewed (Mantel and Broxmeyer 2008; Goligorsky et al. 2009). In fibroblasts, secretion of IGFBP7 and expression of CXCR2 are necessary for induction of replicative and oncogenic senescence (Acosta et al. 2008; Wajapeyee et al. 2008). Sirtuin 1 inhibition by sirtinol or siRNA in endothelial cells results in increased PAI-1 and decreased eNOS expression and promotes SIPS, whereas the opposite effects occur upon overexpression of sirtuin 1 (Ota et al. 2007). In genetically engineered mice lacking sirtuin 1 in endothelial cells and developing endothelial SIPS, the abundance of myocardial and renal mRNA for sirtuins 3 and 6 is reduced (Maizel, unpublished observations).

A host of other pathways regulate SIPS. Decreased c-Myc expression induces senescence via a telomere-independent mechanism driven by a member of polycomb group histone methyltransferase, Bmi-1, and a member of gate-keeping tumor suppressor, p16^{INK4a}, in human fibroblasts and endothelial cells (Guney et al. 2006), and the same mechanism is involved in oxidative stress-induced premature senescence. It was reported (Guo et al. 2007) that a ring finger protein Mel-18 downregulates Bmi-1 resulting in accelerated senescence. This effect of Mel-18 is linked to downregulation of c-Myc. Sharpless and DePinho (2007) argue that senescence promoting pathways, like p16^{INK4a}, ARF-p53, or FOXO-reactive oxygen species, are repressed by the activity of Polycomb group proteins, thus preserving stem cells from developing senescence, but at the same time, limiting their ability to divide asymmetrically. These regulators are linked with the Forkhead box transcription factor FoxM1c, deficiency of which leads to premature cell senescence (Li et al. 2008): induction of FoxM1c expression inhibits p53, p19^{Arf} and p21 and induces expression of its downstream target, Bmi-1, resulting in retardation of senescence program. Elevated levels of the cyclin-dependent kinase inhibitor p16^{INK4a} represent the hallmark of both replicatively and prematurely senescent cells. In fact, genetically triggered elimination p16^{INK4a}-expressing cells delays the onset of aging-associated disorders (Baker et al. 2011). In addition, reduced levels of the Notch ligand Delta-like 1, which is necessary for stem cell activation, is among other characteristic findings in aged cells, as exemplified by muscle cells (Conboy and Rando 2002; Conboy et al. 2005). As opposed to the p16^{INK4a}-dependent cell-intrinsic (cell-autonomous) mechanism of senescence, the Delta-like 1-induced senescence represents an example of the cell-extrinsic, non-cell-autonomous mechanism. In this latter category of pro-senescence pathways, senescent cells reinforce and propagate signals, such as inflammatory cytokines, TGF-β, Dikkopff-1 antagonist of Wnt signaling, insulin, insulin-like growth factor-1 to engage increasing numbers of neighboring cells in cell cycle arrest and functional demise of the organ (Coppe et al. 2010).

Persistent activation of the Wnt pathway represents another mechanism of inducing premature or accelerated aging by enforcing uncontrollable cell cycling and leading to its exhaustion in the stem cell compartment (Liu et al. 2007). Activation of mTOR pathway by persistent cell proliferation serves as a mediator of Wnt/β-catenin pathway (Castilho et al. 2009), thus exhibiting a dualistic nature of this pathway: β-catenin-induced uncontrolled proliferation, which in turn activates mTOR and suppresses it. This paradigm represents another example of reciprocal relations between tumor growth and senescence. Notably, activation of mTOR pathway is downstream of insulin and insulin-like growth factor (IGF-1) signaling systems which contribute to premature senescence.

Wnt pathway is also chronically activated in a mutant mouse strain deficient in Klotho, a model of premature senescence. Alternative splicing of Klotho gene results in the synthesis of a membrane form of the protein (a co-receptor for FGF23, a regulator of mineral metabolism) and a secreted form, which regulates activities of insulin and IGF-1. Recent work identified the retinoic acid-inducible gene-1 (RIG-1) as a target suppressed by the intracellular form of Klotho. Moreover, RIG-1 is a potent inducer of interleukins 6 and 8 (IL-6 and IL-8); therefore defective Klotho signaling re-activates RIG-1 and enhances secretion of IL-6 and IL-8, both pro-inflammatory components of the senescence-associated secretory products (Liu et al. 2011).

Caveolin-1 plays a role in cell senescence. The study of oxidative stress-induced premature senescence in fibroblasts (Dasari et al. 2006) resulted in identification of oxidative stress response elements of the mouse caveolin-1 promotor to the sequences −244/−222 and −124/−101, both within the boundaries of Sp1 binding. Mitogen-activated protein kinase p38 is an upstream regulator of Sp1-mediated activation of the caveolin-1 promotor and its inhibition prevents caveolin-1 upregulation and counteracts development of SIPS. Notably, the combination of elevated expression of caveolin-1 and activated AKT/mTOR pathway represents a powerful predictor of cancer progression and vascularization, as has been demonstrated for renal cell carcinoma (Campbell et al. 2008).

The role of miR-22, an inducer of cell senescence, in repressing cancer progression (Xu et al. 2011) has recently been identified. miR-22 is overexpressed in senescent cells, but downregulated in cancer cell lines and its pharmacological overexpression leads to induction of senescence. Moreover, synthetic miR-22 suppresses tumor growth and metastasis in vivo by facilitating cell senescence. The latter is also associated with an immotile phenotype and reduced metastatic disease in a mouse model of breast carcinoma. It is not clear at this time whether all these pathways inducing SIPS are mechanistically integrated at some higher level or whether they represent separate individual routes eventuating in producing a common state of premature senescence.

Lysosomal Dysfunction Heralds Sirtuin Depletion and SIPS

One of the goals of cancer therapy is to reduce the blood supply to tumor cells. Premature senescence of endothelial cells represents, therefore, one of potential targets. Our studies into SIPS of endothelial cells illustrate several paradigms of this process. Endothelial cell dysfunction is induced by diverse "cardiovascular risk factors" and follows a relatively standard sequence of transformations. Dysfunctional endothelial cells exhibit downregulation of key mitochondrial enzymes – enoyl-coA-hydratase and aconitase-2, responsible for the entry and processing, respectively, of substrates in the Krebs cycle – thus leading to a switch from the oxidative metabolism toward the normoxic glycolysis, the Warburg-type metabolism (Addabbo et al. 2009). In fact, by-passing this enzymatic bottleneck by providing cells with the downstream metabolite, α-ketoglutarate, rescues the dysfunctional endothelial cells. This Warburg-type metabolic switch is followed by cell cycle arrest, reversible at the early stages, if the initiating factors are removed or counteracted, but becoming irreversible after subversion of autophagy leading to the subsequent apoptotic cell death and microvascular rarefaction. These processes, together with the ongoing endothelial-to-mesenchymal transition, lead to obliteration and ablation of vascular beds, rendering tissue chronically hypoxic. The switch from cell senescence to apoptosis may be governed in part by developing lysosomal dysfunction, as detailed below.

Lysosomal dysfunction and resulting impaired autophagy are associated with and causally contribute to aging. In SIPS, oxidative stress leads to lysosomal permeabilization. Patschan et al. (2007, 2008) demonstrated that an AGE-modified long-lived protein, collagen I, leads in endothelial cells to a surge of ROS production, collapse of lysosomal pH gradient within 30–60 min, concomitant lysosomal permeabilization, and eventual subversion of autophagy manifesting in non-fusion of autophagosomes and lysosomes and/or accumulation of giant autophagolysosomes, perhaps, as a result of a defective digestion of their cargo. Furthermore, Chen et al. (2012) showed that (a) pretreatment of endothelial cells with an antioxidant reduces lysosomal permeabilization and SIPS, (b) pretreatment with a broad inhibitor of cathepsins prevents SIPS; and (c) several cathepsins can directly cleave SIRT1. This pathway for SIPS is depicted in Fig. 13.1.

Relevance of Sirtuins to Cancer

The role of sirtuins in multitude of cell functions has been comprehensively reviewed (Haigis and Sinclair 2010). In cancer cells, the role of SIRT1, SIRT2, SIRT3, SIRT6 and SIRT7 has been suggested. SIRT 1,2 and 3 appear to act as tumor suppressors (Bruzzoni et al. 2013) and mice with deficiency of these sirtuins are predisposed to oncogenic transformation. The proposed mechanisms of tumor suppression by sirtuins include deacetylation of RelA/p65 subunit of NF-kB, deacetylation of β-catenin (SIRT1); tubulin deacetylation (SIRT2); regulation

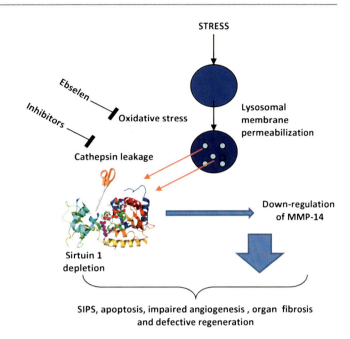

Fig. 13.1 A schematic depiction of the sequence of events triggered by cell stress and leading to, among other targets, production of reactive oxygen species, permeabilization of lysosomal membrane, and local release of cathepsins which cleave sirtuins. Developing sirtuin 1 deficiency in turn suppresses the expression of the membrane type 1 matrix metalloproteinase (MMP-14), thus resulting in the accumulation of the extracellular matrix and development of fibrosis

of mitochondrial function (SIRT3); negative regulation of NF-kB and HIF-1α (SIRT6). Paradoxically, the same members of sirtuin family under certain conditions can facilitate oncogenesis. SIRT1 and possibly SIRT2 accomplishes this by deacetylation and inactivation of p53 and FOXO transcription factors; SIRT3 may exert pro-oncogenic activity in certain squamous cell carcinomas; whereas SIRT6, which is responsible for the maintenance of telomeres, DNA repair and genome stability, may be involved by increasing the threshold for tumor cell injury. The above actions of sirtuins could increase resistance of cancer cells to stress and suppress senescence and apoptotic programs. SIRT7 has been suggested to promote oncogenic transformation by acting on histone 3 and consequently repressing tumor suppressor genes (Barber et al. 2012). The existing dichotomy has been well-illustrated for sirtuin-1 actions as a tumor promoter and suppressor (Fang and Nicholl 2011) (Fig. 13.2). Collectively, these findings justify the search for specific and selective sirtuin inhibitors and activators.

There is strong clinical evidence for the role of sirtuins in carcinogenesis. SIRT1 expression is elevated in gastric cardiac carcinoma, as examined using tissue microarray technique and immunohistochemical stains in a large cohort of patient samples (Feng et al. 2011). Similarly, expression of SIRT1 is significantly elevated in 56% of the hepatocellular carcinoma tissues compared to non-tumor tissues (Choi et al. 2011. Silencing SIRT1 results in cell growth arrest in hepatocellular carcinomatous cells, thus suggesting an association of SIRT1 expression with hepatocellular development and cancer cell growth. Observations by Chen et al. (2011) are also consistent with these findings and demonstrate a link between SIRT1 expression and genomic stability. The dualism of sirtuins action in tumors has been summarized (Fang and Nicholl 2011). Sirtuins are also implicated in drug resistance of tumors. Olmos et al. (2011) argue that SIRT1 is overexpressed in many drug-resistant cancers. The role of SIRTs in drug resistance may be related to their ability to target and modulate the activity of tumour suppressors, including p53, p73, E2F1, and FOXO3a. While in normal cells SIRT-dependent deacetylation of transcription factors is used to fine-tune gene expression, in cancer cells it helps to evade proliferative arrest and cell death in response to chemotherapy.

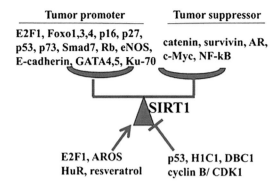

Fig. 13.2 Dual role of sirtuin 1 in cancer. Sirtuin 1 has a potential to activate several tumor promoters and tumor suppressors, depending on the biological content and type of the tumor (Reprinted with permission of Cancer Letters from the reference Fang and Nicholl 2011)

A novel sirtuin-dependent pathway regulating cellular lifespan has recently been proposed (Dang et al. 2009). In the yeast, Sir2 deacetylase is responsible for chromatin silencing by preserving H4 lysine 16 in deacetylated form. In aged cells, Sir2 becomes depleted leading to increased acetylation of H4 and loss of histones at subtelomeric regions. This results in compromised transcriptional silencing at these loci. . A histone acetyltransferase Sas2 antagonizes effects of Sir2 on this process, and reciprocal relations between Sir 2 and Sas2 determine cellular lifespan. It remains to be examined whether similar processes take place in mammalian cells, and whether SIRT1-Sas2 relations are relevant to tumor cell evasion of senescence.

Senescence Induces a Sessile Cell Phenotype and Encapsulation of These Cells by ECM: Role of Collagen XVIII and Endostatin

Invasiveness of tumor cells depends in part on remodeling of the extracellular matrix by specific metalloproteinases, among which the membrane type 1 enzyme, MMP-14 (MT1-MMP), plays a key role. The process of cancer cell invasion of the extracellular matrix has recently been monitored in great detail (Wolf et al. 2007). Coordination of mechanotransduction and collagen remodeling are accomplished by force generation at the leading edge and proteolysis at the trailing edge of fibrosarcoma and breast cancer cells. At the leading edge, the presence of β1-integrins and MMP-14 insures the proper adhesion and re-alignment of collagen fibers along the direction of cell migration. Subsequently migrating cells utilize the same tracks as did the leaders, thus turning them into highways for a large group of invading cells.

It is interesting that a relatively modest number of cells that do develop SIPS is capable of affecting tissue functions. Krtoloca et al. (2001) have shown that co-culture of SIPS fibroblasts with tumor epithelial cells accelerates growth of the latter. This phenomenon cannot be reproduced using normal fibroblasts. Furthermore, a similar phenomenon has been demonstrated in hepatocytes (Zhang and Cuervo 2008). Such a disseminated effect of SIPS cells is explained by generation of secretory signals by dysfunctional cells, thus affecting their neighbors. These senescence-associated secretory products (SASP) are represented by TGF-alpha, galectin-3, IGFBP-3,-4, and -6, MIC-1 (Suzuki and Boothman 2008). In dysfunctional senescent endothelial cells, one of the detectable messages is represented by collagen XVIII and its C-terminal antiangiogenic fragment, endostatin (O'Riordan et al. 2007). Gene microarray analysis of cultured dysfunctional endothelial cells (treatment with an inhibitor of nitric oxide synthase, NOS) revealed upregulation of collagen XVIII, the finding confirmed in vivo in mice chronically treated with NOS inhibitor. Enhanced generation of endostatin in these animals led to the development of endothelial-mesenchymal transition and eventual rarefaction of renal microvasculature, thus further compounding vascular and parenchymal pathology. Endothelial cells exposed to diverse stressors respond with lysosomal dysfunction, leakage of cathepsins and degradation of SIRT1. In turn, SIRT1 depletion leads to down-regulation of MMP-14. This mechanism results in accumulation of ECM and encapsulation of cells, making them sessile, and potentially retarding tumor cell invasiveness and metastatic behavior.

Future Perspectives of Employing SIPS for Cancer Therapies

As briefly summarized above, mechanistic studies of SIPS have uncovered the diversity of pathways leading to cell cycle arrest and promoting cellular senescence, reduced resistance to stressors, and death. Many of these pathways, however, have a large footprint that makes them operational in many cells and tissues, therefore, systemic use of the gained knowledge on induction of SIPS would be predicted to have multiple untoward effects. Under the circumstances, the old problem of targeted drug delivery to tumor cells resurfaces again to optimize any translational findings in the field of SIPS to efficient therapy of tumor cells.

Assuming that this hurdle will be removed by successful targeting of tumor cells, several rational strategies to induce premature senescence in tumor compartment can be envisaged, as bulleted below.

1. Overexpression of the ring finger protein Mel-18, which downregulates BMI-1 and c-Myc.
2. Downregulation of Polycomb group proteins to induce p16INK4a locus, p53 and FOXO
3. Induction of the Forkhead transcription factor FoxM1c
4. Downregulation of Klotho or upregulation of RIG-1
5. Local delivery or induction of miR-22
6. Downregulation of sirtuins expression or their activity.

Each of these strategies should theoretically accelerate tumor cell senescence and could have therapeutic applicability. As to the integration of such approaches into the existing therapies, it would make sense to use them for the intermittent pulse therapy interspersed with the pulses of anti-angiogenic therapy. Such a "zebra" chronological approach could potentially mutually amplify each: therapeutic premature senescence predisposing tumor and vascular cells to stress-intolerance and therapeutic anti-angiogenesis facilitating the state of dormancy of cell cycle-arrested cells.

References

Acosta J, O'Loghlen A, Banito A, Guijarro M, Augert A, Raguz S, Fumagalli M, DaCosta M, Brown C, Popov N (2008) Chemokine signaling via the CXCR2 receptor reinforces senescence. Cell 133:1006–1018

Addabbo F, Ratliff B, Park HC, Kuo MC, Ungvari Z, Csiszar A, Krasnikov B, Sodhi K, Zhang F, Nasjletti A, Goligorsky MS (2009) The Krebs cycle and mitochondrial mass are early victims of endothelial dysfunction: proteomic approach. Am J Pathol 174(1):34–43

Baker D, Wijshake T, Tchkonia T, LeBrasseur N, Childs B, van de Sluis B, Kirkland J, van Deursen J (2011) Clearance of p16Ink4a-positive senescent cells delays ageing-associated disorders. Nature 479:232–236

Barber M, Michishita-Kioi E, Xi Y et al (2012) SIRT7 links H3K18 deacetylation to maintenance of oncogenic transformation. Nature 487(7405):114–118

Bruzzoni S, Parenti M, Grozio A, Ballestrero A, Bauer I, Del Rio A, Nencioni A (2013) Rejuvenating sirtuins: the rise of a new family of cancer drug targets. Curr Pharmacol Des 19(4):614–623

Campbell L, Jasani B, Edwards K, Gumbleton M, Griffith D (2008) Combined expression of caveolin-1 and an activated AKT/mTOR pathway predicts reduced disease-free survival in clinically confined renal cell carcinoma. Br J Cancer 98:931–940

Castilho R, Squarize C, Chodosh L, Williams B, Gutkind S (2009) mTOR mediates Wnt-induced epidermal stem cell exhaustion and aging. Cell Stem Cell 5:279–289

Chen J, Zhang B, Wong N, Lo AW, To KF, Chan AW, Ng MH, Ho CY, Cheng SH, Lai PB, Yu J, Ng HK, Ling MT, Huang AL, Cai XF, Ko BC (2011) Sirtuin 1 is upregulated in a subset of hepatocellular carcinomas where it is essential for telomere maintenance and tumor cell growth. Cancer Res 71(12):4138–4149

Chen J, Xavier S, Moskowitz-Kassai E, Chen R, Lu CY, Sanduski K, Špes A, Turk B, Goligorsky MS (2012) Cathepsin cleavage of sirtuin 1 in endothelial progenitor cells mediates stress-induced premature senescence. Am J Pathol 180:973–983

Choi HN, Bae JS, Jamiyandorj U, Noh SJ, Park HS, Jang KY, Chung MJ, Kang MJ, Lee DG, Moon WS (2011) Expression and role of SIRT1 in hepatocellular carcinoma. Oncol Rep 26(2):503–510

Conboy I, Rando T (2002) The regulation of Notch signaling controls satellite cell activation and cell fate determination in postnatal myogenesis. Dev Cell 3:397–409

Conboy I, Conboy M, Wagers A, Girma E, Weissman I, Rando T (2005) Rejuvenation of aged progenitor cells by exposure to a young systemic environment. Nature 433:760–764

Coppe J, Desprez P, Krtolica A, Campisi J (2010) The senescence-associated secretory phenotype: the dark side of tumor suppression. Annu Rev Pathol 5:99–118

Dang W, Stefen K, Perry R, Dorsey J, Johnson B, Shilatifard A, Kaeberlein M, Kennedy B, Berger S (2009) Histone H4 lysine 16 acetylation regulates cellular lifespan. Nature 459:802–807

Dasari A, Bartholomew J, Volonte D, Galbiati F (2006) Oxidative stress induces premature senescence by stimulating caveolin-1 gene transcription through p38 mitogen-activated protein kinase/Sp1-mediated activation of two GC-rich promoter elements. Cancer Res 66:10805–10814

Debacq-Chainiaux F, Pascal T, Boilan E, Bastin C, Bauwens E, Toussaint O (2008) Screening of senescence-associated genes with specific DNA arrays reveals the role of IGFBP-3 in premature senescence of human diploid fibroblasts. Free Radic Biol Med 44:1817–1832

Fang Y, Nicholl M (2011) Sirtuin 1 in malignant transformation: friend or foe? Cancer Lett 306:10–14

Feng AN, Zhang LH, Fan XS, Huang Q, Ye Q, Wu HY, Yang J (2011) Expression of SIRT1 in gastric cardiac cancer and its clinicopathologic significance. Int J Surg Pathol 19(6):743–750

Goligorsky MS, Chen J, Patschan S (2009) Stress-induced premature senescence of endothelial cells: a perilous state between recovery and point of no return. Curr Opin Hematol 16(3):215–219

Guney I, Wu S, Sedivy J (2006) Reduced c-Myc signaling triggers telomere-independent senescence by regulating Bmi-1 and p16(INK4a). Proc Natl Acad Sci U S A 103:3645–3650

Guo W, Datta S, Band V, Dimri G (2007) Mel-18, a polycomb group protein, regulates cell proliferation and senescence via transcriptional repression of Bmi-1 and c-Myc oncoproteins. Mol Biol Cell 18:536–546

Haigis MC, Sinclair DA (2010) Mammalian sirtuins: biological insights and disease relevance. Annu Rev Pathol 5:253–295

Hayflick L, Moorhead P (1961) The serial cultivation of human diploid cell strains. Exp Cell Res 25:585–621

Krtoloca A, Parrinello S, Lockett S, Desprez P, Campisi J (2001) Senescent fibroblasts promote epithelial cell growth and tumorigenesis: a link between cancer and aging. Proc Natl Acad Sci U S A 98:12072–12077

Li S, Smith D, Leung W, Cheung A, Lam E, Dimri G, Yao K (2008) FoxM1c counteracts oxidative stress-induced senescence and stimulates Bmi-1 expression. J Biol Chem 283:16545–16553

Liu H, Fergusson M, Castilho R, Liu J et al (2007) Augmented Wnt signaling in a mammalian model of accelerated aging. Science 317:803–806

Liu F, Wu S, Ren H, Gu J (2011) Klotho suppresses RIG-1-mediated senescence-associated inflammation. Nat Cell Biol 13:254–262

Mantel C, Broxmeyer H (2008) Sirtuin 1, stem cells, aging, and stem cell aging. Curr Opin Hematol 15:326–331

Olmos Y, Brosens JJ, Lam EW (2011) Interplay between SIRT proteins and tumour suppressor transcription factors in chemotherapeutic resistance of cancer. Drug Resist Updat 14(1):35–44

O'Riordan E, Mendelev N, Patschan S, Chander P, Goligorsky MS (2007) Chronic NOS inhibition actuates endothelial-mesenchymal transformation. Am J Physiol 292:H285–H294

Ota H, Akishita M, Eto M, Iijima K, Kaneki M, Ouchi Y (2007) Sirt1 modulates premature senescence-like phenotype in human endothelial cells. J Mol Cell Cardiol 43:571–579

Patschan S, Chen J, Gealekman O, Krupincza K, Wang M, Shu L, Shayman JA, Goligorsky MS (2007) Mechanisms of premature cell senescence: lysosomal dysfunction and ganglioside accumulation in endothelial cells. Am J Physiol Renal 294:100–109

Patschan S, Chen J, Polotskaia A, Mendelev N, Cheng J, Patschan D, Goligorsky MS (2008) Lipid mediators of autophagy in stress-induced premature senescence of endothelial cells. Am J Physiol Heart 294:H1119–H1129

Sharpless N, DePinho R (2007) How stem cells age and why this makes us grow old. Nat Rev Mol Cell Biol 8:703–713

Suzuki M, Boothman D (2008) Stress-induced premature senescence (SIPS) – influence of SIPS on radiotherapy. J Radiat Res 49:105–112

Swarbrick A, Roy E, Allen T, Bishop M (2008) Id1 cooperates with oncogenic Ras to induce metastatic mammary carcinoma by subversion of the cellular senescence response. Proc Natl Acad Sci U S A 105:5402–5407

Wajapeyee N, Serra R, Zhu X, Mahalingam M, Green M (2008) Oncogenic BRAF induces senescence and apoptosis through pathways mediated by the secreted protein IGFBP7. Cell 133:363–374

Wolf K, Wu Y, Liu Y, Geiger J, Tam E, Overall C, Stack M, Friedl P (2007) Multi-step pericellular proteolysis controls the transition from individual to collective cancer cell invasion. Nat Cell Biol 9:893–904

Xu D, Takeshita F, Hino Y, Fukunaga S, Kudo Y, Tamaki A, Matsunaga J, Takahashi R, Takata T, Shimamoto A, Ochiya T, Tahara H (2011) miR-22 represses cancer progression by inducing cellular senescence. J Cell Biol 193:409–424

Zhang C, Cuervo AM (2008) Restoration of chaperone-mediated autophagy in aging liver improves cellular maintenance and hepatic function. Nat Med 14(9):959–965

Part II
Tumor and Cancer

Nuclear Protein Pirin Negates the Cellular Senescence Barrier Against Cancer Development

14

Silvia Licciulli and Myriam Alcalay

Contents

Abstract	131
Introduction	132
Pirin: Structure and Function	132
PIR in Cancer	133
Melanoma and the Senescence Barrier	134
Alteration of Pirin Expression in Melanoma	135
Pirin Controls the Senescence Response in Melanoma	136
Discussion	139
References	141

S. Licciulli (✉)
Kissil Lab, Department of Cancer Biology,
The Scripps Research Institute, 130 Scripps Way,
Jupiter, FL, USA
e-mail: SLicciul@scripps.edu

M. Alcalay
Department of Experimental Oncologia, Istituto
Europeo di Oncologia, Via Adamello 16, 20139,
Milan, Italy
e-mail: myriam.alcalay@ifrom-ieo-campus.it

Abstract

Pirin (PIR) is a highly conserved protein whose biological role has not yet been fully elucidated. Several studies reported its involvement in cancer progression, proposing a function in apoptosis. We have shown that PIR is primarily expressed in melanocytes and melanoma cells and displays a complex pattern of expression and localization. High levels of PIR protein are found in normal melanocytes whereas low or undetectable levels are present in nevi. Additionally, PIR expression is found in a subset of melanoma cases with increasing levels correlating with tumor progression.

Knock-down experiments performed in melanoma cells with high PIR expression have shown a role for PIR in controlling cellular senescence. In this model, PIR ablation results in impairment of cell proliferation, morphological changes characteristic of cellular senescence and expression of senescence markers. Furthermore, oncogene activation and other senescence stimuli induce PIR downregulation.

Based on our data, we propose here two alternative models to explain PIR expression pattern in nevi and melanoma and its involvement in the control of senescence. We propose that PIR plays a prominent role in negatively controlling senescence in melanocytic cells and that it could represent a novel marker for melanoma progression and a potential therapeutic target.

> **Keywords**
>
> Acute myeloid leukemia (AML) • Cancer • Gene expression studies • Melanoma • Molecular pathways • Pirin (PIR) protein • Prokaryotic orthologs • Ras-Raf-MEK-ERK signalling pathway • Senescence barrier

Introduction

Pirin: Structure and Function

Pirin (PIR) protein was first described in 1997 as the predominant interactor of the replication/transcription factor NFI/CTF1 (Wendler et al. 1997). The original study also analysed the expression pattern of *PIR* mRNA in human tissues and found ubiquitously low expression with higher transcript levels in heart and liver.

Human PIR protein is composed of 290 amino acids with a predicted molecular mass of 32 kDa. It has been described as a nuclear protein by means of different experimental approaches (Wendler et al. 1997).

PIR protein is highly conserved throughout evolution. It is present in mammals, plants, fungi and even prokaryotic organisms. Computer alignments demonstrated significant homologies of 29 residues at the N-terminus of PIR with putative proteins from all aligned species, with the sole exception of the yeast *Saccharomyces cerevisiae* (Wendler et al. 1997).

The biochemical and structural properties of PIR protein have been extensively described. PIR was assigned to the cupin superfamily of proteins (Pang et al. 2004), which is among the most functionally diverse of any described to date, comprising both enzymatic and non-enzymatic members. The crystal structure of human PIR has been resolved and has suggested an enzymatic activity involved in biological redox reactions (Pang et al. 2004; Zeng et al. 2003), confirmed by comparison with a putative PIR homologue from Escherichia coli (Adams and Jia 2005). Both bacterial and human PIR have quercetinase activity and can catalyze the quercetin 2,3-dioxygenase reaction releasing carbon monoxide as a product. Quercetin is a naturally occurring, ubiquitous flavonoid, involved in the regulation of many cellular processes (Adams and Jia 2005). In another study, the PIR homologue of the bacterium *Serratia marcescens* was shown to regulate pyruvate catabolism by interacting with the pyruvate dehydrogenase E1 subunit and modulating its activity (Soo et al. 2007). Therefore, PIR possesses one or more enzymatic activities in different eukaryotic and prokaryotic cells, suggesting that the protein has pleiotropic functions depending on the organism, cell type, or functional status of the cell.

The first report on PIR biological function came from the tomato homologue *Le-pirin*, which shares 56% homology with human PIR. *Le-pirin* mRNA levels dramatically increase during camptothecin-induced Programmed Cell Death (PCD) (Orzaez et al. 2001), which is a physiological plant process similar to animal apoptosis, employed for selective removal of cells that are no longer needed or potentially dangerous for the plant organism. The authors also found that yellowing leaves, which are in an advanced stage of senescence, display the highest levels of *Le-pirin*.

PIR function has been linked to apoptosis also in human cells. *PIR* mRNA was found up-regulated in bronchial epithelial cells of chronic cigarette smokers and this condition was specifically associated with apoptosis (Gelbman et al. 2007). Therefore, PIR up-regulation could represent one mechanism associated with the disruption of the airway epithelial barrier by cigarette smoke.

Prokaryotic orthologs of PIR have been correlated to stress response. Specifically, two adjacent genes encoding for PIR orthologs in the cyanobacterium *Synechocystis* sp. PCC 6803, *pirA* and *pirB*, are highly induced under high salinity and other stress conditions (Hihara et al. 2004). Notably, induction of the *pirAB* genes was not related to cell death in this organism, reinforcing the idea that PIR function may vary in different species and in different contexts.

Molecular Pathways Involving PIR

Different studies have proposed the involvement of PIR in transcriptional processes, based on its interaction with transcription regulators. Nuclear factor I (NFI), which was the first PIR interactor identified (Wendler et al. 1997), consists of a family of sequence-specific DNA binding proteins whose cellular function has not yet been completely elucidated (Wendler et al. 1997). NFI/CTF1 might be involved in DNA replication and it can stimulate RNA-polymerase II-driven transcription. A number of cellular promoters contain NFI/CTF1 binding sites, and it has been shown that NFI/CTF1 can modulate transcription of the corresponding genes (Santoro et al. 1988). As an interactor of a broad acting nuclear factor, PIR itself could be involved in several different molecular processes.

Another report described interaction of PIR with the proto-oncoprotein Bcl-3 (Dechend et al. 1999). Bcl-3 is a member of the IkB multiprotein family, which modulates the activities of NF-kB/Rel transcription factors (Baldwin 1996). The reported interaction with PIR gives rise to a quaternary complex including the p50-NF-kB transcription factor, Bcl-3 and DNA, and can strongly increase the amount of DNA-bound p50-Bcl-3 complex (Dechend et al. 1999). The biological consequences of the interaction between PIR and Bcl-3 are difficult to predict, since Bcl-3 can either activate NF-kB target gene transcription or dissociate the transcription factor from DNA to inhibit gene expression. Therefore, the effects of PIR binding on the pathway activation could be elucidated only by identifying the target genes to which the Bcl-3-NF-kB-PIR complex binds.

PIR in Cancer

Different *in vitro* studies based on proteomic and genomic approaches have suggested a role for PIR in transformation and metastasis. Expression levels of PIR were correlated with metastatic potential in tumor cell lines of Adenoid cystic carcinoma (ACC) (An et al. 2004). Based on the interaction of PIR with Bcl-3 (Dechend et al. 1999), the proposed mechanism is that PIR mediates the anti-apoptotic effects of NF-kappaB activation.

Gene expression studies conducted on transformed cell lines treated with different antineoplastic compounds revealed opposite regulation of *PIR* expression, suggesting that PIR may participate to more than one cellular pathway depending on the cell-type and functional context. An analysis of the gene expression profile of neuroglioma cells treated with curcumin (diferuloyl methane), which can induce antiproliferative effects, showed that the biological pathways mostly affected by the treatment were oxidative stress response and cell cycle/apoptosis. Coherently with its previously described function in apoptosis, *PIR* gene was up-regulated by curcumin treatment and was classified and discussed by the authors as an apoptosis gene (Panchal et al. 2008).

Another gene expression study identified PIR as one of the most strongly regulated genes after treatment with Polysaccharide-K (PSK) (Yoshikawa et al. 2004), a chemo-immunotherapeutic agent that prevents distant metastases in colorectal cancer (Mitomi et al. 1992). *PIR* expression resulted significantly repressed in two human colorectal adenocarcinoma cell lines independent of p53 mutational status. This result, in apparent contrast with the functions thus far associated to PIR, may suggest a different role in metastasis-associated functions, such as cellular adhesion or migration.

We have shown that PIR expression is significantly repressed in a large proportion of acute myeloid leukemia (AML) cases, regardless of subtype or genetic abnormalities. Since AML is characterized by block of differentiation and accumulation of myeloid precursors, we investigated the role of PIR in this process and found that PIR expression increases during in vitro myeloid differentiation of primary mouse hematopoietic precursor cells. Furthermore, ablation of PIR in the U937 myelomonocytic cell line or in murine primary hematopoietic precursors resulted in impairment of terminal myeloid differentiation. Gene expression profiling of U937 cells after knockdown of PIR revealed increased expression of genes associated with the early phases of hematopoiesis, in particular homeobox A (HOXA) genes. Our results suggest that PIR is required for terminal myeloid maturation and that PIR downregulation may contribute to the differentiation arrest associated with AML (Licciulli et al. 2010a).

Melanoma and the Senescence Barrier

Melanoma Overview

Melanoma constitutes the most dangerous type of skin cancer. It arises from the malignant transformation of melanocytes when they escape from the tight regulation provided by surrounding keratinocytes, proliferate and spread, leading to formation of a naevus. This precursor lesion can progress to *in situ* melanoma, which grows laterally and remains confined to the epidermis, so that this stage is defined as the radial-growth phase (RGP). RGP melanoma can progress to vertical growth-phase (VGP), a more aggressive stage that can invade both the upper layer of the epidermis and the dermis and subcutaneous tissue after crossing the basement membrane. Finally, local and distant metastasis represent the most advanced step of tumour progression. Not all melanomas evolve through each of these stages, but it is believed that the transition from RGP to VGP is the crucial step to the acquisition of metastatic potential and poor clinical outcome.

The Ras-Raf-MEK-ERK signalling pathway is the most frequently altered in melanoma and constitutes an obligatory step in the initiating stages of melanocytes transformation. The constitutive activation of the pathway is able to stimulate proliferation and survival and to provide essential functions for tumour growth and maintenance (Gray-Schopfer et al. 2005).

Activation of the pathway by binding of extracellular ligands to their cell-surface receptors initiates a cascade of intracellular events, eventually leading to phosphorylation of mitogen-activated protein kinase (MAPK) and extracellular signal-regulated kinase (ERK). The Ras-Raf-MEK-ERK pathway mediates diverse biological functions such as cell growth, survival and differentiation, predominantly influencing regulation of transcription, metabolism and cytoskeletal rearrangements (Wellbrock et al. 2004).

The most commonly altered component of the pathway is *BRAF*, a serine-threonine-specific protein kinase that activates ERK and is mutated in 50–70% of melanomas. One specific mutation, a glutamic acid to valine substitution at position 600 (V600E), accounts for more than 90% of all *BRAF* mutations in melanoma (Davies et al. 2002). Interestingly, *BRAF*^{V600E} is also found with high frequency in benign and dysplastic nevi, suggesting that it represents an early event in melanoma progression (Pollock et al. 2003).

Senescence Overview

Normal somatic cells have a restricted proliferative potential and, after a definite number of cell divisions, they exit from the cell cycle and enter a state of irreversible growth arrest known as cellular senescence (Mathon and Lloyd 2001). Senescent cells undergo morphologic, biochemical and chromatin changes that can be used to identify the senescence state: cells become enlarged, flattened and enriched in cytoplasmic vacuoles besides displaying senescence-associated β-galactosidase (SA-β-gal) reactivity as a blue perinuclear staining. The inactive proliferative status reflects in lack of BrdU incorporation. Furthermore, senescent cells accumulate senescence-associated heterochromatic foci (SAHF) that are responsible for epigenetic silencing of cell cycle genes, such as E2F target genes (Narita et al. 2003).

The best-known form of senescence is called replicative senescence and depends on the progressive shortening of telomeres that activates p53 and p21 (Campisi and d'Adda di Fagagna 2007). Other senescence-like phenotypes can occur in the absence of telomere attrition, as a response to a variety of stimuli such as activation of oncogenes (Oncogene-Induced Senescence, OIS), oxidative stress, DNA damage, and drug treatment. Premature senescence is morphologically indistinguishable from replicative senescence but cannot be bypassed by overexpression of telomerase.

The senescence phenomenon has been extensively studied *in vitro* for decades; more recently, several groups have also reported on the presence of senescent cells *in vivo* at premalignant stages of tumor development in mouse tumor models and in lesions from human patient samples, supporting the idea that senescence might constitute a barrier mechanism against tumor development. According to this model, in order for tumor cells to become malignant, they need to bypass or escape senescence by establishing and selecting for events that counteract the barrier and rescue proliferation, such as mutations or epigenetic silencing of DNA damage-response and/or cell cycle checkpoint genes.

Senescence induction and maintenance are primarily mediated by the Arf/p53/p21 and p16/pRb tumor suppressor pathways. p53 and Rb proteins are considered the central players in the induction of cellular senescence: in response to various stimuli, p53 protein is stabilized and proceeds to activate its transcriptional targets, such as $p21^{CIP1/WAF1}$ (Kulju and Lehman 1995); activated (dephosphorylated) pRb binds to E2F-family transcription factors to repress their transcriptional targets, thus inhibiting cell cycle progression (Narita and Lowe 2004). The relative importance of p53 and Rb pathways in the control of senescence is cell-type dependent. A pivotal role in the control of senescence dependent on p53 and Rb pathways is played by the *CDKN2* locus (in mouse *Ink4-Arf*), which encodes for the tumor suppressors CDKN2a (p16), CDKN2b (p15) and CDKN2d ($p14^{Arf}$) (Kim and Sharpless 2006).

Melanoma Is an In Vivo Model of the Senescence Barrier to Tumor Progression

Melanocytic nevi of mice carrying endogenous BrafV600E (Dhomen et al. 2009) and human nevi (Michaloglou et al. 2005; Gray-Schopfer et al. 2006) were among the first *in vivo* reports of senescence.

Melanocytic nevi represent an intriguing borderline situation in which an activated oncogene can co-exist with long term-arrest of proliferation. In fact, even if nevi frequently harbour the V600E *BRAF* mutation (Pollock et al. 2003), after an initial phase of growth, proliferation is halted (Bennett 2003). Following this observation, it was demonstrated that the growth arrest of nevi results from oncogene-induced senescence counteracting $BRAF^{V600E}$-mediated oncogenic signalling (Michaloglou et al. 2005). Furthermore, $BRAF^{V600E}$ was found to induce p16 expression and senescence in primary human melanocytes *in vitro* (Gray-Schopfer et al. 2006; Michaloglou et al. 2005), and p16 defects are found in dysplastic but not in benign nevi (Papp et al. 2003). $p16^{INK4a}$ is inactivated in 30–70% of human melanomas (Sharpless and Chin 2003) and disruption of the p16/Rb pathway is required for immortalization (Gray-Schopfer et al. 2006).

p21 and p53 upregulation are not common events in nevi (Michaloglou et al. 2005), suggesting that p16 is the major regulator of BRAF-induced senescence in melanocytes. However, both *in vitro* and *in vivo*, growth-arrested melanocytes display a mosaic pattern of p16 expression, despite a homogeneus pattern of senescence-associated markers, which may reflect that senescence can also be induced by p16-independent pathways in melanocytes (Michaloglou et al. 2005).

Although *TP53* mutation rate in melanoma is considerably lower than in other tumour types and inactivation of p53 is not a critical event in the progression of melanoma, other genes of the p53 pathway may be involved in the induction of p16-independent senescence. Much attention has been focused on $p14^{Arf}$: since most of the genetic events leading to inactivation of *$p16^{INK4a}$* also abolish the coding sequence of *$p14^{ARF}$*, one explanation for the lack of *TP53* mutations could be that p53 function is already abrogated due to inactivation of $p14^{ARF}$ (Dahl and Guldberg 2007).

Alteration of Pirin Expression in Melanoma

PIR expression was thoroughly studied in multiple normal and transformed human tissues and cell lines both at the messenger and at the protein level (Licciulli et al. 2011). In accordance with previous data (Wendler et al. 1997), PIR levels were found variably low in most of the samples analysed, with the sole exception of nevi, primary and metastatic melanomas.

An in-depth analysis of a melanoma-specific tissue microarray (TMA) showed a complex and diverse pattern of PIR expression at different stages of melanoma progression: moderate and high PIR levels were detected in a small fraction of nevi, in approximately one third of primary melanomas and almost half of metastatic melanomas and a significant correlation was observed between PIR staining and disease progression from nevus to metastatic melanoma. Furthermore, PIR expression significantly increases with the thickness of primary lesions (Breslow index) and with Clark staging, which indicates the level of skin invasion (Licciulli et al. 2011). Scattered normal melanocytes present in the basal layer of normal skin areas showed intense PIR staining. Taken together, these results indicate that PIR is expressed at high levels in melanocytes, virtually absent in mature nevi and highly expressed in a subset of primary and metastatic melanomas.

Adding a further level of complexity to the picture, PIR subcellular localization is not homogeneous across all samples. PIR has been described as a nuclear protein, predominantly localized within sub-nuclear dot-like structures (Wendler et al. 1997). Concordantly, normal melanocytes from healthy tissue sections showed PIR expression exclusively in the nucleus, whereas melanoma samples positive for PIR staining showed a varied pattern of protein localization, ranging from nuclear to prevalently cytoplasmic (Licciulli et al. 2010b). Interestingly, cytoplasmic PIR appeared to correlate with melanoma progression: the percentage of cases with cytoplasmic localization increases in the transition from Radial Growth Phase (RGP) melanoma, that represents an early step of tumor progression, to more infiltrating Vertical Growth Phase (VGP) melanoma and to metastatic melanoma, where virtually no PIR is found in the nucleus. Data obtained by IHC analysis of primary tumours are highly informative and reliable as to protein expression and localization in vivo. Furthermore, the observations have been confirmed in several melanoma cell lines and primary cultures by means of high throughput immunofluorescence analysis. We also analysed the complete coding sequence for PIR in melanoma cell lines with different patterns of nuclear or cytoplasmic staining to search for mutations that might be specifically associated to modifications in its localization pattern. No variations in *PIR* nucleotide sequence were detected, demonstrating that PIR localization pattern does not depend on mutations in the coding sequence of the gene.

These data show that an abnormal pattern of PIR sub-cellular localization is a characteristic feature of a subset of melanoma cases and suggest it may represent a marker associated with disease progression.

Pirin Controls the Senescence Response in Melanoma

Given the specific and complex pattern of PIR expression in melanoma, we decided to investigate its function in this tumor model. For the purpose of further analyses, we employed a knock-down approach in melanoma cell lines with high PIR expression. WM266-4 metastatic melanoma cell line was transduced with lentiviral shRNAs to induce stable knockdown of PIR expression (Fig. 14.1a). As a result, the proliferation rate was significantly reduced, as observed in growth curve experiments and in colony formation assays (Fig. 14.1b, c), and cells exhibited a flattened and enlarged morphology. Furthermore, PIR ablation induced SA-β-galactosidase reactivity, which was negligible in parental cells (Fig. 14.1d). Taken together, these results, which were confirmed in a second melanoma cell line (not shown), suggest that ablation of PIR is associated with induction of cellular senescence in PIR-expressing melanoma cells.

Recent data have shown that senescent cells are characterized by the production of a variety of cytokines and other secretory proteins referred to as Senescence-Associated Secretory Phenotype (SASP). Multiple functions have been proposed for the SASP: stimulate clearance of senescent cells by the immune system, promote tissue repair, as well as deleterious effects such as promotion of

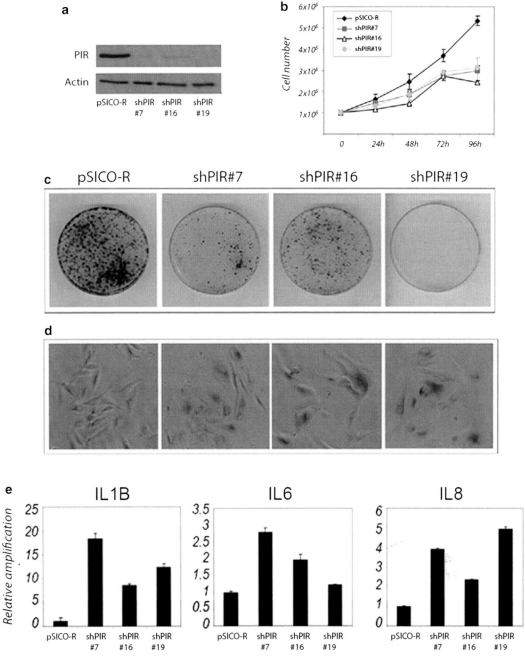

Fig. 14.1 Induction of senescence after PIR ablation in WM266-4 melanoma cells. (**a**) Western blot shows high efficiency of knock-down achieved with lentiviral constructs. (**b**) Growth curve. (**c**) Colony forming assay. (**d**) SA-β-gal staining. (**e**) Quantitative RT-PCR showing transcriptional effect of PIR ablation on *IL1B*, *IL6* and *IL8* expression (Adapted from Licciulli et al. 2011 and unpublished data)

tumor progression and aging (Rodier and Campisi 2011). Provided that PIR is involved in regulation of transcription, we investigated the transcriptional effects of its ablation. Interestingly, among the genes up-regulated after PIR knock-down in melanoma cells there are some cytokine-encoding genes, in particular *IL1B*, *IL6* and *IL8* (Fig. 14.1e). Therefore, besides the classic marks of senescence, PIR ablation is able to induce the characteristic SASP in melanoma cells.

Other genes whose expression was altered after PIR knock-down are involved in proliferation, cell cycle check-point as well as DNA repair. These transcriptional changes could be causative for the establishment of the senescence phenotype or could be incidental and further studies are necessary to clarify this aspect.

The senescent phenotype resulting from ablation of PIR could either be specific of malignant cells or represent a general feature of melanocytic cells. To clarify this we analysed the effect of PIR knockdown in normal cultured melanocytes and observed a decrease in proliferation and an increase in the percentage of SA-β-galactosidase positive cells. It appears, therefore, that PIR function in controlling cellular senescence is a general feature of melanocytic cells and is in place both in malignant and in non-transformed melanocytes (Licciulli et al. 2011).

In order to assess the biological relevance of our findings, we investigated whether PIR expression is affected by the pathways that typically induce senescence in melanoma cells. To this aim we employed an experimental model of oncogenic BRAF-induced senescence represented by human diploid fibroblasts expressing a conditional form of the constitutively activated BRAF oncogene. In these cells, mutant BRAF is fused to the ligand-binding domain of the estrogen receptor (ER), and activity of the fusion protein is rapidly induced by 4-hydroxytamoxifen (OHT) (Woods et al. 1997). We measured PIR levels after induction of senescence by OHT treatment and showed that PIR expression was significantly repressed both at the messenger and at the protein level, suggesting that the establishment of OIS is associated with PIR downregulation (Licciulli et al. 2011).

Chemotherapy is another well-established senescence stimulus in cancer cells (Gewirtz et al. 2008). We therefore measured PIR protein and mRNA levels in WM266-4 melanoma cells treated with low doses of camptothecin (CPT), an antineoplastic agent that induces a senescence-like phenotype in colon cancer cells (Han et al. 2002). PIR mRNA levels showed a fivefold decrease after 24h of CPT treatment, and a decrease in protein expression was clearly detectable after 48h. In addition, increased size, flattened morphology, and staining for SA-β-galactosidase activity after 96h of CPT treatment confirmed that cells had undergone senescence. Thus, rather than a specific consequence of OIS, PIR down-regulation seems to be a general aspect of the induction of senescence in melanoma cells (Licciulli et al. 2011).

Taken together, our data point to a relevant role for PIR in controlling the onset of senescence in melanoma but the molecular mechanism through which PIR exerts this function remains largely unknown. We have analyzed some of the key players in senescence pathways after PIR knock down and induction of senescence, namely p53 and p16, but we failed to observe significant changes in expression levels. It is possible that PIR functions as an effector in the establishment of senescence, therefore acting downstream of the classic senescence pathways. We hypothesize that its transcriptional regulation activity can direct gene expression changes necessary to execute the senescence program.

Provided that PIR expression is required to repress senescence in PIR-expressing melanoma, we wondered what would be the effect of its forced expression in the melanoma cell lines that present low or no PIR. To this aim, PIR negative IGR39 cells were infected with a retroviral construct expressing the entire PIR coding sequence. No significant effect on cellular proliferation and/or viability was observed (Licciulli et al. 2011).

Similarly, overexpression of PIR in TIG3-BRAF cells failed to rescue the induction of senescence by OHT treatment and oncogenic BRAF expression, although it was sufficient to revert some of the molecular effects associated with BRAF-induced senescence. Specifically, oncogenic BRAF induced dramatic upregulation of *IL1A*, *IL1B*, and *IL8* genes within 48h of OHT treatment, as previously described (Kuilman et al. 2008). Co-expression of PIR significantly reduced the extent of induction of all genes.

These data suggest that PIR is required but not sufficient to prevent the establishment of senescence, similarly to what observed in PIR-negative melanoma.

Discussion

PIR is a highly conserved protein whose expression is deregulated in different tumor types. We found that expression of PIR is abundant in melanocytes, and analysis of PIR expression in different stages of melanoma progression revealed a complex and intriguing pattern. PIR is virtually absent in mature nevi, but is detectable at high levels in a fraction of advanced primary and metastatic melanomas. The expression pattern we observed suggests that PIR expression is switched off in nevi and reactivated in a subset of melanomas, in which PIR levels might increase as the disease progresses.

Gain and loss of function experiments were performed respectively in PIR-negative or PIR-positive melanoma cells to study the role of PIR in melanocytic cells. PIR ablation in normal melanocytes and in PIR-positive melanoma cell lines induced growth arrest and a characteristic senescence-like phenotype. Senescence is an anticancer barrier of particular relevance in melanoma, which often stems from a nevus composed of senescent melanocytes. It is therefore conceivable that in the formation of a nevus decreased PIR expression is required for the onset of cellular senescence. In accordance with this view, we observed that residual PIR expression in intradermal nevi is restricted to the superficial portion of the lesion, whereas deeper sections, which are rich in senescent cells, do not show any PIR staining. Furthermore, PIR levels rapidly and significantly decrease on expression of activated BRAF, which is an extremely frequent event in nevi and melanoma and has been directly associated with the onset of oncogene-induced senescence in nevi (Michaloglou et al. 2005). PIR down-regulation is not specific to oncogene-induced senescence, as shown by its rapid decrease in response to camptothecin treatment (Licciulli et al. 2011), and might therefore represent a more general feature of the senescence program induced by diverse stimuli in melanocytic cells. In summary, PIR expression appears to be required to overcome the senescence barrier in PIR-positive melanomas, given that its ablation in this model rapidly results in a senescence-like phenotype.

Based on these data, we envision two alternative models to explain PIR involvement in melanoma progression (Fig. 14.2). One possibility is that PIR-negative melanomas arise from nevi in which PIR has been switched off during the senescence phase put in place to counteract the transforming effects of oncogene activation. Subsequently, in the progression from nevus to primary melanoma, cells by-pass senescence and reacquire the ability to proliferate through mechanisms that do not require PIR expression. In this view, PIR-positive melanomas have a different origin than nevi and do not go through the senescence stage and the associated PIR downregulation (Fig. 14.2a). Alternatively, it is possible that both PIR positive and PIR-negative melanomas evolve from the nevus stage, after which cells may by-pass senescence and progress to the malignant stage either by re-acquiring PIR expression or through alternative mechanisms that do not involve PIR (Fig. 14.2b). In both cases, it would be very informative to compare the mutational status of genes involved in the overcome of senescence, such as cell cycle check-point and DNA repair genes, between PIR-negative and PIR-positive melanoma. This would help clarify whether PIR expression can represent an alternative mechanism to by-pass senescence in the absence of the classic alterations that have been shown so far to drive tumor progression beyond the senescence barrier.

The observation that PIR overexpression concomitant with oncogenic BRAF activation failed to rescue the induction of OIS in human fibroblasts suggests that PIR expression is required but not sufficient to overcome senescence. On the other hand, it is possible that PIR function for senescence control is specific to melanocytic cells. To answer this, the same experiment would have to be repeated in a model of BRAF-induced OIS in normal melanocytes. Supporting the idea

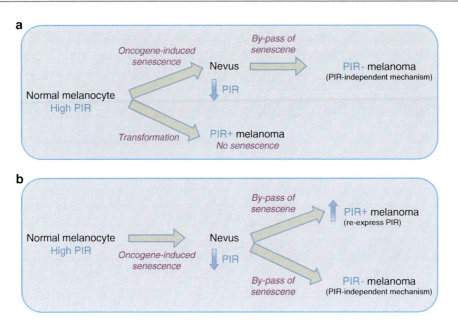

Fig. 14.2 Proposed models for involvement of PIR in control of senescence and melanoma progression (See text for details)

of tissue-type specific roles of PIR, we described a very different effect of PIR ablation in the hematopoietic compartment. We showed that PIR knockdown in hematopoietic precursors impairs terminal myeloid differentiation, and its downregulation in leukaemia may therefore contribute to the maturation arrest characteristic of acute myeloid leukemia (Licciulli et al. 2010a).

We also observed that PIR ablation impairs migration of melanoma cells (unpublished). Others have shown that PIR controls SNAI2 (Slug) expression (Miyazaki et al. 2010) thus providing a molecular mechanism for the control of cell migration. Interestingly, growing evidence points to the existence of intertwining relation between senescence and Epithelial to Mesenchymal Transition (EMT), another crucial process contributing to cancer progression (Ansieau et al. 2008). The observation that PIR transcriptionally regulates Slug could provide another proof that PIR controls senescence by regulating the transcriptional program required for senescence to take place. Overall, our results suggest that PIR expression levels are critical for the establishment of the senescence response in melanocytic cells and that PIR is one of multiple molecular switches that come into play in this cell type in the balance between senescence and proliferation.

Our cohort of human melanoma samples was a very valuable resource to establish the correlation between PIR expression and melanoma progression and allowed us to draw observations that are relevant to the human disease. Nonetheless, it does not allow us to conduct longitudinal study of PIR expression in melanoma progression because all the samples derive from different patients at different stages of disease progression. The appropriate tool to perform such studies would be a mouse model of PIR ablation, which would allow for analysis of PIR levels in the same lesion over time. This analysis could also prove very useful in studying the origin of melanoma relative to a pre-existing nevus or as a *de-novo* lesion.

Besides modulating the transcription of key effectors of senescence, PIR may also exert its function through other mechanisms. Our detailed analysis of PIR subcellular localization in melanoma revealed that advanced stage melanomas display increasing amount of cytoplasmic PIR (Licciulli et al. 2010b), suggesting the existence of additional functions that could confer yet unknown characteristics to the tumor cells.

In summary, our results reveal a complex pattern of expression of PIR in melanoma and demonstrate that PIR is a negative regulator of senescence in melanocytic cells. It is now well established that senescence is a robust physiological anti-tumor mechanism, which is able to halt tumor progression and to elicit a tissue response so that senescent cells can be cleared in vivo through an innate immune response (Xue et al. 2007). In the past few years the concept of pro-senescence therapy has emerged in the field as a novel therapeutic approach to exploit induction of senescence as a strategy to treat cancers. Inhibition of PIR function could prove useful in eliciting a senescence response in melanoma in vivo. A small molecule inhibitor of PIR has already been described, which is able to inhibit melanoma cell migration (Miyazaki et al. 2010) and could possibly be tested for its ability to induce senescence in melanoma cells. In conclusion, we suggest that PIR has a significant potential as a tumor marker and as a therapeutic target in the treatment of melanoma.

Acknowledgements This work was supported by grants from the Italian Association for Cancer Research (Associazione Italiana per la Ricerca sul Cancro, AIRC) and Fondazione Cariplo to M.A.

References

Adams M, Jia Z (2005) Structural and biochemical analysis reveal pirins to possess quercetinase activity. J Biol Chem 280:28675–28682

An J, Sun JY, Yuan Q, Tian HY, Qiu WL, Guo W, Zhao FK (2004) Proteomics analysis of differentially expressed metastasis-associated proteins in adenoid cystic carcinoma cell lines of human salivary gland. Oral Oncol 40:400–408

Ansieau S, Bastid J, Doreau A, Morel AP, Bouchet BP, Thomas C, Fauvet F, Puisieux I, Doglioni C, Piccinin S, Maestro R, Voeltzel T, Selmi A, Valsesia-Wittmann S, Caron de Fromentel C, Puisieux A (2008) Induction of EMT by twist proteins as a collateral effect of tumor-promoting inactivation of premature senescence. Cancer Cell 14:79–89

Baldwin AS Jr (1996) The NF-kappa B and I kappa B proteins: new discoveries and insights. Annu Rev Immunol 14:649–683

Bennett DC (2003) Human melanocyte senescence and melanoma susceptibility genes. Oncogene 22:3063–3069

Campisi J, d'Adda di Fagagna F (2007) Cellular senescence: when bad things happen to good cells. Nat Rev Mol Cell Biol 8:729–740

Dahl C, Guldberg P (2007) The genome and epigenome of malignant melanoma. APMIS 115:1161–1176

Davies H, Bignell GR, Cox C, Stephens P, Edkins S, Clegg S, Teague J, Woffendin H, Garnett MJ, Bottomley W, Davis N, Dicks E, Ewing R, Floyd Y, Gray K, Hall S, Hawes R, Hughes J, Kosmidou V, Menzies A, Mould C, Parker A, Stevens C, Watt S, Hooper S, Wilson R, Jayatilake H, Gusterson BA, Cooper C, Shipley J, Hargrave D, Pritchard-Jones K, Maitland N, Chenevix-Trench G, Riggins GJ, Bigner DD, Palmieri G, Cossu A, Flanagan A, Nicholson A, Ho JW, Leung SY, Yuen ST, Weber BL, Seigler HF, Darrow TL, Paterson H, Marais R, Marshall CJ, Wooster R, Stratton MR, Futreal PA (2002) Mutations of the BRAF gene in human cancer. Nature 417:949–954

Dechend R, Hirano F, Lehmann K, Heissmeyer V, Ansieau S, Wulczyn FG, Scheidereit C, Leutz A (1999) The Bcl-3 oncoprotein acts as a bridging factor between NF-kappaB/Rel and nuclear co-regulators. Oncogene 18:3316–3323

Dhomen N, Reis-Filho JS, da Rocha Dias S, Hayward R, Savage K, Delmas V, Larue L, Pritchard C, Marais R (2009) Oncogenic Braf induces melanocyte senescence and melanoma in mice. Cancer Cell 15:294–303

Gelbman BD, Heguy A, O'Connor TP, Zabner J, Crystal RG (2007) Upregulation of pirin expression by chronic cigarette smoking is associated with bronchial epithelial cell apoptosis. Respir Res 8:10

Gewirtz DA, Holt SE, Elmore LW (2008) Accelerated senescence: an emerging role in tumor cell response to chemotherapy and radiation. Biochem Pharmacol 76:947–957

Gray-Schopfer VC, da Rocha Dias S, Marais R (2005) The role of B-RAF in melanoma. Cancer Metastasis Rev 24:165–183

Gray-Schopfer VC, Cheong SC, Chong H, Chow J, Moss T, Abdel-Malek ZA, Marais R, Wynford-Thomas D, Bennett DC (2006) Cellular senescence in naevi and immortalisation in melanoma: a role for p16? Br J Cancer 95:496–505

Han Z, Wei W, Dunaway S, Darnowski JW, Calabresi P, Sedivy J, Hendrickson EA, Balan KV, Pantazis P, Wyche JH (2002) Role of p21 in apoptosis and senescence of human colon cancer cells treated with camptothecin. J Biol Chem 277:17154–17160

Hihara Y, Muramatsu M, Nakamura K, Sonoike K (2004) A cyanobacterial gene encoding an ortholog of Pirin is induced under stress Conditions. FEBS Lett 574:101–105

Kim WY, Sharpless NE (2006) The regulation of INK4/ARF in cancer and aging. Cell 127:265–275

Kuilman T, Michaloglou C, Vredeveld LC, Douma S, van Doorn R, Desmet CJ, Aarden LA, Mooi WJ, Peeper DS (2008) Oncogene-induced senescence relayed by an interleukin-dependent inflammatory network. Cell 133:1019–1031

Kulju KS, Lehman JM (1995) Increased p53 protein associated with aging in human diploid fibroblasts. Exp Cell Res 217:336–345

Licciulli S, Cambiaghi V, Scafetta G, Gruszka AM, Alcalay M (2010a) Pirin downregulation is a feature of AML and leads to impairment of terminal myeloid differentiation. Leukemia 24:429–437

Licciulli S, Luise C, Zanardi A, Giorgetti L, Viale G, Lanfrancone L, Carbone R, Alcalay M (2010b) Pirin delocalization in melanoma progression identified by high content immuno-detection based approaches. BMC Cell Biol 11:5

Licciulli S, Luise C, Scafetta G, Capra M, Giardina G, Nuciforo P, Bosari S, Viale G, Mazzarol G, Tonelli C, Lanfrancone L, Alcalay M (2011) Pirin inhibits cellular senescence in melanocytic cells. Am J Pathol 178:2397–2406

Mathon NF, Lloyd AC (2001) Cell senescence and cancer. Nat Rev Cancer 1:203–213

Michaloglou C, Vredeveld LC, Soengas MS, Denoyelle C, Kuilman T, van der Horst CM, Majoor DM, Shay JW, Mooi WJ, Peeper DS (2005) BRAFE600-associated senescence-like cell cycle arrest of human naevi. Nature 436:720–724

Mitomi T, Tsuchiya S, Iijima N, Aso K, Suzuki K, Nishiyama K, Amano T, Takahashi T, Murayama N, Oka H et al (1992) Randomized, controlled study on adjuvant immunochemotherapy with PSK in curatively resected colorectal cancer. The cooperative study group of surgical adjuvant immunochemotherapy for cancer of colon and rectum (Kanagawa). Dis Colon Rectum 35:123–130

Miyazaki I, Simizu S, Okumura H, Takagi S, Osada H (2010) A small-molecule inhibitor shows that pirin regulates migration of melanoma cells. Nat Chem Biol 6:667–673

Narita M, Lowe SW (2004) Executing cell senescence. Cell Cycle 3:244–246

Narita M, Nunez S, Heard E, Narita M, Lin AW, Hearn SA, Spector DL, Hannon GJ, Lowe SW (2003) Rb-mediated heterochromatin formation and silencing of E2F target genes during cellular senescence. Cell 113:703–716

Orzaez D, de Jong AJ, Woltering EJ (2001) A tomato homologue of the human protein PIRIN is induced during programmed cell death. Plant Mol Biol 46:459–468

Panchal HD, Vranizan K, Lee CY, Ho J, Ngai J, Timiras PS (2008) Early anti-oxidative and anti-proliferative curcumin effects on neuroglioma cells suggest therapeutic targets. Neurochem Res 33:1701–1710

Pang H, Bartlam M, Zeng Q, Miyatake H, Hisano T, Miki K, Wong LL, Gao GF, Rao Z (2004) Crystal structure of human pirin: an iron-binding nuclear protein and transcription cofactor. J Biol Chem 279:1491–1498

Papp T, Pemsel H, Rollwitz I, Schipper H, Weiss DG, Schiffmann D, Zimmermann R (2003) Mutational analysis of N-ras, p53, CDKN2A (p16(INK4a)), p14(ARF), CDK4, and MC1R genes in human dysplastic melanocytic naevi. J Med Genet 40:E14

Pollock PM, Harper UL, Hansen KS, Yudt LM, Stark M, Robbins CM, Moses TY, Hostetter G, Wagner U, Kakareka J, Salem G, Pohida T, Heenan P, Duray P, Kallioniemi O, Hayward NK, Trent JM, Meltzer PS (2003) High frequency of BRAF mutations in nevi. Nat Genet 33:19–20

Rodier F, Campisi J (2011) Four faces of cellular senescence. J Cell Biol 192:547–556

Santoro C, Mermod N, Andrews PC, Tjian R (1988) A family of human CCAAT-box-binding proteins active in transcription and DNA replication: cloning and expression of multiple cDNAs. Nature 334:218–224

Sharpless E, Chin L (2003) The INK4a/ARF locus and melanoma. Oncogene 22:3092–3098

Soo PC, Horng YT, Lai MJ, Wei JR, Hsieh SC, Chang YL, Tsai YH, Lai HC (2007) Pirin regulates pyruvate catabolism by interacting with the pyruvate dehydrogenase E1 subunit and modulating pyruvate dehydrogenase activity. J Bacteriol 189:109–118

Wellbrock C, Karasarides M, Marais R (2004) The RAF proteins take centre stage. Nat Rev Mol Cell Biol 5:875–885

Wendler WM, Kremmer E, Forster R, Winnacker EL (1997) Identification of pirin, a novel highly conserved nuclear protein. J Biol Chem 272:8482–8489

Xue W, Zender L, Miething C, Dickins RA, Hernando E, Krizhanovsky V, Cordon-Cardo C, Lowe SW (2007) Senescence and tumour clearance is triggered by p53 restoration in murine liver carcinomas. Nature 445:656–660

Yoshikawa R, Yanagi H, Hashimoto-Tamaoki T, Morinaga T, Nakano Y, Noda M, Fujiwara Y, Okamura H, Yamamura T (2004) Gene expression in response to anti-tumour intervention by polysaccharide-K (PSK) in colorectal carcinoma cells. Oncol Rep 12:1287–1293

Zeng Q, Li X, Bartlam M, Wang G, Pang H, Rao Z (2003) Purification, crystallization and preliminary X-ray analysis of human pirin. Acta Crystallogr D Biol Crystallogr 59:1496–1498

Defects in Chromatin Structure and Diseases

15

Umberto Galderisi and Gianfranco Peluso

Contents

Abstract	143
Introduction	143
Chromatin Remodeling Complexes in Senescence and Cancer	146
Chromatin Remodeling Complexes in the Williams Syndrome and the Schimke Immuno-osseous Dysplasia	148
Chromatin Remodeling Complexes in the ATR-X Syndrome, Juberg Marsidi Syndrome, Sutherland-Haan Syndrome and Smith Fineman Myers Syndrome	148
Chromatin Regulation, Histone Acetylation and Diseases	148
Histone Acetylation and Rubinstein Taybi Syndrome	149
Histone Acetylation and Cancer	149
Histone Acetylation and Ageing	149
Chromatin Regulation, DNA Methylation and Diseases	150
ICF Syndrome and DNA Methylation	151
Ageing and DNA Methylation	151
Conclusions	152
References	152

U. Galderisi (✉)
Department of Experimental Medicine, Section of Biotechnology and Molecular Biology, Second University of Naples, Naples, Italy
e-mail: umberto.galderisi@unina2.it

G. Peluso
Institute of Biochemistry of Proteins and Enzymology, C.N.R., Naples, Italy

Abstract

Chromatin is the combination of DNA and other proteins that make up the contents of the nucleus. Chromatin influences all DNA-templated processes including transcription activation and repression, DNA replication and repair. It is now widely appreciated that perturbation of mechanisms regulating chromatin status may have a great impact on cellular physiopathology. In this chapter we describe some genetic diseases related to mutations in genes involved in regulation of chromatin status and also "acquired pathologies" ascribed to defects of ATP-dependent chromatin remodeling and histone post-translational modifications.

Keywords

ATR-X and juberg marsidi syndrome • Chromatin structure and diseases • Histone acetyltransferases (HAT) • ICF syndrome and DNA methylation • Mesenchymal stem cell (MSC) • Rubinstein taybi syndrome and histone acetylation • Schimke immuno-osseous dysplasia • Senescence and cancer • Sutherland-haan and smith fineman myers syndrome • Williams syndrome

Introduction

Chromatin is the combination of DNA and other proteins that make up the contents of the nucleus. The primary protein components of chromatin are histones, which are part of the nucleosome core particles. Each nucleosome consists of approximately 147 base pairs of DNA wrapped around a histone octamer consisting of two copies each of the core histones H2A, H2B, H3, and H4. Nucleosomes are folded through a series of successively higher order structures, such as euchromatin and heterochromatin. The euchromatin is a lightly packed form of chromatin, while the heterochromatin is a tightly packed form of DNA. Chromatin influences all DNA-templated processes including transcription activation and repression, DNA replication and repair (Groth et al. 2007).

The role of chromatin on transcription relies on the fact that gene expression is regulated at multiple levels: sequence, chromatin, and nuclear level. General transcription factors, activators, and repressors act on DNA sequences (promoters, enhancers, silencers, insulators) to modulate gene expression. Many studies on the regulation of gene expression focused on factors acting at sequence level. Nevertheless, regulation of gene expression at chromatin level is hierarchically higher than the sequence level; hence modifications of chromatin status may have a great impact on cell biology. The lightly packed euchromatin is often under active transcription since allows the binding of activators and other factors on gene regulating DNA sequences. At the opposite, the tightly packed heterochromatin silences gene expression by impairing the binding of RNA polymerases and associated factors on gene promoters (Groth et al. 2007).

Chromatin status influences also DNA replication process. The DNA replication fork moves through chromatin without apparent impediment. Replication fork progression disrupts preexisting nucleosomes that have to be properly re-associated on daughter DNA strands this because not only DNA sequence fidelity but also the associated chromatin structure has to pass on to the next generation to ensure that both the genetic and the epigenetic information remain unaltered over generations. Lack of accuracy in this information affects processes such as cell proliferation, commitment, differentiation (Groth et al. 2007).

The integrity of the genome is continuously challenged by both endogenous and exogenous DNA damaging agents. These damaging agents can induce a wide variety of lesions in the DNA, such as double strand breaks, single strand breaks, oxidative lesions and pyrimidine dimers. Wrapping of DNA along nucleosomes to form chromatin may be considered a double edges swords: it protects DNA from damaging agents but can also impair the access of DNA repair proteins to sites of DNA damage. Emerging evidences suggest that a tight regulation of relaxed and compact chromatin structures play a key role in implementing a correct DNA repair (Groth et al. 2007).

It is evident that a proper regulation of chromatin structure have a great impact on cellular functions. Changes in chromatin status depends mainly on covalent histone modifications, on ATP-dependent chromatin remodeling complexes and on methylation of DNA. The post-translational modifications of the histonetails through acetylation, methylation, phosphorylation, and other modifications enable tight regulation of the chromatin structure by modifying the ratio between hetero- and eu-chromatin (Fig. 15.1). This can obtained also through ATP dependent chromatin remodeling factors, which alter histone–DNA interactions, such that nucleosomal DNA becomes more accessible to interacting proteins. ATP-dependent chromatin remodeling complexes can use the energy supplied by ATP hydrolysis to affect nucleosomal organizations by either sliding the nucleosomes along the DNA or by displacing or replacing histones within nucleosomes (Fig. 15.2). Therefore, ATP-dependent chromatin remodeling can actively and directly modify chromatin structures, while histone modifications are thought to mainly play signaling roles through the so called "histone code," which defines specific interactions between chromatin and its interacting partners (Wu et al. 2009).

Change in the hetero-/euchromatin ratio relies also upon methylation/demethylation of DNA, in

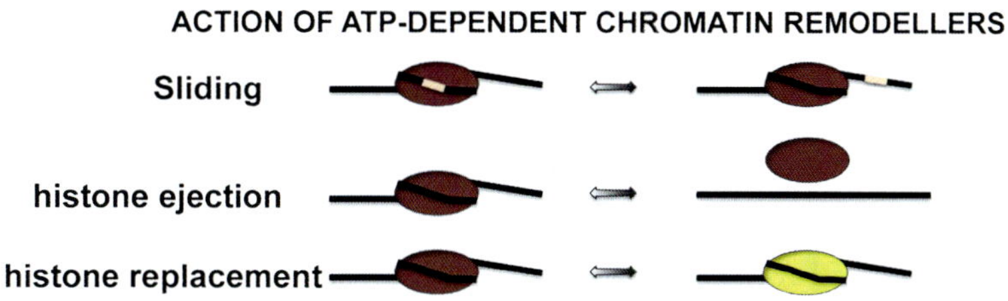

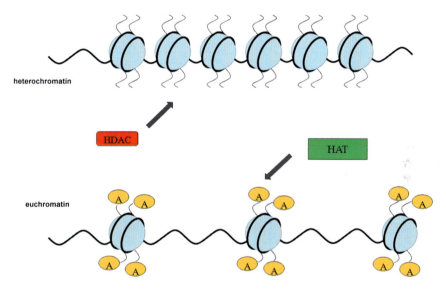

Fig. 15.1 Covalent histone modifications. The N terminus of histone represents a highly conserved domain that is likely to be exposed or extend outwards from the chromatin fibre. A number of distinct post-translational modifications are known to occur at the N terminus of histones including acetylation, phosphorylation, and methylation. Other modifications are known and may also occur in the globular domain. These specific modifications result in changes of chromatin structure and regulate DNA accessibility. In the figure is depicted the activity of histone acetyltransferases (HAT) and histone deacetylase (HDAC), which catalyze the addition or removal of acetyl groups on histones, respectively. Acetylated histones promote euchromatin formation, whereas heterochromatin is associated with lack of acetylation

Fig. 15.2 Chromatin remodeling by ATP-dependent remodelers. Chromatin remodelers can affect gene expression in several ways. Nucleosomes can slide in a new position, which may expose hidden promoters and other gene regulatory regions. Nucleosomes can be ejected to expose DNA or histone dimers can be replaced. Replacement can promote, for example, the exchange of the resident H2A-H2B dimers with dimers containing H2B and the histone H2A variant H2A.Z, which promote gene activation and efficient DNA repair

addition to histone modifications and ATP-chromatin remodeling. DNA methylation refers to the addition of a methyl group to a DNA molecule. The addition of methyl groups to cytosines at CpG sites in mammalian cells is catalyzed by the enzymatic activity of the DNA methyltransferase (DNMT) family, composed of the maintenance DNA methyltransferase DNMT1 and the de novo DNA methyltransferases DNMT3A and DNMT3B (Goll and Bestor 2005) (Fig. 15.3).

Although the frequency of CpG dinucleotides is relatively low in the human genome, they are enriched in the so-called CpG islands, which are DNA regions with a high content of cytosine and guanine. The CpG islands are often associated with genes and are distributed in different locations in the genome, including in promoters, exons, 5′ flanking regions, and 3′ terminal areas (Takai and Jones 2002). DNA methylation is generally associated with chromatin compaction and gene silencing, whereas when regions of a

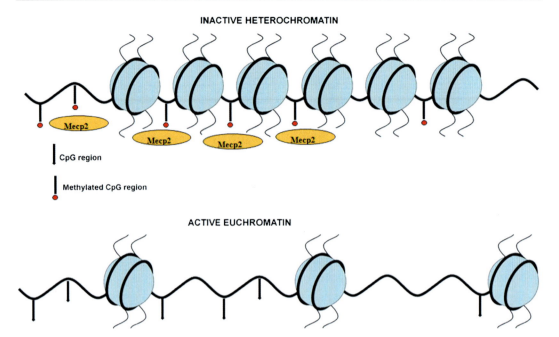

Fig. 15.3 DNA methylation and chromatin status. Change in the hetero-/euchromatin ratio relies also upon methylation/demethylation. The addition of methyl groups to cytosines at CpG sites in mammalian cells is catalyzed by the enzymatic activity of the DNA methyltransferase (DNMT) family. DNA methylation is generally associated with chromatin compaction and gene silencing. Methylated DNA is a docking site for proteins that possess a methyl-CpG binding domain (MBD), such as Mecp2

gene that can be methylated are undermethylated, the gene is transcriptionally active or can be activated.

It is now widely appreciated that perturbation of mechanisms regulating chromatin status may have a great impact on cell biology. Modification of DNA and chromatin is integral to the correct control of gene expression, DNA replication and repair. Deregulation of chromatin functions may impair the correct repair of DNA. This may lead to trigger of apoptosis and/or senescence to block damaged cells. Alternatively, cells may undergo neoplastic transformation. Therefore one would expect that there are diseases in humans that result from mutations/alterations in the components of chromatin or in the enzymes that modify chromatin structure. Some diseases are the direct consequence of a specific gene mutation, others are related to a general impairment in regulation of chromatin function that may have as final result the onset of tumor formation or of degenerative disorders. In the following paragraphs we describe some genetic diseases related to mutations in genes involved in regulation of chromatin status and also "acquired pathologies" ascribed to defects of ATP-dependent chromatin remodeling and histone post-translational modifications.

Chromatin Remodeling Complexes in Senescence and Cancer

Cells are the fundamental structure composing our bodies and hence the cellular decline (called senescence) contributes to the ageing. Endogenous and exogenous stresses may induce cellular senescence. Stressors are mainly macromolecule damage events, which include: shortening of chromosome telomeres; non-telomeric DNA damage; excessive mitogenic signals, which may cause DNA damage; and non-genotoxic stress, such as perturbations to chromatin organization. For many years the analysis of chromatin perturbation as a leading event in triggering senescence has been overlooked. Now, it is well recognized that chromatin DNA packaging is not immune to the ravages of time. All eukaryotes experience changes in chromatin organization and gene-expression patterns as they age. This can be due to perturbation in the function of chromatin modifiers (Campisi and d'Adda di Fagagna 2007; Symonds et al. 2009).

ATP-dependent remodeling complexes seem to be crucial for both the assembly of chromatin structures and their dissolution. These complexes contain SWI2/SNF2-like ATPases and regulate the presence and position of nucleosomes on DNA, enabling or blocking the binding of transcription factors to nucleosomes. The vertebrate genome contains around 30 genes encoding proteins similar to the yeast SWI2/SNF2 ATPases, which are essential for mating type switching and nutrient responses in yeast. Besides the ATPases, the remodeling complexes comprise nine or more proteins, including both conserved (core) and non-conserved components. Several studies had shown that SWI2/SNF2 ATPases possess remodeling activities. Nevertheless, in addition to SWI2/SNF2 proteins, there is evidence that other core components are required for chromatin-remodeling activity (Kingston and Narlikar 1999).

The evolutionarily conserved ATP-dependent chromatin remodeling complexes can be broadly divided into four main families on the basis of the sequence and structure of the ATPase subunit: SWI/SNF, ISWI, CHD and INO80 complexes. However, many of the predicted SWI/SNF-like ATPases do not fit any of these classes and await characterization. Understanding the mechanism of remodeling complexes has been a major area of investigation. Biochemical analysis has demonstrated several different outcomes of remodeling activity in vitro, including nucleosome sliding, octamer transfer to another DNA molecule, dinucleosome formation, and altered nucleosome structure (Kingston and Narlikar 1999).

The mammalian SWI/SNF family includes several members that share most of the same subunits, i.e., the ATPase enzyme, either the BRG1 or the BRM proteins, and/or the presence of tissue specific isoforms. Complexes containing BRG1 have been shown to be required for cell cycle control, apoptosis and differentiation in several biological systems. In particular, in recent years, it has become clear that stem cells have a specific chromatin organization that distinguishes them from more differentiated cells. This directed attention toward chromatin remodeling factors as key players in the regulation of stem cell identity. Our group demonstrated that BRG1 appeared to be actively involved in regulation of mesenchymal stem cell (MSC) biology (Napolitano et al. 2007; Alessio et al. 2010). Forced BRG1 expression in MSC induced significant cell cycle arrest. This was associated with a large increase in apoptosis accompanied by the senescence process. At the molecular level, these phenomena were related to activation of the Rb- and p53-related pathways. In MSC cultures, BRG1 downregulation did affect cell cycle profile, induced a decrease in apoptosis and a significant augmentation of senescent cells. Notably, following BRG1 silencing we detected an increase in senescent cells as observed in MSCs overexpressing BRG1. These data imply that each type of perturbation of mechanisms regulating chromatin status, as occurs either with up- or downregulation of BRG1 activity, may impair cellular physiology (Napolitano et al. 2007; Alessio et al. 2010).

Other researchers showed that BRG1 can act as tumor suppressor gene. Indeed, BRG1 inactivating mutations can be found in lung, breast, and prostate cancer. The BRG1 mutations are biallelic and the reintroduction of the wild-type protein into tumor cell lines alters expression of genes involved in cell-cycle regulation, cell adhesion, and cell migration (Hendrich and Bickmore 2001).

Overall, our data and those of other groups suggest that BRG1 belongs to the class of genes displaying tightly regulated expression, where even subtle alterations may disrupt the normal functioning of cells. One explanation for why certain genes require precise control is their potential to regulate, or be involved in balancing, disparate downstream pathways possessing mutually opposing activities. This may be the case with BRG1, which can modulate gene expression in either a positive or a negative manner.

Other components of the SWI/SNF complexes appear to play a role in senescence process. SWI/SNF complexes have 10 to 12 conserved members ranging in size from 250 kDa (BAF250) to 47 kDa (hSNF5, also known as INI1/BAF47/SmarcB1).

Several scientists evidenced that loss of the hSNF5 gene occurs in virtually 100% of the pediatric rhabdoid tumors. Further studies have shown that re-expression of hSNF5 in human rhabdoid tumor cell lines causes a G1 cell cycle arrest, flattened cell morphology, activation of senescence-associated proteins and expression acid-beta galactosidase, a marker of senescent cells. Chai and collaborators showed that hSNF5 induces replicative senescence by directly regulating the expression of p21CIP1/WAF1 and p16INK4a transcription, which are key genes of senescence related pathways. Authors also showed that both retinoblastoma (Rb) and p53 pathways were involved in this process. These results implicate a role for hSNF5 in control of cell proliferation and senescence and a potential mechanism to account for the embryonic lethality of SNF5-null mice and the development of rhabdoid tumors in infant patients (Chai et al. 2005).

Chromatin Remodeling Complexes in the Williams Syndrome and the Schimke Immuno-osseous Dysplasia

Williams syndrome (OMIM 194050) is caused by heterozygosity for a microdeletion at 7q11.2. It is an autosomal dominant disorder characterized by aortic stenosis, dysmorphism, mental and growth deficiency, aberrant vitamin D metabolism and hypercalcemia. The transcription factor BAZ1B is deleted in Williams syndrome. The encoded protein recruits BRG1 and BRM with their associated chromatin-remodeling factors to vitamin D regulated promoters. When reintroduced into cell lines from Williams syndrome patients, BAZ1B restores vitamin D-responsive transcription (Kitagawa et al. 2003). These findings suggest that aberrant chromatin remodeling might play a key role in the pathophysiology of Williams syndrome.

Schimke immuno-osseous dysplasia (OMIM 242900) is related with mutations in SMARCAL1 (SWI/SNF-related, matrix-associated, actin-dependent regulator of chromatin, subfamily A-like 1). Mutations cause an autosomal recessive disorder of T-cell immunodeficiency, spondyloepiphyseal dysplasia, renal failure hypothyroidism, episodic cerebral ischemia, and bone marrow failure. Until now the role of SMARCAL1 has not been fully elucidated but clinical studies suggest that SMARCAL1 may regulate a subset of genes involved in the proliferation of affected tissues (Boerkoel et al. 2002).

Chromatin Remodeling Complexes in the ATR-X Syndrome, Juberg Marsidi Syndrome, Sutherland-Haan Syndrome and Smith Fineman Myers Syndrome

ATR-X gene encodes a putative ATP-dependent type II helicase of the SNF2 family. The ATRX chromatin-remodeling complex contains the Daxx transcription cofactor that targets ATRX-mediated chromatin remodeling to specific promoters. Other studies demonstrated that the ATRX protein interacts with two different heterochromatic proteins, HP1-alpha and EZH2. Moreover, ATR-X resides predominantly in repetitive DNA and that mutation of ATRX causes aberrant methylation of repetitive DNA elements (Xue et al. 2003).

The ATR-X syndrome (OMIM 301040), the Juberg Marsidi syndrome (OMIM 309590), the Sutherland-Haan syndrome (OMIM 309470) and the Smith Fineman Myers syndrome (309580) result from mutation in the ATRX gene. Mutations in this gene give rise to mental retardation and can also result in facial and skeletal abnormalities, urogenital abnormalities, mild alpha-thalessemia and microcephaly (Hendrich et al. 2001). All these studies evidences that some genetic diseases are related to specific mutations in genes regulation chromatin status.

Chromatin Regulation, Histone Acetylation and Diseases

Histone acetyltransferases (HAT) and histone deacetylase (HDAC) catalyze the addition or removal of acetyl groups on histones, respectively. Acetylated lysine residues serve as binding sites for proteins that contain the bromo domain, which is present in some transcription factors, as well as in subunits of chromatin-remodeling complexes. HAT activities are usually found in large multisubunit complexes. Overall, acetylation of lysine residues on histones is associated with an open chromatin state which may induce transcriptional activation and DNA repair (Grunstein 1997; Dhalluin et al. 1999). Conversely, histone deacetylation promotes the formation of compact chromatin. HDAC enzymes deacetylate all four histones and can act on non-histone substrates as well. Histone deacetylation is associated with chromatin condensation and transcriptional repression.

Histone Acetylation and Rubinstein Taybi Syndrome

The CREB binding protein (CBP) is a HAT that can acetylate histones and other proteins, including p53 and components of basal transcription machinery. As a HAT, CBP promotes the decondensation of chromatin and facilitates transcription through recruitment of several transcriptional regulators including c-Myc, c-Fos, c-Jun, and CREB (Hendrich et al. 2001).

Rubinstein Taybi syndrome (OMIM 180849) is a congenital malformation and a mental retardation disease that is inherited in autosomal dominant manner. Deletions in the 16p13.3 region induce a haploinsufficiency of CBP function with impaired acetylation activity. The mechanisms by which decreased CBP function causes malformations remain obscure, although the mental retardation in may be explained partially by the role of CBP-mediated histone acetylation in synapse plasticity and long-term memory (Guan et al. 2002). Overall, these findings suggest that Rubinstein Taybi syndrome may be considered a "chromatin disease" that is associated to defective acetylation of histones and other proteins.

Histone Acetylation and Cancer

Human cancers are quite often associated with a deregulation of DNA methylation and posttranslational histone modifications, in particular histone acetylation, which leads to deregulation of gene expression. Studies of a panel of normal tissues, primary tumors, and human cancer cell lines indicate that a loss of acetylated Lys16 (K16-H4) and trimethylated Lys20 (K20-H4) of histone H4 is a common event in human cancer that is associated with the hypomethylation of repetitive sequences. Moreover, these changes occur early in tumorogenesis suggesting that that the global loss of monoacetylated and trimethylated forms of histone H4 is a crucial event in cancer development. Other findings on gastrointestinal tumors concluded that the decrease in histone acetylation is not only involved in tumorigenesis but also in tumor invasion and metastasis (Ropero and Esteller 2007).

Currently there are no definitive answers on the mechanisms involved in histone hypoacetylation. These changes can be explained by a decrease in HAT activity due to the mutations or chromosomal translocations characteristic of leukemias, or to changes that result in the increased activity of HDACs.

Regarding this last issue, there are a number of studies showing altered expression of individual HDACs in tumor samples. For example, there is an increase in HDAC1 expression in gastric, prostate, colon and breast carcinomas. Overexpression of HDAC2 has been found in cervical and gastric cancers, and in colorectal carcinoma with loss of APC expression. Other studies have reported high levels of HDAC3 and HDAC6 expression in colon and breast cancer specimens, respectively (Ropero et al. 2007). The findings described above suggest that alteration in chromatin status with the transcriptional repression of tumor-suppressor genes by overexpression and aberrant recruitment of HDACs to their promoter region could be a common phenomenon in tumor onset and progression.

Histone Acetylation and Ageing

Levels of histone acetylation changes during ageing. The phenomenon seems associated mainly to modification in the activity of HDAC rather than in the function of HAT.

Several HDAC enzymes have been identified and classified in homo sapiens. Class I HDAC include HDAC1, 2, 3 and 8, which are related to yeast RPD3 deacetylase and have high homology in their catalytic sites. Class II HDAC are related to yeast Hda1 and include HDAC4, -5, -6, -7, -9 and -10. There are also sirtuins, which are HDAC classified as a third class of deacetylase enzymes. Among them, SIRT1 is orthologous to yeast silent information regulator 2 (sir2) (Marks and Dokmanovic 2005).

The sirtuins represent one of the best examples of changes in protein acetylation status that is linked to ageing. Pioneer studies showed a clear correlation between expression of proteins of the Sir complex, Sir2/Sir3/Sir4, and longevity in yeast. Other studies, evidenced that a deletion of Sir4 abolished silencing of yeast-mating loci and telomeres and caused a decrease in lifespan. In contrast, a gain-of-function mutation in Sir4 induced the relocation of the Sir complex from mating loci and telomeres to the nucleolus and extended lifespan. Further researches evidenced that the increase in Sir2 activity prolong lifespan in yeast, C. elegans, and drosophila.

Following these seminal findings, studies on sirtuins and ageing were extended to mammals. SIRT1 is the closest mammalian ortholog of Sir2 and also this protein has been correlated with ageing. SIRT1 deacetylates histone H4 at lysine 16 (H4K16) and histone H3 at lysine 9 (H3K9). The role of SIRT1 in ageing is complex since this enzyme deacetylates also non-histone substrates. SIRT1, in fact, can deacetylate p53 leading to its inactivation (Chua et al. 2005).

Changes in the level of HDAC1 appear to be associated with ageing. In vitro senescence of human fibroblasts is accompanied by a progressive decrease in the level of HDAC1. In agreement with this analysis, the treatment of human fibroblasts with HDAC inhibitors (HDACi) induce a senescent phenotype in vitro (Bandyopadhyay et al. 2007). In our recent studies, we evidenced that in mesenchymal stem cells the inhibition of class I HDAC with the drug MS-275, a class I selective inhibitor, promoted senescence (Di Bernardo et al. 2009, 2010). Whether the effects of HDAC inhibition results from global changes in histone acetylation and hence in chromatin status or from the transcriptional activation of discrete gene loci remain to be determined. In contrast with the above findings is the observation that down-regulation of p300/CBP histone acetyltransferase activates a senescence checkpoint in human melanocytes. This is in agreement with data showing that in over-expression of HDAC1 in melanoma cells may induce an irreversible senescent phenotype (Bandyopadhyay et al. 2007).

These contradictory observations can be reconciled considering that histone acetylation/deacetylation may have complex modulation during ageing. It is not known how senescence can be triggered both by heterochromatin disruption (HDAC inhibition) and by activities that are associated with heterochromatin formation (HDAC over-expression). Both manipulations cause extensive but incomplete changes in chromatin organization, so each may alter the expression of different critical genes, and the response may be cell-type specific. For example, it could be hypothesized that a decrease in HDAC activity may induce the expression of pro-senescence (ageing) genes, such as those involved in cell cycle arrest, production of senescence-associated cytokines and secreted proteins, etc. On the other side, an increase in HDAC function may repress the expression of anti-senescence (ageing) genes, such as those that promote cell cycle progression, DNA repair, scavenging of reactive oxygen species, etc. These different outcomes may be depend on cell-type and on other still unknown exogenous and endogenous factors.

Chromatin Regulation, DNA Methylation and Diseases

In recent years, numerous studies have demonstrated that a close correlation exists between alteration in DNA methylation and diseases, such as cancer. We will not describe alteration of DNA methylation in cancer since there are exhaustive reviews on this topic. We will focus our attention on some other diseases/disorders that may be related to altered chromatin status due to altered regulation of DNA methylation.

ICF Syndrome and DNA Methylation

ICF syndrome (Immunodeficiency, Centromere instability and Facial anomalies syndrome) is a very rare autosomal recessive immune disorder. Moreover, it is becoming evident that also ageing is associated with deregulated control of gene expression due to impairment in DNA methylation/demethylation processes. It is characterized by variable reductions in serum immunoglobulin levels which cause most ICF patients to succumb to infectious diseases before adulthood. ICF syndrome patients exhibit facial anomalies which include hypertelorism, low-set ears, epicanthal folds and macroglossia. ICF syndrome can be caused by a mutation in the DNMT3b gene, located on chromosome 20q11.2. The mutations identified in ICF patients interfere, but not completely abolish, the methyltransferase activity. This leads to demethylation of satellite DNA, undercondensation and chromosome instability as observed in T lymphocytes of ICF patients (Hendrich et al. 2001). Several other findings suggest that failure in proper DNA methylation in ICF syndrome deregulates the expression of key genes that perturb craniofacial, cerebral and immunological development.

Ageing and DNA Methylation

Pioneering studies, evidenced the age-dependent loss of genomic methylation in DNA isolated from the various organs of humpback salmon was significantly decreased during ontogenesis. These findings were confirmed also in mammals. The maximal amount of methylated cytosine was observed in DNA isolated from tissues of embryos and newborn animals and gradually decreased upon aging. Mice and human fibroblasts evidenced a decline in DNA methylation during in vitro cultivation (Wilson and Jones 1983).

Decline in DNA methylation may be associated with change in DNA methyltransferase activity. Several studies have demonstrated that the activity of DNMT1 substantially decreases with aging. On this premise, it has been hypothesized that during ageing the genomic hypomethylation is a result of passive demethylation, especially of highly methylated GC-rich DNA domains caused by the functional inability of DNMT1. This hypothesis is further supported by the results of in vivo and in vitro studies demonstrating that also expression of DNMT3A decreased with ageing and by the fact that the deletion of Dnmt3a gene substantially shortens the mouse life span (Nguyen et al. 2007).

Is the decrease of DNA methylation a cause or a consequence of ageing? The answer is not simple. Both possibilities may be true. Several studies have proved that accumulation of DNA damage may affect the normal status of DNA methylation. Every living organism is exposed to various genomic insults on a daily basis caused by many endogenous and exogenous factors. Several findings evidenced that the presence of unrepaired lesions in DNA substantially alters the methylation capacity of DNA methyltransferases, leading to DNA hypomethylation. The presence of unrepaired DNA lesions is due both to progressive accumulation of DNA damage and to an age-dependent decrease in DNA repair proficiency. These studies emphasized the interdependence between the function of DNA methyltransferases and the integrity of the genome in the maintenance of a normal level of DNA methylation. Thus, disruption of any of these mechanisms during aging may result in a subsequent loss of DNA methylation (Imai and Kitano 1998).

Despite a general decrease in DNA methylation during ageing, specific DNA regions become hypermethylated as organism ages. This increase in DNA methylation occurs mainly in some CpG islands that are typically undermethylated in normal tissues. A similar phenomenon (global DNA demethylation along with hypermethylation at specific loci) has been observed also in cancer cells. The significance of these modifications in DNA methylation pattern is not clear. Some scientists state that these peculiar changes in DNA methylation may be explained taking into account that during ageing or cancer there is a redistribution of chromatin modifiers (DNA methyltransferases; histone modifiers; ATP-dependent chromatin remodelers). According to this hypothesis endogenous or exogenous stressors may induce a shift of chromatin modifiers from normal targets to other site on the DNA (Imai et al. 1998).

Methylated DNA is a docking site for proteins that possess a methyl-CpG binding domain (MBD). Human proteins MECP2, MBD1, MBD2, MBD3, and MBD4 comprise a family of nuclear proteins related by the presence of MBD domain that allows their specific binding to methylated DNA. These proteins regulate gene expression by causing changes in chromatin structure.

Among the different methylated DNA binding proteins, MECP2 is of particular interest since its mutation is associated with Rett syndrome, which is one of the most common genetic causes of mental retardation in young females. Our research group has shown that mesenchymal stem cells obtained from Rett patients are prone to senescence in comparison with wild-type cells. These data were confirmed in an in vitro model of partial MECP2 silencing (Squillaro et al. 2008, 2010). We observed that senescence induced by reduced level of MECP2 is associated with failure in the maintenance of DNA and occurs through canonical Rb- and p53-related pathways.

Conclusions

The studies of chromatin structure have demonstrated the critical roles of chromatin remodeling in sequential gene activation and inactivation during the entire life of a cell. Defects in chromatin components that cell uses to regulate gene expression are found in human inherited and acquired diseases. The "chromatin diseases" may affect multiple biological systems with huge phenotypic variations. This reflects the number and the broad range of genes that are involved in regulation of chromatin status, whose expression must be perturbed in these pathologies. An unresolved issue is why is mental retardation a prominent feature of inherited "chromatin disorders"? Until now we have only a simplistic explanation. Brain functions depend on the integrated actions of hundreds genes and hence they are more sensitive to perturbation of key regulator of gene expression, such as chromatin remodelers and histone modifiers.

Among the acquired "chromatin diseases" cancer and ageing play a prominent role. The long-term maintenance of the nuclear architecture is vital for the normal functioning of cells and tissues over a lifetime. Cells undergo significant chromatin changes in response to exogenous and endogenous stresses. These stimuli may damage DNA and can alter the cellular identity and function by triggering neoplastic transformation or senescence process.

It should be born in mind that chromatin reorganization is a main contributor to cancer and ageing. Understanding the role of chromatin alteration in tumor onset and senescence may help in understanding, treating, and ultimately slowing these phenomena.

References

Alessio N, Squillaro T, Cipollaro M, Bagella L, Giordano A, Galderisi U (2010) The BRG1 ATPase of chromatin remodeling complexes is involved in modulation of mesenchymal stem cell senescence through RB-P53 pathways. Oncogene 29(40):5452–5463

Bandyopadhyay D, Curry JL, Lin Q, Richards HW, Chen D, Hornsby PJ, Timchenko NA, Medrano EE (2007) Dynamic assembly of chromatin complexes during cellular senescence: implications for the growth arrest of human melanocytic nevi. Aging Cell 6(4):577–591

Boerkoel CF, Takashima H, John J, Yan J, Stankiewicz P, Rosenbarker L, Andre JL, Bogdanovic R, Burguet A, Cockfield S, Cordeiro I, Frund S, Illies F, Joseph M, Kaitila I, Lama G, Loirat C, McLeod DR, Milford DV, Petty EM, Rodrigo F, Saraiva JM, Schmidt B, Smith GC, Spranger J, Stein A, Thiele H, Tizard J, Weksberg R, Lupski JR, Stockton DW (2002) Mutant chromatin remodeling protein SMARCAL1 causes Schimke immuno-osseous dysplasia. Nat Genet 30(2):215–220

Campisi J, d'Adda di Fagagna F (2007) Cellular senescence: when bad things happen to good cells. Nat Rev Mol Cell Biol 8(9):729–740

Chai J, Charboneau AL, Betz BL, Weissman BE (2005) Loss of the hSNF5 gene concomitantly inactivates p21CIP/WAF1 and p16INK4a activity associated with replicative senescence in A204 rhabdoid tumor cells. Cancer Res 65(22):10192–10198

Chua KF, Mostoslavsky R, Lombard DB, Pang WW, Saito S, Franco S, Kaushal D, Cheng HL, Fischer MR, Stokes N, Murphy MM, Appella E, Alt FW (2005) Mammalian SIRT1 limits replicative life span in response to chronic genotoxic stress. Cell Metab 2(1):67–76

Dhalluin C, Carlson JE, Zeng L, He C, Aggarwal AK, Zhou MM (1999) Structure and ligand of a histone acetyltransferase bromodomain. Nature 399(6735):491–496

Di Bernardo G, Squillaro T, Dell'Aversana C, Miceli M, Cipollaro M, Cascino A, Altucci L, Galderisi U (2009) Histone deacetylase inhibitors promote apoptosis and senescence in human mesenchymal stem cells. Stem Cells Dev 18(4):573–581

Di Bernardo G, Alessio N, Dell'Aversana C, Casale F, Teti D, Cipollaro M, Altucci L, Galderisi U (2010) Impact of histone deacetylase inhibitors SAHA and MS-275 on DNA repair pathways in human mesenchymal stem cells. J Cell Physiol 225(2):537–544

Goll MG, Bestor TH (2005) Eukaryotic cytosine methyltransferases. Annu Rev Biochem 74:481–514

Groth A, Rocha W, Verreault A, Almouzni G (2007) Chromatin challenges during DNA replication and repair. Cell 128(4):721–733

Grunstein M (1997) Histone acetylation in chromatin structure and transcription. Nature 389(6649):349–352

Guan Z, Giustetto M, Lomvardas S, Kim JH, Miniaci MC, Schwartz JH, Thanos D, Kandel ER (2002) Integration of long-term-memory-related synaptic plasticity involves bidirectional regulation of gene expression and chromatin structure. Cell 111(4):483–493

Hendrich B, Bickmore W (2001) Human diseases with underlying defects in chromatin structure and modification. Hum Mol Genet 10:2233–2242

Imai S-I, Kitano H (1998) Heterochromatin islands and their dynamic reorganization: a hypothesis for three distinctive features of cellular aging. Exp Gerontol 33:555–570

Kingston RE, Narlikar GJ (1999) ATP-dependent remodeling and acetylation as regulators of chromatin fluidity. Genes Dev 13(18):2339–2352

Kitagawa H, Fujiki R, Yoshimura K, Mezaki Y, Uematsu Y, Matsui D, Ogawa S, Unno K, Okubo M, Tokita A, Nakagawa T, Ito T, Ishimi Y, Nagasawa H, Matsumoto T, Yanagisawa J, Kato S (2003) The chromatin-remodeling complex WINAC targets a nuclear receptor to promoters and is impaired in Williams syndrome. Cell 113(7):905–917

Marks PA, Dokmanovic M (2005) Histone deacetylase inhibitors: discovery and development as anticancer agents. Expert Opin Investig Drugs 14(12):1497–1511

Napolitano MA, Cipollaro M, Cascino A, Melone MA, Giordano A, Galderisi U (2007) Brg1 chromatin remodeling factor is involved in cell growth arrest, apoptosis and senescence of rat mesenchymal stem cells. J Cell Sci 120(Pt 16):2904–2911

Nguyen S, Meletis K, Fu D, Jhaveri S, Jaenisch R (2007) Ablation of de novo DNA methyltransferase Dnmt3a in the nervous system leads to neuromuscular defects and shortened lifespan. Dev Dyn 236(6):1663–1676

Ropero S, Esteller M (2007) The role of histone deacetylases (HDACs) in human cancer. Mol Oncol 1(1):19–25

Squillaro T, Hayek G, Farina E, Cipollaro M, Renieri A, Galderisi U (2008) A case report: bone marrow mesenchymal stem cells from a Rett syndrome patient are prone to senescence and show a lower degree of apoptosis. J Cell Biochem 103(6):1877–1885

Squillaro T, Alessio N, Cipollaro M, Renieri A, Giordano A, Galderisi U (2010) Partial silencing of methyl cytosine protein binding 2 (MECP2) in mesenchymal stem cells induces senescence with an increase in damaged DNA. FASEB J 24(5):1593–1603

Symonds CE, Galderisi U, Giordano A (2009) Aging of the inceptive cellular population: the relationship between stem cells and aging. Aging (Albany NY) 1(4):372–381

Takai D, Jones PA (2002) Comprehensive analysis of CpG islands in human chromosomes 21 and 22. Proc Natl Acad Sci U S A 99(6):3740–3745

Wilson VL, Jones PA (1983) DNA methylation decreases in aging but not in immortal cells. Science 220(4601):1055–1057

Wu JI, Lessard J, Crabtree GR (2009) Understanding the words of chromatin regulation. Cell 136(2):200–206

Xue Y, Gibbons R, Yan Z, Yang D, McDowell TL, Sechi S, Qin J, Zhou S, Higgs D, Wang W (2003) The ATRX syndrome protein forms a chromatin-remodeling complex with Daxx and localizes in promyelocytic leukemia nuclear bodies. Proc Natl Acad Sci U S A 100(19):10635–10640

The Role of Fibrosis in Tumor Progression and the Dormant to Proliferative Switch

Lara H. El Touny, Dalit Barkan, and Jeffrey E. Green

Contents

Abstract	155
Introduction	156
Fibrosis	156
Mechanisms of Fibrosis	156
Fibrosis in Breast Cancer	157
Fibrosis and Breast Cancer: Risk, Initiation and Progression	157
Fibrosis and Breast Cancer Metastasis	159
Fibrosis, Collagen-I and Breast Cancer Dormancy	161
Discussion	162
References	163

L.H. El Touny • J.E. Green (✉)
Laboratory of Cancer Biology and Genetics,
National Cancer Institute, Bethesda, MD, USA
e-mail: jegreen@mail.nih.gov

D. Barkan
Department of Biology, Faculty of Sciences,
Haifa University, Haifa, Israel

Abstract

The extracellular matrix is known to play a pivotal role in normal breast development as well as tumorigenesis and breast cancer progression. Several lines of clinical evidence have associated the presence of fibrotic-like, activated stroma with poor therapeutic response and prognosis in breast cancer patients. Recent evidence suggests that extracellular changes are requisite for the formation of a pre-metastatic niche that provides a permissive environment for disseminated breast cancer cells to survive and proliferate. It is also thought that in the absence of favorable environmental cues at a metastatic site, disseminated tumor cells can be maintained in a dormant, metabolically active state until they encounter or modulate their surroundings into an environment that supports their proliferation. We have shown *in vivo* that the induction of lung fibrosis via adenoviral instillation of TGFß, which results in collagen-I accumulation, can induce the proliferation of an otherwise dormant breast cancer cell line (D2.0R). We have recapitulated this dormant-to-proliferative switch by collagen-I supplementation in a three dimensional *in vitro* model of dormancy, suggesting that collagen-I is a major contributor to the overtly proliferative integrin β1-dependent state of the D2.0R cells in fibrotic lungs. This work has highlighted the importance of the integrin β1 pathway and its downstream effectors as principal players in sensing microenvironmental changes by dormant breast cells and activating pro-proliferative pathways resulting in overt metastases.

Keywords

Breast cancer • Cancer-associated fibroblast (CAF) • Collagen-I and breast cancer dormancy • Extracellular matrix (ECM) • Fibrosis • Hypoxia inducible factor (HIF) • Lysyl oxidase (LOX) • Proliferative switch • Tumor progression and dormancy • VEGF receptor 1$^+$ (VEGFR1$^+$)

Introduction

It is projected that more than 200,000 women will develop breast cancer in 2012 in the United States, with more than 15% of these patients succumbing to the disease (Siegel et al. 2012). Despite major clinical advances and available therapies, breast cancer mortality due to disseminated breast cancer cells and subsequent metastasis remains very high, second only to deaths from lung cancer among women. Therefore, a better understanding of the mechanisms underlying the progression of confined and resectable breast tumors into a metastatic phenotype would provide better targets in the fight against breast cancer. Interestingly, advances in basic research have determined that alterations in the extracellular matrix (ECM) surrounding the tumors can, in fact, contribute significantly to the acquisition of tumor growth and metastatic properties including uncontrolled proliferation, resistance to apoptosis, angiogenesis and tissue invasion.

The extracellular matrix as part of the microenvironment of the breast epithelial cells is undoubtedly one of the principal determinants of the behavior of mammary epithelial cells. In fact, defined cell–ECM interactions are a prerequisite for the structural integrity and specialized function of the normal breast epithelium. Initially thought to only signal through the specific growth factor reservoirs and extracellular components present in this "niche", methodological advances have resulted in the recognition of a mechanical role for the ECM in controlling mammary epithelial behavior. Most glandular epithelial tissues including the breast are in a state of tensional homeostasis so that their normal state is highly mechanically compliant. The ECM in a compliant tissue is composed of a relaxed meshwork of type I and III collagens and elastin that, together with fibronectin, form a relaxed network of fibers (Paszek and Weaver 2004). Additionally, ECM proteins interact with receptors on the epithelial cells, activating signaling cascades that alter cellular proliferation.

Based on the dynamic interaction between breast epithelial cells and their surrounding stroma, it is not surprising that alterations in the ECM of breast tissue can have a role in the initiation or progression of breast cancer.

Several clinical observations point toward a relationship between fibrosis, resulting in major alterations in the structure and composition of the ECM, and breast cancer progression. Fibrosis is the result of a wound healing response gone awry which results in a "stiffened" matrix and a more rigid microenvironment that is sensed by the mammary cells. This mechano-transduction process converts mechanical stimuli into chemical signals and allows the cell to adapt to its microenvironment, often contributing to tumor progression.

Collagen is the most abundant component of the ECM and it accumulation is a hallmark of fibrosis. This chapter will focus on the evidence supporting a role for a "fibrotic-like" collagen-rich microenvironment in breast cancer progression with emphasis on the modulation of the metastatic phenotype and specifically, tumor cell dormancy.

Fibrosis

Mechanisms of Fibrosis

Upon epithelial/endothelial cell injury; a multifaceted response is initiated that culminates in wound healing (Wynn 2008). This takes place when damaged cells and vasculature promote the release of inflammatory mediators that initiate an anti-fibrinolytic coagulation cascade and the establishment of a provisional matrix composed of fibrin and plasma fibronectin. Concomitant with platelet-mediated vasodilation and increased blood vessel permeability, fibroblasts produce matrix metallopreoteases (MMPs), which disrupt the basement membrane, allowing recruitment of neutrophils and macrophages to the site of injury. Upon exposure to pro-fibrotic cytokines and growth factors, such as transforming growth factor-β (TGFβ), interleukin-13 and platelet-derived growth factor (PDGF) (secreted by immune cells in the wound microenvironment), myofibroblasts, bone marrow-derived fibrocytes and fibroblasts as well as epithelial cells that have become activated induced to undergo an epithelial to mesenchymal transition (EMT).

This "activation" is defined by the expression of a "myofibroblast" phenotype characterized by the concurrent expression of α-smooth muscle actin (αSMA) and vimentin, stress fibers, and a prominent rough endoplasmic reticulum. These activated myofibroblasts promote the formation of an extracellular matrix rich in fibronectin, hyaluronic acid, tenascin, and collagen and laminin types representative of an immature matrix and reminiscent of a tumor invasion ECM (Herrick et al. 1992). This "stiffened" and remodeled ECM promotes the proliferation and the directional migration of cells within the tissue towards the wound site resulting in wound contraction. In contrast to a tumor progression scenario, this metaplastic phase is short-lived and followed by a remodelling phase that favors epithelial and stromal cell apoptosis, dissolution of immature vessels and the degradation of the immature ECM, while shifting the balance towards the reconstruction of a collagen-I rich mature matrix.

In a successful healing process, scar resolution follows with collagen levels returning to normal tissue levels. Evidently, wound healing, tissue remodeling and repair are protective mechanisms activated in response to injury in order to maintain functional integrity of organs. However, the normal ECM building/degrading processes may be disrupted through continued exposure to injury resulting in chronic inflammation and repair, defects in the resolution phase of wound healing as well as a number of underlying diseases or genetic predispositions that can accentuate deregulated healing or activate scarring processes in the absence of insult.

This disruption is exemplified by shifting the balance towards tissue inhibitors of metalloproteases (TIMP) production over MMP synthesis, resulting in excessive accumulation of ECM components. Specifically, collagen, as part of this excessive ECM, coalesces into fibrous bundles resistant to degradation, enhancing the stiffness of the tissue and supporting a continuous mesenchymal cell proliferation, migration and ECM accumulation, resulting in fibrosis and subsequent disruption of function in the affected organ (Yates et al. 2011).

Fibrosis in Breast Cancer

Fibrosis and Breast Cancer: Risk, Initiation and Progression

An interest in the potential role of fibrosis in tumor formation and development has been sparked even before the elucidation of the epithelial-stromal interactions by the identification of tumors near scar tissue. In fact, reports have indicated the incidence of breast cancers in surgical scars of women who had undergone non-breast related surgical procedures an average of 10 years earlier than their initial breast cancer diagnosis (Freund et al. 1976).

During that same time period, a number of studies had suggested a link between the common phenomenon of high density regions on mammograms and breast cancer risk. Increased breast density could be due to enhanced deposition, reduced remodeling or increase post-translational crosslinking of ECM components, especially type I collagen. In fact, histological analyses of radiographically opaque areas of the breast have revealed increased fibrous tissue, collagen deposition, and fibroblast accumulation, characteristic markers of fibrosis. Increased mammary density has subsequently been characterized as one of the greatest risk factors for the development of breast cancer (Boyd et al. 2002), representing a two- to six-fold increase in tumor susceptibility among women with dense breasts. Increased mammary density has also been established as a prognostic indicator of breast cancer treatment and response in young female patients. The converse association has been observed where softening of breast tissue after tamoxifen treatment is considered a good response prognostic indicator for breast cancer treatment (Cuzick et al. 2011).

Clinical studies have also demonstrated the presence of fibrosis in breast cancer patients. This fibrosis could manifest itself as the "activated stroma" that occurs during breast cancer progression and enhances the selection and expansion of developing neoplasms, particularly that of late-stage metastatic tumors, which will be discussed in more detail below. However, it also manifests as fibrotic foci characterized by stromal fibrosis and cyst formation, which are found in approximately 50% of infiltrating ductal carcinoma (IDC) cases (Hasebe et al. 2002). In this scenario, the fibrotic focus is a scar-like area in the center of carcinomas whose presence predicts poor clinical outcome in invasive breast carcinoma (Colpaert et al. 2003).

It is well established that the stroma present in the normal mammary gland is strikingly different from carcinoma-associated stroma. This observation has been highlighted by transcriptome wide analyses that have reported significant changes in stromal gene expression (mainly components of the ECM and MMPs (Ma et al. 2009) associated with tumor development. A gene expression signature indicative of activated wound responses is common to more than 90% of non-neoplastic tissues adjacent to breast cancer (Troester et al. 2009). When compared to normal tissue, stroma surrounding/accompanying breast tumors is characterized by an increased number of fibroblasts and immune cell infiltrates, increased collagen I and fibrin deposition, elevated smooth muscle actin (SMA) expression among other changes, resulting in both altered physical tensile properties and signaling cascades of stromal cells (Cukierman 2004). In addition to differences in its composition, the architecture of this stroma is also strikingly different from its normal counterpart; rather than relaxed non-oriented fibrils, collagen I in breast tumors is often highly linearized and either oriented adjacent to the epithelium or projecting perpendicularly into the tissue (Leventhal et al. 2009). This "activated stroma" phenotype is accompanied by disruption of the basement membrane and the onset of a fibrotic response, reminiscent of a wound healing signature.

Studies using mouse models have established the causal relationship between "activated stroma" and breast cancer initiation. In fact, injecting either normal or oncogenically-transformed breast cells into irradiated "activated, desmoplastically reactive stroma" of murine mammary glands (Barcellos-Hoff 1998) resulted in the promotion of ductal carcinoma in situ-like lesions, adenomas and poorly differentiated tumors. Similarly, a human-in-mouse (HIM) model, in which immortalized mammary fibroblasts expressing (TGFβ) or hepatocyte growth factor (HGF) are injected into cleared fat pads led to tumor formation (Kuperwasser et al. 2004). This direct effect of the stroma on the malignant phenotype of breast cells is further supported by the discovery that it could be reversed by blocking cues from the tumor microenvironment, such as the inhibition of integrin β1 (Kenny and Bissell 2003).

A direct relationship between collagen-I specifically and breast cancer initiation has also been established in the MMTV-PyVT/Col1a1^{tmJae} mouse model where collagen-I degradation is compromised (Provenzano et al. 2008). These mice presented with a threefold increase in the number of mammary tumors that exhibited a very invasive phenotype. This collagen-dense stroma also resulted in a significantly larger number of lung metastases in comparison to wild type PyVT counterparts.

Several studies linking breast cancer incidence to the extracellular matrix have focused on manipulating the stromal expression of genes that are well known fibrosis modulators. Knockdown of caveolin-1 (Cav-1), a negative regulator of TGFb, induces extensive fibrosis in several organs. Gene-profiling analysis shows that the Cav-1$^{-/-}$ mammary fibroblast gene signature significantly overlaps with the cancer-associated fibroblast (CAF) signature and predicts poor clinical outcome in breast cancer patients treated with tamoxifen therapy (Sotgia et al. 2012). Furthermore, the role of Cav-1 deficient stroma in supporting tumor growth was shown by the reconstitution of Cav-1$^{-/-}$ mammary fat pads with wild-type mammary epithelial cells or orthotopic transplantation of mammary tumors in Cav-1$^{-/-}$ mice which resulted in ductal hyperplasia and greater tumor growth, respectively, as compared with wild type hosts.

Another mediator of ECM stiffness, lysyl oxidase (LOX) has generated a lot of interest in its effect on breast cancer initiation and progression. LOX is a copper-dependent amine oxidase that initiates the process of covalent intra- and intermolecular crosslinking of collagen and is frequently elevated in tumors (Erler et al. 2009). LOX overexpression increases ECM stiffness and promotes invasion and progression of Ha-Ras overexpressing mammary organoids; conversely, inhibition of LOX reduces tissue fibrosis and tumor incidence in the Neu breast cancer model (Leventh al. 2009). These observations suggest that collagen deregulation and, therefore, ECM stiffness play a direct causative role in cancer pathogenesis. Interestingly, overexpression of LOX alone is insufficient to cause tumor formation suggesting that deregulation of ECM dynamics is a "co-conspirator," rather than a primary inducer of tumorigenesis in the breast.

Opposing the co-conspirator theory, a more recent report has suggested that mammary gland fibrosis, observed upon the transgenic replacement of E-cadherin expression with mesenchymal N-cadherin in the adult mammary gland can result in tumor formation, albeit in only 10% of the mice analyzed at an advanced age of 12 months (Kotb et al. 2011). Furthermore, in concordance, with the previous studies, this transgenic manipulation of cadherin expression that results in progressive fibrosis of the mammary gland also increased the incidence of tumor formation when p53 expression was reduced. Finally, in a model of post-partum involution, premalignant cells, exposed to increasing amounts of fibrillar collagen, acquire malignant phenotypes which could be reversed by the inhibition of COX-2 (Lyons et al. 2011). Several mechanisms by which a fibrosis-like microenvironment can mediate the initiation and progression of breast cancer have been proposed. In fact, increases in collagen deposition or ECM stiffness, alone or in combination, have been shown to upregulate integrin signaling, and thus promote cell survival and proliferation via crosstalk with growth factor receptors and PI3 kinase (Egeblad et al. 2010). Furthermore, various ECM components or their functional fragments have pro- or anti-apoptotic effects. Therefore, it is conceivable that deregulation of ECM remodeling can reduce the apoptotic response of mutant cells. Finally, studies using live imaging have shown that cancer cells migrate rapidly on collagen fibers in areas enriched in collagen (Condeelis and Segall 2003), suggesting an enhanced migratory/invasive potential. Together, deregulation of ECM dynamics plays an essential role in cancer progression through the promotion of cancer cell proliferation, loss of cell differentiation, and cancer cell invasion.

Fibrosis and Breast Cancer Metastasis

The metastatic site offers new challenges for circulating tumor cells. In a distant site, cells must lodge, survive, extravasate, and grow before they are considered of clinical relevance. Although the early steps of invasion and metastatic spread might be frequent, efficient, experimental models and clinical observations indicate that successful metastatic outgrowth is quite an inefficient and rate limiting process. Indeed, even after invasion into the parenchyma, disseminated metastatic cells may become quiescent with relatively few reactivated later that develop into proliferating lesions. The steps or changes necessary for this "switch to proliferation" have yet to be completely elucidated. The importance of interactions between metastatic tumor cells and the microenvironment was evident from the earliest studies in this field, and most clearly enunciated by Paget (1889) in the 'seed and soil' hypothesis. However, during the past decade research efforts have intensified to elucidate how the microenvironment and particularly the ECM affect metastatic growth and breast cancer patient outcome.

This section will only focus on the evidence linking the effects of the extracellular matrix and collagen, in particular, on the ability of metastatic breast cancer cells to survive and grow in distant organs. The clinical evidence supporting a causal link between the effect of the ECM at a metastatic site and the ability of cancer cells to establish clinically relevant macrometastases has been limited owing to the difficulty in obtaining clinical samples from metastases. Nevertheless, Loya et al. (2009) have reported that occult lymph node metastases are enriched with fibrosis. This observation has been expanded to include the upregulation of TGFβ signaling and coordinate overexpression of the TGFβ-inducible ECM proteins collagen-I, fibronectin, periostin and versican, as well as the formation of elaborate fibrillar networks ensheathing nests of breast cancer cells in all LN, liver, and lung macrometastases (Soikkeli et al. 2010).

Experimentally, the majority of our understanding of the effect of the ECM in the metastatic sites comes from experiments defining and describing the "pre-metastatic niche". A successful metastasis requires not only a local niche to support cancer cell growth at the primary site but also a metastatic niche to allow invading cancer cells to survive, colonize, and expand to form macrometastasis.

For example, increased fibronectin expression is essential for VEGF receptor 1^+ (VEGFR1$^+$) hematopoietic progenitor cells, which also express the fibronectin receptor integrin α4β1, to migrate and adhere to the niche in the lung (Kaplan et al. 2005). Once there, VEGFR1$^+$ hematopoietic progenitor cells secrete MMP9, which is known to play a role in lung-specific metastasis, and thus further modulate and deregulate the pre-metastatic niche. In addition to fibronectin, other ECM components also play a role in the pre-metastatic niche. Mammary carcinoma cells expressing the hyaluronan receptor CD44 survive better than cells with low levels of CD44.

Another signaling cascade reminiscent of fibrotic changes that takes place in the pre-metastatic niche is the recruitment of granulin-expressing bone marrow cells to sites where incipient tumors reside, which then deregulate behaviors of the distant stromal cells (Elkabets et al. 2011). Granulin (GRN) mediates the assembly of a desmoplastic tumor stroma by stimulating fibroblasts. Fibroblasts stimulated with GRN elaborate proteins that define cancer-associated fibroblasts (CAFs), including αSMA and a host of pro-inflammatory and matrix remodeling cytokines, thereby supporting the growth of micrometastases.

The most convincing evidence of the role of collagen in enhancing breast cancer in distant sites is available from studies examining the effects of LOX expression in tumors. Hypoxia inducible factor (HIF)-induced LOX, LOXL2 and LOXL4 secretion from breast cancer cells in the primary tumor mediate the crosslinking of collagens in the lungs, facilitating the invasion of CD11b+ bone marrow-derived cells (BMDCs). These BMDCs can bestow a growth advantage to metastatic cells by increasing their recruitment via chemokines, by promoting the formation of blood vessels in the incipient metastatic lesion, thereby supporting cancer cell survival and proliferation. Additionally, they suppress natural killer cell dependent immune responses in tumor-bearing animals (Wong et al. 2012). Thus, the BMDCs that are recruited to the metastatic niche create a favorable microenvironment for lung colonization of metastatic cancer cells and their proliferation. In fact, knockdown of LOX or LOXL4 or LOXL2 results in suppressed collagen crosslinking, BMDC recruitment, and breast cancer metastasis. The LOX-mediated changes may be similar to the ones at the primary niche and may further trigger the angiogenic switch and lead to cancer cell expansion from micrometastasis to macrometastasis. However, this notion remains to be tested experimentally.

Thus, these data suggest that deregulation of ECM dynamics is an important step during the formation of a pre-metastatic niche. However, from these studies it is hard to determine whether the observed enhancement of metastasis is due to a direct role for the extracellular matrix on the proliferation of micrometastases and their progression to clinically relevant disease.

Fibrosis, Collagen-I and Breast Cancer Dormancy

We therefore attempted to shed some light on this question by using a hyperplastic nodule-derived mammary cancer cell line (D2.0R) that has been shown to efficiently reach and survive in the lungs of nude mice, but fails to proliferate and remains dormant (Naumov et al. 2002). We have resorted to inducing fibrosis in the lungs of nude mice, using an adenoviral vector expressing a mutant TGFb[223/225], which does not associate with the latency associated peptide (LAP) and is therefore activated independently of cleavage. Intranasal instillation of adenoviral TGFb[223/225] results in lung fibrosis and collagen-I accumulation. In order to determine the effect of lung fibrosis on the proliferation of disseminated breast cancer cells, tail vein injection of breast cancer cells was used as the delivery method to bypass any effect mediated by lung fibrosis on the preceding steps in the metastatic cascade, if any. Mice with lung fibrosis exhibited a shift towards overtly proliferating metastatic lesions in the lungs as opposed to control mice without fibrosis (Barkan et al. 2010) (Fig. 16.1). This proliferation was also shown to be dependent on integrin β1 expression as integrin β1 shRNA expressing D2.0R cells

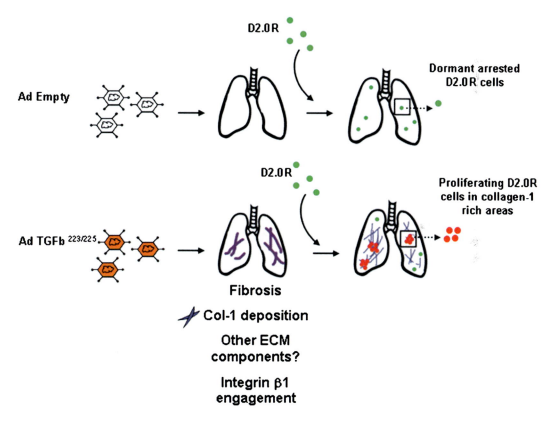

Fig. 16.1 *In vivo* fibrosis model. The intranasal instillation of adenoviral TGFb[223/225] prior to the tail vein injection of D2.0R cells results in fibrosis and collagen-I deposition. Tail-vein injected D2.0R cells readily proliferate in collagen-I rich areas in the fibrotic lung in an integrin β1-dependent fashion but remain growth arrested and dormant in the absence of fibrosis

failed to proliferate in the fibrotic lungs. Importantly in our study, both parental, D2.0R non-target shRNA as well as integrin β1 shRNA expressing D2.0R cells failed to proliferate in the normal, non-fibrotic lungs but remained viable suggesting that integrin β1 expression was specifically required for the proliferation of the mammary cancer cells and not their viability in the fibrotic lung.

Given that collagen-I is enriched in the fibrotic environment but is not the only player in the fibrotic response, we aimed at determining whether collagen-I can directly affect the proliferation of the otherwise dormant D2.0R cells. Towards this aim, we supplemented the in vitro three-dimensional model which recapitulates the behavior of dormant breast cancer cells in vivo with collagen-I (Barkan et al. 2008). In the presence of collagen-I, the otherwise dormant D2.0R cells readily proliferated in an integrin β1-dependent manner within 24 h. Further strengthening the link between collagen-I enrichment and proliferation of dormant cell, we made use of the behavior of the D2A1 cell line. This cell line is characterized by a transient dormant phenotype followed by a spontaneous switch to proliferation in vitro and in vivo. Upon tail vein injections of D2A1 cells in a normal lungs, the macrometastases that were observed at the end of the experiment, were surrounded by collagen-I expression which wasn't seen in areas devoid of macrometastases. The collagen-I induced proliferation was subsequently shown to be dependent on the integrin β1 downstream activation of the focal adhesion kinase (FAK)/SRC pathway that resulted in the activation of the extracellular receptor kinase (ERK1/2), the phosphorylation of the myosin light chain kinase (MLCK) and stress fiber formation. Interestingly, FAK phosphorylation/activation was also shown to be reduced by transfection with LOX antisense oligonucleotides and rescued by LOX re-expression in MDA-MB 231 cells, showing an enzymatic role for hypoxia and LOX in the regulation of FAK through β1 integrin (Erler et al. 2006) but also highlighting the importance of FAK activity in the metastatic niche.

Discussion

Our work has demonstrated that fibrosis in vivo or the supplementation of collagen-I in a 3D model of dormancy in vitro, induces important changes that affect the behavior of dormant cells and induces their proliferation in metastatic sites. This work has also identified integrin β1, FAK and SRC as potential targets in the context of dormancy. Integrins have played a crucial role in strengthening the connection between the extracellular matrix and metastasis. For example, α3β1 integrin is required for adhesion to laminin during pulmonary metastasis (Wang et al. 2004) whereas blocking αvβ1 integrin inhibited metastasis through suppression of expression of the protease urokinase plasminogen activator (Upa) which is consistent with the inhibition of metastasis associated with uPA deficiency in the MMTV-PyMT model of breast cancer (Almholt et al. 2005). However, in this system it is possible that the reduction in lung metastasis could be due to a decreased number of metastases (suggesting an effect on survival in the lung) or decreased growth of the metastases.

In our work, we have demonstrated that integrin β1 is needed for the proliferation of the dormant cells in the fibrotic environment, that in part, may be mediated by the presence of collagen-I. On the other hand, our studies are corroborated by the observation that FAK is an important player in the dormant to proliferative switch in other dormancy models (Aguirre-Ghiso 2002). In this work, we have shown that proliferating breast cancer cells can be induced back to dormancy by interfering with the cues that they receive from their collagen-I rich surroundings. However, we cannot be sure that this interruption is sufficient to keep this dormant population in check with further dynamic remodelling of the ECM or upon a possible angiogenic switch. Therefore, these conclusions open the door to research aimed at finding therapies that would interfere with the viability of this dormant population of cells.

It is also very important to mention that fibrosis is a very relevant phenotype for breast cancer patients. In fact, osteosclerotic breast cancer bone metastases exhibit marrow fibrosis and new bone formation (Kamby and Sengeløv 1997). On the other hand, certain breast cancer therapies have been shown to induce fibrosis in breast cancer patients as a side effect. Radiotherapy has been shown to induce a long-term increase in the turnover of type I collagen and leads to the accumulation of cross-linked type I collagen in the skin of breast cancer patients (Sassi et al. 2001). However these observations are not limited to the skin, but also involve radiotherapy-induced pulmonary fibrosis (Vågane et al. 2008), thereby affecting one of the principal sites suspected to contain dormant breast cancer cells. These findings highlight the importance of determining which patients have a higher predisposition for tumor dissemination when considering radiotherapy or certain other regimens that have been shown to induce fibrosis in breast and other cancer patients. Alternatively, these findings warrant the examination of possible therapies aimed at a rapid reversal of the therapy-induced fibrosis in high risk patients.

References

Aguirre Ghiso JA (2002) Inhibition of FAK signaling activated by urokinase receptor induces dormancy in human carcinoma cells in vivo. Oncogene 21:2513–2524

Almholt K, Lund LR, Rygaard J, Nielsen BS, Danø K, Rømer J, Johnsen M (2005) Reduced metastasis of transgenic mammary cancer in urokinase-deficient mice. Int J Cancer 113:525–532

Barcellos-Hoff MH (1998) The potential influence of radiation-induced microenvironments in neoplastic progression. J Mammary Gland Biol Neoplasia 3:165–175

Barkan D, Kleinman H, Simmons JL, Asmussen H, Kamaraju AK, Hoenorhoff MJ, Liu Z-y, Costes SV, Cho EH, Lockett S, Khanna C, Chambers AF, Green JE (2008) Inhibition of metastatic outgrowth from single dormant tumor cells by targeting the cytoskeleton. Cancer Res 68:6241–6250

Barkan D, El Touny LH, Michalowski AM, Smith JA, Chu'I, Davis AS, Webster JD, Hoover S, Simpson RM, Gauldie J, Green JE (2010) Metastatic growth from dormant cells induced by a col-I-enriched fibrotic environment. Cancer Res 70:5706–5716

Boyd NF, Dite GS, Stone J, Gunasekara A, English DR, McCredie MRE, Giles GG, Tritchler D, Chiarelli A, Yaffe MJ, Hopper JL (2002) Heritability of mammographic density, a risk factor for breast cancer. N Engl J Med 347:886–894

Colpaert CG, Vermeulen PB, Fox SB, Harris AL, Dirix LY, Van Marck EA (2003) The presence of a fibrotic focus in invasive breast carcinoma correlates with the expression of carbonic anhydrase IX and is a marker of hypoxia and poor prognosis. Breast Cancer Res Treat 81:137–147

Condeelis J, Segall JE (2003) Intravital imaging of cell movement in tumours. Nat Rev Cancer 3:921–930

Cukierman E (2004) A visual-quantitative analysis of fibroblastic stromagenesis in breast cancer progression. J Mammary Gland Biol Neoplasia 9:311–324

Cuzick J, Warwick J, Pinney E, Duffy SW, Cawthorn S, Howell A, Forbes JF, Waren RML (2011) Tamoxifen-induced reduction in mammographic density and breast cancer risk reduction: a nested case-control study. J Natl Cancer Inst 103:744–752

Egeblad M, Rasch MG, Weaver VM (2010) Dynamic interplay between the collagen scaffold and tumor evolution. Curr Opin Cell Biol 22:697–706

Elkabets M, Gifford AM, Scheel C, Nilsson B, Reinhardt F, Bray M-A, Carpenter AE, Jirström K, Magnusson K, Ebert BL, Pontén F, Weinberg RA, McAllister SS (2011) Human tumors instigate granulin-expressing hematopoietic cells that promote malignancy by activating stromal fibroblasts in mice. J Clin Invest 121:784–799

Erler JT, Bennewith KL, Nicolau M, Dornhöfer N, Kong C, Le Q-T, Chi J-TA, Jeffrey SS, Giaccia AJ (2006) Lysyl oxidase is essential for hypoxia-induced metastasis. Nature 440:1222–1226

Erler JT, Bennewith KL, Cox TR, Lang G, Bird D, Koong A, Le Q-T, Giaccia AJ (2009) Hypoxia-induced lysyl oxidase is a critical mediator of bone marrow cell recruitment to form the premetastatic niche. Cancer Cell 15:35–44

Freund H, Biran S, Laufer N, Eyal Z (1976) Breast cancer arising in surgical scars. J Surg Oncol 8: 477–480

Hasebe T, Sasaki S, Imoto S, Mukai K, Yokose T, Ochiai A (2002) Prognostic significance of fibrotic focus in invasive ductal carcinoma of the breast: a prospective observational study. Mod Pathol 15: 502–516

Herrick SE, Sloan P, McGurk M, Freak L, McCollum CN, Ferguson MW (1992) Sequential changes in histologic pattern and extracellular matrix deposition during the healing of chronic venous ulcers. Am J Pathol 141:1085–1095

Kamby C, Sengeløv L (1997) Pattern of dissemination and survival following isolated locoregional recurrence of breast cancer. A prospective study with more than 10 years of follow up. Breast Cancer Res Treat 45:181–192

Kaplan RN, Riba RD, Zacharoulis S, Bramley AH, Vincent L, Costa C, MacDonald DD, Jin DK, Shido K, Kerns SA, Zhu Z, Hicklin D, Wu Y, Port JL, Altorki N, Port ER, Ruggero D, Shmelkov SV, Jensen KK, Rafii S, Lyden D (2005) VEGFR1-positive haematopoietic bone marrow progenitors initiate the pre-metastatic niche. Nature 438:820–827

Kenny PA, Bissell MJ (2003) Tumor reversion: correction of malignant behavior by microenvironmental cues. Int J Cancer 107:688–695

Kotb AM, Hierholzer A, Kemler R (2011) Replacement of E-cadherin by N-cadherin in the mammary gland leads to fibrocystic changes and tumor formation. Breast Cancer Res 13:R104

Kuperwasser C, Chavarria T, Wu M, Magrane G, Gray JW, Carey L, Richardson A, Weinberg RA (2004) Reconstruction of functionally normal and malignant human breast tissues in mice. Proc Natl Acad Sci U S A 101:4966–4971

Levental KR, Yu H, Kass L, Lakins JN, Egeblad M, Erler JT, Fong SFT, Csiszar K, Giaccia A, Weninger W, Yamauchi M, Gasser DL, Weaver VM (2009) Matrix crosslinking forces tumor progression by enhancing integrin signaling. Cell 139:891–906

Loya A, Guray M, Hennessy BT, Middleton LP, Buchholz TA, Valero V, Sahin AA (2009) Prognostic significance of occult axillary lymph node metastases after chemotherapy-induced pathologic complete response of cytologically proven axillary lymph node metastases from breast cancer. Cancer 11:1605–1612

Lyons TR, O'Brien J, Borges VF, Conklin MW, Keely PJ, Eliceiri KW, Marusyk A, Tan A-C, Schedin P (2011) Postpartum mammary gland involution drives progression of ductal carcinoma in situ through collagen and COX-2. Nat Med 17:1109–1115

Ma X-J, Dahiya S, Richardson E, Erlander M, Sgroi DC (2009) Gene expression profiling of the tumor microenvironment during breast cancer progression. Breast Cancer Res 11:R7

Naumov GN, MacDonald IC, Weinmeister PM, Kerkvliet N, Nadkarni KV, Wilson SM, Morris VL, Groom AC, Chambers AF (2002) Persistence of solitary mammary carcinoma cells in a secondary site: a possible contributor to dormancy. Cancer Res 62:2162–2168

Paget S (1889) The distribution of secondary growths in cancer of the breast. Cancer Metastasis Rev 8:98–101

Paszek MJ, Weaver VM (2004) The tension mounts: mechanics meets morphogenesis and malignancy. J Mammary Gland Biol Neoplasia 9:325–342

Provenzano PP, Inman DR, Eliceiri KW, Knittel JG, Yan L, Rueden CT, White JG, Keely PJ (2008) Collagen density promotes mammary tumor initiation and progression. BMC Med 6:11

Sassi M, Jukkola A, Riekki R, Höyhtyä M, Risteli L, Oikarinen A, Risteli J (2001) Type I collagen turnover and cross-linking are increased in irradiated skin of breast cancer patients. Radiother Oncol 58:317–323

Siegel R, Naishadham D, Jemal A (2012) Cancer statistics, 2012. CA Cancer J Clin 62:10–29

Soikkeli J, Podlasz P, Yin M, Nummela P, Jahkola T, Virolainen S, Krogerus L, von Heikkilä P, Smitten K, Saksela O, Hölttä E (2010) Metastatic outgrowth encompasses COL-I, FN1, and POSTN up-regulation and assembly to fibrillar networks regulating cell adhesion, migration, and growth. Am J Pathol 177:387–403

Sotgia F, Martinez-Outschoorn UE, Howell A, Pestell RG, Pavlides S, Lisanti MP (2012) Caveolin-1 and cancer metabolism in the tumor microenvironment: markers, models, and mechanisms. Annu Rev Pathol 7:423–467

Troester MA, Lee MH, Carter M, Fan C, Cowan DW, Perez ER, Pirone JR, Perou CM, Jerry DJ, Schneider SS (2009) Activation of host wound responses in breast cancer microenvironment. Clin Cancer Res 15:7020–7028

Vågane R, Bruland ØS, Fosså SD, Olsen DR (2008) Radiological and functional assessment of radiation-induced pulmonary damage following breast irradiation. Acta Oncol 47:248–254

Wang H, Fu W, Im JH, Zhou Z, Santoro SA, Iyer V, DiPersio CM, Yu Q-C, Quaranta V, Al-Mehdi A, Muschel RJ (2004) Tumor cell alpha3beta1 integrin and vascular laminin-5 mediate pulmonary arrest and metastasis. J Cell Biol 164:935–941

Wong CC-L, Zhang H, Gilkes DM, Chen J, Wei H, Chaturvedi P, Hubbi ME, Semenza GL (2012) Inhibitors of hypoxia-inducible factor 1 block breast cancer metastatic niche formation and lung metastasis. J Mol Med (Berl) 90(7):803–815

Wynn TA (2008) Cellular and molecular mechanisms of fibrosis. J Pathol 214:199–210

Yates CC, Bodnar R, Wells A (2011) Matrix control of scarring. Cell Mol Life Sci 68:1871–1881

Diagnosis of Branchial Cyst Carcinoma: Role of Stem Cells and Dormancy

17

Athanassios Kyrgidis

Contents

Abstract	165
Introduction	166
Branchial Arches, Pouches, Clefts	166
Branchial Cleft Cyst	166
Branchial Cleft Cyst Carcinoma	168
Genomic Instability: Stem Cells: Niche	169
The Processes Involved in Malignant Transformation	169
The Hallmarks for Malignant Transformation	170
Stem Cells	171
Stem Cell Niche	172
Dormancy	172
Angiogenic Dormancy	173
Immunosurveillance	173
Cellular Dormancy-Autophagy	174
Dominant Tumor Dormancy Escape	174
HPV Associated Head and Neck Cancer	175
Discussion	176
Epilogue	177
References	177

A. Kyrgidis (✉)
Department of Oral Maxillofacial Surgery,
Aristotle University of Thessaloniki, 3 Papazoli St,
Thessaloniki, 546 30, Greece
e-mail: akyrgidi@gmail.com

Abstract

Branchial cleft cysts are among the commonest causes for a congenital neck mass. Branchial cleft cyst carcinoma (BCCC) is a type of cancer that arises from cells within these cysts. Congenital branchial cysts and BCCC tumours may result from progenitor cell rests of the embryological branchial development.

In the stem-cell carcinogenesis model, a niche of stem-cells, maintains a distinct cellular population. Based on this model, one could anticipate that an explicit stem cell population, remnants of the branchial embryogenesis, would be required to generate any branchial cleft cyst. These cells (TSCs) would need to survive in dormancy for some time before their awakening could be trailed by the clinical presentation of the brachial cleft cyst. For the brachial cleft cyst to continue existing rather than involute, the progenitor stem cells would need to proliferate. With time and while the cyst remains unresected, stem cells could acquire mutations due to genomic instability. During the following years, this population of stem cells would proliferate, maintain and enlarge the cyst and acquire further mutations. With accumulation of further mutations in the stem cells, dysplasia and malignant transformation to TSCs might occur. These mutations might by the driving force for malignant transformation of some progenitor cell clone to TSCs. Such clone TSC populations could be responsible for BCCC tumours. Inadvertent microbial, viral or traumatic inflammation of the region, could result in amplification of neck lump signs described above. From the stem-cell perspective, host monocytes and macrophages approaching the area, along with regional fibroblasts and mesenchymal normal stem cells would modify the cyst microenvironment. This might result in advanced proliferation, possible new mutations and further novel epigenetic changes. Such a process would augment the machinery available to TSCs to overcome the hurdles proposed by Hanahan and Weinberg.

Keywords

Apoptosis and epigenetics • Branchial cyst carcinoma • Head and neck cancer • Malignant transformation • Meiomitosis • Nemosis • Neosis • Senescence • Stem cells and dormancy • Tumor dormancy

Introduction

Branchial Arches, Pouches, Clefts

In the development of vertebrate animals, the pharyngeal arches (which develop into the branchial arches or gill arches in fish) are the origin for a multitude of structures. In humans, they develop during the fourth week in utero as a series of mesodermal outpouchings on the left and right sides of the developing pharynx. In fish, the branchial arches support the gills (branchia is Greek word for gills). Phylogenetically, the branchial apparatus is related to gill slits. In fish and amphibians, these structures are responsible for the development of the gills, hence the name branchial.

Branchial arches grow and join in the ventral midline. The first arch, as the first to form, separates the mouth pit or stomodeum from the pericardium. By differential growth the neck elongates and new arches form, so the pharynx has six arches ultimately. Each pharyngeal arch has a cartilaginous stick, a muscle component which differentiates from the cartilaginous tissue, an artery, and a cranial nerve. Each of these is surrounded by mesenchyme. Arches do not develop simultaneously, but instead possess a development spread over a period of time (Bhattacherjee et al. 2007).

Pharyngeal pouches (or branchial pouches) form on the endodermal side between the arches, and pharyngeal grooves (or clefts) form from the lateral ectodermal surface of the neck region to separate the arches. The pouches line up with the clefts, and these thin segments form the gills in fish. In mammals the endoderm and ectoderm not only remain intact, but continue to be separated by a mesoderm layer. At the fourth week of embryonic life, the development of four branchial (or pharyngeal) clefts results in six ridges known as the branchial (or pharyngeal) arches, which contribute to the formation of various structures of the head, the neck, and the thorax. Although there are six pharyngeal arches, in humans the fifth arch only exists transiently during embryologic growth and development. Since no human structures result from the fifth arch, the arches in humans are I, II, III, IV, and VI. More is known about the fate of the first arch than the remaining four. Arches I, II and III contribute to structures above the larynx, while arches IV and VI contribute to the larynx and trachea. The second arch grows caudally and, ultimately, covers the third and fourth arches. The buried clefts become ectoderm-lined cavities, which normally involute around week 7 of development. If a portion of the cleft fails to involute completely, the entrapped remnant forms an epithelium-lined cyst with or without a sinus tract to the overlying skin (Bhattacherjee et al. 2007).

Branchial Cleft Cyst

Branchial cleft cysts (also known as "Pharyngeal fistulas" or "Branchial cysts" or "Branchiomas") are congenital, oval, moderately movable epithelial cystic masses that develop under the skin in the neck between the sternocleidomastoid muscle and the pharynx. They are considered to from the failure of the obliteration of the second branchial cleft or pouch. They are among the most common causes for a congenital neck mass and are located anterior to the sternocleidomastoid muscle and posterior to the angle of the mandible. Cyst walls are composed of either squamous or columnar epithelial cells with lymphoid infiltrate – a characteristic feature- with often prominent germinal centers. The cyst may contain granular and/or keratinized cellular debris. Similar to some cysts of odontogenic origin, cholesterol crystals may also be found in the branchial cyst content. Branchial cleft cysts are remnants of embryonic development and result from a failure of obliteration of the branchial clefts (which in fish develop into gills – Greek word: branchia) The exact incidence of branchial cleft cysts in the US population is unknown. Branchial cleft cysts are the most common congenital cause of a neck mass. An estimated 2–3% of cases are bilateral (Ozolek 2009; Papadogeorgakis et al. 2009).

About two-thirds of first branchial anomalies are cysts with the remaining divided equally between sinuses and fistulas (Banikas et al. 2011). There are distinct types of branchial cysts, according to their origin.

- First or primary branchial cleft cysts typically originate in the angle of the mandible and extend to the external auditory canal. They are often associated with the facial nerve, being the nerve of the second or hyoid branchial arch. With regard to their relation to the facial nerve, Work classified first branchial cleft anomalies into types I and II (Work 1972). Type I lesions tend to course lateral to the facial nerve, near the external auditory canal and the parotid region. Type II lesions track medial to the nerve presenting as preauricular, infra-auricular, or postauricular swellings near the angle of the mandible.
- Second branchial cleft cysts are the most common. Branchial cysts are smooth, nontender, fluctuant masses, which occur along the lower one third of the anteromedial border of the sternocleidomastoid muscle between the muscle and the overlying skin. They are found along the anterior border of the sternocleidomastoid muscle, oftenly pass through the carotid bifurcation and discharge into the tonsillar pillar.
- Third branchial cleft cysts are uncommon. The external opening occurs about 2/3 of the way down the sternocleidomastoid muscle anteriorly similar to second branchial cleft cysts. The tract ascends along the carotid sheath posteriorly to the internal carotid artery, under the glossopharyngeal nerve (structure of the fourth arch), and over the vagus nerve and hypoglossal nerve to open into the piriform sinus (structures of the fourth arch). The lesion may be tender if secondarily inflamed or infected. When associated with a sinus tract, mucoid or purulent discharge onto the skin or into the pharynx may be present (Ozolek 2009; Work 1972).

Branchial cleft cysts are congenital in nature, but they may not present clinically until later in life, usually by early adulthood. A family history of branchial cleft cysts may be present. Many branchial cleft cysts are asymptomatic. They may become tender, enlarged, or inflamed, or they may develop abscesses, especially during periods of upper respiratory tract infection, due to the lymphoid tissue located beneath the epithelium. Spontaneous rupture of an abscessed branchial cleft cyst may result in a purulent draining sinus to the skin or the pharynx.

Contrast-enhanced CT scan will show a cystic and enhancing mass in the neck. MRI allows for finer resolution during preoperative planning. It may aid preoperative planning and identify compromise of local structures. Ultrasonography helps to delineate the cystic nature of these lesions. Fine-needle aspiration may be helpful to distinguish branchial cleft cysts from malignant neck masses. Fine-needle aspiration and culture may help guide antibiotic therapy for infected cysts. HPV-DNA genotyping is also important as there has been a reported tendency for HPV+ SCCs to present with cystic neck nodes (Banikas et al. 2011).

Depending on the size and the anatomical extension of the mass, locally compressive symptoms, such as dysphagia, dysphonia, dyspnea, and stridor, may occur. Furthermore, untreated branchial cleft cyst lesions are prone to recurrent infection and abscess formation with resultant scar formation and possible compromise to local structures. For that, surgical excision has long been suggested. Notably, complications of surgical excision of branchial cleft cysts result from damage to nearby vascular or neural structures, which include carotid vessels and the facial, hypoglossal, vagus, and lingual nerves; cysts can recur following surgery as well. Following surgical excision of branchial cleft cysts, recurrence is uncommon, with a risk estimated at 3%, unless previous surgery or recurrent infection has occurred, in which case, it may be as high as 20%.

Branchial Cleft Cyst Carcinoma

Branchial cleft cyst carcinoma (BCCC) is a type of cancer that arises from cells within the cyst, and is considered to be a controversial entity. Fewer than 40 cases have been reported in the literature (Banikas et al. 2011). Similar to branchiogenic cysts, branchial cleft cyst carcinoma presents as a mass in the superior lateral neck more commonly in men between the fifth and sixth decade, with symptoms lasting from 2 weeks to decades (Maturo et al. 2007; Khafif et al. 1989). Lateral neck masses are frequently treated by ENT and maxillofacial surgeons. Knowledge and clinical suspicion of the characteristic features of each lesion is necessary for accurate diagnosis and surgical treatment (Al-Khateeb and Al Zoubi 2007). The possibility of a primary malignancy in a congenital branchial cyst was first introduced in 1882 by Von Volkman who used the term branchial cleft cyst carcinoma. In 1950, Martin et al. (1950) reviewed 250 published cases and formulated four diagnostic criteria for branchial cleft cyst carcinoma:

1. The tumour must originate on a line extending from the tragus along the anterior border of the sternocleidomastoid muscle to the clavicle, in a region where branchial cysts are naturally found.
2. The histological aspect of the tumour must be compatible with tissue originating from a branchial cyst.
3. No primary tumour must be registered after a survival period of 5 years.
4. Demonstration of a squamous cell carcinoma (SCC) developing in the epithelial lining of a cyst.

The third criterion was subsequently challenged as the 5-year follow-up was considered too theoretical and of limited diagnostic and therapeutic value (Bhanote and Yang 2008; Khafif et al. 1989). Besides two alternate criteria were set forth by Khafif et al. (1989) which are considered more practical and are more widely used. These two criteria include:

5. Absence of any identifiable primary tumour after exhaustive evaluation of the patient (including endoscopy, x-ray examinations, complete CT scan of the head and neck, and appropriate biopsy).
6. Identification of a cystic structure partially lined by normal squamous or pseudo-stratified columnar epithelium with transition to carcinoma. Lymphoid elements such as germinal elements may be present, but not peripheral lobulation, internodular trabeculae, or perinodal sinuses as would be seen in lymph nodes.

BCCC is a diagnosis of exclusion (Devaney et al. 2008). It has been reported that further to the criteria proposed by Martin et al. (1950) and modified by Khafif et al. (1989) surgical excision of the cervical mass should provide definite pathological evidence of both a pre-existing benign developmental cyst and a malignant tumour arising therein (Devaney et al. 2008).

For over a century head and neck surgeons have both defended and attacked the notion of a squamous carcinoma arising in a pre-existing branchial cleft cyst (Pai et al. 2009; Hudgins and Gillison 2009; Goldenberg et al. 2006). This is probably more than a mere academic debate, as treatment of a localized carcinoma arising within a developmental cyst and still entirely confined within that cyst would differ from treatment of a metastatic deposit derived from a (presumably occult) head and neck primary tumour situated elsewhere (Devaney et al. 2008).

It has been suggested that the missing link in BCCC aetiopathogenesis could be the demonstration of in situ BCCC (Bhanote and Yang 2008). In a recent case report demonstrated histopathology was able to detect areas of dysplasia within the cystic epithelium, which coexisted with areas of infiltrating tumour. The authors suggested that these evidence of dysplastic epithelia in the cystic wall, coexisting with cancer transformed epithelia are confirmative of the existence of BCCC.

In this chapter we shall attempt to rationalize the existence of this controversial tumour and discuss its aetiopathogenesis through the context of stem cell biology and the role of tumor dormancy in cancer relapse and regression and immune surveillance of tumors. To achieve this goal, we systematically reviewed the common medical databases (PubMed, EMBASE, Scopus and Cochrane CENTRAL) for studies reporting on stem cell, dormancy which could be applicable to branchial cleft cyst carcinoma diagnosis. Articles were not reviewed on the grounds of outcome or methodology. Rather, the systematic review process aimed to identify eligible articles to support the aim of this study. Results were critically appraised and classified according to the genetic basis of cancer, the stem cell cancer theory and the current knowledge of tumour dormancy.

Genomic Instability: Stem Cells: Niche

The Processes Involved in Malignant Transformation

Apoptosis is the process of programmed cell death that occurs in multicellular organisms. Programmed cell death involves a series of biochemical events leading to a characteristic cell morphology and death. Apoptosis occurs via two interconnected pathways: the extrinsic (death receptor) and intrinsic (mitochondrial-dependent) pathways, both of which ultimately lead to the activation of effector-caspases (caspases-3, -6, -7), the final mediators of cell death, PTEN, CDKN2A and APAF-1 (Kyrgidis and Antoniades 2010).

Epigenetics refer to changes in phenotype (appearance) or gene expression caused by mechanisms other than changes in the underlying DNA sequence, hence the name *epi-*(Greek: over; above)-*genetics*. These changes may remain through cell divisions for the remainder of the cell's life and may also last for multiple generations. However, there is no change in the underlying DNA sequence of the organism; instead, non-genetic factors cause the organism's genes to behave differently. Epigenetic alterations include both DNA modification (methylation of CpG islands) and alteration in DNA packaging (histones). Over the past 10 years, more than 20 genes in oral have been shown to be dysregulated through epigenetic changes (Kyrgidis and Antoniades 2010).

Nemosis is a novel term referring to fibroblast activity. Nemosis involves an increased production of growth factors and proinflammatory and proteolytic proteins, while at the same time cytoskeletal proteins are degraded. The cells that produce these cytokines are cancer associated fibroblasts (CAF), in plain English these are peritumoral fibroblasts. The expression of CAF markers differs between normal fibroblasts and CAFs in nemosis. This emphasizes the heterogeneity of fibroblasts and the evolving tumor-promoting properties of CAFs (Kyrgidis and Antoniades 2010).

Senescence is a characteristic of normal cells that refers to their limited proliferative potential, that is, they permanently cease proliferating ('senesce') after about 50 population doublings. During normal cell proliferation the ends of chromosomes (telomeres) become progressively shortened due to a phenomenon known as the 'end replication problem'. Senescence is thought to be triggered by signals generated by one or more telomeres reaching a critical length (Kyrgidis and Antoniades 2010).

Neosis is a novel type of cell division, characterized by karyokinesis via nuclear budding followed by asymmetric, intracellular cytokinesis, producing several small mononuclear cells, termed the Raju cells, with extended mitotic life span. Normally, multinucleate/polyploid giant cells (MN/PGs) formed due to DNA damage are supposed to die via mitotic catastrophe. Before they die, some MN/PGs may undergo neosis. In the latter case, mitotic derivatives of Raju cells would give rise to transformed cell lines, which inherit may genomic instability, display a phenotype and transcriptome different from the maternal cell population, and anchorage-independent growth (Kyrgidis and Antoniades 2010). Some authors suggested that stemness may be a transient, cyclic property afforded by de-polyploidisation of senescing cells which have undergone polyploidisation. Furthermore, it has been suggested that mitotic catastrophe may represent a mechanism through which the cell can switch from the usual mitotic cell-cycle to this novel but evolutionarily conserved life cycle. Intriguingly, some most recent data suggest that cell senescence may be reversible and that stem cells are tolerant to polyploidy caused by genotoxic stress (Kyrgidis and Antoniades 2010). Senescence phenotype can be induced by (i) telomere attrition-induced senescence at the end of the cellular mitotic life span and (ii) also by replication history-independent, accelerated senescence due to inadvertent activation of oncogenes or by exposure of cells to genotoxins. Tumor suppressor genes p53/RB/p16INK4A and related senescence checkpoints are involved in effecting the onset of senescence. However, senescence as a tumor suppressor mechanism is a leaky process and senescent cells with mutations or epigenetic alterations in these genes escape mitotic catastrophe-induced cell death by becoming polyploid cells. These polyploid giant cells, before they die, give rise to several cells with viable genomes via neosis and the immediate neotic offsprings are the Raju cells. The latter inherit genomic instability and transiently display stem cell properties in that they differentiate into tumor cells and display extended, but, limited mitotic life span, at the end of which they enter senescent phase and can undergo secondary/tertiary neosis to produce the next generation of Raju cells. Neosis could be repeated several times during tumor growth in a non-synchronized fashion, is the mode of origin of resistant tumor growth and contributes to tumor cell heterogeneity and continuity. The main event during neosis might be the production of mitotically viable daughter genome following epigenetic alterations from the non-viable polyploid genome of neosis maternal cell. This could allow for the growth of resistant tumor cells. Because spindle checkpoint is not activated during neosis, aneuploidy can result (Kyrgidis and Antoniades 2010). Thus, tumor cells may also be destined to die due to senescence, but may be able to escape senescence due to mutations or epigenetic alterations in the senescent checkpoint pathway.

Meiomitosis is a mitotic division of cancer cells where these cells also express meiotic machinery during their division. Normal meiosis is a cellular division pathway in germ cells that promotes the recombination of large expanses of chromosomal DNA, potentially similar in size to those lost or gained in cancers. A major characteristic of meiosis is chiasma or sister chromatid exchange. In this process, homologous chromosomes are held together by cohesions, and after resolution of chiasma (the recombination crossover points with the other chromosome pair), the homologous chromosomes are segregated together. In normal mitosis, there are no chiasma and sister chromatids are directly separated. In meiomitosis, the presence of chiasma, cohesins, and an aberrantly assembled mitotic spindle would be expected to result in significant shearing and/or mis-segregation of DNA sequences. Cells surviving this event would likely be aneuploid (Kyrgidis and Antoniades 2010).

The Hallmarks for Malignant Transformation

Back in 2000, Hanahan and Weinberg reported the hallmarks of cancer. In recent years, cancer research is focussing on finding new biomarkers of potential prognostic value and on better understanding the pathophysiological processes of the disease, on improving current therapies or introducing novel treatment approaches (Kyrgidis et al. 2005).

Human cells, in order to transform to cancer cells, need to (1) provide growth signals-obtain growth self-sufficiency, (2) ignore growth-inhibitory signals, (3) avoid apoptosis, (4) replicate without limit, (5) sustain angiogenesis and (6) invade and proliferate (Kyrgidis et al. 2005).

In this chapter we will try to focus on branchial cleft cyst carcinoma and dormancy through a cell cycle perspective. For this, the hallmarks proposed by Hanahan and Weinberg offer the required alterations for cancer and metastasis development. Of course, this segregation is only aimed to provide better understanding of the various mechanisms by researchers. In cell cycle reality, all these molecular and genetic changes interplay to give birth to a population of cells with enhanced properties. Cancer may be considered a process of "natural selection." If these new cells acquired properties that give them advantage over the remainder of cells in a multicellular organism, they would proliferate. Darwin's law states "survival of the fittest"; cancer cells are only fittest than normal cells and this gives them the advantage to proliferate. This proliferation takes place at the expense of the community. Once these cells reach a significant number (clinically manifesting as cancer) they could attempt a "mutiny" and subsequently destroy the host organism-cell community (Kyrgidis and Antoniades 2010).

Stem Cells

While extensive progress has been made on the molecular mechanics of cancer, the exact population of cancer cells where these changes will occur has not been defined. Along with the progress of stem-cell biology, the possible existence of "cancer stem-cells" that is rare cancer cells with indefinite potential for self-renewal that drive carcinogenesis has been proposed (Kyrgidis and Antoniades 2010). The recognition of a subpopulation of cancer stem cells (CSCs) in solid cancers has reinvigorated the field. These cells have the capacity for self-renew and give rise to more differentiated cell forms (Kyrgidis and Antoniades 2010). Importantly, these cells have the ability to survive long enough so that the required number of "hits" for malignant transformation occur (Kyrgidis et al. 2010).

CSCs have access to embryologic developmental programs, including the capacity to differentiate along multiple cell lineages. To sustain and repair tissue, a considerable number of daughter cells, CSCs must maintain significant plasticity into adulthood. Carcinogenesis is currently considered to be an aberrant developmental process (Kyrgidis and Antoniades 2010).

According to the stem cell model, the initial mutations accumulate in a quiescent stem cell. When eventually, environmental signaling would activate the stem cell, due to these mutations, proliferation would not be appropriately controlled. The neoplastic cells produced would attempt to follow normal differentiation pathways Although CSCs are likely to play a role in the cause of a variety of cancers, there have been no research reports proving the successful isolation of specific CSCs from SCC lesions to date. Immunohistochemistry, transcriptional studies and micro-morphological data have all been employed to demonstrate or refute the CSC hypothesis in oral carcinomas. Treatment strategies of SCC are based on the idea that oral cancer is generated by a population of cells of equal proliferative and aggressive potential More recent reports suggest that the CSC hypothesis may likely also apply to SCC. SCC CSCs could either originate directly from malignant stem cells or the tumorigenic potential may be promoted in the supra-basal compartment, where differentiating cells deviate from their predetermined differentiation route and transform into malignant intraepithelial cells. The latter possibility is supported by the existence of the Tis-stage (carcinoma in situ) with the presence of typical malignant intraepithelial cells and an intact basal membrane. In Tis-lesions, malignant potential seems to develop in the stratum spinosum, but not in the stratum basale of the squamous epithelium. The dedifferentiation cascade in SCC is typically associated with p53-mutations and loss of DNA-repair mechanisms (Kyrgidis and Antoniades 2010). In addition, oncogene activation, tumor suppressor gene inactivation, and the downregulation of cell-cell contact proteins, such as E-cadherin and β-catenin (which is also an intracellular mediator (Kyrgidis et al. 2010)), were described in SCC. SCC is comprised from malignant keratinocytes and their corresponding malignant stem cells might be found in the stratum basale of the oral and dermal epithelium, which has been identified in immunohistochemical studies to harbor epithelial stem cells (Kyrgidis and Antoniades 2010).

Stem Cell Niche

Both normal and cancer stem cells need a supporting niche. This niche is composed of tissue cells and extracellular matrix creating very specific microenvironment that hosts stem cells and controls their self-renewal and progeny production (Favaro et al. 2008). Notably, the niche is not merely a docking site; Niche supporting cell population and extracellular matrix modulate stem cell homeostasis by either maintaining them in a quiescent state or helping them to proliferate when required. Recent evidence suggests that tumor cells target and parasitize the haemopoietic stem cell (HSC) niche during metastasis (Chen and Pienta 2011). Normal cells recruited during the angiogenic process, or chemotactically attracted to metastatic site paracrine mediators produced by cancer cells, have been shown to create an appropriate niche for TSCs proliferation and survival (Favaro et al. 2008; Chen and Pienta 2011). Angiogenic stimulants as those produced during inflammation, are useful modifying niche properties and striving towards proliferation of normal cells. However, the same is very true for TSCs and the same mechanisms could lead to re-activation of poorly angiogenic micrometastases seeded into tissues (Favaro et al. 2008). Microenvironmental control of tumor cell proliferation is essential in the stem cell niche, but intrinsic properties of the tumor cells – such as their heterogeneous growth potential – could also contribute to determining the final outcome of the metastatic process. The release of tumor cells from quiescence has been classically attributed to the promoting role of growth factors produced by the host, i.e. following surgical removal of the primary tumor (see below).

Dormancy

The Australian pathologist Rupert Willis originally coined the term 'dormant tumour cells' having analyzed the metastatic spread of human cancers in autopsy studies (Klein 2011). Hadfield who introduced the idea of a 'temporary mitotic arrest' first employed the term "dormant" to describe the malignant cells which, although remaining viable for relatively long periods, show no evidence of multiplication during this time, yet retain all their former and vigorous capacity to multiply (Alsabti 1979; Klein 2011).

While the original meaning of the term clearly links tumor dormancy to cancer recurrence as metastatic disease, tumor dormancy is, however, sometimes used for the latency period of an undetected primary tumor until its clinical diagnosis (Klein 2011). Since disseminated tumor cells (DTCs) can be routinely detected in the bone marrow or lymph nodes of cancer patients long before manifestation of metastases, they have been considered as metastasis precursor/founder cells and as material correlate of minimal residual disease and clinical tumor dormancy (Klein 2011). DTCs are widespread both locally and systematically in a cancer patient who endured primary tumour excision. However, the notion of those cells that is able to cause metastasis is unknown, Also unknown is the notion of those cells that will eventually give origin to metastases.

Patients with cancer can develop recurrent metastatic disease with latency periods that range from years even to decades. This pause can be explained by cancer dormancy, a stage in cancer progression in which residual disease is present but remains asymptomatic. Cancer dormancy is poorly understood, resulting in major shortcomings in our understanding of the full complexity of the disease (Aguirre-Ghiso 2007).

Understanding of tumour dormancy would necessitate identification of the host factors that act to transiently restrain residual tumour cells from outgrowth and the mechanisms involved in the final breakthrough to overt neoplasia. Tumor dormancy is a state of co-existence in which tumor cells persist in a clinically normal host for prolonged periods of times. This state of dormancy is characterized by two remarkable features:
1. Tumor cells are not destroyed by the host's natural defence mechanisms.

2. Tumor cells do not grow out rapidly to form a clinically evident tumour.

The aetiology behind dormancy is not yet clear. Aquirre-Ghiso proposed three distinct types of tumor dormancy:
1. Angiogenic dormancy,
2. Cellular dormancy-autophagy and (Aguirre-Ghiso 2007; Klein 2011)
3. Immunosurveillance (Aguirre-Ghiso 2007).

Other studies have shown that removal of the primary cancer reactivates proliferative and metastatic pathways in residual tumor (Woolgar et al. 2011). On theory explaining dormancy following tumor resection may be that the primary tumor was "dominant", preventing or suppressing growth of the nodal deposit (s) and that the rapid growth follows loss of the primary tumor's inhibitory influence. Loss of the suppressive effect following surgery on the primary tumor removes the suppressive effect and promotes metastatic growth.

Therefore, a fourth category may be added:
4. Dominant tumor dormancy escape (Woolgar et al. 2011).

It is possible that some primary tumors produce angiostatin that inhibits angiogenesis (Woolgar et al. 2011); hence, these mechanisms may well be combined to induce dormant state, rather than each one of them induce dormancy on its own.

Angiogenic Dormancy

Tumor dormancy can result from arrested angiogenesis in which tumor mass expansion is blocked due to inability to recruit new and functional blood vessels. Tumor size during dormancy results from a balance of proliferation and apoptosis of tumour cells. Escape from dormancy is associated with generation of neovasculature and decrease of apoptotic tumour cells (Almog 2010). Cancer cells produce their own angiogenesis stimulators. This way they can provoke the release of stimulators bound to the surrounding extracellular matrix and induce macrophages to secrete more angiogenesis stimulators, thereby further promoting angiogenesis. Metastatic cell dormancy state is characterized by an equilibrium between cell proliferation and apoptosis which is thought to be controlled by increased apoptosis, circuitously induced by angiogenesis inhibitors. Patients with solid tumours often have occult micrometastases at the time of detection and surgical removal of their primary tumors. Primary tumor resection is thought to arouse angiogenesis, initiating the proliferation of dormant micrometastases. Dormant micrometastases may have already acquired angiogenic potential before they left the maternal tumour or not (Park et al. 2011; Klein 2011). In the latter case, while the primary tumor cells may be capable of inducing angiogenesis, DTCs may have left the primary site before acquisition of this trait and would therefore be required to independently acquire it to develop metastasis. In this regard, the provision of additional therapy to inhibit angiogenesis after surgery has been suggested to be a rational oncological treatment approach (Park et al. 2011).

Immunosurveillance

A plausible mechanism is immunesurveillance in which tumor mass growth is blocked by active immune response (Almog 2010). The ability to suppress tumor formation had been shown to involve the action of a number of host immune and non-immune effector mechanisms. Maintenance of tumor dormancy may involve the loss of exposed cell surface targets for host effector mechanisms. Tumor escape may also result from impairments in immune mechanisms due to senescence (Alsabti 1979). Tumor dormancy can also be induced by potent immunosuppression of tumor growth by the immune system, where equilibrium between the immune system and the tumor cells results in long-term persistence of tumor dormancy. There is accumulating evidence that cellular as well as humoral responses are required for immune surveillance and the maintenance of tumor dormancy. Indeed, various bone marrow derived cells and inflammatory cells had been implicated in promoting tumor progression and development by inducing inflammation and angiogenesis Both cellular components (mainly macrophages and fibroblasts) and humoral factors associated with inflammation have been shown to enhance tumor growth in numerous preclinical studies (Woolgar et al. 2011).

Cellular Dormancy-Autophagy

Tumour dormancy can result from cell cycle arrest or cellular dormancy in which blockage of tumour mass expansion results from quiescence of tumour cells. The escape from dormancy is associated with increased cell proliferation (Almog 2010). The initial concept of dormancy conceived by Willis theorizes that the restraining ectopic environment is different to that of the site of origin. There is growing evidence that extracellular matrix (ECM), signaling through integrins, is regulating dormancy and is responsible for any flee from it. The bone marrow is the most studied metastatic site (Kyrgidis et al. 2011). It has been long recognized that the establishment of osteolytic and osteoblastic metastases requires an intricate communication between tumor cells, osteoclasts, osteoblasts and other cells (Kyrgidis et al. 2009; Klein 2011). Osteoclast activation – which is dependant on osteoblast RANK signaling- leads to the release of several growth factors and cytokines immobilized by the mineralized bone matrix and thereby promotes metastatic colonization, while osteoblasts may directly induce cancer cell dormancy (Klein 2011).

Aguirre-Ghiso and Ossowski used the human HEp3 head and neck cancer cell line to demonstrate the extracellular signal-regulated kinase (ERK) MAPK/p38SAPK activity ratio predicts whether the cells will proliferate or enter a state of dormancy in vivo (Aguirre-Ghiso et al. 2001). This stress dependent activation of p38 provides a survival signal via the upregulation of the endoplasmic reticulum stress responses a fact that could also confer chemotherapeutic drug resistance to those dormant cells (Klein 2011). mTOR signaling has also been implicated in the survival of dormant cells (Schewe and Aguirre-Ghiso 2008). The team by Aguirre-Ghiso elucidated the fundamental role of p38 and they further attempted to identify specific transcription factor networks which supposedly activated by p38 would launch the quiescence program. Of note findings are that quiescence is activated via a number of different mechanisms such as growth factor withdrawal and loss of adhesion. Moreover, quiescence is not simply equivalent to exit from the cell cycle like in G1 or G2 arrest but there is suppression of cell differentiation; the latter suppression of differentiation is reversible, in contrast to previously reported senescence (Schewe and Aguirre-Ghiso 2008; Adam et al. 2009; Aguirre-Ghiso et al. 2001, 2003; Klein 2011). This quiescence-related de-differentiation provides the basis for speculations about a possible link between cancer stem cells and dormancy.

Human normal cells that encounter a variety of stresses undergo an evolutionarily conserved process of self-digestion termed autophagy. The importance of this intracellular damage response for pathophysiology has been established across multiple fields, including infectious disease, neurodegeneration, heart failure, and cancer (Amaravadi 2008). In cancer in particular, the debate continues to rage as to whether or not autophagy is primarily a mechanism of cell death or cell survival. In addition to quiescence, autophagy, which can produce the death or permit the survival of tumor cells (Mathew et al. 2007) may be an alternative but relevant dormancy mechanism. The tumor suppressor ARHI was identified as an effector of autophagy; (Amaravadi 2008). Evidence suggests that cell destiny, which could be either autophagic death – apoptosis or autophagic dormancy and growth resumption may depend on the surrounding microenvironment (Mathew et al. 2007; Jang et al. 2011; Amaravadi 2008).

Dominant Tumor Dormancy Escape

Angiogenic and immune mechanisms of dormancy have in common that they may regulate primary and metastatic growth. In fact, the molecular genetic analysis of DTCs revealed that DTCs display often different and also fewer genetic alterations than primary tumor cells,

suggesting early dissemination of human cancers and divergent and possibly protracted somatic progression. In a large proportion of early disseminated tumor cells a small number of genetic alterations are found. This fact is may mean two things:
- Either undetected genetic alterations are taking place or
- the lack of growth of these lacking alterations DTCs may be driven by further epigenetic alterations.

The general consensus that emerges from these data is that two stages might explain tumor dormancy:
- one that involves the induction of growth arrest and prolonged survival of disseminated single or small groups of tumor cells and
- a second one that occurs in a small tumor mass where a constant balance between apoptosis and proliferation keeps the lesion small and undiagnosed (Aguirre-Ghiso 2006).

What is more, reported genetic disparities of primary tumors, DTCs and metastases obliges the immune system to react continuously to new emerging variants of cancer cells. Because the immune system controls tumor growth not only by direct cytotoxic attack but also by the induction of growth arrest it may also act as a dormancy mechanism in a manner of cell-cycle inhibition-arrest. Thereby, any segregation of dormancy mechanisms can only be valid in an effort to analyze them; rather the process is complete with all mechanisms described above interplaying and counter parting each other.

In this regard, an experimental model of tumor dormancy therapy for advanced head and neck carcinoma has been reported. Following transplantation of KB cells into nude mice, the mice were given tiracoxib, a selective cyclooxygenase (COX)-2 inhibitor, probucol, an antioxidant, and S-1, an oral pro-drug of 5-fluorouracil (5-FU), or combinations of two of them. The combined administration of tiracoxib with probucol significantly inhibited the tumor growth. The angiogenesis in this group was markedly reduced. Tiracoxib and probucol did not affect the intratumoral concentration of 5-FU when coadministered with S-1. The combined use of tiracoxib and probucol is thus a candidate for use in maintenance therapy after the primary therapy for patients with advanced head and neck carcinoma (Nishimura et al. 2000).

HPV Associated Head and Neck Cancer

HPV infection and activation of the H-ras gene has been firstly reported in oral verrucous carcinomas. This fact is in accordance with the multiple hit carcinogenesis theory and suggest that in – at least-some cases, oral cancer may be a sequelae of the latter events. The association between H&N SCC and human papilloma virus (HPV) has drawn much light in the past decade (Kyrgidis and Antoniades 2010). The expression of the E6 and E7 oncogenes coupled with inactivation of RB and p53 proteins being the major molecular alterations associated with the presence of HPV-16. RB protein levels have been reported to be down regulated in HPV-16-positive SCC, a fact clearly indicative of E7 activity. With regard to E6 and p53, the HPV-16-positive tumors could be classified into two groups. Those tumors (i) exhibiting E6 gene expression and lacking p53 mutations and those (ii) lacking E6 expression and carrying p53 mutations, the latter subgroup having a worse prognosis (Kyrgidis and Antoniades 2010). HPVs are small oncogenic viruses, which are implicated in epithelial carcinogenesis, and p53 is a tumor suppressor gene with a central role in the prevention of genomic injury. The E6 oncoproteins of these high risk HPVs are known to bind and induce degradation of p53 tumor suppressor protein through the ubiquitin pathway. It has been reported that HPV is not associated with the recent surge in the incidence of biologically aggressive oral cavity cancer in young populations nor is HPV presence in oral leukoplakia is a prognostic indicator of malignant transformation.

Detection of those HPV – positive head and neck cancers allow for the segregation of this subset of tumours having better prognosis. This subset of HPV-related SCC typically occurs in younger patients, rather than non-HPV-related SCC of the head and neck, and a neck-cyst like clinical appearance is common (Pai et al. 2009). It has been proposed that occult head and neck tumours could more likely appear as cystic node metastases (Hudgins and Gillison 2009) and that controversial entities like BCCC should be better considered metastases from occult carcinomas, since cancer treatment approaches would be sufficient (Banikas et al. 2011). Although, it might be tempting to adopt such an over-simplistic approach, the wide recognition TSCs and their role in carcinogenesis provides a new context to rationalize the aetiopathogenesis of BCCC.

Discussion

In the stem-cell carcinogenesis model, a niche of stem-cells, maintains a distinct cellular population. Based on this model, one could anticipate that an explicit stem cell population, remnants of the branchial embryogenesis, would be required to generate any brachial cleft cyst. These cells would need to survive in dormancy for some time before their awakening could be trailed by the clinical presentation of the brachial cleft cyst. For the brachial cleft cyst to continue existing rather than involute, the progenitor stem cells would need to proliferate. With time and while the cyst remains unresected, stem cells could acquire mutations due to genomic instability. During the following years, this population of stem cells would proliferate, maintain and enlarge the cyst and acquire further mutations. With accumulation of further mutations in the stem cells, dysplasia and malignant transformation to TSCs might occur. These mutations might by the driving force for malignant transformation of some progenitor cell clone to TSCs. Such clone TSC populations could be responsible for BCCC tumours. Inadvertent microbial, viral or traumatic inflammation of the region, could result in amplification of neck lump signs described above. From the stem-cell perspective, host monocytes and macrophages approaching the area, along with regional fibroblasts and mesenchymal normal stem cells would modify the cyst microenvironment. This might result in advanced proliferation, possible new mutations and further novel epigenetic changes. Such a process would augment the machinery available to TSCs to overcome the hurdles proposed by Hanahan and Weinberg.

Authors who oppose the existence of BCCC argue that occult head and neck tumours could more likely appear as cystic node metastases which are much more common, and therefore controversial entities like BCCC should be better considered metastases from occult carcinomas. They further suggest that in any latter case, cancer treatment approaches for cancer of unknown primary would be sufficient. Cancer of unknown primary (CUP) is getting more uncommon as diagnostic modalities advance. It is now accepted that a significant proportion of tumours diagnosed in the past as CUP are tumors of the nasopharynx. Still, there are patients who are HPV-negative, in whom modern imaging modalities failed to detect a primary SCC, who present with a cystic neck lump, which on histology is SCC. One cannot exclude the possibility that some of these cases are branchial cleft cyst carcinomas, arising from remnants of branchial epithelia. In light of novel findings about tumor stem cells, niches and tumor dormancy, such a scenario appears plausible. But even if one chooses to forgo this scenario, the stem cell – tumor dormancy theory may still apply: It has been reported that CUP may be the clinical presentation of occult micrometastases, following primary tumors that involute (Briasoulis and Pavlidis 1997). Keratoacanthoma is a common skin tumor in the head and neck region which is known to frequently involute. Notably. distinction between keratoacanthoma and well-differentiated squamous cell carcinoma remains a problem, even with modern methods, (Putti et al. 2004) a fact that might implicate both tumours are manifestations of a single entity. Keratoacanthoma stem cells will likely be quite similar to SCC stem cells, apart from infiltrative properties, as reported by Hanahan and Weinberg. In the stem cell niche environment, quiescence or apoptosis are regulated by a myriad of factors, thereby making probable some type of SCC involution. In the latter case, involution of the primary tumour could lead to awakening of micrometastasis via dominant tumour escape dormancy and manifestation of a solid or cystic neck mass.

Epilogue

To endure in time, a literature review in a book chapter, should not provide just a re-hash of the literature. Rather, the author of the chapter needs to shape the literature of a field into a story in order to enlist the support of readers to continue that story (Kyrgidis and Antoniades 2010). In this chapter, we have tried our best to select from a wide list of research papers, put them side by side and create a story that combines them, a theatre play with actors and events but still without an ending. We hope that this way the current chapter attracts the reader into the deeper understanding or oral cancer biology and the implications for novel biomarkers. Ideally, by reaching this point, the reader should be able to understand the complex mechanisms involved in dormancy, quiescence and metastasis in a straightforward manner and also foresee what can be done in the future. The aim of this chapter was not to merely provide information on what is currently known and accepted with regard to carcinogenesis but also to allow the reconstruction of carcinogenesis knowledge.

It is evident that cancer will eventually end being a "black box" for researchers. Cancer precursor cells – stem cells more likely – undergo several genome changes. These changes could be either mutations or epigenetic. These changes provide the stem cells with the ability to self-invoke growth signals, the ability to suppress anti-growth signals, to avoid apoptosis, to replicate without limit, to invade, proliferate and sustain blood supply. Stem cells are able to collect progressively these changes in their genome. Mutations, epigenetic changes, nemosis, neosis and meiomitosis all have roles in the machinery that moves precursor stem cells round the hallmarks of cancer.

References

Adam AP, George A, Schewe D, Bragado P, Iglesias BV, Ranganathan AC, Kourtidis A, Conklin DS, Aguirre-Ghiso JA (2009) Computational identification of a p38SAPK-regulated transcription factor network required for tumor cell quiescence. Cancer Res 69:5664–5672

Aguirre-Ghiso JA (2006) The problem of cancer dormancy: understanding the basic mechanisms and identifying therapeutic opportunities. Cell Cycle 5:1740–1743

Aguirre-Ghiso JA (2007) Models, mechanisms and clinical evidence for cancer dormancy. Nat Rev Cancer 7:834–846

Aguirre-Ghiso JA, Liu D, Mignatti A, Kovalski K, Ossowski L (2001) Urokinase receptor and fibronectin regulate the ERK (MAPK) to p38 (MAPK) activity ratios that determine carcinoma cell proliferation or dormancy in vivo. Mol Biol Cell 12:863–879

Aguirre-Ghiso JA, Estrada Y, Liu D, Ossowski L (2003) ERK (MAPK) activity as a determinant of tumor growth and dormancy; regulation by p38 (SAPK). Cancer Res 63:1684–1695

Al-Khateeb TH, Al Zoubi F (2007) Congenital neck masses: a descriptive retrospective study of 252 cases. J Oral Maxillofac Surg 65:2242–2247

Almog N (2010) Molecular mechanisms underlying tumor dormancy. Cancer Lett 294:139–146

Alsabti EAK (1979) Tumor dormancy. J Cancer Res Clin Oncol 95:209–220

Amaravadi RK (2008) Autophagy-induced tumor dormancy in ovarian cancer. J Clin Invest 118:3837–3840

Banikas V, Kyrgidis A, Koloutsos G, Sakkas L, Antoniades K (2011) Branchial cyst carcinoma revisited: stem cells, dormancy and malignant transformation. J Craniofac Surg 22:918–921

Bhanote M, Yang GC (2008) Malignant first branchial cleft cysts presented as submandibular abscesses in fine-needle aspiration: report of three cases and review of literature. Diagn Cytopathol 36:876–881

Bhattacherjee V, Mukhopadhyay P, Singh S, Johnson C, Philipose JT, Warner CP, Greene RM, Pisano MM (2007) Neural crest and mesoderm lineage-dependent gene expression in orofacial development. Differentiation 75:463–477

Briasoulis E, Pavlidis N (1997) Cancer of unknown primary origin. Oncologist 2:142–152

Chen K-W, Pienta K (2011) Modeling invasion of metastasizing cancer cells to bone marrow utilizing ecological principles. Theor Biol Med Model 8:36

Devaney KO, Rinaldo A, Ferlito A, Silver CE, Fagan JJ, Bradley PJ, Suarez C (2008) Squamous carcinoma arising in a branchial cleft cyst: have you ever treated one? Will you? J Laryngol Otol 122:547–550

Favaro E, Amadori A, Indraccolo S (2008) Cellular interactions in the vascular niche: implications in the regulation of tumor dormancy. APMIS 116:648–659

Goldenberg D, Sciubba J, Koch WM (2006) Cystic metastasis from head and neck squamous cell cancer: a distinct disease variant? Head Neck 28:633–638

Hudgins PA, Gillison M (2009) Second branchial cleft cyst: NOT!! AJNR Am J Neuroradiol 30:1628–1629

Jang DW, Avivar-Valderas A, Banach A, Aguirre-Ghiso J (2011) Autophagy and tumor cell dormancy in head and neck cancer. Laryngoscope 121:S125–S125

Khafif RA, Prichep R, Minkowitz S (1989) Primary branchiogenic carcinoma. Head Neck 11:153–163

Klein CA (2011) Framework models of tumor dormancy from patient-derived observations. Curr Opin Genet Dev 21:42–49

Kyrgidis A, Antoniades K (2010) Oral cancer: biomarkers, hallmarks, stem cell origin, and clinical implications. In: Kristoff HC (ed) Cancer biomarkers. Nova, Hauppauge

Kyrgidis A, Kountouras J, Zavos C, Chatzopoulos D (2005) New molecular concepts of Barrett's esophagus: clinical implications and biomarkers. J Surg Res 125:189–212

Kyrgidis A, Triaridis S, Vahtsevanos K, Antoniades K (2009) Osteonecrosis of the jaw and bisphosphonate use in breast cancer patients. Expert Rev Anticancer Ther 9:1125–1134

Kyrgidis A, Tzellos TG, Triaridis S (2010) Melanoma: stem cells, sun exposure and hallmarks for carcinogenesis, molecular concepts and future clinical implications. J Carcinog 9

Kyrgidis A, Tzellos TG, Toulis K, Antoniades K (2011) The facial skeleton in patients with osteoporosis: a field for disease signs and treatment complications. J Osteoporos 2011:147689

Martin H, Morfit HM, Ehrlich H (1950) The case for branchiogenic cancer (malignant branchioma). Ann Surg 132:867–887

Mathew R, Karantza-Wadsworth V, White E (2007) Role of autophagy in cancer. Nat Rev Cancer 7:961–967

Maturo SC, Michaelson PG, Faulkner JA (2007) Primary branchiogenic carcinoma: the confusion continues. Am J Otolaryngol 28:25–27

Nishimura G, Yanoma S, Satake K, Ikeda Y, Taguchi T, Nakamura Y, Hirose F, Tsukuda M (2000) An experimental model of tumor dormancy therapy for advanced head and neck carcinoma. Jpn J Cancer Res 91:1199–1203

Ozolek JA (2009) Selective pathologies of the head and neck in children: a developmental perspective. Adv Anat Pathol 16:332–358

Pai RK, Erickson J, Pourmand N, Kong CS (2009) p16 (INK4A) immunohistochemical staining may be helpful in distinguishing branchial cleft cysts from cystic squamous cell carcinomas originating in the oropharynx. Cancer Cytopathol 117:108–119

Papadogeorgakis N, Petsinis V, Parara E, Papaspyrou K, Goutzanis L, Alexandridis C (2009) Branchial cleft cysts in adults. Diagnostic procedures and treatment in a series of 18 cases. Oral Maxillofac Surg 13:79–85

Park Y, Kitahara T, Takagi R, Kato R (2011) Does surgery for breast cancer induce angiogenesis and thus promote metastasis? Oncology 81:199–205

Putti TC, Teh M, Lee YS (2004) Biological behavior of keratoacanthoma and squamous cell carcinoma: telomerase activity and COX-2 as potential markers. Mod Pathol 17:468–475

Schewe DM, Aguirre-Ghiso JA (2008) ATF6α-Rheb-mTOR signaling promotes survival of dormant tumor cells in vivo. Proc Natl Acad Sci U S A 105:10519–10524

Woolgar JA, Ferlito A, Takes RP, Rodrigo JP, Silver CE, Devaney KO, Rinaldo A (2011) The sudden presentation and progression of overt cervical metastases following treatment of head and neck cancers. Eur Arch Otorhinolaryngol 268:1–4

Work WP (1972) Newer concepts of first branchial cleft defects. Laryngoscope 82:1581–1593

Function of the ING Proteins in Cancer and Senescence

18

Uyen M. Tran*, Uma Rajarajacholan*, and Karl Riabowol

Contents

Abstract	179
Introduction	180
Discovery of the INhibitors of Growth (INGs)	180
Structural Components of ING Proteins	180
Post-translational Modification of the ING Proteins	182
INGs as Chromatin Regulators	182
INGs and Short Non-coding RNAs	184
INGs in Cancer	184
Cell Cycle Regulation	184
DNA Damage Response and Genomic Stability	185
Cell Motility and Angiogenesis	185
Apoptosis	185
Mutation of *Ing* Genes	186
INGs and the p53 Tumour Suppressor	186
INGs in Aging	187
Telomeres and Replicative Senescence	187
Stress Induced Premature Senescence (SIPS)	188
Oncogene Induced Senescence	188
Changes Associated with Senescence	189
Morphological Changes	189
Genetic Changes	189
Epigenetic Changes	190
Senescence Markers	190
Tumour Suppressor Networks and Senescence	191
The ARF-p53-p21 Pathway	191
The p16-pRb-E2F Pathway	191
The ING-HAT/HDAC Pathway	192
Conclusions and Future Perspectives	193
References	193

*These authors made equal contributions

U.M. Tran • U. Rajarajacholan • K. Riabowol (✉)
Department of Biochemistry and Molecular Biology and Oncology, University of Calgary, 3330 Hospital Drive NW, T2N4N1 Calgary, AB, Canada
e-mail: karl@ucalgary.ca

Abstract

Age is one of the strongest correlates to the incidence of cancers known, suggesting that the two processes are linked. Cell aging (senescence) is increasingly being linked to epigenetic pathways, many of which have recently been found to be markedly altered in precancerous and cancer cells. Thus, misregulation of epigenetic pathways may impact both cancer and aging by influencing genetic and biochemical pathways common to both processes. Similar to the p53 and retinoblastoma (Rb) tumor suppressors that affect chromatin structure by genetic and epigenetic mechanisms, the INhibitor of Growth (ING) type II tumour suppressors affect pathways that contribute to cell aging and cancer. In particular, the INGs have been demonstrated to act as readers of the histone code by virtue of interacting specifically with the histone H3 residue H3K4Me3 and as targeting subunits of the writers of the histone code by being stoichiometric members of histone acetyltransferase (HAT) and histone deacetylase (HDAC) complexes. The ING proteins are frequently deregulated or mislocalized in various cancer types and certain ING variants are upregulated in replicative senescence, affecting epigenetic mechanisms and altering chromatin structure. The *ing* gene that is best studied, *ing1*, encodes two major splicing variants that encode ING1a, a regulator of cell senescence, and ING1b, a protein that promotes apoptosis. In this chapter, we provide an overview of the domains of the ING proteins, descriptions of the processes affected by the INGs and describe in more detail their functions in cancer and cell aging. Those characteristics of the INGs that allow them to impact both processes are highlighted.

Keywords

ARF-p53-p21 pathway • Cancer and senescence • Cell cycle regulation • Cell motility and angiogenesis • Chromatin regulators • DNA damage response and genomic stability • Inhibitors of growth (INGs) • p16-Rb-E2F and ING-HAT/HDAC pathway • Stress-induced premature and oncogene-induced senescence • Telomeres and replicative senescence

Introduction

Discovery of the INhibitors of Growth (INGs)

The first *ing* gene was discovered using a subtractive hybridization approach to detect genes that were preferentially expressed in normal mammary epithelial cells but not in several breast cancer cell lines. Screening of a senescent cell library and an *in vivo* selection process was subsequently developed and conducted to isolate growth suppressor genes downregulated in cancer cell lines (Garkavtsev et al. 1996). Since then, five *ings* encoding ING1-ING5 have been identified based upon sequence homology. They are classified as type II tumour suppressors because they are frequently downregulated or mislocalized, rather than mutated, in disease (Soliman and Riabowol 2007). The *ing* genes are evolutionarily well conserved and have been identified in rats, frogs, fish, mosquitoes, fruit flies, worms, fungi and plants. The five *ing* genes map to chromosomes 13q34, 4q35, 7q31, 12p13.3, and 2q37.3 for *ing1* to *ing5*, respectively. With the exception of *ing3*, which might represent the "primordial" *ing* gene based upon bioinformatic analyses, all *ings* are found near telomeric regions, making them potentially vulnerable to the effects of telomere erosion. As might be expected for epigenetic regulators, ING proteins have been found to function in a broad range of biological processes including apoptosis, DNA repair, cell cycle regulation, chromatin remodelling, tumorigenesis and angiogenesis (Soliman and Riabowol 2007).

Structural Components of ING Proteins

Most of the *ing* genes encode multiple splicing variants. The *ing1* gene encodes two major protein isoforms, ING1a and ING1b, which are generated by differential splicing of the first exon of the gene. These two proteins appear to have opposing biological functions. ING1b is the shorter, dominant isoform and it is highly expressed in young (low passage) primary cultured fibroblasts and epithelial cells. In addition, upon overexpression and in response to DNA damage, ING1b promotes apoptosis (Helbing et al. 1997). In contrast, ING1a is only significantly expressed during cell aging and induces a senescent cellular phenotype when ectopically expressed (Soliman et al. 2008). All other *ings*, with the exception of *ing5*, encode at least two splicing isoforms from differential incorporation of internal exons or alternative promoter usage (Fig. 18.1).

Bioinformatic analysis revealed that ING members consist of seven major protein domains. Every ING protein contains a nuclear localization signal (NLS), a lamin interaction domain (LID) and a plant homeodomain (PHD). However, several other domains are found in subsets of the INGs. ING1b contains a partial bromodomain (PBD), whose function remains unknown, and a PCNA-interacting-protein (PIP) motif, which helps to mediate ING1b binding with the prolif-

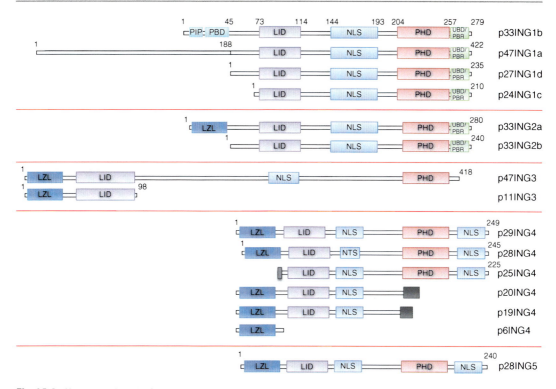

Fig. 18.1 Domains of the ING proteins

erating cell nuclear antigen (PCNA) in a highly DNA damage-inducible manner (Soliman and Riabowol 2007; Thalappilly et al. 2011). In ING2 to ING5, a leucine zipper-like (LZL) motif occupies the regions comparable to those of the PIP and PBD of ING1. This motif consists of 4–5 conserved leucine or isoleucine residues, promoting the formation of a hydrophobic face near the helical N-termini of ING proteins. The LZL was reported to affect ING functions in DNA repair and apoptosis. Lastly, ING1 and ING2 contain a UBD or ubiquitin-binding domain that was found to be crucial for stabilizing p53 by affecting its ubiquitin-mediated proteosomal degradation (Thalappilly et al. 2011). This domain overlaps with the polybasic region (PBR) found at the C-termini of ING1 and ING2 that was determined to be necessary and sufficient for the binding of stress-inducible phosphatidylinositol 5′-monophosphate signalling lipids, resulting in the promotion of apoptosis (Li et al. 2011).

Thus, this short region links stress-induced phospholipid signaling to protein turnover via the proteosomal system.

The PHD is the most highly conserved region of the ING proteins and is found in approximately 150 other members of the human proteome. The PHD contains a C_4-H-C_3 form of zinc finger. It allows INGs to act as chromatin regulators by recognizing and binding to chromatin at lysine 4 of histone 3 (H3K4) in a methylation sensitive manner (Soliman and Riabowol 2007). Subsequent to chromatin binding, INGs recruit chromatin-modifying complexes including histone deacetylases (HDACs) such as Sin3A, and histone acetyltransferases (HATs) such as MOZ/MORF, HBO1 and Tip60, targeting them to regulate local chromatin structure. Since the ING proteins are also stoichiometric members of these HAT and HDAC complexes, altered levels of INGs necessarily affect the acetylation status of targets of the complexes.

Unlike the PHD, the LID domain is unique to the ING family. It is the site through which INGs bind to lamin A, a type V intermediate filament that helps define the physical structure of the nucleus and organize chromatin (Han et al. 2008). The interaction that ING1 has with lamin A is thought to reinforce its nuclear localization, augmenting its ability to carry out its functions in HAT and HDAC complexes. Coincidently, the expression of ING1b is reduced at both the RNA and protein levels in certain laminopathies and in blood cancers that lack lamin A. In addition to fulfilling a role as a consequence of lamin A binding, the LID domain also contains the sequence KIQI/KVQL, which mediates ING binding to components of HAT and HDAC complexes such as the SAP30 protein subunit of the Sin3 HDAC complex, suggesting that there may be competition between lamin A, HAT and HDAC complexes for ING protein binding and that ING proteins may target HAT and HDAC complexes as a consequence of interactions with both methylated H3K4 and lamin A.

The NLS of the INGs also promotes their nuclear localization through the use of the importin pathway. The ING1 protein binds the karyopherin-α and karyopherin–β importin proteins, particularly in the presence of stressors, to target ING proteins to the nucleus (Soliman and Riabowol 2007).

Post-translational Modification of the ING Proteins

While localization of ING1 to the nucleus occurs as a result of binding to karyopherins, relocalization of ING1 to the cytoplasm happens when a region between the NLS and PHD, known as the 14-3-3 binding motif, is phosphorylated at serine 199 (S199) and is bound by 14-3-3 family proteins (Soliman and Riabowol 2007). As a result, the ability of ING1 to regulate genes such as p21 is inhibited because of reduced levels of ING1 in the nucleus. ING1 is also phosphorylated on serine 126 (Ser-126) by both the cyclin-dependent kinase 1 (Cdk1) and the damage checkpoint effector kinase Chk1 (Garate et al. 2007). The former protein maintains phosphorylation of Ser-126 under basal conditions while the latter increases phosphorylation of this residue after DNA damage. This post-translational modification acts to increase the stability of ING1, thus, regulating the expression of cyclin B1 and proliferation in melanoma cells. ING1 has also been reported to be ubiquitinated as a mechanism of turnover via the proteasome, but it also binds ubiquitin through its UBD (Thalappilly et al. 2011).

Lysine 195 found between the NLS and PHD of ING2 can be sumoylated. This modification enhances the interaction of ING2 with the Sin3A HDAC complex and is required for ING2 to efficiently bind to the promoter of genes, such as transmembrane protein 71 (TMEM71), to regulate their transcription (Ythier et al. 2008). Recent findings suggest that similar to ING2, ING1 may also contain a sumoylation consensus motif, ψKxE, in which ψ is a hydrophobic residue, K is the sumoylated lysine, x is any residue and E is a glutamic acid (Satpathy et al. 2013). If sumoylated, the protein could be affected in various ways, including changes in protein-protein interactions that could alter the stability of ING1. Furthermore, additional data indicates that ING1 could also be a target of phosphorylation by Src tyrosine kinases (Yu et al. 2013), which have been shown to be involved in the development, progression and growth of several cancers.

INGs as Chromatin Regulators

In the nucleus, DNA exists in a chromatin DNA-protein complex. The most fundamental unit of this structure is the nucleosome, which is composed of 147 base pairs of DNA wrapped around an octamer of core histone proteins. Each octamer consists of two molecules each of H2A, H2B, H3 and H4. Broadly defined, epigenetics refers to the heritable regulation of gene expression without changing the primary DNA sequence per se. It involves mechanisms that alter the chromatin, and hence gene expression, through DNA methylation, histone modifications, variant histone substitutions, energy-dependent nucleosome rearrangement and noncoding RNAs (ncRNAs). All ING proteins are crucial stoichiometric components of

18 Function of the ING Proteins in Cancer and Senescence

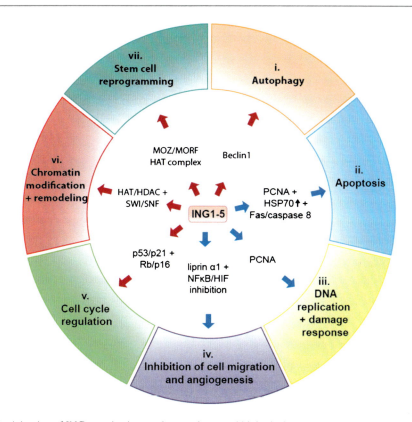

Fig. 18.2 Participation of ING proteins in protein complexes and biological processes

histone acetylase and deacetylase complexes, which modify histones. ING1b-mediated gene repression has been attributed to two domains located at either the N-terminus or the C-terminus. The former is HDAC-dependent and is sensitive to the histone deacetylase inhibitor trichostatin A (TSA) while the latter is HDAC-independent. Paradoxically, overexpression of ING1b also increases histone H3 and H4 acetylation levels, both of which are marks of increased transcription activities (Soliman and Riabowol 2007). This is believed to be due to the disruption of the ING1-containing Sin3A HDAC complex, and therefore the loss of HDAC activity due to altered stoichiometry. ING1a expression, on the other hand, modestly decreases acetylation of the same histones, which is consistent with a weaker association with HDAC complexes or with factors outside HAT and HDAC protein complexes. More strikingly, another study found that the HDAC inhibitor TSA can enhance ING1b expression in glioblastoma cells to facilitate TSA-induced apoptosis in the absence of p53 even though ING1b is found to bind to HDACs (Tamannai et al. 2010). Of note, ING1b was also found to associate with components of the SWI/SNF complex (Han et al. 2008), suggesting that ING1b can regulate gene expression through influencing chromatin-remodelling by this mechanism as well.

Similar to ING1, ING2 was found to be a member of Sin3 HDAC1 and HDAC2 complexes. On the other hand, INGs 3–5 are components of HAT complexes: ING3 is found in the NuA4/Tip60 HAT complex, which is functionally conserved in yeast; ING4 associates with the HBO1 HAT; and ING5 complexes with either HBO1 HAT or nucleosome H3-specific MOZ/MORF HATs (Doyon et al. 2006). These HAT and HDAC complexes have all been shown to be important for cell cycle regulation and oncogenic transformation, consistent with the biological functions of INGs in senescence and oncogenesis, as outlined in Fig. 18.2.

INGs and Short Non-coding RNAs

ING1 has been reported to regulate another epigenetic pathway by affecting early microRNA biogenesis. Specifically, ectopic expression of ING1 resulted in decreased promoter activity of the DGCR8 gene, which encodes a microRNA processor that cleaves pre-miRNAs involved in cell proliferation for export to the cytoplasm (Gomez-Cabello et al. 2010). Expression of INGs can also be inhibited by microRNAs. Of note, ING4 is targeted by microRNA-650, a short non-coding RNA that promotes the proliferation and growth of gastric cancer cells and ING1 is a direct target of microRNA-622 in gastric cancer and is inhibited, promoting cell invasion and tumour metastasis (Guo et al. 2011). Thus, the addition of microRNAs to the network involving INGs implies that these proteins, like other tumour suppressors, have functions that can be highly multidimensional.

INGs in Cancer

The ING proteins are often deregulated or mislocalized in cancer. For example, ING1 levels are reduced in approximately 40% of primary breast cancers (Toyama et al. 1999) and in many types of blood and solid tumours. The lack of ING1 availability contributes to decreased apoptosis and cell cycle control. Intriguingly, knocking out ING1 in mice resulted in the animals becoming susceptible to spontaneous B-cell lymphoma formation, and having increased sensitivity to repetitive low-dose DNA damage. This result is consistent with studies showing that INGs are involved in DNA repair and genome maintenance. In human non-small cell lung carcinoma, the expression of ING1, along with p53 and the autophagy protein Beclin1, decreased with increasing pathological stage. A similar effect with ING4 was observed in glioblastomas. In addition, ING proteins were found to be deregulated in multiple other cancers including melanoma, basal cell carcinoma, myeloid leukemia, ovarian cancer, bladder cancer, hepatocellular carcinoma, colorectal cancer, head and neck cancer, oral cancer and pancreatic cancer. Thus, detection of levels of INGs and associated proteins may act as a potential approach for early cancer diagnosis and determination of patient prognosis.

Various studies have demonstrated that INGs function in numerous pathways that counter carcinogenesis. Firstly, the ectopic expression of INGs alone or in conjunction with stress can cause cells to undergo programmed cell death or apoptosis (Helbing et al. 1997). Secondly, INGs have the ability to induce cell cycle arrest, further preventing cells from overcoming regulatory networks and acquiring uncontrolled growth (Garkavtsev et al. 1996). Consistently, loss of ING1 (Garkavtsev and Riabowol 1997) or ING2 (Pedeux et al. 2005) inhibits the onset of senescence. Moreover, ING4 can repress processes that give cells enhanced motility for migration and support blood vessel growth for a continuous supply of nutrients in tumour development (Moreno et al. 2010), an ability shared with ING1. Therefore, since the levels of INGs are often deregulated or mislocalized in various cancer types, their loss or dysregulation could potentially promote cancer progression by multiple mechanisms including apoptosis, senescence, cell cycle control, DNA damage response, cell migration and angiogenesis.

Cell Cycle Regulation

The ability of ING proteins to influence the cell cycle depends primarily on their ability to regulate gene expression. Consequently, the localization of INGs to the nucleus is crucial and the induction of specific genes is inhibited when INGs are triggered to translocate to the cytoplasm. Several INGs activate the promoter region of p21, a transcriptional target of p53. The p21cyclin-dependent kinase inhibitor is a mediator of the G1 cell cycle checkpoint and thus, it helps to trigger cell cycle arrest. One mechanism by which INGs induce p21 is by binding to hSIR2, a negative regulator of p53 activity (Kataoka et al. 2003). This prevents the deacetylation of p53 by hSIR2amd thus maintains p53 in an active state. As noted previously, ING1 also regulates p21 levels by inhibiting ubiquitin-mediated proteosomal degradation of p53 through physically interacting with herpesvirus-associated ubiquitin-specific protease (HAUSP), a p53 and MDM2 deubiquitinase (Thalappilly et al. 2011).

The ability of ING2 to enhance transcription and inhibit cell proliferation also appears to be contingent upon its association with the transcription factor SnoN and R-Smad2, and activation of the TGF-β signalling pathway (Sarker et al. 2008).

DNA Damage Response and Genomic Stability

INGs have been reported to participate in the nucleotide excision repair (NER) process. In melanoma cells, expression of either ING1b or ING2 was needed for efficient DNA repair after UV irradiation. The ability of ING1b to function in DNA damage could partly be due to its UV-inducible physical association with proliferating cell nuclear antigen (PCNA). Recent findings also suggest that ING1b stabilizes PCNA through E3 ligase Rad-18-mediated monoubiquitination of PCNA to maintain genomic integrity during replication stress (Wong et al. 2011). In addition, for NER to occur, both ING1b and ING2 rely heavily on chromatin relaxation by rapid induction of histone H4 acetylation and on the recruitment of the damage recognition factor xeroderma pigmentosum group A protein (XPA) to sites of DNA damage. Similar to the above observation, studies in yeast have shown that the yeast homolog of ING4 and ING5, called Yng2, requires functional histone H3 and H4 acetylation in order to effectively maintain genome stability during chromosome replication in S phase.

Cell Motility and Angiogenesis

Cell spreading, cell migration, contact inhibition and angiogenesis are also processes affected by certain members of the ING tumour suppressor family. Specifically, ING4 and all of its variants have been shown to be major players in these processes. One of the first reports that hinted at the potential suppressive role of ING4 revealed that it could halt loss of contact inhibition evoked by the protooncogene MYCN and hinder breast cancer cell growth in soft agar (Kim et al. 2004). Furthermore, overexpression of ING4 in glioblastomas showed marked reduction of angiogenesis. This effect was attributed to the ability of ING4 to bind to the p65/RelA subunit of the nuclear factor NF-κB, and prevent transcription of angiogenesis-related genes in this pathway via histone deacetylation (Garkavtsev et al. 2004).

In melanoma cells, ING4 was observed to be a downstream target of BRMS1, a metastasis suppressor. Upon stimulation, ING4 simultaneously repressed IL-6 expression and NF-κB activity in these cells to inhibit angiogenesis (Li and Li 2010). Concomitantly, ING4 was shown to inhibit the activation of hypoxia inducible factor (HIF) in low oxygen conditions (Moreno et al. 2010). In addition, ING4 can affect cell spreading and migration by binding to liprin α1, a protein involved in the formation and disassembly of focal adhesions. Liprin α1 subsequently directs ING4 to lamellipodia where it performs its repressive functions (Shen et al. 2007). However, the effects of ING4 are not entirely repressive. It can facilitate the expression of IκB, an NF-κB inhibitor, by increasing histone H4 acetylation at the IκB promoter region (Coles et al. 2010). ING1 can also play a significant role in angiogenesis and migration. For example, in glioblastoma multiforme, ING1 can prohibit neoangiogenesis by regulating the expression of angiopoietins, which are proteins that promote blood vessel growth.

Apoptosis

ING proteins have been noted in many studies to induce apoptosis in cells when overexpressed. Overexpression of ING1 can also increase sensitivity of cells to DNA damaging agents such as UV irradiation or Adriamycin, a chemical that causes DNA double strand breaks by interfering with replication at the level of the topoisomerase II enzyme. The function of ING1 in this process can be attributed, in part, to its interaction with PCNA. In melanoma cells, ING3 proteins may contribute to apoptosis via the extrinsic Fas/caspase 8 apoptosis pathway. Also, of note, past studies have shown that ING1b could sensitize cells to extrinsic apoptosis via the upregulation of the heat shock protein HSP70 (Soliman and Riabowol 2007).

Mutation of *Ing* Genes

As noted previously, the ING proteins are type II tumor suppressors. While they are frequently downregulated and show aberrant subcellular localization, they are not commonly mutated in cancers. However, some studies have identified silent and/or point mutations and loss of heterozygosity (LOH) in different members of the ING family. Microsatellite analysis in ameloblastomas showed high frequency of loss of heterozygosity at the locus of each *ing* gene, with particularly high occurrence in the *ing5* locus (Borkosky et al. 2010). Another study reported three missense mutations in the leucine zipper like (LZL) region of the *ing5* gene in oral squamous cell carcinoma. This study had identified a high frequency of allelic loss at chromosome 2q37 that corresponds to candidate tumor suppressor genes like PPP1R7, ILKAP, DTYMK and *ing5* (Cengiz et al. 2010). A similar deletion mapping study in head and neck cancers identified four frequently deleted regions corresponding to the genes Caspase 9, SAP30, *ing2* and SMARCA2 [reviewed in Li et al. 2011]. Ythier and colleagues have reported that ING2 protein levels were reduced in adenocarcinoma but they did not identify any mutations or loss of heterozygosity of the *ing2* gene in about 120 non-small cell lung cancer tissues that were used in their study. However, they observed a silent single nucleotide polymorphism (SNP) in about 95% of their samples (Ythier et al. 2010).

Mutational analysis of 83 primary melanomas and 55 melanoma cell lines did not find any mutations in the *ing1* gene, concluding that mutation in ING1 is very rare in the case of melanoma (Stark et al. 2006). Another study reported that about 25% of the 111 ovarian cancer tissues that they used showed an allelic loss of *ing1* but none carried a mutated *ing1* gene (Shen et al. 2005). When ING4 mutations were analyzed in 50 head and neck squamous cell carcinomas, there was a significant rate of deletion and a reduction in expression, but no other mutations or loss of heterozygosity was found [reviewed in Li et al. 2011]. In contrast to the majority of reports, an independent study using 46 human cutaneous melanoma samples reported that at least 20% of the tissues examined carried a missense mutation of the *ing1* gene, in a region corresponding to the SAP30 interacting domain and PHD finger region and this mutation had a significant impact on ING1 function in nucleotide excision repair of UV induced DNA damage (Ythier et al. 2008). A common point mutation in the PHD region of ING4 (N214D) was also reported in lung cancers, which affected ING4 protein stability, and blocked the function of ING4 in inhibiting proliferation, cell migration, anchorage-independent growth, angiogenesis and the induction of apoptosis (Moreno et al. 2010).

INGs and the p53 Tumour Suppressor

ING-induced apoptosis has a complex relationship with p53. While the function of ING proteins in apoptosis can be affected by the presence of p53, many reports have suggested that the status of p53 is not crucial for ING1 activity. For example, knocking out the ING1b isoform in mouse fibroblasts increased cell proliferation in both p53 positive and p53-null backgrounds. The converse is also true; that is, p53 can function without ING1b (Coles et al. 2010).

Although INGs localize to the nucleus in response to stress to alter chromatin structure, they can also alter gene expression indirectly through affecting the protein stability of p53. Previously, there has been evidence that ING1 can affect MDM2 activity by physically interacting with ARF, a protein that can sequester MDM2 in the nucleolus to increase p53 activity (Gonzalez et al. 2006). This interaction is important for the nucleolar localization of ING1b and is also required for its functions in cell cycle arrest and transactivation of p53-mediated genes. Additionally, a recent report has shown that ING1b can help to stabilize p53 proteins via its ubiquitin binding domain or UBD. ING1b binds monoubiquitinated p53 to prevent polyubiquitination and subsequent degradation of p53 by the proteasome. Moreover, ING1b can also act as the middleman, bringing p53 and the mitochondrial herpesvirus-associated ubiquitin-specific protease (HAUSP) together so that HAUSP could deubiquitinate p53 into more stable forms (Thalappilly et al. 2011).

While ING1 prolongs the half-life of p53 proteins, ING2 can increase the levels of its active forms through enhancement of p53 acetylation at lysine 382 by the acetyltransferase p300 (Li et al. 2011). Additionally, ING4 can also promote p53 acetylation. This ability is abrogated when the p53-binding region in the NLS of ING4 is citrullinated by peptidylarginine deaminase 4 (PAD4). PAD4 neutralizes arginine side chains in the NLS to eliminate the interaction of ING4 with p53 and negatively affects the half-life of ING4 in this process (Guo and Fast 2011). Despite the identification of pathways by which INGs can enhance p53 activities, there is still much to be uncovered about the relationship between the two proteins. This includes the question of how ING1, in many cases, can influence p53 gene targets irrespective of p53 status in the cell.

INGs in Aging

The finding that normal human somatic cells in culture have a finite proliferative capacity was first reported by Hayflick and Moorhead (Hayflick and Moorhead 1961). This limited replicative potential has led many to speculate that it may act as a first-line tumour suppressive mechanism to block the growth of transformed cells, preventing them from becoming fully cancerous. Consistent with this idea, in both mouse models and human tissues, a number of markers of senescence have been noted in pre-malignant stages of tumourigenesis in the lung, pancreas, breast and in melanocytic nevi. Characteristics of senescence were absent in their malignant counter-parts suggesting that loss of senescence is a requirement for gaining tumour malignancy (Collado and Serrano 2010). Furthermore, mice lacking major tumour suppressors like p53, Rb, ING1, ING2, PTEN, CHK2 and p16, all of which have characterized functions in the induction of senescence, are readily cancer-prone and can overcome senescence much more rapidly than the normal cells. However, loss or mutation of these genes is not believed to be sufficient for cancer cell immortalization. In addition to this, a mitogenic stimulus such as the aberrant activation of the Ras or MAPK kinase pathway is required for tumorigenesis (Campisi and d'Adda di Fagagna 2007).

Telomeres and Replicative Senescence

The factor that enforces the "Hayflick limit" at which cells stop dividing and undergo permanent growth arrest and senescence was first speculated to be telomere length when it was noted that telomeres shortened with passage of cells in culture. Telomeres are stretches of hexanucleotide DNA repeats (TTAGGG in mammals) that cap the ends of chromosomes and prevent them from recombining and fusing, thereby maintaining genome stability. During replication there is a loss of about 100–200 bp of telomeric DNA per cell division due to the end replication problem of DNA polymerase. Human somatic cells have an average of ~ 15 kb of telomeric repeats during infancy, which allows cell division to proceed until average telomere length becomes critically short. In vertebrates, telomeric sequences have G rich 3' single stranded overhangs that form a telomere- or t-loop that complexes with six major proteins that form the Shelterin complex. This complex protects the telomere from chromosomal fusions and prevents it from being recognized as DNA damage by several different pathways. As a consequence of cell division and telomeres becoming critically short, a DNA damage signalling cascade is initiated, activating proteins like ATM/ATR and CHK1/CHK2, resulting in a signaling cascade that culminates in cell senescence. In germline cells, the enzyme telomerase, which consists of a catalytic subunit (telomerase reverse transcriptase; TERT), an RNA template component (TERC) and many additional regulatory proteins, adds telomeric repeats to the ends of the chromosomes after each replication cycle. This machinery, however, is not active in most human somatic cells and is insufficient to compensate for DNA lost due to the end replication problem in cells such as lymphocytes and various forms of stem cells that retain varying amounts of telomerase activity. In contrast, the great majority of cancer cells gain indefinite proliferative ability by increasing expression of the *tert* gene by an unknown mechanism to maintain telomere length by maintaining active telomerase.

Loss of telomeric sequence increases the frequency of alternative splicing of many genes including lamin A (Cao et al. 2011) and ING1 (Soliman et al. 2008). While disruption of either protein induces premature senescence (Han et al. 2008), the mechanisms by which telomere erosion induces alternative splicing and how the alternative splice products enforce a senescent phenotype, remain important questions.

Stress Induced Premature Senescence (SIPS)

Following Hayflick's seminal description of the limited replicative potential of primary human fibroblasts, the subsequent definition of cellular senescence and discovery of the loss of telomeric DNA sequences with age, the initial dogma was that senescence occurred only as a consequence of telomere attrition. Despite this idea still dominating thoughts about the nature of normal cell senescence, many studies have subsequently found that various features of replicative senescence were observed in young and replication competent cells, but as a consequence of different stress stimuli. For example, altered oxygen tension or treatment with DNA damaging agents, among other stresses, result in many phenotypes similar to telomere-induced senescence, but in the absence of short telomeres. This gave rise to the concept of stress induced premature senescence or SIPS. The process of SIPS occurs independent of telomere length and telomerase activity and a number of stresses result in relocalization of ING1 to the nucleoli of cells (Soliman and Riabowol 2007) where it may contribute to the initiation of SIPS by affecting protein synthesis.

Oncogene Induced Senescence

Protooncogenes are genes that, when activated by mutation to form oncogenes or by overexpression, have the ability to transform normal cells into cancer cells in conjunction with additional mutations (Campisi and d'Adda di Fagagna 2007). Based on this ability, oncogenes were previously associated only with tumorigenesis. The first observation that oncogenes could induce senescence as part of a tumor suppressive program came from studies with oncogenic HRAS, a cytoplasmic signal transducer, that is upstream of many growth factor signalling pathways including the PI3K, AKT, MAPK pathways. Normal primary cells transfected with Ras showed many features of replicative senescence, but these are independent of telomere length. Subsequently, other proteins that are downstream of Ras signalling were also studied to examine whether they had an ability to induce senescence and these reports showed that RAF, MEK, MOS and BRAF were all capable of inducing senescence in young fibroblasts when overexpressed (Campisi and d'Adda di Fagagna 2007). Other forms of stress-induced senescence include DNA damaging chemotherapeutic drugs and oxidative stress. The DNA damage caused by chemotherapeutic drugs elicits a DNA damage response that has a number of signal transduction intermediates in common with telomere-initiated senescence. This form of senescence is more prevalent in cancers with intact p53 rather than mutant p53 protein since p53 is a key transducer of the DNA damage signal and is activated in senescence as well. This suggests that this form of SIPS is mainly dependent on the p53-p21 axis for senescence induction. Drugs that elicit this response include doxorubicin, camptothecin, etoposide and many other topoisomerase poisons.

Oxidative stress induced senescence was first identified in murine cells. Mice have approximately 30–150 kb of telomeric DNA repeats at the ends of their chromosomes depending upon the strain examined, with an average length that is several-fold longer than in humans, yet, mouse cells senesce within just a few population doublings in culture. This was determined to be due mainly to a higher sensitivity to oxidative stress in murine, compared to human cells since cells were typically cultured in conditions of ambient (21%) oxygen. When the same cells are cultured under "normoxic" conditions of 2–3% oxygen that they are normally exposed to *in vivo*, mouse fibroblasts proliferate much longer than human fibroblasts.

Changes Associated with Senescence

Since the initial description of cellular senescence, its relevance in biological aging has been debated. However, with reports of numerous senescence markers such as hyperactivity of genes like p53, Rb and p16, the production of senescence-associated beta-galactosidase (SA-β-Gal) and others in normal but aged individuals, the link between cellular senescence and biological aging has become more evident and convincing (Campisi and d'Adda di Fagagna 2007). As individuals age, the stem cell pools stop replenishing, leading to failure in regeneration. Consistent with this idea, many senescence markers are also found in the stem cell populations of aged individuals.

Morphological Changes

Senescent cells are viable and metabolically active, but are unresponsive to mitogenic growth factor stimuli and are characterized by a permanent cell cycle arrest at the G_0/G_1 phase of the cell cycle. They typically assume a large, flattened morphology and frequently become multinucleated. Senescent cells also mislocalize a number of proteins such as actin and lamin A. The basis for mislocalization of proteins is unknown, but it may be a function of alternative isoforms being expressed (Cao et al. 2011; Soliman et al. 2008).

Genetic Changes

One well documented change that occurs as cells become senescence is the increased expression of cell cycle inhibitors. The cyclin-dependent kinase inhibitors (CDKi) that are best documented to be upregulated in senescent cells are p16 and p21. While p16 functions primarily in the Rb pathway by inhibiting cyclin D-CDK4 and cyclin D-CDK6 complexes, p21 functions downstream of p53 and can inhibit cyclin A-CDK2 and cyclin E-CDK2 complexes to induce cell cycle arrest and senescence. These two CDK inhibitors play an active role in arresting cells at the G_0/G_1 phase of the cell cycle, enforcing a senescence phenotype by maintaining Rb growth inhibitory activity, preventing the initiation of DNA replication inspite of the availability of growth factors. This irreversible growth arrest distinguishes senescent cells from quiescent cells, where the cells undergo a temporary arrest at the G_0/G_1 phase but enter the cell cycle normally as soon as the growth conditions become favourable.

Other genes that are differentially affected during cellular senescence include different cyclins, the proliferating cell nuclear antigen (PCNA), several histones, FOS, Egr1 and extracellular matrix associated proteins. Many E2F target genes that are required for growth are repressed by Rb during senescence since E2F transcription factors play a key role in cell cycle progression (Campisi and d'Adda di Fagagna 2007). Another important event that happens most extensively during senescence is alternative splicing of many genes. A recent study (Cao et al. 2011) identified more than 80 genes that are differentially alternatively spliced in senescent cells. This splicing change was dependent on telomere attrition. Among the genes that were identified in this study were those with known roles in cytoskeletal function and organization, filamentous microtubular proteins like actin, tubulin and vimentin and lamin A (LMNA). The splice variant of LMNA that is produced gives rise to a protein called progerin, which plays a causal role in inducing the Hutchinson Gilford Progeria Syndrome (HGPS) form of premature aging (Cao et al. 2011).

Epigenetic Changes

Recent advancements in microscopy and imaging techniques have helped us to better understand chromatin organization and cell nuclear architecture. The importance of maintaining nuclear architecture, the failure of which often results in pre-mature aging syndromes, is evident from the diseases like Werners' syndrome, HGPS and other laminopathies. The cell nuclei derived from patients with Progeria, for example, show an anomalous nuclear structure, with alterations in the distribution of heterochromatin that may result from the expression of progerin, the truncated form of lamin A formed due to an altered splicing of the *lmna* gene. Consistent with this observation, over-expression of progerin in young fibroblasts results in senescence (Cao et al. 2011) further emphasising the importance of maintaining nuclear morphology.

Senescent cells contain more condensed heterochromatin than young cells. Specific regions of heterochromatin form during senescence called senescence associated heterochromatic foci that are enriched in methylated H3K9 histone marks and show few signs of euchromatin signatures like H3K9 acetylation or H3K4 methylation (Narita et al. 2003). The two histone chaperones HIRA and ASF1 also physically interact with, and may help mediate the formation of SAHF (Narita et al. 2003). Additionally, Rb localization to SAHF and HIRA/ASF1 translocation to promyelocytic leukemia (PML) bodies have also been shown to help initiate the process of SAHF formation (Narita et al. 2003). Other epigenetic changes that occur with senescence include activation of the MTS1 locus by the JMJD3 DNA demethylase. The polycomb proteins keep this locus suppressed during development and JMJD3 demethylates H3K27 to remove this repression mark to activate the expression p14^{ARF} and p16^{INK4A} (Campisi and d'Adda di Fagagna 2007).

Senescence Markers

One of the most widely used markers of senescence is senescence-associated beta-galactosidase (SA-β-Gal) staining, which results from increased activity of lysosomal beta galactosidase (Campisi and d'Adda di Fagagna 2007). While widely used as a simple colorimetric biochemical test, SA-β-Gal activity is also induced by a variety of stresses, some of which induce stress-induced premature senescence. Another widely used, and perhaps the most important characteristic marker of senescence is blockage of the cell cycle resulting in the failure to replicate DNA. Methods to measure this include testing the incorporation of 5-bromodeoxyuridine and by determining the expression levels of Ki67, a proliferation marker that is downregulated in senescent cells. Senescence induction can also be studied by checking the expression levels of various CDK inhibitors induced during senescence like p16, p21, p15, p27 and also by measuring levels of cyclins. While levels of cyclins A, B and E are downregulated during senescence, the levels of cyclins D1 and D2 are reproducibly increased in senescent cells (Meyyappan et al. 1998), perhaps since they serve as mitogen sensors that are upregulated in response to reduced mitogenic signaling.

As aluded to above, one major limitation with these markers of senescence is that they are not entirely specific to senescent cells. For instance, senescence associated beta galactosidase staining also shows positive staining when cells are serum starved and when they are stressed by growing them to confluence in culture. Furthermore, proliferation assays like BrdU and Ki67 markers do not distinguish senescent from quiescent cells. The gene expression changes that are noted above as senescence markers can be cell type and stress type dependent and thus cannot be applied to all senescence-inducing conditions in all cell types. For example, p16 cannot be used as a marker for oncogene induced senescence since OIS acts primarily through the p53/p21 pathway while replicative senescence appears to be mediated to a greater degree by the Rb/p16 pathway. Thus, the need for markers that are specific for identification of the state of senescence, both *in vitro* and *in vivo*, still remains, although ING1a appears to be a promising candidate (Soliman et al. 2008).

Tumour Suppressor Networks and Senescence

Although many tumor suppressor genes are now being studied for their role in senescence induction, two key players that have been extensively studied are the p53 and pRb tumor suppressors. These two proteins, depending upon the stress stimuli and the cell type, drive cells into senescence by regulating the expression of their target genes. Although there is cross talk between both of these pathways, these tumour suppressors also appear to be able to work independent of each other in many cases.

The ARF-p53-p21 Pathway

The multifunctional p53 tumor suppressor protein plays a key role in senescence induction. Loss of p53 by knockout, expressing dominant negative forms, using siRNA or by over-expressing the oncoprotein E6, results in immortalization of many cell types, irrespective of the expression and function of Rb. Loss of p53 in these cells also makes them highly susceptible to malignant transformation. The p14^{ARF} protein encoded by the MTS1 locus is responsible for the activation of p53 in senescent cells. This protein inhibits MDM2 protein, the E3 ubiquitin ligase for p53 that ubiquitinylates and promotes p53 degradation. Thus, by inhibiting MDM2, p14^{ARF} stabilizes and activates p53. Activation of p53 can result in either apoptosis or cell cycle arrest and senescence, depending on the cellular context. While many of the pro-apoptotic targets of p53 are known, the only senescence-promoting target gene of p53 is the CDK inhibitor p21. However, loss of p21 in humans does not bypass senescence nor extend their replicative life span, suggesting roles for additional pathways in human cell senescence.

The p16-pRb-E2F Pathway

The Rb family of pocket binding proteins consists of three members; p105Rb, p107 and p130. These proteins bind to and inactivate the E2F family of transcription factors. When Rb proteins are inactivated by phosphorylation, the E2F factors are released and activated, leading to the expression of genes involved in cell cycle progression, DNA synthesis and replication, DNA repair and checkpoints, proliferation and maintenance of chromosomal structure. Consistent with this, senescent cells contain Rb in its active, hypophosphorylated state, tightly bound to E2F transcription factors. The Rb tumour suppressor proteins are phosphorylated and inactivated by cyclin E/Cdk2 and subsequently by cyclin D/Cdk4/Cdk6. In addition, inactivation of Rb by the oncoprotein E7 combined with expression of the hTERT protein results in the immortalization of human epithelial cells. Interestingly, Rb disruption in senescent cells can facilitate their cell cycle re-entry, suggesting that Rb is a key player in maintaining the post-mitotic nature of senescent cells, perhaps by promoting the formation of SAHF (Narita et al. 2003).

Both p16^{INK4a} and p14^{ARF} (p19^{ARF} in mouse) are encoded in the same locus on chromosome 9q21. This locus is called the INK4A/ARF locus or the multiple tumor suppressor-1 (MTS1) locus. This locus is frequently deleted in many tumours, resulting in the loss of function of both the Rb and p53 pathways. The p16^{INK4a} and p14^{ARF} promoters are activated depending upon the stress stimuli to induce senescence via the Rb and p53 pathways, respectively. How these two promoters are activated or what the factors are that decide which promoter is activated is still not clear. It has been reported that Bmi-1, a polycomb group protein family repressor, keeps p16 inhibited during development and that epigenetic mechanisms during senescence abrogates this inhibition, leading to the activation of p16. In support of this argument, overexpression of Bmi-1 in cells leads to a delay in senescence.

The ING-HAT/HDAC Pathway

Knockdown of ING1 using antisense constructs extends the *in vitro* lifespan of human fibroblasts by ~10% (Garkavtsev and Riabowol 1997). This study also reported an increase in ING1 levels in senescent fibroblasts and that the loss of ING1 promoted neoplastic transformation, suggesting a role for ING1 in the initiation of cellular replicative senescence as part of its intrinsic tumor suppressor function. Recent observations further suggest that it is the ING1a isoform of ING1 that might transduce senescence signaling. During senescence, the ING1a:ING1b ratio increases by 20–30-fold and ectopic expression of ING1a, but not ING1b was associated with a number of senescence-like phenotypes (Soliman et al. 2008). ING1a expression in cells increased the number of cells staining positive for SA-β–Gal, arrested cells at the G_0/G_1 phase of the cell cycle and increased the levels of Rb and p16. In addition to these markers of senescence, expressing ING1a also resulted in the accumulation of SAHF in the nucleus that contained heterochromatin protein 1 gamma (HP1γ) and the amount of chromatin immunoprecipitated using anti-ING antibodies increased with cell age, indicating that ING bound to chromatin more avidly in senescent, compared to young cells. This suggests a role for ING1a in maintaining the relatively heterochromatic state of senescent cells. In addition, ectopic expression of ING1a did not induce any signs of apoptosis, which is the well-characterized response of many cell types to ING1b. Taken together, these observations suggest that ING1b, the major isoform expressed in young cells, functions in the regulation of apoptosis while ING1a, the senescence specific isoform of ING1 functions in the initiation of replicative senescence and might possibly be involved in the p16/Rb pathway. Further studies are needed in order to understand how these isoforms are differentially expressed in cells and to understand the mechanism behind their seemingly antagonistic functions.

Abad and colleagues have recently reported a role for ING1 in oncogene-induced senescence in young fibroblasts. They noted that ING1 accumulates in the chromatin of senescent cells and it is induced by oncogenic Rasv12 activation, while loss of ING1 abrogates senescence triggered by certain oncogenes (Abad et al. 2011). This is similar to observations made previously where the amount of ING1 associated with chromatin, increased in senescent cells.

Currently, it is unclear which tumour suppressor pathways the INGs are associated with, in terms of their roles in inducing senescence. As noted above, with studies using young fibroblasts, the ectopic expression of the longest isoform of ING1 increases the Rb and p16 protein levels and also shows signs of SAHF, suggesting that ING proteins function through the Rb pathway in inducing growth arrest. But studies with ING2 suggests the INGs might also function through the p53/p21 senescence pathway (Ythier et al. 2008). Ectopic expression of ING2 in young replication-competent fibroblasts showed senescent phenotypes in a p53-dependent manner and loss of ING2 extends the replicative life span of cells in culture. It was also suggested that ING2 associates with and recruits p300, a histone acetyl transferase complex to increase the acetylation of p53 and thus mediates p53 dependent senescence signaling (Ythier et al. 2008). Identification of an ING2 isoform that lacks the p53 binding region in its promoter shows no effect in p53 dependent senescence in cells, whereas, ING2a which is regulated by p53, is downregulated in senescent fibroblasts (Unoki et al. 2008). This observation further supports the idea that the senescence function of ING2 is dependent on the p53 pathway and not so much on the Rb pathway. More detailed mechanistic studies will help to confirm which pathways that the INGs affect, in determining cell fate during senescence.

Conclusions and Future Perspectives

It is becoming increasingly clear that the p53 and Rb tumour suppressor networks strongly influence the processes of tumorigenesis and senescence through common signal transduction intermediates that impact chromatin structure. The first member of the ING family of epigenetic regulators was discovered as a candidate tumor suppressor since knockdown promoted tumorigenesis whereas overexpression blocked cell growth and induced apoptosis (Garkavtsev et al. 1996). More recently ING1 has been shown to influence cell senescence, making the similarities between p53, Rb and the ING family more pronounced. Due to the broad spectrum of processes each of the tumor suppressors influence, it is not surprising that they also affect each other, since all three have been linked to histone deacetylase activity, which can broadly affect gene expression. Understanding how these major tumor suppressor axes interact should prove increasingly useful in the development of cancer treatments based upon individual genetic and epigenetic traits of the patient's tumors being targeted.

Acknowledgements We apologize to our colleagues for the many relevant papers we were unable to cite due to length constraints. UMT is a recipient of an Alberta Cancer Foundation (ACF) graduate award and KR holds a Scientist Award from Alberta Innovates-Health Solutions (AI-HS). This work was supported by grants from AI-HS and the Canadian Breast Cancer Foundation.

References

Abad M, Moreno A, Palacios A, Narita M, Blanco F, Moreno-Bueno G, Palmero I (2011) The tumor suppressor ING1 contributes to epigenetic control of cellular senescence. Aging Cell 10:158–171

Borkosky SS, Gunduz M, Beder L, Tsujigiwa H, Tamamura R, Gunduz E, Katase N, Rodriguez AP, Sasaki A, Nagai N, Nagatsuka H (2010) Allelic loss of the ING gene family loci is a frequent event in ameloblastoma. Oncol Res 18:509–518

Campisi J, d'Adda di Fagagna F (2007) Cellular senescence: when bad things happen to good cells. Nat Rev Mol Cell Biol 8:729–740

Cao K, Blair CD, Faddah DA, Kieckhaefer JE, Olive M, Erdos MR, Nabel EG, Collins FS (2011) Progerin and telomere dysfunction collaborate to trigger cellular senescence in normal human fibroblasts. J Clin Invest 121:2833–2844

Cengiz B, Gunduz E, Gunduz M, Beder LB, Tamamura R, Bagci C, Yamanaka N, Shimizu K, Nagatsuka H (2010) Tumor-specific mutation and downregulation of ING5 detected in oral squamous cell carcinoma. Int J Cancer 127:2088–2094

Coles AH, Gannon H, Cerny A, Kurt-Jones E, Jones SN (2010) Inhibitor of growth-4 promotes IkappaB promoter activation to suppress NF-kappaB signaling and innate immunity. Proc Natl Acad Sci U S A 107:11423–11428

Collado M, Serrano M (2010) Senescence in tumours: evidence from mice and humans. Nat Rev Cancer 10:51–57

Doyon Y, Cayrou C, Ullah M, Landry AJ, Côté V, Selleck W, Lane WS, Tan S, Yang XJ, Côté J (2006) ING tumor suppressor proteins are critical regulators of chromatin acetylation required for genome expression and perpetuation. Mol Cell 21:51–64

Garate M, Campos EI, Bush JA, Xiao H, Li G (2007) Phosphorylation of the tumor suppressor p33(ING1b) at Ser-126 influences its protein stability and proliferation of melanoma cells. FASEB J 21:3705–3716

Garkavtsev I, Riabowol K (1997) Extension of the replicative life span of human diploid fibroblasts by inhibition of the p33ING1 candidate tumor suppressor. Mol Cell Biol 17:2014–2019

Garkavtsev I, Kazarov A, Gudkov A, Riabowol K (1996) Suppression of the novel growth inhibitor p33ING1 promotes neoplastic transformation. Nat Genet 14:415–420

Garkavtsev I, Kozin SV, Chernova O, Xu L, Winkler F, Brown E, Barnett GH, Jain RK (2004) The candidate tumour suppressor protein ING4 regulates brain tumour growth and angiogenesis. Nature 428: 328–332

Gomez-Cabello D, Callejas S, Benguria A, Moreno A, Alonso J, Palmero I (2010) Regulation of the microRNA processor DGCR8 by the tumor suppressor ING1. Cancer Res 70:1866–1874

Gonzalez L, Freije JM, Cal S, Lopez-Otin C, Serrano M, Palmero I (2006) A functional link between the tumour suppressors ARF and p33ING1. Oncogene 25:5173–5179

Guo Q, Fast W (2011) Citrullination of inhibitor of growth 4 (ING4) by peptidylarginine deiminase 4 (PAD4) disrupts the interaction between ING4 and p53. J Biol Chem 286:17069–17078

Guo XB, Jing CQ, Li LP, Zhang L, Shi YL, Wang JS, Liu JL, Li CS (2011) Down-regulation of miR-622 in gastric cancer promotes cellular invasion and tumor metastasis by targeting ING1 gene. World J Gastroenterol 17:1895–1902

Han X, Feng X, Rattner JB, Smith H, Bose P, Suzuki K, Soliman MA, Scott MS, Burke BE, Riabowol K (2008) Tethering by lamin A stabilizes and targets the

ING1 tumour suppressor. Nat Cell Biol 10:1333–1340

Hayflick L, Moorhead PS (1961) The serial cultivation of human diploid cell strains. Exp Cell Res 25:585–621

Helbing CC, Veillette C, Riabowol K, Johnston RN, Garkavtsev I (1997) A novel candidate tumor suppressor, ING1, is involved in the regulation of apoptosis. Cancer Res 57:1255–1258

Kataoka H, Bonnefin P, Vieyra D, Feng X, Hara Y, Miura Y, Joh T, Nakabayashi H, Vaziri H, Harris CC, Riabowol K (2003) ING1 represses transcription by direct DNA binding and through effects on p53. Cancer Res 63:5785–5792

Kim S, Chin K, Gray JW, Bishop JM (2004) A screen for genes that suppress loss of contact inhibition: identification of ING4 as a candidate tumor suppressor gene in human cancer. Proc Natl Acad Sci U S A 101:16251–16256

Li J, Li G (2010) Cell cycle regulator ING4 is a suppressor of melanoma angiogenesis that is regulated by the metastasis suppressor BRMS1. Cancer Res 70:10445–10453

Li X, Kikuchi K, Takano Y (2011) ING Genes Work as Tumor Suppressor Genes in the Carcinogenesis of Head and Neck Squamous Cell Carcinoma. J Oncol 2011:963614

Meyyappan M, Wong H, Hull C, Riabowol KT (1998) Increased expression of cyclin D2 during multiple states of growth arrest in primary and established cells. Mol Cell Biol 18:3163–3172

Moreno A, Palacios A, Orgaz JL, Jimenez B, Blanco FJ, Palmero I (2010) Functional impact of cancer-associated mutations in the tumor suppressor protein ING4. Carcinogenesis 31:1932–1938

Narita M, Nunez S, Heard E, Lin AW, Hearn SA, Spector DL, Hannon GJ, Lowe SW (2003) Rb-mediated heterochromatin formation and silencing of E2F target genes during cellular senescence. Cell 113:703–716

Pedeux R, Sengupta S, Shen JC, Demidov ON, Saito S, Onogi H, Kumamoto K, Wincovitch S, Garfield SH, McMenamin M, Nagashima M, Grossman SR, Appella E, Harris CC (2005) ING2 regulates the onset of replicative senescence by induction of p300-dependent p53 acetylation. Mol Cell Biol 25:6639–6648

Sarker KP, Kataoka H, Chan A, Netherton SJ, Pot I, Huynh MA, Feng X, Bonni A, Riabowol K, Bonni S (2008) ING2 as a novel mediator of transforming growth factor-beta-dependent responses in epithelial cells. J Biol Chem 283:13269–13279

Satpathy S, Nabbi A, Riabowol K (2013) RegulatING chromatin regulators: post-translational modification of the ING family of epigenetic regulators. Biochem J Rev 450:433–442

Shen DH, Chan KY, Khoo US, Ngan HY, Xue WC, Chiu PM, Ip P, Cheung AN (2005) Epigenetic and genetic alterations of p33ING1b in ovarian cancer. Carcinogenesis 26:855–863

Shen JC, Unoki M, Ythier D, Duperray A, Varticovski L, Kumamoto K, Pedeux R, Harris CC (2007) Inhibitor of growth 4 suppresses cell spreading and cell migration by interacting with a novel binding partner, liprin alpha1. Cancer Res 67:2552–2558

Soliman MA, Riabowol K (2007) After a decade of studyING, a PHD for a versatile family of proteins. Trends Biochem Sci 32:509–519

Soliman MA, Berardi P, Pastyryeva S, Bonnefin P, Feng X, Colina A, Young D, Riabowol K (2008) ING1a expression increases during replicative senescence and induces a senescent phenotype. Aging Cell 7:783–794

Stark M, Puig-Butille JA, Walker G, Badenas C, Malvehy J, Hayward N, Puig S (2006) Mutation of the tumour suppressor p33ING1b is rare in melanoma. Br J Dermatol 155:94–99

Tamannai M, Farhangi S, Truss M, Sinn B, Wurm R, Bose P, Henze G, Riabowol K, von Deimling A, Tallen G (2010) The inhibitor of growth 1 (ING1) is involved in trichostatin A-induced apoptosis and caspase 3 signaling in p53-deficient glioblastoma cells. Oncol Res 18:469–480

Thalappilly S, Feng X, Pastyryeva S, Suzuki K, Muruve D, Larocque D, Richard S, Truss M, von Deimling A, Riabowol K, Tallen G (2011) The p53 tumor suppressor is stabilized by inhibitor of growth 1 (ING1) by blocking polyubiquitination. PLoS One 6:e21065

Toyama T, Iwase H, Watson P, Muzik H, Saettler E, Magliocco A, DiFrancesco L, Forsyth P, Garkavtsev I, Kobayashi S, Riabowol K (1999) Suppression of ING1 expression in sporadic breast cancer. Oncogene 18:5187–5193

Unoki M, Kumamoto K, Robles AI, Shen JC, Zheng ZM, Harris CC (2008) A novel ING2 isoform, ING2b, synergizes with ING2a to prevent cell cycle arrest and apoptosis. FEBS Lett 582:3868–3874

Wong RP, Lin H, Khosravi S, Piche B, Jafarnejad SM, Chen DW, Li G (2011) Tumour suppressor ING1b maintains genomic stability upon replication stress. Nucleic Acids Res 39:3632–3642

Ythier D, Larrieu D, Brambilla C, Brambilla E, Pedeux R (2008) The new tumor suppressor genes ING: genomic structure and status in cancer. Int J Cancer 123:1483–1490

Ythier D, Brambilla E, Binet R, Nissou D, Vesin A, de Fraipont F, Moro-Sibilot D, Lantuejoul S, Brambilla C, Gazzeri S, Pedeux R (2010) Expression of candidate tumor suppressor gene ING2 is lost in non-small cell lung carcinoma. Lung Cancer 69:180–186

Yu L, Thakur S, Leong-Quong RYY, Suzuki K, Pang A, Bjorge JD, Riabowol K, Fujita DJ (2013) Src regulates the activity of the ING1 tumor suppressor. PLoS One 8(4):e60943

Premalignancy and Cellular Senescence

19

Hussein A. Abbas and Raya Saab

Contents

Abstract	195
Introduction	196
Premalignant Lesions	196
Cellular Senescence	197
Inducers of Cellular Senescence	198
Characteristics and Markers of Senescent Cells	199
Pathways Involved in Senescence	200
Senescence as a Barrier of Malignant Transformation	201
Tumor Progression from Premalignant Lesions	202
Senescence as a Tumor Promoting Factor	203
Clinical Implications	204
References	205

H.A. Abbas • R. Saab (✉)
Children's Cancer Center of Lebanon, American University of Beirut, Beirut, Lebanon
e-mail: haa125@aub.edu.lb; rs88@aub.edu.lb

Abstract

Tumorigenesis is s a multistage process characterized by multiple genetic and molecular insults. The progression from premalignancy to malignancy requires bypass of tumor suppressor mechanisms such as apoptosis and cellular senescence. While apoptosis has long been considered a tumor suppressor mechanism, senescence has recently been verified to be a major impediment for cancer progression, especially in premalignant lesions. In this chapter, we discuss the characteristics of premalignant in comparison to malignant tumors, the role of senescence in impeding the transition, and the mechanisms by which tumor cells are able to escape cellular senescence during cancer progression. This chapter also summarizes what is currently known about the process of cellular senescence in tumor suppression, including the inducers, markers, and molecular pathways involved. Finally, we discuss implications of induction of cellular senescence as a therapeutic strategy in clinical management of premalignant and malignant tumors.

Keywords

Cellular senescence • DNA damage response (DDR) pathway • Malignant transformation • Premalignant lesions • Retinoblastoma (RB) • Senescence-associated heterochromatin foci (SAHF) • Senescent cells • Tumors suppressor protein 53 (TP53)

Introduction

During the normal lifespan of the individual, cells are continuously exposed to a variety of metabolic, oncogenic, oxidative and other stresses. Such stressors lead to a constant risk of transformation and tumorigenesis. However, this is impeded by cell-autonomous tumor suppressor mechanisms, including two well-described responses: apoptosis and cellular senescence. For cancer to occur, one or both of these tumor suppressor mechanisms have to be abrogated at the cellular level.

Cellular senescence is a relatively recently characterized tumor suppressor response. As discussed further below, features of senescence are found in many premalignant lesions, and it is thought to contribute to cessation of progression of such lesions into invasive cancer. The growing interest in this field stems from the hope that understanding the molecular pathways underlying cellular senescence, and the evasion strategies of cancer cells, may lead to development of therapeutic interventions to inhibit tumor progression from a premalignant state, or even reverse the tumorigenic phenotype and arrest tumor growth.

In this review, we will attempt to shed light on cellular senescence as a tumor suppressor mechanism in premalignant lesions, and we will summarize what is known regarding mechanisms of escape or evasion of senescence, which leads to tumor progression from premalignancy into invasive cancer.

Premalignant Lesions

Ample data supports that, in most instances, cancer development is a multistep process requiring several environmental, molecular, genetic and cellular alterations. One of the first and most well characterized models of the multi-step progression of cancer is in adenomatous colorectal carcinoma, where tumors progress from benign adenomas into invasive malignant carcinomas (Fearon and Vogelstein 1990). This transformation is well documented, at least in a subgroup of colon carcinomas, to occur through sequential cellular acquisition of mutations in well-characterized oncogenic and tumor suppressor pathways (Fig. 19.1 upper panel). Specifically, hyperplastic adenomas occur after genetic loss of *APC*, and increased expression of EGFR and COX2. Invasive and malignant properties are acquired when cells further develop *Ras* mutations, then loss of SMAD3/4 and DCC, as well as aberrations in mismatch repair genes such as *MLH1/2* and *MSH2/3/6*. The combination of all of these sequential genetic events then leads to the phenotype of invasive cancer. In order for metastasis to occur, further mutations and genetic alterations are acquired, such as loss of TGFβ and the tumor suppressor p53 (Hanahan and Weinberg 2011).

Another well-characterized model of sequential genetic events leading to tumor progression is seen in a subtype of malignant gliomas (Sanson et al. 2004). In the so-called secondary glioblastoma multiforme, the first lesion is a biologically "low-grade" astrocytoma, which has a mutation

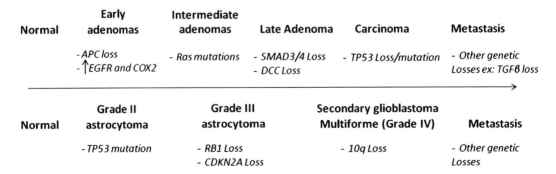

Fig. 19.1 The molecular changes resulting in progression from normal, to pre-malignant, and eventually malignant tumors, during onset of colorectal cancer (*upper panel*) and malignant glioma (*lower panel*)

in the *p53* tumor suppressor gene, and is at that stage classified as Grade II, with low malignant potential. Loss of RB1 and CDKN2A, and amplification of CDK4 instigate the progression from Grade II astrocytoma to Grade III astrocytoma. Finally, the loss of chromosome *10q* leads to secondary gliobastoma multiforme, classified as Grade IV because of its highly invasive and malignant behavior (Fig. 19.1 lower panel).

The above examples demonstrate that several genetic and biologic events need to take place before a normal cell can become a cancerous cell. In 2001, Hanahan and Weinberg outlined the six hallmarks of a cancer cell: sustaining proliferative signaling and independence of growth factors, evasion of cell death, induction of angiogenesis, replicative immortality, insensitivity to anti-growth signals, and finally activating invasion and metastasis pathways (Hanahan and Weinberg 2011). In 2011, and based on accrued evidence, reprogramming of energy metabolism and evading immune destruction were added as key elements in cancer development and progression (Hanahan and Weinberg 2011). The exact sequence and specific genetic mutations needed for acquisition of these properties may differ in each tissue and particular cancer subtype, but the end result is phenotypically similar.

During development of a sporadic premalignant lesion, the first insult in most instances involves an event that leads to increased proliferation, such as the expression of an oncogene or loss of a cell cycle checkpoint gene. The accelerated or enhanced proliferation can then lead to acquisition of further genetic insults that allow the cell to acquire the rest of the properties listed above, and transform into cancer. Examples of such lesions include skin nevi commonly heralded by acquisition of b-raf mutations, neurofibromas that form due to Ras pathway activation because of loss of Nf1, and lung hyperplastic adenomas driven by mutations in Kras or EGFR, among many others. However, these hyperplastic lesions in the vast majority of cases stay "dormant" and do not progress to cancer, due to cell-autonomous mechanisms that are activated by the instigating oncogenic signals, to prevent tumor development.

Whether cells exposed to oncogenic stress induce senescence or apoptosis as a tumor suppressor mechanism seems to be context-and cell-type dependent, and is the subject of continuing research and investigation. In addition, whether cells "sense" an oncogenic event and activate tumor suppression seems to depend on the type and intensity of oncogenic stimulation. For example, in transgenic mice where expression of Ras in mammary tissue can be induced by doxycycline, low levels of Ras allow persistence of proliferation and tumor formation; however, when Ras levels are high, the oncogenic stress is "sensed" by the Arf-p53 pathway, which then induces the senescence checkpoint, inhibiting further cellular proliferation (Sarkisian et al. 2007). Thus, the level of activation of the initiating insult, and its ability (or lack of it) to engage cell autonomous tumor suppressor pathways, may account for the observed dichotomy between tumors that are preceded by pre-malignant lesions that can remain stable for a long time (where tumor suppressor pathways have been engaged), versus those that directly progress to cancer without an obvious growth-arrested, premalignant condition.

Cellular Senescence

Cellular senescence is defined as a state of irreversible cell cycle arrest that can be induced by extrinsic and intrinsic stress signals. Even in the presence of abundant nutrients in a mitogen-rich milieu, senescent cells are unable to proliferate. The first observation of cellular senescence came from experiments by Hayflick and Moorhead (1961) on cultured fibroblasts. They recognized that cells lack the ability to undergo mitotic divisions indefinitely; rather, cells exited mitosis after a defined number of replications. This phenomenon was dubbed *replicative senescence*, and has since been termed the Hayflick limit, referring to the limited ability of cells to divide indefinitely. Notably, the cells remained viable for several weeks, but without sustained growth in size or number, despite nurturing growth conditions.

It was not until 1997 that cellular senescence was recognized to be a possible tumor suppressor mechanism, induced in response to a variety of tumor promoting insults (Serrano et al. 1997). In lung tumors, atypical adenomatous hyperplasia (AAH) are early hyperplastic lesions, thought to progress to preinvasive adenocarcinoma in situ, before developing into invasive adenocarcinoma. Bronchiolo-alveolar adenomas of the lung that are transformed by KRAS^{G12V} have features of oncogene induced senescence as shown by the increased expression of the markers p15, p16, HP1ϒ, Dec1, DcR2 and β-galactosidase activity (Collado et al. 2005). Conversely, adenocarcinomas circumvent the senescence pathway and continue proliferating and invading surrounding tissues (Collado et al. 2005). Evidence from K-ras^{G12V} knock-in mice confirms that senescence is a major hallmark of premalignant lung adenomas (Collado et al. 2005). Likewise, pancreatic intraductal neoplasias (PIN) have senescent characteristics compared to the pancreatic ductal adenocarcinomas in K-ras^{G12V} mice (Collado et al. 2005). Other known premalignant lesion, such as nevi (as opposed to their malignant counterpart melanoma), pituitary adenomas (as opposed to carcinomas), and prostatic intraepithelial neoplasia (as opposed to prostatic carcinoma), among others, also show features of cellular senescence that are not present in their malignant counterparts (reviewed in Saab 2010).

Inducers of Cellular Senescence

Cellular senescence has been shown to be induced by multiple oncogenic insults, including telomere attrition, DNA damage, oxidative stress, oncogenic stimuli, loss of key tumor suppressors, as well as metabolic stress. All of these insults seem to converge upon activation of one or both of two tumor suppressor pathways: the p53 and RB pathways, which then lead to irreversible cell cycle exit and features of senescence (Campisi and d'Adda di Fagagna 2007).

Among the pathways involved in activating cellular senescence, the DNA damage response (DDR) pathway has been found to be a central player. The DDR pathway is activated by phosphorylation of the ataxia-telangiectasia mutated (ATM) protein, which in turn phosphorylates a number of downstream effectors, including histone H2AX and p53. Via a positive feedback loop, phosphorylated H2AX (ϒH2AX) further induces ATM recruitment and formation of the DNA damage foci, which typically include p53-binding protein 1 (53BP1) and mediator of DNA-damage checkpoint (MDC1) (d'Adda di Fagagna 2008). The presence of DDR signaling during the induction and maintenance of cellular senescence has been documented in multiple models in-vitro and in-vivo, where DDR foci are seen, characterized by focal staining for ϒH2AX, and co-localization of DNA damage checkpoint regulators such as NBS1, 53BP1and MDC1. These DDR foci are referred to as senescence-associated DNA damage foci (SDF), and have since been used in identifying senescent lesions in vivo and in vitro, as discussed further below.

The DDR pathway can be activated by DNA damage due to direct insults such as exposure to radiation, but can also occur downstream of other insults such as oncogenic stimulation and stalled replication forks, telomere shortening, or oxidative and metabolic stress (Campisi and d'Adda di Fagagna 2007). Continuous activation of oncogenic Hras, Myc, Braf, Cyclin E and E2F1 lead to excessive cellular proliferation, which can lead to increased DNA replication origins and stalling of replication forks, activating DDR signaling and downstream tumor suppressor pathways (Di Micco et al. 2006). Also, reactive oxygen species can induce DNA damage and eventually senescence, as does metabolic stress. For instance, stimulating normal human fibroblasts with interferon-β induces the production of reactive oxygen species, which leads to DNA damage and ϒH2AX foci, ultimately resulting in cellular senescence (Moiseeva et al. 2006). Inactivation of DDR components, such as by shRNA targeting ATM or CHK2, abrogates senescence both in vitro and in vivo, and increases susceptibility to tumor formation in vivo, proving the central role of this pathway in inducing cellular senescence. Continued active signaling through the DDR pathway, and persistence of SDF in senescent cells, has been proposed as a mechanism for maintenance of senescence and cell cycle exit. Recent elegant work showed that persistence of a few DDR foci, primarily located near telomeric regions, is seen in senescent cells and is necessary for maintaining cell cycle exit (Fumagalli et al. 2012).

Characteristics and Markers of Senescent Cells

Unlike reversible growth arrest and unlike quiescence, cells that undergo senescence are thought to be irreversibly arrested (Campisi and d'Adda di Fagagna 2007). One of the basic definitions of senescence, therefore, is a state of irreversible cell cycle exit, shown by absence of DNA replication and cell division, in the presence of ongoing oncogenic signaling or growth stimulation. In vitro, senescent cells are characterized morphologically by large size, flattened shape, and increased vacuolation. Despite the presence of growth signals, senescent cells are resistant to proliferation. While growth factor deprivation, oxidative stress and ceramide may induce apoptosis in non-senescent human fibroblasts, the senescent fibroblasts resist apoptosis (Campisi and d'Adda di Fagagna 2007).

The most cited marker associated with senescence in vitro is positive histochemical staining for β-galactosidase activity at acidic pH. This is due to increased accumulation of lysosomal β-galactosidase in senescent cells, turning the substrate blue and coloring the cells when examined under a light microscope (Lee et al. 2006). Although this marker has become widely used and approved as modality to identify senescent cells, it can be falsely positive, and its absence does not necessarily rule out senescence. Moreover, this technique is highly amenable to false negatives especially in tissue sections, where fresh frozen tissue is needed and prompt staining of fresh sections is essential.

Senescent cells also show increased expression of specific cell cycle inhibitors, such as p21^{Cip1} and p16^{Ink4a}, which are components of the p53 and RB pathways, respectively. The expression of these proteins directly contributes to the irreversible growth arrest in most cases (reviewed in Saab 2010). Other proteins have been shown to be expressed in senescent cells, even though their function is not as clear. These include cell cycle inhibitors such as p15^{Ink4b} and p18^{Ink4c}, decoy death receptor-2 (DCR2), differentiated embryochondrocyte expressed-1 (DEC1), and promyelocytic leukemia protein (PML), among others (reviewed in Saab 2010).

Epigenetic changes are also seen in senescent cells, as condensed heterochromatin foci designated as senescence-associated heterochromatin foci (SAHF); and the compacted DNA shows resistance to nuclease digestion (Adams 2007). Detection of SAHF requires the binding of certain dyes to DNA, such as DAPI. These foci also contain repressive protein marks such as heterochromatin-associated protein 1-gamma (HP1γ), macroH2A histone, and histone-3 trimethylated at lysine 9 (H3K9me3). Other non-histone components of SAHF include high-mobility group A (HMGA) proteins, anti-silencing function 1a (ASF1a) and histone repressor A (HIRA). These proteins seem to be essential for proper formation of SAHF and induction of a senescence phenotype (Adams 2007). SAHF are thought to contribute to maintenance of senescence by suppressing transcription of cell-cycle genes such as E2F, Cyclin A, and others. Formation of SAHF seems to be dependent on the retinoblastoma tumor suppressor protein (pRB) in most cases, although p53 might contribute in certain contexts (Adams 2007).

Interestingly, recent evidence has shown that SAHF formation is dependent on DNA replication, and SAHF formation in response to cellular stressors seems to occur in a cell-type and context dependent manner. For example, SAHF are formed primarily in the setting of oncogene-induced senescence in human fibroblasts, but not in replicative senescence or other senescence-inducing stimuli such as ionizing radiation, telomere shortening, and H_2O_2 and hydroxyurea treatment (Di Micco et al. 2011; Kosar et al. 2011). In addition, heterochromatin induction can also be seen in proliferating oncogene-expressing cells and in tumor tissue, but in such cases are not associated with E2F and Cyclin A promoters (Di Micco et al. 2011). This suggests that the role and function of such foci differs depending on cellular context, and that such chromatin condensation may be tumor-suppressive only if it inhibits the correct cell cycle genes, among other yet undetermined genes or pathways. The details and complexity of function of SAHF in the setting of cellular senescence, and the reasons for its specificity to oncogenic signaling, still awaits further characterization.

Another characterizing feature of senescent cells is their contribution to the extracellular milieu, by secreting a number of proteins collectively referred to as the Senescence-Associated Secretory Phenotype (SASP) or Secretory Messaging System (SMS) (Kuilman and Peeper 2009). These include growth factors such as insulin-like growth factor 1 (IGF1), pro-inflammatory cytokines, such as GRO1 and IL-6, and other proteins that include components of the TGFβ, WNT and IGF1 signaling cascades. Interestingly, while acting in a cell-autonomous manner to induce senescence, such chemokines have been shown in some settings to contribute to non-autonomous transformation of neighboring cells. This has been suggested as a mechanism by which cancer incidence is increased in aging and senescent tissue (Campisi and d'Adda di Fagagna 2007).

Despite the plethora of senescence markers discussed above, none of these is exclusive for a senescent cell. Depending on the cell type and the initiating insult, a senescent cell may express some of those markers, but not others. Moreover, some markers, including p14Arf, p15Ink4b, p16Ink4c and DCR2, can be expressed in cancer cells, as seen in prostate and breast cancers, even to higher levels than in their benign counterparts (Evan and d'Adda di Fagagna 2009; Zhang et al. 2006). However, such tumors have a high proliferation index, and therefore it is likely that signaling pathways downstream of these proteins are impaired in these settings. The quest for definitive senescent markers is still ongoing. In the meantime, senescent cells are identified by a combination of the above-mentioned markers, when they occur in the appropriate setting of cell cycle exit under pressure of oncogenic signaling or other pro-tumorigenic insults.

Pathways Involved in Senescence

Retinoblastoma (RB) and tumor suppressor protein 53 (TP53) proteins have been shown to be cornerstone elements in the senescence induction pathways. These two proteins orchestrate a variety of downstream effectors. RB is required to inhibit the E2F family of proteins, which functions to activate S phase-specific cell cycle genes. In order for E2F proteins to function, cyclin dependent kinases 4 and 6 (CDK4 and CDK6) phosphorylate RB and render it incapable of inhibiting E2F. However, CDK-inhibitors such as $p15^{INK4b}$, $p16^{INK4a}$, and $p18^{INK4c}$ inhibit the activity of Cyclin D-CDK4/6 complexes, and result in activation of RB. Although $p16^{INK4a}$ seems to be the most prevalent activator of RB in inducing senescence, a role for $p15^{INK4b}$ and $p18^{INK4c}$ has been shown in certain contexts (Solomon et al. 2008; Zalzali et al. 2012; Zhang et al. 2006). Importantly, loss of p16INK4a or loss of RB prevents induction of senescence, and prevents formation of SAHF (Campisi and d'Adda di Fagagna 2007; Narita et al. 2003). Notably, although P16 is required for the formation of SAHF, it does not seem to be necessary for SAHF maintenance (Campisi and d'Adda di Fagagna 2007).

P53 is a tumor suppressor that acts by altering transcription of genes involved in cell cycle arrest, apoptosis, senescence and differentiation. Loss of p53 abrogates oncogene-induced senescence, as well as senescence induced by DNA damage, loss of tumor suppressors, or telomere attrition (Campisi and d'Adda di Fagagna 2007; Hanahan and Weinberg 2011). P53 is stabilized and activated by ATM, which is a component of the DDR pathway. P53 can also be activated by the tumor suppressor p14ARF (p19Arf in mice), which is induced in response to oncogenic stimuli. ARF is a positive regulator of senescence, and it stabilizes TP53 by inhibiting the latter's negative regulator MDM2. RAS can also activate p38MAPK via MKK3/6 pathway, and p38MAPK then phosphorylates p53, resulting in senescence (Kuilman et al. 2010; reviewed in Saab 2010). P53 seems to be essential for senescence induction in almost all models of cellular senescence studied to date. Activation of p53 can lead to induction of p21CIP1, which acts to inhibit the Cyclin A-CDK2 complex, resulting in activation of the RB tumor suppressor protein. Notably, senescent human fibroblasts are seen to express either p21 or p16, but not both simultaneously when replicative senescence is attained.

The cellular decision regarding whether to activate the P53 or RB senescence pathway likely depends on the cell type and the particular tumorigenic stimulus; and in some contexts, both pathways are activated (reviewed in Evan and d'Adda di Fagagna 2009). For instance, ionizing radiation induced-DNA damage activates the P53/P21 pathway, and seems to act primarily through the p53 pathway. Interestingly, telomere disruption in mice activates the P53/P21 senescence pathway while both P53/p21 and RB/p16 are activated in human cells (reviewed in Campisi and d'Adda di Fagagna 2007). Cell type is also an important determinant of which pathway is activated. Epithelial cells tend to induce the RB/p16 pathway primarily when undergoing senescence, and senescence could also occur in a P53- and RB-independent manner. The definitive players in this setting are still obscure (reviewed in Campisi and d'Adda di Fagagna 2007).

Senescence as a Barrier of Malignant Transformation

Accumulating evidence over the past 10 years now clearly demonstrates that cellular senescence, in addition to prominently inhibiting cell proliferation in vitro, is also clearly a tumor suppressor response in vivo. The strongest evidence for the involvement of senescence in tumor suppression in vivo comes from mouse models, where features of senescence are found in pre-invasive lesions, are associated with induction of tumor suppressors, and are lost upon tumor progression (reviewed in Saab 2011).

Collado et al. (2005) used the K-ras^{G12V} conditional tamoxifen-inducible oncogenic mouse model to induce premalignant lung adenomas, malignant lung adenocarcinomas, premalignant pancreatic intraductal neoplasias and malignant ductal adenocarcinomas. They showed that premalignant lesions, but not malignant ones, stain intensely for SABG, and show a number of senescence markers that include low proliferation index (by Ki67 staining), and increased expression of markers such as p16, p15, DcR2, Dec1. In addition, cells in premalignant lesions show staining for markers of SAHF, such as HP1ϒ. Importantly, tumor progression and loss of senescence markers occurs upon loss of p16 or p53. These findings have since been underscored by multiple other mouse models of premalignant lesions inhibited from further transformation by activation of a senescence program. Such models include $Braf^{V600E}$-driven nevi, N-Ras^{G12D} driven lymphoma, E2F-driven pituitary adenoma, PTEN loss-driven prostatic intraepithelial neoplasia, pre-malignant hepatocytes, Cyclin D1-driven pinealoma, among others (reviewed in Saab 2011). In all these models, features of senescence are observed in the premalignant lesions; this phenotype is abrogated upon disruption of pathways involved in senescence induction. Depending on the particular model, such disruptions typically involve the p53 pathway (Chen et al. 2005), components of the RB pathway (Lazzerini Denchi et al. 2005), or effectors of SAHF formation (such as the histone methyltransferase Suv39h1 in Ras-driven lymphoma) (Braig et al. 2005), underscoring the importance of these pathways in senescence induction.

In addition to mouse models, features of senescence have also been widely documented in human premalignant lesions, and are lost in invasive malignant counterparts of these lesions. For example, human nevi, which commonly have $BRAF^{V600E}$ mutations, tend to remain dormant and rarely progress into malignant melanomas. Human nevi specimens were shown to have elevated p16 expression and SA-β-galactosidase activity, features not present in malignant melanoma specimens (Michaloglou et al. 2005). Similarly, human prostate intraepithelial hyperplasia (premalignant lesion) have higher levels of SA-β-galactosidase activity compared to malignant prostate cancer, and advanced human prostate cancers have lost p53 as compared to pre-invasive lesions (Chen et al. 2005). In human colon adenomas, and in human precancerous lesions of the urinary bladder, SA-β-galactosidase activity as well as other senescence markers such as HP1α and HP1β strongly correlated with DDR activation while carcinomas lacked this phenotype (Bartkova et al. 2006). Interestingly, the decreased DNA damage checkpoint activation and absence of senescence in colon carcinomas is strongly associated with mutations of p53 but not with p16 expression (Bartkova et al. 2006).

Collectively, these findings propose a model where oncogenic stimulation and enhanced DNA replication leads to metabolic disruption and activation of the DDR pathway, which then activates the downstream p53 and possibly RB pathways, leading to senescence. Since p53 and ATM are key regulators of the DDR, there is a selective pressure to abrogate these pathways for the premalignant to malignant transformation in these lesions.

In vivo evidence suggests that the immune system may also play a role in clearing senescent cells and contributing to tumor suppression. This has been demonstrated in premalignant NRas^{G12V}-transduced hepatocytes, which undergo senescence, and subsequently are associated with an inflammatory response and an infiltration by immune cells (neutrophils, CD4+ and CD8+ lymphocytes and monocytes/macrophages) in vivo (Kang et al. 2011). Premalignant hepatocytes and liver-infiltrating immune cells were found to secrete various cytokines involved in the senescence surveillance inflammatory response such as CTACK, leptin R, MCP1, RANTES, IL-2 and TECK. Interestingly, premalignant hepatocytes were eventually cleared about 60 days after Nras^{G12V}-transduction (Kang et al. 2011). The investigators further showed that, in immune-deficient (SCID) mice, cells continued to proliferate and levels of p21 and p16 were significantly decreased, suggesting bypass of senescence. Further analysis revealed that CD4+ T lymphocytes were merely responsible for the senescence surveillance in hepatocytes and required the presence of antigen presenting cells (APC), and responded to a specific NRas^{G12V} epitope. In *MYC* and *BCR-ABL*-induced T-cell acute lymphoblastic lymphoma and pro-B-cell leukemia, respectively, CD4+ T-lymphocytes are also required to induce senescence and inhibit angiogenesis (Rakhra et al. 2010).

These models suggest that senescence surveillance via the adaptive immune system and the activation of the innate immunity can induce clearance of premalignant senescent cells, and may play a role in suppression of tumor progression in humans.

Tumor Progression from Premalignant Lesions

While senescence is largely considered an irreversible tumor suppressor mechanism, premalignant lesions by definition present a risk for malignancy, and are associated with a finite lifetime risk of progression into cancer. For cells within a premalignant lesion to proliferate and result in invasive cancer, it is reasonable to assume that such cells must bypass the senescence process by acquiring mutations in key genes responsible for induction of senescence. Such cells then lose the ability to undergo senescence and are at a high likelihood of further proliferation and acquisition of gene mutations that lead to transformation and a malignant phenotype (Fig. 19.2a). This is seen in the majority of mouse models and cell culture models of cellular senescence, where absence of components of the p53 pathway, the RB pathway, the DDR pathway, or proteins involved in formation of SAHF, results in bypass of cell cycle exit, absence of senescence, and continued proliferation (in vitro) and tumor formation (in vivo) (Campisi and d'Adda di Fagagna 2007; Evan and d'Adda di Fagagna 2009; Zhang et al. 2006).

Another plausible but yet unproven scenario, is for a few cells within the senescent lesion to escape the senescent state, through genetic or epigenetic aberrations that may interfere with maintenance of the senescent phenotype (Fig. 19.2b, c). In a model of Cyclin D1-induced

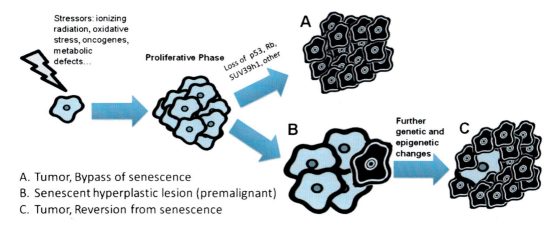

A. Tumor, Bypass of senescence
B. Senescent hyperplastic lesion (premalignant)
C. Tumor, Reversion from senescence

Fig. 19.2 A schematic model for malignant tumor progression from pre-malignant lesions, by either primary bypass of senescence (**A**), or the theoretical possibility of reversion of senescent cells into the cell cycle (**C**), from within a senescent hyperplastic lesion (**B**)

senescence in the mouse pineal gland, genetic loss of p53 leads to invasive tumors without a premalignant senescent state, while loss of p18Ink4c permits senescence in the majority of cells, but a few cells continue to proliferate and eventually lead to tumor progression, albeit at a delayed age, in 100 % of mice (Zalzali et al. 2012). Whether a subset of the p18Ink4c −/− pineal cells escape induction of senescence, or fail to maintain it, is difficult to determine using currently available techniques. However, evidence using in vitro models suggests that escape from an established senescent state may be possible. For instance, abrogation of interleukin-6 or of p53 in senescent cells can allow reentry into cell cycle (Kuilman et al. 2010); shRNA-mediated suppression of p53 in senescent MEFs allows reversal of senescence (Dirac and Bernards 2003), and acute ablation of RB function in senescent MEFS leads to re-entry into the cell cycle (MacPherson et al. 2003). The effect of p53 disruption on the senescent phenotype seems, however, to be context dependent, and seems to depend on the status of the RB pathway in the senescent cells (Beausejour et al. 2003). Senescence induced by oncogenic Ras can also be reversed by inactivation of one of several DDR components (Di Micco et al. 2006). Similarly, targeting proteins involved in formation of SAHF, such as HMGA proteins, leads to resumption of DNA synthesis (Narita et al. 2006).

Finally, senescent epithelial cells left in culture for prolonged periods of time have been observed to result in resumption of proliferation, at least in a subset of cells (Gosselin et al. 2009; Romanov et al. 2001).

All of the evidence for senescence reversion detailed above has so far been shown only in cultured cells. Evaluating whether cells revert from a senescent phenotype in vivo using mouse models has been challenging. Future work is needed to determine which of the two scenarios occurs in vivo, and which one is the predominant mechanism of tumor progression of premalignant lesions. Such a distinction is important because of its implications on therapeutic approaches for prevention of tumor progression, by identifying the need to target the induction of senescence, the maintenance of senescence, or both, in premalignant cells.

Senescence as a Tumor Promoting Factor

Interestingly, senescence in fibroblasts has also been implicated in providing a pro-tumorigenic milieu for surrounding epithelial cells. Earlier experiments from 1950s showed that the milieu created by wound healing can induce proliferation in dormant nevi (reviewed in Kuilman et al. 2010). Senescent fibroblasts secrete MMP-3 and can transform premalignant (immortal but not tumorigenic) mammary epithelial cells to invasive cancer cells (Parrinello et al. 2005). Similarly, senescent human fibroblasts can promote the transition from premalignancy to malignancy in skin epithelial cells, prostate, ovary, mammary glands and skin (Shan et al. 2009). Paracrine signals, such as insulin-like growth factor-1 (IGF-1), stromal-derived factor 1 (SDF-1), and hepatocyte growth factor (HGF) secreted by senescent fibroblasts that are associated with premalignant cells potentially contribute to this transformation (Shan et al. 2009). These findings increase the complexity of considering cellular senescence as a goal in tumor therapy, and make it imperative to better understand the mechanisms underlying such paracrine signaling, in order to minimize the risk of secondary cancer progression in neighboring non-senescent cells.

Clinical Implications

While much has been learned about the process of senescence and its contribution to tumor suppression in humans, many questions remain regarding its possible utilization as a mechanism to treat established cancers, or prevent cancer progression in premalignant lesions. As mentioned above, its clinical utility is further complicated by the finding of tumor promoting properties of senescent fibroblasts, via secretion of various cytokines.

The effectiveness of senescence in inducing tumor suppression can be seen in several in vivo models. Restoration of p53 in mouse models of p53-null driven sarcomas and liver carcinomas induces p53-dependent senescence and inflammatory markers in cancer cells but not normal cells, and potently inhibits tumor progression (Ventura et al. 2007; Xue et al. 2007). Similarly, reactivation of wild-type p53 in sarcomas in vivo induces senescence in tumors but not in normal tissues (Ventura et al. 2007). These experiments suggest that restoration of the relevant tumor suppressor pathways can activate an anti-tumorigenic senescent response in cancer cells. However, the effectiveness of such an approach for long term tumor control has not been addressed yet, as the potential of the persistent (albeit senescent) tumor cells to re-activate proliferative pathways, is not yet clear. In other words, inducing a senescent state in established tumors may arrest tumor progression for a while, but whether such cells can later revert from a senescent phenotype remains to be investigated.

Several studies have shown that senescence can be induced in tumor cells after treatment with radiation, or with chemotherapy such as cisplatin, hydroxyurea, butyrate, doxorubicin, and others (Vergel et al. 2010). In fact, chemotherapy-induced senescence in residual tumor tissue has been seen in human tumor samples after treatment, but no correlation has yet been made to risk of progression or recurrence (te Poele et al. 2002). Interestingly, DNA damaging agents produced the strongest senescent response in cancer cells in comparison to other agents (Vergel et al. 2010). Mammary tumor cells treated with neoadjuvant chemotherapy have a significant senescent response induced by p53/p21 axis and possibly maintained by elevated p16Ink4a (te Poele et al. 2002; Vergel et al. 2010). In other studies, chemotherapeutic induction of senescence was shown not to be merely p53 and p16-dependent. For instance, guanylate cyclase inhibitors induce senescence in a p21-dependent, but p53-independent, manner and act in a DNA-damage-independent process (Vergel et al. 2010).

The question remains as to how the senescent phenotype can be maintained, and how to assess subsequent risks for bypass or reversion from senescence. A recent study showed that, in *MMTV-Wnt1* mammary tumors in mice, doxorubicin treatment leads to features of p53-dependent senescence in mammary tumor cells, but is insufficient to maintain tumor suppression (Jackson et al. 2012). Whether that is due to heterogeneous effects of chemotherapy leading to incomplete induction of senescence due to, for example, heterogeneity in cellular drug delivery or drug effect; or whether this is due to an unsustainable phenotype of senescence in a subset of cancer cells, for example through susceptibility to neighboring pro-tumorigenic cytokines released by senescent fibroblasts, remains unclear. Thus, one may conclude that, while senescence is clearly important in prevention of de novo tumorigenesis, and may be useful in control of progression of established tumor cells, its role in achieving sustained responses in established cancers is yet unclear and awaits further investigation.

References

Adams PD (2007) Remodeling of chromatin structure in senescent cells and its potential impact on tumor suppression and aging. Gene 397:84–93

Bartkova J, Rezaei N, Liontos M, Karakaidos P, Kletsas D, Issaeva N, Vassiliou LV, Kolettas E, Niforou K, Zoumpourlis VC, Takaoka M, Nakagawa H, Tort F, Fugger K, Johansson F, Sehested M, Andersen CL, Dyrskjot L, Orntoft T, Lukas J, Kittas C, Helleday T, Halazonetis TD, Bartek J, Gorgoulis VG (2006) Oncogene-induced senescence is part of the tumorigenesis barrier imposed by DNA damage checkpoints. Nature 444:633–637

Beausejour CM, Krtolica A, Galimi F, Narita M, Lowe SW, Yaswen P, Campisi J (2003) Reversal of human cellular senescence: roles of the p53 and p16 pathways. EMBO J 22:4212–4222

Braig M, Lee S, Loddenkemper C, Rudolph C, Peters AH, Schlegelberger B, Stein H, Dorken B, Jenuwein T, Schmitt CA (2005) Oncogene-induced senescence as an initial barrier in lymphoma development. Nature 436:660–665

Campisi J, d'Adda di Fagagna F (2007) Cellular senescence: when bad things happen to good cells. Nat Rev Mol Cell Biol 8(9):729–740

Chen Z, Trotman LC, Shaffer D, Lin HK, Dotan ZA, Niki M, Koutcher JA, Scher HI, Ludwig T, Gerald W, Cordon-Cardo C, Pandolfi PP (2005) Crucial role of p53-dependent cellular senescence in suppression of Pten-deficient tumorigenesis. Nature 436:725–730

Collado M, Gil J, Efeyan A, Guerra C, Schuhmacher AJ, Barradas M, Benguria A, Zaballos A, Flores JM, Barbacid M, Beach D, Serrano M (2005) Tumour biology: senescence in premalignant tumours. Nature 436:642

d'Adda di Fagagna F (2008) Living on a break: cellular senescence as a DNA-damage response. Nat Rev Cancer 8:512–522

Di Micco R, Fumagalli M, Cicalese A, Piccinin S, Gasparini P, Luise C, Schurra C, Garre M, Nuciforo PG, Bensimon A, Maestro R, Pelicci PG, D'Adda D, Fagagna F (2006) Oncogene-induced senescence is a DNA damage response triggered by DNA hyper-replication. Nature 444:638–642

Di Micco R, Sulli G, Dobreva M, Liontos M, Botrugno OA, Gargiulo G, Dal Zuffo R, Matti V, D'Ario G, Montani E, Mercurio C, Hahn WC, Gorgoulis V, Minucci S, D'Adda Di Fagagna F (2011) Interplay between oncogene-induced DNA damage response and heterochromatin in senescence and cancer. Nat Cell Biol 13:292–302

Dirac AM, Bernards R (2003) Reversal of senescence in mouse fibroblasts through lentiviral suppression of p53. J Biol Chem 278:11731–11734

Evan GI, D'Adda Di Fagagna F (2009) Cellular senescence: hot or what? Curr Opin Genet Dev 19:25–31

Fearon ER, Vogelstein B (1990) A genetic model for colorectal tumorigenesis. Cell 61:759–767

Fumagalli M, Rossiello F, Clerici M, Barozzi S, Cittaro D, Kaplunov JM, Bucci G, Dobreva M, Matti V, Beausejour CM, Herbig U, Longhese MP, D'Adda Di Fagagna F (2012) Telomeric DNA damage is irreparable and causes persistent DNA-damage-response activation. Nat Cell Biol 14:355–365

Gosselin K, Martien S, Pourtier A, Vercamer C, Ostoich P, Morat L, Sabatier L, Duprez L, T'Kint de Roodenbeke C, Gilson E, Malaquin N, Wernert N, Slijepcevic P, Ashtari M, Chelli F, Deruy E, Vandenbunder B, De Launoit Y, Abbadie C (2009) Senescence-associated oxidative DNA damage promotes the generation of neoplastic cells. Cancer Res 69:7917–7925

Hanahan D, Weinberg RA (2011) Hallmarks of cancer: the next generation. Cell 144:646–674

Hayflick L, Moorhead PS (1961) The serial cultivation of human diploid cell strains. Exp Cell Res 25:585–621

Jackson JG, Pant V, Li Q, Chang LL, Quintas-Cardama A, Garza D, Tavana O, Yang P, Manshouri T, Li Y, El-Naggar AK, Lozano G (2012) p53-mediated senescence impairs the apoptotic response to chemotherapy and clinical outcome in breast cancer. Cancer Cell 21:793–806

Kang TW, Yevsa T, Woller N, Hoenicke L, Wuestefeld T, Dauch D, Hohmeyer A, Gereke M, Rudalska R, Potapova A, Iken M, Vucur M, Weiss S, Heikenwalder M, Khan S, Gil J, Bruder D, Manns M, Schirmacher P, Tacke F, Ott M, Luedde T, Longerich T, Kubicka S, Zender L (2011) Senescence surveillance of premalignant hepatocytes limits liver cancer development. Nature 479:547–551

Kosar M, Bartkova J, Hubackova S, Hodny Z, Lukas J, Bartek J (2011) Senescence-associated heterochromatin foci are dispensable for cellular senescence, occur in a cell type- and insult-dependent manner and follow expression of p16(ink4a). Cell Cycle 10:457–468

Kuilman T, Peeper DS (2009) Senescence-messaging secretome: SMS-ing cellular stress. Nat Rev Cancer 9:81–94

Kuilman T, Michaloglou C, Mooi WJ, Peeper DS (2010) The essence of senescence. Genes Dev 24:2463–2479

Lazzerini Denchi E, Attwooll C, Pasini D, Helin K (2005) Deregulated E2F activity induces hyperplasia and senescence-like features in the mouse pituitary gland. Mol Cell Biol 25:2660–2672

Lee BY, Han JA, Im JS, Morrone A, Johung K, Goodwin EC, Kleijer WJ, DiMaio D, Hwang ES (2006) Senescence-associated beta-galactosidase is lysosomal beta-galactosidase. Aging Cell 5:187–195

MacPherson D, Sage J, Crowley D, Trumpp A, Bronson RT, Jacks T (2003) Conditional mutation of Rb causes cell cycle defects without apoptosis in the central nervous system. Mol Cell Biol 23:1044–1053

Michaloglou C, Vredeveld LC, Soengas MS, Denoyelle C, Kuilman T, van der Horst CM, Majoor DM, Shay JW, Mooi WJ, Peeper DS (2005) BRAFE600-associated senescence-like cell cycle arrest of human naevi. Nature 436:720–724

Moiseeva O, Mallette FA, Mukhopadhyay UK, Moores A, Ferbeyre G (2006) DNA damage signaling and p53-dependent senescence after prolonged beta-interferon stimulation. Mol Biol Cell 17:1583–1592

Narita M, Nunez S, Heard E, Lin AW, Hearn SA, Spector DL, Hannon GJ, Lowe SW (2003) Rb-mediated heterochromatin formation and silencing of E2F target genes during cellular senescence. Cell 113:703–716

Narita M, Krizhanovsky V, Nunez S, Chicas A, Hearn SA, Myers MP, Lowe SW (2006) A novel role for high-mobility group a proteins in cellular senescence and heterochromatin formation. Cell 126:503–514

Parrinello S, Coppe JP, Krtolica A, Campisi J (2005) Stromal-epithelial interactions in aging and cancer: senescent fibroblasts alter epithelial cell differentiation. J Cell Sci 118:485–496

Rakhra K, Bachireddy P, Zabuawala T, Zeiser R, Xu L, Kopelman A, Fan AC, Yang Q, Braunstein L, Crosby E, Ryeom S, Felsher DW (2010) CD4(+) T cells contribute to the remodeling of the microenvironment required for sustained tumor regression upon oncogene inactivation. Cancer Cell 18:485–498

Romanov SR, Kozakiewicz BK, Holst CR, Stampfer MR, Haupt LM, Tlsty TD (2001) Normal human mammary epithelial cells spontaneously escape senescence and acquire genomic changes. Nature 409:633–637

Saab R (2010) Cellular senescence: many roads, one final destination. Sci World J 10:727–741

Saab R (2011) Senescence and pre-malignancy: how do tumors progress? Semin Cancer Biol 21:385–391

Sanson M, Thillet J, Hoang-Xuan K (2004) Molecular changes in gliomas. Curr Opin Oncol 16:607–613

Sarkisian CJ, Keister BA, Stairs DB, Boxer RB, Moody SE, Chodosh LA (2007) Dose-dependent oncogene-induced senescence in vivo and its evasion during mammary tumorigenesis. Nat Cell Biol 9:493–505

Serrano M, Lin AW, McCurrach ME, Beach D, Lowe SW (1997) Oncogenic ras provokes premature cell senescence associated with accumulation of p53 and p16INK4a. Cell 88:593–602

Shan W, Yang G, Liu J (2009) The inflammatory network: bridging senescent stroma and epithelial tumorigenesis. Front Biosci 14:4044–4057

Solomon DA, Kim JS, Jenkins S, Ressom H, Huang M, Coppa N, Mabanta L, Bigner D, Yan H, Jean W, Waldman T (2008) Identification of p18 INK4c as a tumor suppressor gene in glioblastoma multiforme. Cancer Res 68:2564–2569

te Poele RH, Okorokov AL, Jardine L, Cummings J, Joel SP (2002) DNA damage is able to induce senescence in tumor cells in vitro and in vivo. Cancer Res 62:1876–1883

Ventura A, Kirsch DG, McLaughlin ME, Tuveson DA, Grimm J, Lintault L, Newman J, Reczek EE, Weissleder R, Jacks T (2007) Restoration of p53 function leads to tumour regression in vivo. Nature 445:661–665

Vergel M, Marin JJ, Estevez P, Carnero A (2010) Cellular senescence as a target in cancer control. J Aging Res 2011:725365

Xue W, Zender L, Miething C, Dickins RA, Hernando E, Krizhanovsky V, Cordon-Cardo C, Lowe SW (2007) Senescence and tumour clearance is triggered by p53 restoration in murine liver carcinomas. Nature 445:656–660

Zalzali H, Harajly M, Abdul-Latif L et al. (2012) Temporally distinct roles for tumor suppressor pathways in cell cycle arrest and cellular senescence in Cyclin D1-driven tumor. Mol Cancer 11:28 [Epub ahead of print]

Zhang Z, Rosen DG, Yao JL, Huang J, Liu J (2006) Expression of p14ARF, p15INK4b, p16INK4a, and DCR2 increases during prostate cancer progression. Mod Pathol 19:1339–1343

Loss of Cdh1 Triggers Premature Senescence in Part via Activation of Both the RB/E2F1 and the CLASPIN/CHK1/P53 Tumor Suppressor Pathways

20

Shavali Shaik*, Pengda Liu*, Zhiwei Wang, and Wenyi Wei

Contents

Abstract	207
Introduction	208
Regulation of Cell Cycle by the APC/Cyclosome and the SCF E3 Ubiquitin Ligase Complexes	208
Cdh1 Is a Tumor Suppressor	210
Regulation of Cdh1 by Multiple Mechanisms	210
Cdh1 Loss in Human Cells and MEFS Leads to Premature Senescence	212
Cdh1 Loss Leads to Activation of RB/E2F1 and CLASPIN/CHK1/P53 Pathways to Induce Premature Senescence in Primary Human Fibroblasts	213
Conclusion	215
References	216

*These authors contributed equally to this work

S. Shaik • P. Liu • Z. Wang • W. Wei (✉)
Department of Pathology, Beth Israel Deaconess Medical Center, Harvard Medical School, Boston, MA 02215, USA
e-mail: sshaik@bidmc.harvard.edu;
pliu1@bidmc.harvard.edu; zwang6@bidmc.harvad.edu; wwei2@bidmc.harvard.edu

Abstract

Senescence is recently characterized as one of the evolutionarily conserved protective mechanisms against tumor development. Several upstream factors including oxidative stress, DNA damage and overexpression of certain oncoproteins have been shown to induce premature senescence. Interestingly, it has been discovered that instead of promoting tumorigenesis, loss of certain tumor suppressors such as TSC2, PTEN and NF1 induce premature senescence under certain conditions, presumably by activating the downstream oncoproteins mTORC1/S6k, Akt and Ras, respectively. Interestingly, it has been observed by multiple groups that acute loss of Cdh1 also leads to premature senescence in several cellular settings including mouse embryonic fibroblasts and human primary fibroblasts. This is in part due to the fact that Cdh1 loss leads to stabilization of Ets2, which increases p16 expression and causes premature senescence. Moreover, recent studies from our laboratory further suggested that loss of Cdh1 results in the activation of both the Claspin/Chk1/p53 and the Rb/E2F1 pathways, which ultimately leads to premature senescence in primary human fibroblasts but not in transformed cells with defective p53/Rb pathways. Therefore, our studies support the idea that onset of premature senescence serves as a protection mechanism against sporadic tumorigenesis. It also indicates that loss of Cdh1 tumor suppressor is a relatively late event, which only benefits tumorigenesis for late stage tumors with defective Rb and p53 tumor suppressor pathways. More importantly, our results also indicate that Cdh1 could be an anti-cancer target in certain settings, as complete inactivation of Cdh1 in early stage tumors with wild-type p53 and Rb pathways will lead to induction of premature senescence, thereby aiding tumor regression.

Keywords

Anaphase-promoting complex (APC/C) • Cyclin-dependent protein kinases (Cdks) • DNA binding activity of p53 • DNA damage • Expression of PTEN • Mouse embryonic fibroblasts (MEFs) • Regulation of Cdh1 • Skp1-Cullin1-F-box complex (SCF) • Ubiquitin proteasome system (UPS)

Introduction

Cellular senescence is considered as one of the tumor suppression mechanisms and contributes to overall aging process. Cells undergoing senescence cease proliferation and exhibit several morphological and functional changes (Hwang et al. 2009). Recent evidence suggests that in addition to regulating many cellular processes, the ubiquitin proteasome system (UPS) is also actively involved in regulating senescence (Sitte et al. 2000). Specifically, the two ubiquitin ligase complexes, Anaphase-Promoting Complex (APC/C) and Skp1-Cullin1-F-box complex (SCF) are involved in targeted degradation of many key proteins in cell cycle regulation (Shaik et al. 2012; Vodermaier 2004) as well as senescence (Li et al. 2008). Importantly, APC/C is considered as one of the critical regulators of senescence (Li et al. 2008). It is known that APC/C targets several key substrates including cyclin A, cyclin B, cyclin D, securin and p27 for degradation during M phase to G1 phase. Although Cdh1 and Cdc20 are the two proteins involved in the activation of APC/C, in this review article, we mainly highlight the tumor suppression role of APC/Cdh1 and discussed how loss of Cdh1 leads to onset of premature senescence.

Regulation of Cell Cycle by the APC/Cyclosome and the SCF E3 Ubiquitin Ligase Complexes

The cell cycle progression is under tight control since abnormal cell cycle leads to the development of various deadly diseases such as cancer. The cell cycle events are mainly regulated by both cyclins and cyclin-dependent protein kinases (Cdks) (Doree and Galas 1994). Cdks drive the cell cycle progression by phosphorylating key proteins involved in DNA replication and cell division. The activities of Cdks are regulated largely by their regulatory cyclin partners. Notably, Cdks are constitutively expressed across the cell cycle, whereas cyclins only express at specific cell cycle phases. Thus, recent studies reveal that the cell cycle progression is actually controlled by the periodically expressed cyclins. Furthermore, according to their specific functions, cyclins are classified into four different classes: G1/S cyclins, S cyclins, M cyclins and G1 cyclins (Bloom and Cross 2007). Most cyclins are synthesized and degraded in a cyclic manner. Therefore, the periodical degradation of precious cell phase-specific cyclins and the synthesis of the following cell cycle cyclins are well coordinated by ubiquitination-mediated degradation processes to ensure accurate cell cycle transition. On the other hand, the Cdk inhibitors (CdkIs), which functionally inhibit the kinase activities of various Cdks, serve as the "brake" of cell cycle machinery that mediate the cell cycle arrest in response to various anti-proliferative signals such as growth factor deprivation, DNA damage and contact inhibition. Two families of CdkIs have been identified in mammalian cells based on their sequence homology: the Cip/Kip family, which includes p21, p27 and p57; and the Ink family, which includes p15, p16, p18, and p19. In addition to the above mentioned mechanisms, the activities of Cdks are also regulated by the activating and the inactivating posttranslational modification events including but not restricted to phosphorylation.

The two major ubiquitin E3 ligase complexes, APC/C and SCF complexes regulate cell cycle

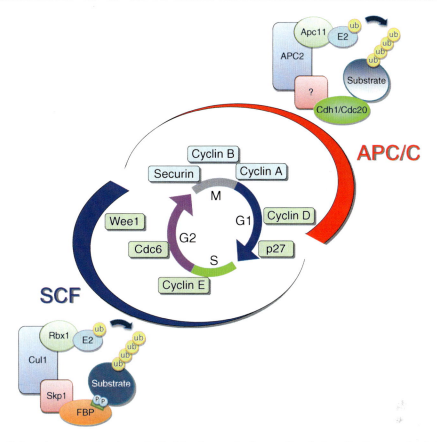

Fig. 20.1 Cell cycle progression is controlled by the APC/C and SCF complexes. The cell cycle progression is precisely regulated by the periodically expressed Cyclins complexed with corresponding Cdks. APC/C and SCF complexes exert their functions by timely targeting many cellular regulatory proteins especially Cyclins and Cdk inhibitors for ubiquitination-mediated degradation. APC/Cdh1 stabilizes G1 phase and controls the G1/S transition by suppressing the accumulation of mitotic as well as S phase Cyclins to maintain low levels of Cdk activities. Inactivation of Cdh1 by Cdh1 phosphorylation leads to the entry of S phase. On the other hand, SCF regulates G1/S through G2/M transitions by degrading Cdk inhibitors such as p27, p21, Wee1 and G1/S Cyclins like Cyclin E, or Cdh1 inhibitors Emi1. Furthermore, APC/Cdh1 also targets Skp2 for degradation, indicating another layer of regulation of cell cycle by APC/Cdh1

progression in a coordinative manner (Vodermaier 2004) (Fig. 20.1). The APC/C's main function is to trigger the transition from metaphase to anaphase by tagging specific proteins including securin and M cyclins for degradation. The two main APC/C activator proteins namely Cdc20 and Cdh1 guide the APC/C to interact with specific sets of downstream ubiquitin substrates for 26S proteasome-mediated degradation at different times in the cell cycle, thus driving the cell cycle forward. On the other hand, SCF complex mainly controls the G1/S and the G2/M transitions. Similar to Cdh1/Cdc20 in APC/C, F-box proteins within the specific SCF complexes mediate the direct and specific interactions between SCF and its substrates. Skp2 and β-TRCP are the two well-studied F-box-proteins. SCF^{Skp2} mainly ubiquitinates and degrades the cyclin-dependent-kinase inhibitors (CdkIs) such as p27 and p21 as well as G1/S cyclins such as cyclin E, while $SCF^{\beta\text{-}TRCP}$ targets a number of substrates including β-catenin, CDC25A, Wee1 and Mdm2 that play key roles in cell cycle regulation (Frescas and Pagano 2008) (Fig. 20.1).

Cdh1 Is a Tumor Suppressor

Cdh1 is one of the substrate adaptor proteins of the APC/C complex. It has seven WD40 repeats in its c-terminus and these repeats play an important role in recognizing the various Cdh1 substrates. Furthermore, most of the APC/C substrates contain a D-Box (RxxL) and/or a KEN box domain that facilitates their interaction with Cdh1 (Kraft et al. 2005). Cdh1 functions as a APC/C activator (Sudo et al. 2001) especially during late mitosis as well as in the early-mid G1 phase, promoting the degradation of many proteins involved in cell cycle progression and other cellular function. Acute depletion of Cdh1 by RNAi leads to an aberrant accumulation of several APC^{Cdh1} substrates including cyclin A, cyclin B, Aurora A, PlK1, Skp2 and Cdc20. One of the major functions of Cdh1 is to suppress the accumulation of mitotic cyclins and other cell cycle regulators in order to maintain cells in the G1 phase. Therefore, Cdh1 is inactive during the early mitotic phase and becomes active during the transition from late mitosis to G1 phase. Cdh1 also plays a major role in the exit from mitosis. During this phase, Cdh1 is dephosphorylated by the protein phosphatase Cdc14, gets activated and subsequently degrades mitotic cyclins that facilitate the cell exit from mitosis (Fig. 20.2).

Importantly, it has been demonstrated recently that Cdh1 functions as a tumor suppressor. To study the role of Cdh1 in cancer, Garcia-Higuera et al. (2008) have generated conditional knockout mice with a targeted mutation in the Fzr1 locus. Complete inactivation of Cdh1 caused embryonic lethality due to specific aberrations in the placenta. However, ablation of Cdh1 causes defects in the proliferation of mouse embryonic fibroblasts (MEFs) in culture, and accumulation of chromosomal defects. On the contrary, it has been observed that *Fzr1*$^{+/-}$ mice develop a variety of tumors late in life without losing the *Fzr1* wild-type allele. These tumors include adenocarcinomas and fibroadenomas of the mammary gland, tumors in the lung, liver, kidney, testis, sebaceous gland and B cell lymphomas (Garcia-Higuera et al. 2008). Interestingly, several Cdh1 substrates including Aurora A, Plk1, cyclin A, cyclin B and Cdc20 were upregulated in *Fzr1*$^{-/-}$ cells and these were part of the reasons for the chromosomal instability observed in human tumors which showed a positive correlation with poor prognosis in cancer (Carter et al. 2006). Moreover, over-expression of several licensing factors that involved in DNA replication is also frequently observed in many human tumors. Consistent with a critical role of Cdh1 in tumor suppression, decreased Cdh1 protein levels were observed in some cancer cell lines and expression of Cdh1 prevented tumor development (Sudo et al. 2001; Wang et al. 2000). Altogether, these findings suggest that Cdh1 may function as a haploinsufficient tumor suppressor in vivo and that loss of Cdh1 may induce gross genomic alterations during cell proliferation, which can ultimately lead to tumor development.

Regulation of Cdh1 by Multiple Mechanisms

As Cdh1 plays a critical role in cell cycle regulation, its activity is tightly regulated in a cell cycle specific manner. Therefore, Cdh1 exists in multiple forms, including free Cdh1 and APC/C-bound active Cdh1 within a given cell. Furthermore, Cdh1 levels were found to fluctuate throughout the cell cycle and Cdh1 activity is regulated by cell-cycle-dependent phosphorylation and dephosphorylation events (Fig. 20.2). Cdh1 is phosphorylated in vivo during S, G2, and M phases by multiple Cdks, and phosphorylation of Cdh1 results in the inactivation of APC/Cdh1, by blocking phospho-Cdh1 to incorporate into functional APC; whereas dephosphorylation by Cdc14B activates APC/Cdh1 by promoting the formation of active APC/C (Fig. 20.2). Specifically, Cdh1 is inactivated during the S phase, G2 phase and the early stages of mitosis via phosphorylation by both cyclin A2-Cdk2 and cyclin B1-Cdk1 (Lukas et al. 1999). Furthermore, at the end of G1 phase Cdh1 is inactivated by E2F1 target, Emi1, which binds Cdh1 to directly turn off the APC^{Cdh1} activity (Hsu et al. 2002). Interestingly, it has been found that Cdh1 may mediate its own degradation

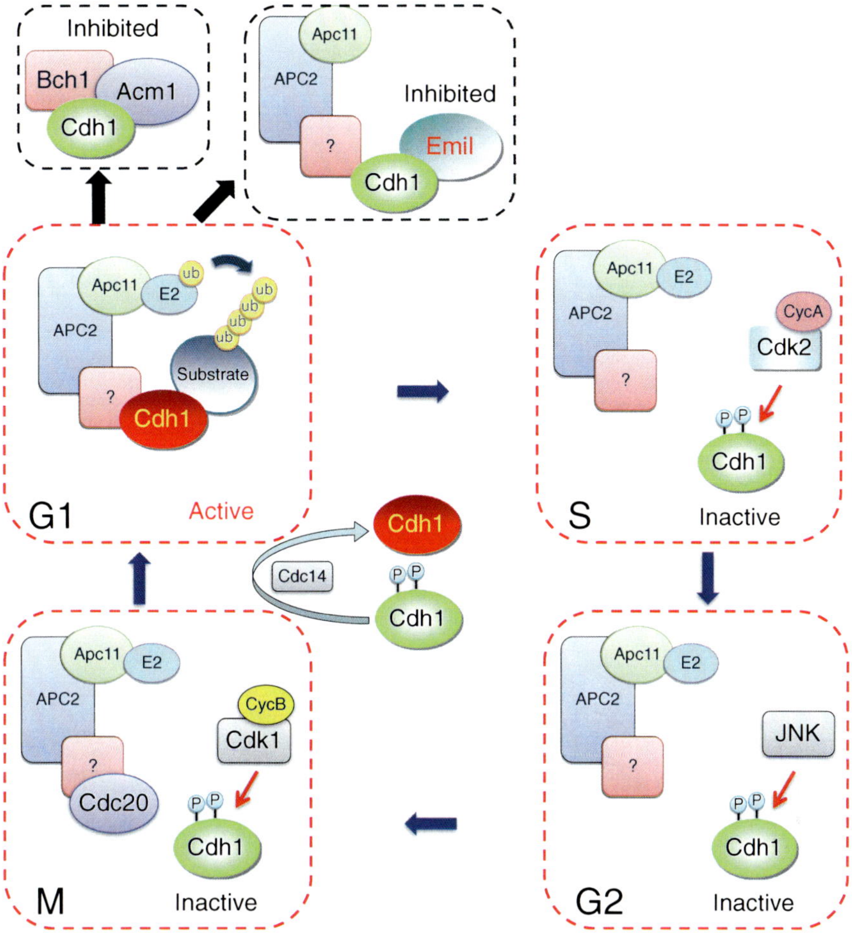

Fig. 20.2 Cell cycle dependent regulation of APC/Cdh1 activity by phosphorylation and dephosphorylation events. When entering the G1 phase, Cdh1 is released from the inhibitory states by dissociation from its inhibitors Emi1 and Acm1 to be incorporated into the functional APC/Cdh1 complex to exert its E3 ligase activity to actively repress the expression of mitotic Cyclins. Cdh1 is phosphorylated by Cdk2/Cyclin A which results in its inactivation, leading to the accumulation of S phase Cyclins to enter S phase. In G2 and M phases, Cdh1 is kept suppressed by phosphorylation by JNK and Cdk1/Cyclin B, respectively. This negative phosphorylation on Cdh1 could be antagonized by the phosphatase Cdc14, resulting the incorporation of unphosphorylated Cdh1 into functional APC/C again

in G1 and G0 phases by Cdh1-activated APC/C complex (Listovsky et al. 2004).

Furthermore, phosphorylation of Cdh1 not only abolishes its ability to activate APC/C, but also determines Cdh1's sub-cellular localization. In this regard, Cdh1 is mainly localized to nuclei during interphase, and to centrosomes during mitosis. Interestingly, a functional nuclear localization signal (NLS) has been identified in Cdh1, and phosphorylation in the vicinity of the NLS led to constitutive cytoplasmic retention of Cdh1 protein (Zhou et al. 2003). In addition, there is at least one additional Cdh1 inactivation mechanism independent of Cdh1 phosphorylation has been identified, in which Acm1 and Bmh1 compete with APC to complex with unphosphorylated Cdh1 (Dial et al. 2007). Interestingly, a recent study demonstrated that PTEN activates APC/C complex by promoting the interaction between APC/C and Cdh1. Expression of PTEN in PC3 cells increased the association between APC and Cdh1, whereas PTEN knockdown reversed the interaction in DU145 cells. Furthermore, findings from in vitro binding assays revealed that PTEN promotes the assembly between APC3 and Cdh1 in a dose-dependent manner. Together, these results indicated that PTEN enhances the activity of APC^{Cdh1} through promoting the association between APC and Cdh1 (Song et al. 2011).

It is interesting to note that some of the Cdh1 substrates are post-translationally modified and such modifications affect the ability of Cdh1 in promoting their destruction. For example, Skp2 is one of the key substrates of APC^{Cdh1} (Wei et al. 2004) and phosphorylated Skp2 cannot be degraded by Cdh1. Previous studies have suggested that a region of Skp2 between amino acids 46 and 90 which contains both the Akt and CKI phosphorylation sites, is required for its interaction with Cdh1 (Wei et al. 2004). Both Ser72 and Ser75 sites in Skp2 were found to be critical for interaction with Cdh1, as substitution of these sites with the phospho-mimetic amino acids disrupted the interaction between Skp2 and Cdh1. Interestingly, when activated Akt1 was overexpressed, the interaction between Skp2 and Cdh1 was dramatically reduced in vivo suggesting that Akt-mediated phosphorylation of Skp2 reduces its association with APC/Cdh1. In line with this hypothesis, our laboratory recently demonstrated that phosphorylation of Skp2 by Akt1 at Ser72 site not only impairs APC/Cdh1-mediated Skp2 destruction but also promotes Skp2 cytoplasmic localization (Gao et al. 2009b). Importantly, this finding expanded our knowledge of how specific Akt-mediated signaling cascades influence proteolysis governed by APC/Cdh1 complexes, and provided evidence that elevated Akt activity and cytoplasmic Skp2 expression may be potential causative factors for development as well as progression of certain cancers. Similarly, phosphorylation of Cdc6 by Cdk2/cyclin E also protects Cdc6 from Cdh1-mediated destruction. In addition to its ability to promote the E3 ligase activity of APC/C by recruiting various substrates to the APC core complex, recent studies have began to reveal APC-independent functions of Cdh1. For example, recent studies from our laboratory indicated that APC-free form of Cdh1 could promote the E3 ligase activity of Smurf1 by disrupting the inhibitory Smurf1 homo-dimers (Wan et al. 2011).

Cdh1 Loss in Human Cells and MEFS Leads to Premature Senescence

Cellular senescence is considered as an essential contributor to the aging process and has been shown to be an important tumor suppression mechanism. After a finite number of divisions, normal somatic cells undergo replicative senescence by withdrawing from the cell cycle and entering into an irreversible growth arrest. In senescent cells, the expression of certain mitogen responsive genes and G1/S specific genes are permanently repressed, whereas the expression and activity of members of the p53/p21 and p16/Rb pathways are changed dramatically in comparison to actively cycling cells (Kuilman et al. 2010). The DNA binding activity of p53 is greatly increased in senescent cells and the expression of cell cycle inhibitors p21 and p16 was found much higher. Furthermore, the senescent cells exhibit increased expression of β-galactosidase enzyme (Dimri et al. 1995) and other cyclin-dependent kinase inhibitors such as p27 and p15 (Kuilman et al. 2010). These cells are also known to accumulate transcriptionally inactive heterochromatic structure known as senescence-associated heterochromatic foci (Campisi 1996).

Interestingly, it has been discovered that Cdh1 is essential for placental development and its complete deficiency causes early lethality. Cdh1 also seems to play a pivotal role in senescence, as Cdh1-deficient mouse embryonic fibroblasts (MEFs) proliferate poorly and enter replicative senescence after only a few passages (Li et al. 2008). In order to investigate the mechanisms by which Cdh1 loss causes senescence, Li et al. (2008) analyzed various Cdh1 substrates in $Cdh1^{-/-}$ MEFs. They found that the Cdh1 substrates Cdc20, Aurora B and Skp2 were stabilized, and p27 (a substrate of Skp2) was decreased as expected. However, p21^{Cip1}, another Skp2's substrate was maintained at the same level in Cdh1-deficient MEFs. Unexpectedly, cyclin B1 levels were found decreased, and no changes in cyclin A levels were observed in $Cdh1^{-/-}$ MEFs. Unfortunately, the decreased p27 levels and no change in p21 levels in $Cdh1^{-/-}$ could not explain how Cdh1 contributes to premature senescence.

Interestingly, further investigations found that p16 protein levels were dramatically increased along with p16 mRNA levels in $Cdh1^{-/-}$ MEFs compared to wild type MEFs. These results indicated that the Cdh1 loss activates p16 transcription, leading to the slow growth and senescence in MEFs. Further studies revealed that Cdh1 deficiency-induced upregulation of p16 was a result of Ets2 stabilization. Transcription factor Ets2 is a characterized Cdh1 substrate, which is activated by the Ras-Raf-MAPK signaling pathway. This pathway is known to stimulate cell proliferation. Therefore, downregulation of Cdh1 leads to Ets2 overexpression, resulting in enhanced cell growth. As overactivation of Ras signaling could also cause senescence in primary cells, it is postulated that Ets2 may mediate in part the senescence effect of Ras signaling. Although, it remains to be determined how Ras signaling interacts with APC/Cdh1, it has been demonstrated that Ets2 could directly activate p16 expression (Ohtani et al. 2001), thus stabilized Ets2 and increased p16 both may responsible for premature senescence in this phenotype.

Cdh1 Loss Leads to Activation of RB/E2F1 and CLASPIN/CHK1/P53 Pathways to Induce Premature Senescence in Primary Human Fibroblasts

Our laboratory made further efforts to characterize the molecular mechanisms by which Cdh1 loss causes premature senescence in various cells. Earlier reports have demonstrated that overexpression of strong oncogenes, such as Ha-Ras or Stat3 led to activation of the ATM/ATR DNA damage pathway and subsequently led to premature senescence (Mallette et al. 2007), and that inactivation of both the p53 and Rb pathways is required for bypassing the growth arrest phenotype (Serrano et al. 1997). Further studies revealed that depletion of Cdh1 in HeLa cells, whose p53 and Rb pathways are defective due to the presence of HPV E6 and E7 proteins, did not cause growth arrest, however retarded growth in U2OS cells, which possess relatively normal p53 pathway, but defective Rb pathway (Gao et al. 2009a). This finding indicated that p53 pathway is partly responsible for the senescent phenotype observed in Cdh1 defective cells. A similar growth retardation phenotype was observed in MEFs when Cdh1 was depleted (Garcia-Higuera et al. 2008; Li et al. 2008). Interestingly, Gao et al. (2009a) observed that normal human fibroblasts responds more severely to acute loss of Cdh1, and undergoes premature senescence as evidenced by elevated β-Galactosidase staining with decreased cell proliferation. Interestingly, increased expression of many senescence-associated proteins especially p21, p27, and to a lesser degree p16, was found after acute loss of Cdh1 in primary human fibroblasts. As p21 is a well-characterized p53 downstream target, the significant induction of p21 provides further evidence for the notion that loss of Cdh1 activates the p53 pathway (Gao et al. 2009a).

Although the exact molecular mechanisms remain unknown for the upregulation of p53 pathway in $Cdh1^{-/-}$ cells, depletion of Cdh1 in HeLa, U2OS, and primary human fibroblasts led to an up-regulation of Claspin protein, and subsequent increase in Chk1 activity. As Chk1 has been shown to activate p53 via phosphorylation at the Ser20 site (Shieh et al. 2000) and an increase in p53-Ser20 phosphorylation in Cdh1 depleted U2OS cells suggested that the Claspin/Chk1 pathway might contribute to the observed increase in p53 activity after Cdh1 loss. To further confirm whether the p53 pathway alone plays a critical role for this senescent phenotype, the matched HCT116 WT and HCT116 $p53^{-/-}$ cells were utilized to examine the significance of this pathway. Surprisingly, depletion of Cdh1 in HCT116 $p53^{-/-}$ cells still resulted in retarded growth arrest. Notably, further results indicated that the growth arrest phenotype was largely abolished after simultaneous inactivation of the p53 and Rb pathways, suggesting that both p53 and Rb pathways are required to induce senescence.

Inactivation of the p53 and Rb pathways allows the human primary fibroblasts to abrogate Cdh1 depletion–induced premature senescence. This further suggested that in addition to the

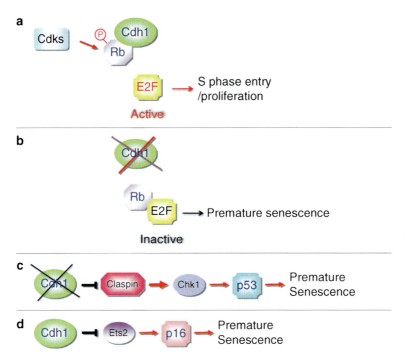

Fig. 20.3 Cdh1 loss leads to premature senescence through activating E2F pathway or inactivating Claspin/Chk1/p53 pathway. (**a**, **b**). Most prominent among the regulators disrupted in cancer cells are two tumor suppressors, the retinoblastoma protein (RB) and the p53 transcription factor. Cdk phosphorylates Rb to prevent its association with E2F thus activating E2F transcriptome (including Cyclin A, Cyclin E and many others) to promote cell cycle progression. (**c**) Loss of Cdh1 results in activation of the Claspin/Chk1 pathway to promote premature senescence. Cdh1 targets Claspin for degradation and loss of Cdh1 leads to the accumulation of Claspin protein. As the upstream activator of Chk1, increased Claspin abundance enhances Chk1 activation to promote p53 activity, leading to premature senescence. (**d**) Loss of Cdh1 leads to increases in the levels of Ets2 to the point that it overcomes Id1's inhibition and accumulation of p16, resulting in senescence

Claspin/Chk1/p53 pathway, Cdh1 also regulates the Rb pathway. In support of this notion, Gao et al. (2009a) found that in addition to the activation of the p53 pathway as evidenced by an enhanced pSer15-p53 signal, inactivation of Cdh1 in both primary lung fibroblasts and primary fore-skin fibroblasts resulted in a marked decrease in E2F1 abundance. Furthermore, they also found that overexpression of Cdh1 in U2OS cells led to a significant increase of E2F1 levels concomitant with Cdh1 induction. Interestingly, Cdh1 and E2F1 are mutually exclusive in interacting with Rb and as a result, overexpression of Cdh1 leads to increased levels of free E2F1, whereas depletion of Cdh1 leads to decreased levels of free E2F1. All together, these findings indicate that loss of Cdh1 results in the activation of Claspin/Chk1/p53 and Rb/E2F1 pathway that ultimately leads to premature senescence (Fig. 20.3). More importantly, these results further suggest that loss of Cdh1 might benefit tumorigenesis in later stages at which point both the p53 and Rb tumor suppressor pathways are inactivated. On the other hand, loss of Cdh1 in early stage with intact p53/Rb pathways will likely retard tumorigenesis in part by triggering premature senescence.

In this regard, the ability of normal cells to undergo premature senescence against various stresses such as DNA damage, oxidative stress and overexpression of oncogenic proteins has been described as one of the major underlying molecular mechanisms against tumor development (Campisi and d'Adda di Fagagna 2007). In line with this hypothesis, recent studies from various laboratories indicated that inactivation of

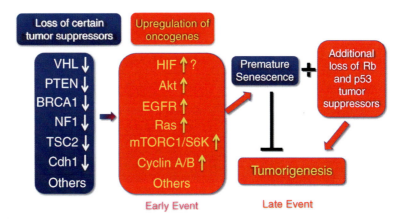

Fig. 20.4 A two step tumor development model. Initial loss of tumor suppressors leads to the upregulated expression of corresponding oncoproteins, resulting in premature senescence to serve as a protection mechanism against tumorigenesis. Further loss of either of the two known gate-keeper tumor suppressors Rb or p53 eventually develops tumor

many tumor suppressors such as VHL (Young et al. 2008), PTEN (Song et al. 2011), BRCA1 (Cao et al. 2003), NF1 (Courtois-Cox et al. 2006), and TSC2 (Zhang et al. 2003) induces premature senescence in variety of cells (Fig. 20.4). It is interesting to note that each tumor suppressor has different mechanisms in causing premature senescence in different cells (Kuilman et al. 2010). For example, VHL loss triggers senescence in an Rb- and p400-dependent manner, whereas PTEN loss-induced senescence is accompanied by p53 induction which was also observed in the case of Cdh1 inactivation (Song et al. 2011). On the other hand, NF1 loss-induced senescence is largely caused by decreases in ERK and AKT activities (Courtois-Cox et al. 2006). All together, these findings indicate that in many cases, loss of certain tumor suppressors will not simply results in onset of tumorigenesis but rather lead to induction of premature senescence. This might be an evolutionarily conserved protection mechanism against sporadic onset of tumorigenesis. On one hand, it provides the molecular explanation for why multiple genetic alterations are required for tumorigenesis. On the other hand, it emphasizes the critical roles for the p53 and Rb tumor suppressor pathways as loss of p53 and Rb tumor suppressor pathways in most cases will impair the onset of premature senescence induced by loss of various tumor suppressors.

Conclusion

There is mounting evidence suggesting that Cdh1 functions as a tumor suppressor by timely degrading the key proteins involved in cell cycle regulation. APC/Cdh1 complex mediates constitutive proteolysis of mitotic cyclins throughout the G1 phase, and regulates Skp2 proteolysis thereby leading to the accumulation of various Cdk inhibitors, such as p27, p21 and p57, which suppresses Cdk activity. It has been suggested that dysfunction of APC/Cdh1 might result in the abnormal accumulation of both mitotic Cdks and non-Cdks, ultimately leading to cancer development. Consistent with a critical role of Cdh1 in tumor suppression, Cdh1 protein levels were found decreased in some cancer cell lines and expression of Cdh1 reduces tumor development (Sudo et al. 2001; Wang et al. 2000). More importantly, it has been demonstrated that overexpression of murine *fzr* (murine counterpart of human Cdh1) in B-lymphoma cell lines increases cell susceptibility to NK (Natural Killer)-cytotoxicity and suppresses tumor formation (Wang et al. 2000). These findings revealed a novel role for Cdh1 in the NK-mediated cell death pathway. Furthermore, Garcia-Higuera et al. (2008) found that $Fzr1^{+/-}$ mice develop a variety of tumors late in life without losing the *Fzr1* wild-type allele. However, the key Cdh1 substrates responsible for this phenotype have not yet been identified. Recent studies from our laboratory as well as many other laboratories coherently indicated that loss of Cdh1 leads to premature senescence in certain cells (Gao et al. 2009a). Mechanistically, it has been described that Cdh1 loss leads to stabilization as well as activation of the key pathways such as Rb/E2F1 and Claspin/Chk1/p53 thus resulting in induction of premature senescence. Notably, as Cdh1 loss leads to induction of premature senescence in normal human fibroblasts, this finding also provides a molecular mechanism for the less frequently observed Cdh1 loss in tumor cells and further suggested that loss of Cdh1 may occurs late in tumor development. More importantly, the novel findings described above provide the rationale to develop new therapeutics that target Cdh1 to treat certain early-stage tumors by inducing senescence to suppress tumorigenesis.

Acknowledgements This work was supported in part from the grants (GM089763 and GM094777) from National Institutes of Health to Wenyi Wei. Shavali Shaik, Pengda Liu and Zhiwei Wang were supported by the institutional NRSA T-32 training grant.

References

Bloom J, Cross FR (2007) Multiple levels of cyclin specificity in cell-cycle control. Nat Rev Mol Cell Biol 8:149–160

Campisi J (1996) Replicative senescence: an old lives' tale? Cell 84:497–500

Campisi J, d'Adda di Fagagna F (2007) Cellular senescence: when bad things happen to good cells. Nat Rev Mol Cell Biol 8:729–740

Cao L, Li W, Kim S, Brodie SG, Deng CX (2003) Senescence, aging, and malignant transformation mediated by p53 in mice lacking the Brca1 full-length isoform. Genes Dev 17:201–213

Carter SL, Eklund AC, Kohane IS, Harris LN, Szallasi Z (2006) A signature of chromosomal instability inferred from gene expression profiles predicts clinical outcome in multiple human cancers. Nat Genet 38:1043–1048

Courtois-Cox S, Genther Williams SM, Reczek EE, Johnson BW, McGillicuddy LT, Johannessen CM, Hollstein PE, MacCollin M, Cichowski K (2006) A negative feedback signaling network underlies oncogene-induced senescence. Cancer Cell 10:459–472

Dial JM, Petrotchenko EV, Borchers CH (2007) Inhibition of APCCdh1 activity by Cdh1/Acm1/Bmh1 ternary complex formation. J Biol Chem 282:5237–5248

Dimri GP, Lee X, Basile G, Acosta M, Scott G, Roskelley C, Medrano EE, Linskens M, Rubelj I, Pereira-Smith O, Peacocke M, Campisi J (1995) A biomarker that identifies senescent human cells in culture and in aging skin in vivo. Proc Natl Acad Sci U S A 92:9363–9367

Doree M, Galas S (1994) The cyclin-dependent protein kinases and the control of cell division. FASEB J 8:1114–1121

Frescas D, Pagano M (2008) Deregulated proteolysis by the F-box proteins SKP2 and beta-TrCP: tipping the scales of cancer. Nat Rev Cancer 8:438–449

Gao D, Inuzuka H, Korenjak M, Tseng A, Wu T, Wan L, Kirschner M, Dyson N, Wei W (2009a) Cdh1 regulates cell cycle through modulating the claspin/Chk1 and the Rb/E2F1 pathways. Mol Biol Cell 20:3305–3316

Gao D, Inuzuka H, Tseng A, Chin RY, Toker A, Wei W (2009b) Phosphorylation by Akt1 promotes cytoplasmic localization of Skp2 and impairs APCCdh1-mediated Skp2 destruction. Nat Cell Biol 11:397–408

Garcia-Higuera I, Manchado E, Dubus P, Canamero M, Mendez J, Moreno S, Malumbres M (2008) Genomic stability and tumour suppression by the APC/C cofactor Cdh1. Nat Cell Biol 10:802–811

Hsu JY, Reimann JD, Sorensen CS, Lukas J, Jackson PK (2002) E2F-dependent accumulation of hEmi1 regulates S phase entry by inhibiting APC(Cdh1). Nat Cell Biol 4:358–366

Hwang ES, Yoon G, Kang HT (2009) A comparative analysis of the cell biology of senescence and aging. Cell Mol Life Sci 66:2503–2524

Kraft C, Vodermaier HC, Maurer-Stroh S, Eisenhaber F, Peters JM (2005) The WD40 propeller domain of Cdh1 functions as a destruction box receptor for APC/C substrates. Mol Cell 18:543–553

Kuilman T, Michaloglou C, Mooi WJ, Peeper DS (2010) The essence of senescence. Genes Dev 24:2463–2479

Li M, Shin YH, Hou L, Huang X, Wei Z, Klann E, Zhang P (2008) The adaptor protein of the anaphase promoting complex Cdh1 is essential in maintaining replicative lifespan and in learning and memory. Nat Cell Biol 10:1083–1089

Listovsky T, Oren YS, Yudkovsky Y, Mahbubani HM, Weiss AM, Lebendiker M, Brandeis M (2004) Mammalian Cdh1/Fzr mediates its own degradation. EMBO J 23:1619–1626

Lukas C, Sorensen CS, Kramer E, Santoni-Rugiu E, Lindeneg C, Peters JM, Bartek J, Lukas J (1999) Accumulation of cyclin B1 requires E2F and cyclin-A-dependent rearrangement of the anaphase-promoting complex. Nature 401:815–818

Mallette FA, Gaumont-Leclerc MF, Ferbeyre G (2007) The DNA damage signaling pathway is a critical mediator of oncogene-induced senescence. Genes Dev 21:43–48

Ohtani N, Zebedee Z, Huot TJ, Stinson JA, Sugimoto M, Ohashi Y, Sharrocks AD, Peters G, Hara E (2001) Opposing effects of Ets and Id proteins on p16INK4a expression during cellular senescence. Nature 409:1067–1070

Serrano M, Lin AW, McCurrach ME, Beach D, Lowe SW (1997) Oncogenic ras provokes premature cell senescence associated with accumulation of p53 and p16INK4a. Cell 88:593–602

Shaik S, Liu P, Fukushima H, Wang Z, Wei W (2012) Protein degradation in cell cycle. In: Yixian Zheng (ed) eLS. Wiley, Chichester, pp 1–8

Shieh SY, Ahn J, Tamai K, Taya Y, Prives C (2000) The human homologs of checkpoint kinases Chk1 and Cds1 (Chk2) phosphorylate p53 at multiple DNA damage-inducible sites. Genes Dev 14:289–300

Sitte N, Merker K, von Zglinicki T, Grune T (2000) Protein oxidation and degradation during proliferative senescence of human MRC-5 fibroblasts. Free Radic Biol Med 28:701–708

Song MS, Carracedo A, Salmena L, Song SJ, Egia A, Malumbres M, Pandolfi PP (2011) Nuclear PTEN regulates the APC-CDH1 tumor-suppressive complex in a phosphatase-independent manner. Cell 144:187–199

Sudo T, Ota Y, Kotani S, Nakao M, Takami Y, Takeda S, Saya H (2001) Activation of Cdh1-dependent APC is required for G1 cell cycle arrest and DNA damage-induced G2 checkpoint in vertebrate cells. EMBO J 20:6499–6508

Vodermaier HC (2004) APC/C and SCF: controlling each other and the cell cycle. Curr Biol 14:R787–R796

Wan L, Zou W, Gao D, Inuzuka H, Fukushima H, Berg AH, Drapp R, Shaik S, Hu D, Lester C, Eguren M, Malumbres M, Glimcher LH, Wei W (2011) Cdh1 regulates osteoblast function through an APC/C-independent modulation of Smurf1. Mol Cell 44:721–733

Wang CX, Fisk BC, Wadehra M, Su H, Braun J (2000) Overexpression of murine fizzy-related (fzr) increases natural killer cell-mediated cell death and suppresses tumor growth. Blood 96:259–263

Wei W, Ayad NG, Wan Y, Zhang GJ, Kirschner MW, Kaelin WG Jr (2004) Degradation of the SCF component Skp2 in cell-cycle phase G1 by the anaphase-promoting complex. Nature 428:194–198

Young AP, Schlisio S, Minamishima YA, Zhang Q, Li L, Grisanzio C, Signoretti S, Kaelin WG Jr (2008) VHL loss actuates a HIF-independent senescence programme mediated by Rb and p400. Nat Cell Biol 10:361–369

Zhang H, Cicchetti G, Onda H, Koon HB, Asrican K, Bajraszewski N, Vazquez F, Carpenter CL, Kwiatkowski DJ (2003) Loss of Tsc1/Tsc2 activates mTOR and disrupts PI3K-Akt signaling through downregulation of PDGFR. J Clin Invest 112:1223–1233

Zhou Y, Ching YP, Chun AC, Jin DY (2003) Nuclear localization of the cell cycle regulator CDH1 and its regulation by phosphorylation. J Biol Chem 278:12530–12536

Suppression of Premature Senescence and Promotion of Metastatic Transformation: Role of Reduced TGF-Beta Signaling in Human Cancer Progression

Shu Lin and Lu-Zhe Sun

Contents

Abstract	219
Introduction	220
Transforming Growth Factor-Beta (TGF-β)	220
Attenuation of TGF-β Signaling in Human Carcinomas	221
TGF-β and Cellular Senescence	222
Disrupted TGF-β Signaling Pathway in Early Stage Tumor Progression	223
TGF-β Signaling Pathway in Later Stage Tumor Progression	223
Discussion	224
References	224

S. Lin • L.-Z. Sun (✉)
Department of Cellular and Structural Biology,
University of Texas Health Science Center,
San Antonio, TX 78229, USA
e-mail: linshu8518@gmail.com; sunl@uthscsa.edu

Abstract

Transforming growth factor-beta (TGF-β) signaling pathway serves as a tumor suppressor by inhibiting cell cycle progression and stimulating senescence and apoptosis in normal and early-stages neoplastic tissues. As tumors progress, TGF-β signaling is often turned to drive multi-step metastasis processes by stimulating cell survival and epithelial-to-mesenchymal transition (EMT) leading to tumor cell migration and invasion. Many human carcinomas including triple-negative breast cancer, often show reduced or loss of key components of TGF-β signaling, indicating reduced tumor-suppressive TGF-β signaling may contribute to the cancer progression. However, molecular mechanisms that drive the switch of TGF-β are not well understood. Few molecular biomarkers have been identified as efficient indicators for the anti-TGF-β cancer therapy. In order to better understand the mechanism mediating the role of TGF-β during cancer progression, we will discuss the question of how the loss of control of cell proliferation and senescence by TGF-β promotes tumor invasion and metastasis and whether a set of transformation/metastasis-related genes are specifically regulated by TGF-β signaling.

Keywords

c-Myc transcription in epithelial cells
• Co-expression of DNRII and H-Ras-V12
• Human carcinomas • Human mammary

epithelial cells (HMECs) • R-Smads and Smad4 • Senescence-like growth arrest (SLGA) • TGF-β1 inhibitor element (TIE) • TGFBR2 mutation • TGF-β's growth inhibition • Transforming growth factor-beta (TGF-β) signaling

Introduction

The cytokine TGF-β was first found as a transforming growth factor stimulating transformation-like phenotypes including cell proliferation under suspension. Later on, it was shown that TGF-β could suppress tumor growth during the early stage of tumorigenesis and could be switched to be a tumor promoter at later stages of tumor progression (Massague 2008). Here, we will introduce the tumor-suppressing role of TGF-β in the progression of human carcinomas, especially triple-negative, basal-like breast cancer. A majority of triple-negative, basal-like breast cell lines possess altered TGF-β signaling and display resistance to TGF-β treatment (Lin et al. 2012). The molecular mechanisms that govern triple-negative, basal-like breast cancer progression are not well understood, resulting in the lack of efficient targeted therapeutics for this aggressive cancer. Identification of TGF-β signaling-involved molecular targets for the prevention or treatment of this deadly disease is urgently needed. In order to unravel the role of reduced TGF-β signaling in human cancer progression, we will introduce how the loss of the control of cell proliferation and senescence by TGF-β promotes tumor invasion and metastasis, and whether a set of transformation/metastasis-related genes are specifically regulated by TGF-β signaling.

Transforming Growth Factor-Beta (TGF-β)

The TGF-β superfamily is composed of TGF-β, bone morphogenetic proteins (BMPs), inhibitins, activins, nodal, anti-Müllerian hormone and related proteins (Derynck and Zhang 2003). As a member of the TGF-β superfamily, TGF-β is a homodimeric polypeptide of 25 kDa essentially expressed in all the tissues of humans and mice. TGF-β plays pivotal roles during body development, as well as in normal physiological and disease processes, by regulating a variety of cellular functions, including cell proliferation, differentiation, migration, apoptosis, extracellular matrix formation, and immune response (Jakowlew 2006). In mammals, three TGF-β isoforms termed TGF-β1, TGF-β2, and TGF-β3 have been identified. The TGF-β precursors undergo proteolytic cleavage in their C-terminal to produce the mature TGF-β. The mature TGF-β is secreted in a latent form by noncovalently interacting with the N-terminal residues of its processed precursor. Upon the interaction with various extracellular proteins, the mature and active TGF-β can be released from the latent complex, allowing TGF-β to engage with the TGF-β receptors on cell membranes (Sun 2004).

Three different cell surface receptors of TGF-β have been identified, called TGF-β type I (RI), type II (RII) and type III (RIII) receptors. RI and RII are serine/threonine kinase receptors (Massague 2008). RIII (also called betaglycan), containing two TGF-β binding sites in its extracellular domain, is believed not to participate directly in TGF-β signal transduction, but present TGF-βs to RII (Sun 2004).

In canonical TGF-β signal transduction, engagement of TGF-β with RII dimer on the cell surface leads to the recruitment, transphosphorylation, and activation of RI dimer. RII dimer and RI dimer form a hetero-tetrameric complex with the TGF-β ligand. Activated RI with conformation change recruits the intracellular mediators of TGF-β signaling, receptor-regulated Smads (R-Smads: Smad2 and Smad3). Adaptor proteins that facilitate the binding of R-Smads to RI include SARA (the SMAD anchor for receptor activation), a zinc double finger FYVE domain containing protein. The kinase activity of RI allows it to phosphorylate the serine residues of R-Smads, which leads to the formation of an oligomeric complex consisting of R-Smads and Smad4. After shuttling into the nucleus, the R-Smads/Smad4 complex binds to specific DNA sequences and cooperates with transcription coactivators or corepressors to regulate the transcription of TGF-β target genes (Massague 2008).

R-Smads and Smad4 possess a highly conserved N-terminal MH1 (Mad homology domains 1) and C-terminal MH2 domain, which are linked by a divergent proline-rich region. The MH1 domains of Smads (except for Smad2) are the site for DNA binding. The MH2 domains essentially regulate Smad-receptor interaction, Smad oligomerization and interaction with transcription factors. The linker domain has been shown to be phosphorylated by MAP Kinases (MAPK). Inhibitory Smads lacking a MH1 domain including Smad6 and Smad7 diminish the activation of R-Smads by RI. In particular, as a decoy of R-Smads, Smad7 binds to RI via its MH2 domains, preventing the recruitment and phosphorylation of R-Smads by RI (Derynck and Zhang 2003).

Attenuation of TGF-β Signaling in Human Carcinomas

Our limited exploration on the complicated gene alteration that governs the cancer progression impedes the efficient treatment for malignant cancers, including breast cancer. On one hand, autocrine TGF-β plays a prominent role in regulating carcinogenesis and cancer cell growth, invasion, and metastasis (Massague 2008). On the other hand, TGF-β inhibits cell cycle proliferation, induces apoptosis, and promotes cellular senescence of most normal epithelial cells, suggesting a tumor suppressor role for TGF-β. Disruption of autocrine TGF-β signaling pathway and/or sensitivity to ectopic TGF-β has been shown to promote early stage tumor progression (Jakowlew 2006).

Deletions and mutations of the genes that encode the components of TGF-β signaling in human tumors support the tumor suppressive role of TGF-β. *TGFBR2*, the gene coding RII, is widely found to be inactivated by mutation in various cancers, such as colon, gastric, ovarian, pulmonary, esophageal, and head and neck cancers. Mutation of *TGFBR2* is mainly due to the mutation of genes involved in DNA mismatch repair, which results in insertion or deletion of repeated nucleotides in genome including *TGFBR2*, a phenomenon called microsatellite instability. *TGFBR2* mutation inactivates RII for TGF-β signal transduction (Markowitz et al. 1995; Lu et al. 1996). Interestingly, RII expression is found to be downregulated during carcinogenesis in other types of cancer without *TGFBR2* mutation including breast, liver, lung, pancreatic, and endometrial carcinomas (Jakowlew 2006). Although less common than in RII, frameshift and missense mutations in the *TGFBR1* gene have been found in some ovarian, esophageal, and neck and head cancers. As a critical mediator of TGF-β signaling, the mutational inactivation of Smad4 occurs in human pancreatic cancer and metastatic colorectal cancer in late stage (Levy and Hill 2006). Smad2 mutation is relatively rare and occurs in a small portion of colon and lung carcinomas (Riggins et al. 1996). Smad3 mutations are detected in a limited number of gastric cancer (Han et al. 2004). Analyses of tumor outcomes in cancers with mutation of RII, RI or Smad genes suggest that they serve as tumor suppressors in the cancer progression.

In multiple experimental mouse models, the growth suppressive role of TGF-β signaling has been demonstrated in normal tissues and in the early stages of tumor outgrowth. Chemical carcinogen-induced breast tumor formation was inhibited by TGF-β1 transgenic expression in the mouse mammary gland (Pierce et al. 1995). Similarly, forced expression of RI or RII in transgenic mice has been shown to inhibit tumor progression (Wang et al. 1996; Minn et al. 2005). In both human cancers and mouse models, low RII expression is positively associated with high grade tumors (Tang et al. 1999; Kim et al. 2000). Loss of the remaining wild-type (wt) Smad4 allele promotes Smad4 heterozygous mice to develop benign gastric and duodenal polyps that are able to progress into invasive tumors (Xu et al. 2000). Thus, epithelial tissues require TGF-β signaling to maintain cellular homeostasis and avoid tumorigenesis.

Particularly, it has been shown that reduced RII and Smad4 levels in human breast tissues are associated with breast cancer development (de Jong et al. 1998; Stuelten et al. 2006). Triple-negative, basal-like breast cancer (estrogen receptor negative, progesterone receptor negative and EGFR2/Her2 negative), accounting for ~15% of all types of breast cancer, is highly prevalent in African American women and short of targets for the treatment or prevention (Carey et al. 2006). Attenuated TGF-β signaling has been demonstrated in a panel of human triple-negative breast cancer cell lines (Lin et al. 2012). Therefore, attenuation of TGF-β signaling may serve as a hallmark of tumorigenesis, including triple-negative, basal-like breast cancer.

TGF-β and Cellular Senescence

The tumor suppressive activity of TGF-β is mediated by its effect on the expression of a number of key proteins controlling cell cycle progression, apoptosis, and senescence. In the G1 phase and the G1 to S phase transition during the cell cycle, cell proliferation and differentiation are specifically regulated. Deregulation of the genes that are involved in the progression from the G1 to S phase of the cell cycle often contributes to tumorigenesis. A proto-oncogene, c-Myc, promotes S phase entry in the cell cycle by regulating the transcription of various cell cycle related genes (Dang 1999). In the promoter region of the c-Myc gene, a Smad-responsive element has been identified, which contains the TGF-β1 Inhibitor Element (TIE), and binding sites of E2F and the co-repressor p107 (Yagi et al. 2002). It has been shown that TGF-β treatment can rapidly inhibit c-Myc transcription in epithelial cells (Pietenpol et al. 1990). TGF-β induces the nuclear translocation of a transcription repression complex, and subsequent docking of this complex in association with Smad4 on the Smad/E2F binding site of c-Myc promoter, resulting in the inhibition of c-Myc transcription (Chen et al. 2002).

In addition, TGF-β can increase the transcription of cyclin-dependent kinase (CDK) inhibitors, such as p15ink4b and p21cip1 (Hannon and Beach 1994; Datto et al. 1995a). p15ink4b interacts with CDK4 and CDK6 to inhibit their kinase activity and association with cyclin D; whereas, p21cip1 mainly inhibits the activities of cyclin A/CDK2 and cyclin E/CDK2. Both cyclin D/CDK4/6 and cycline E/CDK2 can phosphorylate retinoblastoma gene product p-Rb to inactivate its ability to block G1 to S phase transition. Through a synergistic interaction between Smad proteins and the Sp1 transcription factor at the promoter region of p15ink4b and p21cip1 genes, TGF-β stimulates the transcriptional activation of p15ink4b and p21cip1 (Hannon and Beach 1994; Datto et al. 1995b; Feng et al. 2000).

Recently, we have reported that reduced TGF-β signaling suppresses premature senescence in a p21cip1-dependent manner in human mammary epithelial cells (HMECs) (Lin et al. 2012). Triple-negative, basal-like breast carcinomas share a similar basal-like gene expression profile with basal-like HMECs (Sorlie et al. 2001). In order to determine the mechanism of triple-negative, basal-like breast progression, a series of isogenic telomerase-immortalized HMECs with different malignancy through ectopic expression of a dominant-negative RII (DNRII) and an oncogenic Ras, resembling a full spectrum of basal-like breast carcinogenesis (Lin et al. 2012). These engineered HMECs provided good models to determine the molecular mechanisms that contribute to oncogene-induced transformation of basal-like HMECs when TGF-β signaling is attenuated. Abolished autocrine TGF-β signaling in HMECs suppressed oncogenic H-Ras-V12-induced senescence-like growth arrest (SLGA), implicating a tumor-suppressive role of TGF-β signaling in an early neoplastic progression model. Gene profile analysis demonstrated that p21cip1 was a key mediator of Ras-induced SLGA. Attenuated or loss of p21cip1 expression contributed to the escape from SLGA in HMECs with abrogated autocrine TGF-β signaling. In another HMEC model without hTERT-induced immortalization, TGF-β signaling was also reported to be needed for oncogenic Ras-induced senescence (Cipriano et al. 2011). In head and neck squamous cell carcinoma, loss of TGF-β signaling and PTEN in a *Tgfbr1/Pten* double conditional knockout mouse model promotes head and neck squamous cell carcinoma through cellular senescence evasion and cancer-related inflammation (Bian et al. 2012). Therefore, attenuation or loss of TGF-β signaling allows epithelial cells to evade oncogenic gene-induced senescence.

Disrupted TGF-β Signaling Pathway in Early Stage Tumor Progression

Mutational inactivation of TGF-β signaling components appears to occur at late stages of tumorigenesis, as evidenced by the loss of *SMAD4* during the transformation of human pancreatic neoplasm from non-invasive to invasive stage and mutation of *TGFBR2* during the progression of human colon adenomas to malignant carcinomas (Sun 2004). Thus, TGF-β signaling appears necessary for the suppression of tumor invasion and metastasis, in addition to prevention of tumorigenesis and inhibition of the growth of primary tumors. In our recent study, we have demonstrated that attenuation of TGF-β signaling promotes oncogenic Ras-mediated metastatic transformation in HMECs (Lin et al. 2012). Co-expression of DNRII and H-Ras-V12 allowed HMECs to become highly tumorigenic and metastatic when compared with H-Ras-V12-transformed HMECs that spontaneously escaped H-Ras-V12-induced SLGA. It was reported that attenuation of TGF-β increases miR-200 expression to mediate EMT, which promotes metastatic colonization of breast cancer (Korpal et al. 2011). TGF-β/Smad4 signaling has been demonstrated to act as a barrier to the progression of prostate cancer from intraepithelial neoplasia to highly metastatic adenocarcinoma. In the mouse prostate epithelium with Pten loss, additional Smad4 inactivation made prostate epithelial cells bypass oncogene-induced senescence and enhanced tumor cell proliferation and metastasis (Ding et al. 2011). Therefore, from these recent studies, the suppression of premature senescence and promotion of metastatic transformation induced by reduced TGF-β signaling implicates that autocrine TGF-β signaling is an integral part of cellular anti-transformation network in cancer progression.

TGF-β Signaling Pathway in Later Stage Tumor Progression

As we mentioned earlier, tumor cells often evade the growth inhibition by TGF-β during tumor progression. This is because TGF-β signaling is often altered in tumor cells such that the tumor cells become resistant to TGF-β-induced growth inhibition. Loss of TGF-β signaling receptors or Smad4 has been shown to contribute to the insensitivity to TGF-β in certain cancers as presented above. On the other hand, the retention of autocrine TGF-β signaling has been shown to be needed for the progression of certain types of cancer. For example, few or no RII mutations are detected in breast, endometrial, pancreatic and lung carcinomas with microsatellite instability and mutation of Smad4 is rare in breast and ovarian carcinomas (Sun 2004). Thus, some carcinoma cells are resistant to TGF-β's growth inhibitory activity while retaining a functional TGF-β signaling pathway. In this scenario, TGF-β loses its ability to inhibit cell cycle progression due to loss or attenuation of its inhibition of the expression of c-Myc and its induction of the expression of p15ink4b and p21cip1, and consequently becomes incapable of causing hypophosphorylation of p-RB.

Tumor cells resistant to TGF-β's growth inhibition gain a selective growth advantage in comparison with normal cells. Meanwhile, increased expression of TGF-β has been shown to be associated with tumorigenesis and tumor progression. For example, elevated TGF-β isoforms have been found during neoplastic development and progression in breast, colon, prostate, and bladder cancers (Sun 2004). Positive immunohistochemical staining for TGF-β1 significantly correlated with increased tumor progression, metastasis, and death in 74% of patients with breast cancer (Gorsch et al. 1992). Furthermore, multiple studies reported that increased expression of TGF-β could actually promote tumor progression in carcinoma cells. Overexpression of TGF-β1 in human cancer cells caused increased tumor growth and metastasis in xenograft models (Arteaga et al. 1993).

Several mechanisms contribute to the tumor-promoting activity of TGF-β. TGF-β has been shown to promote EMT in various transformed epithelial cells (Massague 2008). During EMT, cells lose epithelial properties and gain the characteristics of mesenchyme, including loss of cell-cell junctions and their associated proteins, increased matrix metalloproteinase activity, induction of actin stress fibers, and acquisition of migration and invasion ability. The switch of TGF-β from tumor-suppression to tumor-promotion may require tumor cells to undergo TGF-β-induced EMT, which contributes to tumor cell migration and invasion. The tumor microenvironment consisting of stromal, immune and vascular cells is considered as the critical determinant of cancer progression. TGF-β acts as a potent immune suppressor that stimulates tumor growth by attenuating host immune surveillance. Additionally, it has been reported that TGF-β serves as an angiogenic stimulator *in vivo*. A wide range of studies demonstrates that together with its role in stimulating EMT, elevated activities of TGF-β signaling in the tumor microenvironment in late stage malignant cancer cells are necessary for its metastatic behavior (Massague 2008).

Induction of EMT by TGF-β is usually through a combination of Smad-dependent and Smad-independent signaling. The activated TGF-β receptor complex activates non-Smad signaling pathways, including Ras/MAPK/ERK signaling in epithelial cells including breast cancer cells and fibroblasts (Massague 2008). During EMT, the activation of ERK is required for disassembly of cell adheren junctions and induction of cell motility by TGF-β. In a gene signature study for TGF-β-induced EMT, TGF-β-stimulated ERK signal has been shown to regulate a subset of target genes, which function in the remodeling of integrin-based cell-matrix adhesion and promote cell motility (Zavadil et al. 2001).

and stimulating apoptosis of epithelial cells. Several models for studying the tumor-suppressing role of TGF-β have been generated. For example, we have generated a series of isogenic basal-like breast cancer progression models that share many similarities with triple-negative, basal-like human breast carcinomas with respect to aberrant regulation of TGF-β and Ras signaling pathways, as well as clinical and pathological features (Lin et al. 2012). From those models, loss of control of cell proliferation and senescence by TGF-β has been shown to be sufficient to promote tumor invasion and metastasis, and a set of transformation/metastasis-related genes including p21cip1 are specifically regulated by TGF-β signaling. TGF-β signaling and its upregulation of p21cip1 were found to be required for the oncogenic Ras-induced senescence-like growth arrest. Attenuation of TGF-β signaling by the expression of DNRII promoted tumorigenesis and metastatic activity of the engineered cells reflecting tumor progression of triple-negative breast cancer. However, with cancer progression, retention of TGF-β signaling has been found to drive tumor progression and metastasis by stimulating EMT, thereby facilitating tumor invasion and metastasis (Massague 2008). The complicated pattern of molecular alterations mediating the switch of TGF-β is not well understood. Anti-TGF-β cancer therapies targeting the tumor-promoting activities of TGF-β have to exercise great caution due to this dual nature of TGF-β signaling. Therefore, further investigation on biomarkers are urgently needed as indicators for anti-TGF-β therapy for the prevention and treatment of metastatic cancer.

Acknowledgements This work was in part supported by National Institutes of Health Grants R01CA75253 and R01CA79683.

Discussion

Studies elucidating the loss of or reduced TGF-β signaling have demonstrated that TGF-β signaling is a potent tumor suppressor by inhibiting cell cycle progression, promoting cell senescence,

References

Arteaga CL, Carty-Dugger T, Moses HL, Hurd SD, Pietenpol JA (1993) Transforming growth factor beta 1 can induce estrogen-independent tumorigenicity of human breast cancer cells in athymic mice. Cell Growth Differ 4:193–201

Bian Y, Hall B, Sun ZJ, Molinolo A, Chen W, Gutkind JS, Waes CV, Kulkarni AB (2012) Loss of TGF-beta signaling and PTEN promotes head and neck squamous cell carcinoma through cellular senescence evasion and cancer-related inflammation. Oncogene 31:3322–3332

Carey LA, Perou CM, Livasy CA, Dressler LG, Cowan D, Conway K, Karaca G, Troester MA, Tse CK, Edmiston S, Deming SL, Geradts J, Cheang MC, Nielsen TO, Moorman PG, Earp HS, Millikan RC (2006) Race, breast cancer subtypes, and survival in the Carolina Breast Cancer Study. JAMA 295:2492–2502

Chen CR, Kang Y, Siegel PM, Massague J (2002) E2F4/5 and p107 as Smad cofactors linking the TGFbeta receptor to c-myc repression. Cell 110:19–32

Cipriano R, Kan CE, Graham J, Danielpour D, Stampfer M, Jackson MW (2011) TGF-beta signaling engages an ATM-CHK2-p53-independent RAS-induced senescence and prevents malignant transformation in human mammary epithelial cells. Proc Natl Acad Sci U S A 108:8668–8673

Dang CV (1999) c-Myc target genes involved in cell growth, apoptosis, and metabolism. Mol Cell Biol 19:1–11

Datto MB, Li Y, Panus JF, Howe DJ, Xiong Y, Wang XF (1995a) Transforming growth factor beta induces the cyclin-dependent kinase inhibitor p21 through a p53-independent mechanism. Proc Natl Acad Sci U S A 92:5545–5549

Datto MB, Yu Y, Wang XF (1995b) Functional analysis of the transforming growth factor beta responsive elements in the WAF1/Cip1/p21 promoter. J Biol Chem 270:28623–28628

de Jong JS, van Diest PJ, van der Valk P, Baak JP (1998) Expression of growth factors, growth inhibiting factors, and their receptors in invasive breast cancer. I: an inventory in search of autocrine and paracrine loops. J Pathol 184:44–52

Derynck R, Zhang YE (2003) Smad-dependent and Smad-independent pathways in TGF-beta family signalling. Nature 425:577–584

Ding Z, Wu CJ, Chu GC, Xiao Y, Ho D, Zhang J, Perry SR, Labrot ES, Wu X, Lis R, Hoshida Y, Hiller D, Hu B, Jiang S, Zheng H, Stegh AH, Scott KL, Signoretti S, Bardeesy N, Wang YA, Hill DE, Golub TR, Stampfer MJ, Wong WH, Loda M, Mucci L, Chin L, DePinho RA (2011) SMAD4-dependent barrier constrains prostate cancer growth and metastatic progression. Nature 470:269–273

Feng XH, Lin X, Derynck R (2000) Smad2, Smad3 and Smad4 cooperate with Sp1 to induce p15(Ink4B) transcription in response to TGF-beta. EMBO J 19:5178–5193

Gorsch SM, Memoli VA, Stukel TA, Gold LI, Arrick BA (1992) Immunohistochemical staining for transforming growth factor beta 1 associates with disease progression in human breast cancer. Cancer Res 52:6949–6952

Han SU, Kim HT, Seong DH, Kim YS, Park YS, Bang YJ, Yang HK, Kim SJ (2004) Loss of the Smad3 expression increases susceptibility to tumorigenicity in human gastric cancer. Oncogene 23:1333–1341

Hannon GJ, Beach D (1994) p15INK4B is a potential effector of TGF-beta-induced cell cycle arrest. Nature 371:257–261

Jakowlew SB (2006) Transforming growth factor-beta in cancer and metastasis. Cancer Metastasis Rev 25:435–457

Kim SJ, Im YH, Markowitz SD, Bang YJ (2000) Molecular mechanisms of inactivation of TGF-beta receptors during carcinogenesis. Cytokine Growth Factor Rev 11:159–168

Korpal M, Ell BJ, Buffa FM, Ibrahim T, Blanco MA, Celia-Terrassa T, Mercatali L, Khan Z, Goodarzi H, Hua Y, Wei Y, Hu G, Garcia BA, Ragoussis J, Amadori D, Harris AL, Kang Y (2011) Direct targeting of Sec23a by miR-200s influences cancer cell secretome and promotes metastatic colonization. Nat Med 17:1101–1108

Levy L, Hill CS (2006) Alterations in components of the TGF-beta superfamily signaling pathways in human cancer. Cytokine Growth Factor Rev 17:41–58

Lin S, Yang J, Elkahloun AG, Bandyopadhyay A, Wang L, Cornell JE, Yeh IT, Agyin J, Tomlinson G, Sun LZ (2012) Attenuation of TGF-beta signaling suppresses premature senescence in a p21-dependent manner and promotes oncogenic Ras-mediated metastatic transformation in human mammary epithelial cells. Mol Biol Cell 23:1569–1581

Lu SL, Zhang WC, Akiyama Y, Nomizu T, Yuasa Y (1996) Genomic structure of the transforming growth factor beta type II receptor gene and its mutations in hereditary nonpolyposis colorectal cancers. Cancer Res 56:4595–4598

Markowitz S, Wang J, Myeroff L, Parsons R, Sun L, Lutterbaugh J, Fan RS, Zborowska E, Kinzler KW, Vogelstein B et al (1995) Inactivation of the type II TGF-beta receptor in colon cancer cells with microsatellite instability. Science 268:1336–1338

Massague J (2008) TGFbeta in cancer. Cell 134:215–230

Minn AJ, Kang Y, Serganova I, Gupta GP, Giri DD, Doubrovin M, Ponomarev V, Gerald WL, Blasberg R, Massague J (2005) Distinct organ-specific metastatic potential of individual breast cancer cells and primary tumors. J Clin Invest 115:44–55

Pierce DF Jr, Gorska AE, Chytil A, Meise KS, Page DL, Coffey RJ Jr, Moses HL (1995) Mammary tumor suppression by transforming growth factor beta 1 transgene expression. Proc Natl Acad Sci U S A 92:4254–4258

Pietenpol JA, Stein RW, Moran E, Yaciuk P, Schlegel R, Lyons RM, Pittelkow MR, Munger K, Howley PM, Moses HL (1990) TGF-beta 1 inhibition of c-myc transcription and growth in keratinocytes is abrogated by viral transforming proteins with pRB binding domains. Cell 61:777–785

Riggins GJ, Thiagalingam S, Rozenblum E, Weinstein CL, Kern SE, Hamilton SR, Willson JK, Markowitz SD, Kinzler KW, Vogelstein B (1996) Mad-related genes in the human. Nat Genet 13:347–349

Sorlie T, Perou CM, Tibshirani R, Aas T, Geisler S, Johnsen H, Hastie T, Eisen MB, van de Rijn M, Jeffrey SS,

Thorsen T, Quist H, Matese JC, Brown PO, Botstein D, Eystein Lonning P, Borresen-Dale AL (2001) Gene expression patterns of breast carcinomas distinguish tumor subclasses with clinical implications. Proc Natl Acad Sci U S A 98:10869–10874

Stuelten CH, Buck MB, Dippon J, Roberts AB, Fritz P, Knabbe C (2006) Smad4-expression is decreased in breast cancer tissues: a retrospective study. BMC Cancer 6:25

Sun L (2004) Tumor-suppressive and promoting function of transforming growth factor beta. Front Biosci 9:1925–1935

Tang B, de Castro K, Barnes HE, Parks WT, Stewart L, Bottinger EP, Danielpour D, Wakefield LM (1999) Loss of responsiveness to transforming growth factor beta induces malignant transformation of nontumorigenic rat prostate epithelial cells. Cancer Res 59:4834–4842

Wang J, Han W, Zborowska E, Liang J, Wang X, Willson JK, Sun L, Brattain MG (1996) Reduced expression of transforming growth factor beta type I receptor contributes to the malignancy of human colon carcinoma cells. J Biol Chem 271:17366–17371

Xu X, Brodie SG, Yang X, Im YH, Parks WT, Chen L, Zhou YX, Weinstein M, Kim SJ, Deng CX (2000) Haploid loss of the tumor suppressor Smad4/Dpc4 initiates gastric polyposis and cancer in mice. Oncogene 19:1868–1874

Yagi K, Furuhashi M, Aoki H, Goto D, Kuwano H, Sugamura K, Miyazono K, Kato M (2002) c-myc is a downstream target of the Smad pathway. J Biol Chem 277:854–861

Zavadil J, Bitzer M, Liang D, Yang YC, Massimi A, Kneitz S, Piek E, Bottinger EP (2001) Genetic programs of epithelial cell plasticity directed by transforming growth factor-beta. Proc Natl Acad Sci U S A 98:6686–6691

Senescence Escape in Melanoma: Role of Spleen Tyrosine Kinase SYK

22

Marcel Deckert and Sophie Tartare-Deckert

Contents

Abstract	227
Introduction	228
Pathogenesis of Melanoma	228
Cellular Senescence: A Barrier to Melanocyte Transformation	229
Spleen Tyrosine Kinase Syk	231
Syk in Hematopoietic Signaling and Diseases	231
Syk in Non-hematopoietic Cancers	231
Syk in Melanomagenesis	232
Expression of Syk in Epidermal Melanocytes and Melanoma Cells	232
Syk and Senescence	232
Discussion and Conclusion	233
References	236

M. Deckert • S. Tartare-Deckert (✉)
C3M, Team Microenvironment, Signaling and Cancer, INSERM, U1065, Nice, France
e-mail: tartare@unice.fr

Abstract

Tissue homeostasis is maintained by appropriate innate cellular responses to various oncogenic or genotoxic stresses. Flaws in pathways controlling these responses can cause cancer. Cellular senescence is a critical tumor suppressor mechanism and a well-recognized failsafe program against melanoma progression. Melanoma is a lethal skin cancer of increasing incidence that is linked to solar ultraviolet (UV) radiation and oncogenic events such as activating mutations in BRAF. Understanding why senescence fails to constraint malignant transformation of epidermal melanocytes is a key question in melanoma biology. Spleen tyrosine kinase (Syk) is a multifunction protein tyrosine kinase critical for immune and hematopoietic signaling that has been implicated in tumor suppression of several carcinomas and skin melanomas. Our recent report indicated that Syk exerts its melanoma suppressive function by inducing p53-dependent premature senescence and stress-activated c-Jun N-terminal kinases (JNKs) activation. We proposed that epigenetic inactivation of Syk that is generally observed in primary and metastatic melanoma cells may contribute to senescence escape and tumorigenicity. In this chapter, we discuss this new aspect of Syk function in melanomagenesis with a focus on cellular circuits controlling BRAFV600E-induced senescence. We also examine the potential implication of Syk in p53-mediated UVB stress signaling in melanocytes.

Keywords

c-Jun N-terminal kinases (JNK) activation • Environmental stress • Genetic disruption of p53 pathway • Immunoreceptor tyrosine-based activation motifs (ITAMs) • KIT, ERBB4 and CDK4 genes • MDM2 activity and PTEN expression • Melanocyte transformation • Pathogenesis of melanoma • Secreted protein acidic and rich in cysteine (SPARC) • Syk in melanomagenesis

Introduction

Pathogenesis of Melanoma

Melanoma is the leading cause of skin cancer-related deaths with an incidence that increases faster than any other cancer, providing a significant health challenge. Melanoma, which represents only 4 % of skin cancers, accounts for 80 % of deaths from dermatologic cancer (Miller and Mihm 2006). In advanced stages, this is one of the most difficult cancers to treat because of its high plasticity, its molecular heterogeneity, its therapeutic resistance and metastatic potential. Treatment of cutaneous melanoma is based on the early excision of the primary tumor. This treatment is effective on localized and thin melanoma. However, melanomas typically progress to metastatic variants resistant to classic radiotherapy and chemotherapies. Thus, the prognosis of melanoma in the metastatic stage of the disease remains extremely poor and the overall 5-year survival for metastatic patients remains less than 10 %. No therapy to date has shown a real long-term effectiveness on the survival of these patients.

Cutaneous melanomas arise from malignant transformation of epidermal melanocytes, specialist pigment cells that emerge out of the neural crest during embryogenesis, located at the basement membrane of the epidermis and involved in photoprotection. The clinical, histological and biological features of melanoma progression are relatively well defined. The first step toward transition to melanoma is uncontrolled proliferation of epidermal melanocytes and dysplasia that may arise within a preexisting benign lesion (nevus or mole) or more frequently directly from a new site. The majority of melanomas evolve in a first phase "horizontally" within the epidermis (RGP or radial growth phase) and in a second phase "vertically" (VGP or vertical growth phase). This latter phase is associated with an invasive penetration into the dermis, the acquisition of an increased survival potential, and a propensity to rapidly spread to the lymph system resulting in regional lymph node metastases followed by distant and visceral metastases (Miller and Mihm 2006). Melanomas develop as a multistep process involving environmental stress such as solar ultraviolet (UV) light exposure, and accumulation of genetic and epigenetic changes in growth and survival pathways. The major oncogenic events during melanomagenesis are mutations affecting the *BRAF* (40–60 % of cases of melanoma) and *NRAS* (5–20 % of melanomas) genes, leading to constitutive activation of the MAP kinase pathway. The V600E point mutation is found in 90 % of mutated BRAF melanoma and is currently the therapeutic target for new anti-melanoma treatments (Flaherty et al. 2012). The most advanced compound targeting the oncogene BRAFV600E is PLX-4032 (Vemurafenib, Zelboraf), which has shown in recent Phase II/III studies improved overall survival and disease-free survival in about half of patients harboring the mutation. However, about 50 % of patients are not responders to Vemurafenib, and the rapid emergence of resistance acquired by melanoma limits its use as monotherapy. Clinical trials are underway with other BRAF inhibitors, in combination or not with Trametinib, a selective MEK1/2 inhibitor (Flaherty et al. 2012). In addition to BRAF and NRAS activating mutations, melanoma formation and progression is also frequently accompanied by mutations in *KIT*, *ERBB4*, and *CDK4* genes. The gene encoding MITF, the melanocyte lineage-specific transcription factor, is amplified in approximately 15 % of sporadic melanomas. Recently, an activating point mutation in MITF has also been identified in familial cases of melanomas (Flaherty et al. 2012).

Failure of tumor suppressor mechanisms also contributes to the molecular pathogenesis of melanoma. Indeed, loss of tumor suppressor genes such as *CDKN2A* and *PTEN* are frequently observed in melanoma (Miller and Mihm 2006). Furthermore, concurrent genetic events involving BRAF mutations and loss of PTEN function are common in human melanomas (Flaherty et al. 2012).

Acquisition of survival pathways promoting resistance to cell death is a common feature of tumor cells. Several such pathways are altered in a large proportion of melanomas, including the PI3-kinase/Akt survival pathway that is turned on following the inactivation of *PTEN* and/or the amplification of *AKT3* (Flaherty et al. 2012). In response to various cellular stresses, p53 transcriptionally induces the expression of specific target genes mediating critical tumor-suppressive functions by regulating cell cycle progression and apoptotic cell death. Consequently, the p53 pathway is disrupted in most human cancers by mutations in p53, or altered expression of downstream effectors or upstream regulators of p53 (Vousden and Lane 2007). Unlike other cancers, the *TP53* locus is intact in approximately 95 % of melanomas. Thus, melanoma cells typically harbor functional p53 protein but acquire capacities to inactivate components of the p53 pathway and to evade stress-induced cell death (Soengas and Lowe 2003). Under normal conditions, intracellular levels of p53 are controlled by the E3 ubiquitin ligase MDM2, which binds to p53 and promotes its ubiquitination and degradation by the proteasome (Vousden and Lane 2007). Several mechanisms have been implicated in the lack of p53-dependent apoptotic demise of melanoma cells, including *MDM2* gene amplification and the loss of the *CDKN2A* gene product ARF, a suppressor of MDM2 activity (Flaherty et al. 2012). Conversely, a major positive regulator of MDM2 activity is Akt, which phosphorylates MDM2 at Ser166 and Ser186. Akt is an important pro-survival molecule, and phosphorylation of MDM2 may thus serve to protect cells from p53-induced apoptotic cell death. Recently, another negative regulator of p53, MDM4, has been identified as a key determinant of impaired p53 function in melanomas (Gembarska et al. 2012). Signals arising from the melanoma microenvironment can also participate to the functional inactivation of the oncosuppressive p53 pathway. We have shown that Secreted Protein Acidic and Rich in Cysteine (SPARC) (also called osteonectin and BM-40), abundantly produced by melanoma cells but not by normal melanocytes (Robert et al. 2006), promotes an autocrine loop of resistance to p53-mediated cell death by activating Akt and MDM2 (Fenouille et al. 2011a, b).

Cellular Senescence: A Barrier to Melanocyte Transformation

Premature cellular senescence is recognized as a potent tumor-suppressive process that constrains the progression of premalignant lesions by stably restricting excessive proliferation driven by oncogenic stimuli. Senescence can also be triggered by various forms of cellular stress, including UV exposure, oxidative stress and DNA damage. Importantly, senescence driven by DNA damage can also participate to the outcome of certain chemotherapies (Collado and Serrano 2010). Cellular senescence is accompanied by a series of changes that distinguish senescence from quiescence or differentiation. Senescent cells display typical features including irreversible proliferation arrest, a flat enlarged morphology with increased granularity, an accumulation of senescence-associated ß-galactosidase (SA ß-gal), the presence of senescence-associated heterochromatin foci (SAHF), and increased production of reactive oxygen species (Kuilman et al. 2010). Stable cell cycle arrest is generally achieved through the activation of the two major tumor suppressor pathways, p53 and RB, respectively involved in the accumulation of cell cycle inhibitors (i.e. p16INK4a, p15INK4b, p21Cip1/Waf1), and in the silencing of S-phase-promoting genes, including E2F-target genes (Collado and Serrano 2010). In addition, senescent cells are characterized by a senescence-associated secretory phenotype (SASP), consisting of secreted factors such as proinflammatory cytokines GRO1, IL-6 and IL-8 that maintain growth arrest, recruit and activate immune cells, and metalloproteases that modify the tissue microenvironment (Kuilman et al. 2010). Studies by Lowe and colleagues have recently shown that NF-kB acts as a master regulator of the SASP, controlling cell-autonomous and non-cell-autonomous mechanisms of the senescence program (Chien et al. 2011). However, premature senescence is greatly dependent on the cellular context and on the type of stress or oncogenic insult, making this process diverse and very complex (Kuilman et al. 2010).

Melanocytic nevi are benign pigmented lesions composed of senescent melanocytes that typically remain growth arrested for years. The most prevalent oncogenic mutation in human melanoma is the V600E mutation in BRAF and remarkably the same mutation is found in a high proportion of nevi. Peeper and colleagues have demonstrated that oncogenic signaling driven by BRAFV600E induces human melanocyte growth arrest and senescence after an initial phase of proliferation. Consistently, early studies in the zebrafish model have shown that the expression of BRAFV600E mutation in the fish melanocyte is sufficient to promote nevus-like lesions (Kuilman et al. 2010). Notably, telomere length in nevi appears similar to what is observed in normal skin, ruling out the involvement of replicative senescence in nevi formation (Kuilman et al. 2010). Altogether, these observations underline the notion that nevi may undergo oncogene-induced senescence (OIS) in an attempt to prevent melanomagenesis. In addition to the clinical relevance of these findings for human melanoma biology, melanocytic cells and nevi thus represent an attractive model to study the mechanisms governing OIS and more generally premature senescence.

Senescence bypass in melanocyte is considered as a major mechanism of melanoma progression, generally occurring following additional alterations in tumor suppressor genes controlling senescence, leading to the acquisition by the pre-malignant cells of a *de novo* capacity of proliferation. Melanocyte senescence was initially shown to be dependent on the expression of the tumor suppressor p16INK4a (Gray-Schopfer et al. 2006). However, cells from senescent nevi do not always express p16INK4a, and NRAS oncogene-mediated senescence is effective in cultured human melanocytes in the absence of INK4a/ARF-encoded proteins (Haferkamp et al. 2009). Consistent with this notion, inducible expression of BRAFV600E in melanocytes leads to nevi formation and melanoma development independently of p16INK4a in mice (Dhomen et al. 2009), suggesting the existence of other pathways regulating the senescence program. The secreted protein IGFBP7 has been proposed as a key player of BRAFV600E-mediated melanocyte senescence, involved both in the initiation and the maintenance of the senescence phenotype (Wajapeyee et al. 2008). However, the relevance of these observations have been recently questioned (Scurr et al. 2010). Alteration of several other pathways can abrogate BRAFV600E-induced senescence to promote nevi progression towards melanoma. For example, enforced activation of the PI3-kinase pathway resulting from decreased PTEN expression was recently involved in senescence escape and melanomagenesis (Vredeveld et al. 2012). Also, inactivation of the p53 pathway has been implicated in the senescence bypass in melanocyte. Genetic disruption of the p53 pathway was shown to prevent the formation of nevus-like hyperplasia and promotes melanoma in mice and zebrafish (Kuilman et al. 2010), supporting a crucial role of inactivating the p53 pathway in the bypass of OIS. Another interesting candidate in regulating the senescence program in melanocyte may be the transcription factor FOXO4, which was described as a potential downstream effector of BRAFV600E-induced senescence (de Keizer et al. 2010). Finally, other parameters such as the normal hypoxic microenvironment of the skin and HIF-1 alpha expression (Bedogni and Powell 2009) or aberrant expression of the microenvironmental protein SPARC that antagonizes p53 function (Fenouille et al. 2011a, b) are likely to contribute to senescence escape and melanoma development. Together, these studies indicate that the establishment of melanocyte senescence driven by stress or oncogenic signaling is a much more complex phenomenon than initially envisioned.

Spleen Tyrosine Kinase Syk

Syk in Hematopoietic Signaling and Diseases

Spleen tyrosine kinase (Syk) is a 72-kDa non-receptor cytoplasmic protein tyrosine kinase, originally identified as a 40 kDa proteolytic fragment in spleen. Syk contains two tandem NH2 terminal Src homology 2 (SH2) domains, multiple tyrosine phosphorylation sites in a central linker domain and a COOH terminal tyrosine kinase domain. This structural organization typical of the Syk family of tyrosine kinases is only shared by zeta-activated protein of 70 kDa (ZAP-70) in human. In leukocytes, Syk exerts crucial functions in proximal signaling activation downstream immunoreceptors (i.e. TCR, BCR, FcR). When immunoreceptors are engaged by their ligands, SH2 domains of Syk bind phosphorylated immunoreceptor tyrosine-based activation motifs (ITAMs) within the cytoplasmic tails of the immunoreceptor, leading to the activation of Syk catalytic activity, phosphorylation of enzymes and adaptor proteins and downstream signaling pathways (Mocsai et al. 2010). Syk has been also implicated in signal transduction downstream a variety of other membrane receptors, including integrins and C-type lectins. Consistent with the variety of receptor signaling pathways in which Syk has been involved, genetic studies have demonstrated the functions of hematopoietic Syk in critical cellular responses including adaptive and innate immunity, lymphoid cell proliferation and differentiation, phagocytosis, platelet activation and bone formation (Mocsai et al. 2010). Increasing evidence now point to a role of Syk in diverse pathological states, including autoimmune diseases and haematological malignancies. A Syk inhibitor (R788/fostamatinib) has recently been shown effective in the treatment of rheumatoid arthritis (Weinblatt et al. 2010). We and others have shown that targeting Syk in neoplastic B cells affects their growth and survival (Baudot et al. 2009; Chen et al. 2008). Furthermore, a recent clinical study has demonstrated the potential of Syk inhibition in the treatment of non-hodgkin lymphoma (Friedberg et al. 2009). Finally, genetic studies have shown that homozygous Syk mutant mice suffered severe haemorrhaging as embryos, due to a failure of the vascular and lymphatic systems to separate during development (Abtahian et al. 2003). These early studies hence suggested an additional role for Syk during vascular development. It appears now that this function of Syk is linked to its activity downstream the C-type lectin receptor CLEC-2 in the megakaryocyte/platelet lineage, a hematopoietic-derived cell population that is required for brain vasculature and lymphatic development (Finney et al. 2012).

Syk in Non-hematopoietic Cancers

The first evidence for a biological activity of Syk in non-hematopoietic cells came from studies by Coopman and colleagues showing its expression in normal mammary epithelial cells (Coopman et al. 2000). Syk expression was then detected in a variety of tissues and cells, including breast, lung and gastric epithelial cells, endothelial and neuron-like cells, nasal fibroblasts, hepatocytes and epidermal melanocytes. Although the exact function of Syk in non-hematopoietic cells is still unclear, the presence or absence of Syk expression in those cell lineages have been associated either with tumor promotion or tumor suppression. Accumulating evidence suggests that Syk can function as a tumor suppressor in breast cancer. Syk is detected in normal breast epithelial cells and benign breast lesions, but its expression is low or absent in metastatic breast cancer cells (Coopman and Mueller 2006). In poor prognosis breast cancers, the absence of Syk was associated with epigenetic silencing of its promoter. Reexpression of Syk in breast cancer cells reduced tumor growth and metastasis in mouse xenografts (Coopman et al. 2000), and regulated integrin and E-cadherin-mediated cell adhesion (Larive et al. 2009; Zhang et al. 2009). The tumor suppressive activity of Syk in breast cancer cells was also associated with abnormal mitotic progression and a non-apoptotic cell death similar to mitotic catastrophe (Coopman and Mueller

2006). Recently, the tumor suppressive function of Syk in the mammary gland was supported by genetic studies showing that loss-of-heterozygosity of the *Syk* gene in mouse resulted in loosened breast epithelium cell-cell contacts and aberrant cell polarization, two processes preventing the loss of morphological integrity and quiescence observed in neoplastic cells. In addition, knockdown of Syk in breast epithelial cells increased proliferation, anchorage independent growth, cell motility and induced an epithelial to mesenchymal transition (EMT)-like process associated with formation of extracellular matrix-degrading invadopodia (Sung et al. 2009). Consistent with a role of Syk as a tumor suppressor in several other non-hematopoietic cancers, epigenetic loss of Syk expression associated with promoter hypermethylation was also found in bladder, hepatocellular, gastric, pancreatic, lung and cervical carcinomas.

In contrast to this tumor suppressive activity, Syk appears to exert a tumor promoting activity in retinoblastoma, an aggressive childhood cancer initiated by biallelic loss of the tumor suppressor RB1 (Zhang et al. 2012). Syk is upregulated in poor prognosis retinoblastomas, where it is required for tumor cell survival. Pharmacological and genetic targeting of Syk in retinoblastoma cell lines activates a cell death program involving the degradation of the anti-apoptotic protein Mcl-1 (Zhang et al. 2012), in a manner reminiscent of what occurs in leukemia (Baudot et al. 2009).

Thus, in non-hematopoietic cells, Syk activity may depend on the cellular context and may not be restricted to a suppressive function.

Syk in Melanomagenesis

Expression of Syk in Epidermal Melanocytes and Melanoma Cells

Syk also appears as a candidate tumor suppressor in malignant melanoma. We and others have found that Syk is expressed and tyrosine phosphorylated in normal skin melanocytes, but lost in a majority of RGP and VGP primary melanoma tumors and metastatic cell lines harboring the activating V600E mutation in BRAF. Of note, we found that whereas Syk protein is detected in immortalized mouse melanocytes, it is not expressed in B16 mouse melanoma cell line (unpublished observation). Sequencing and methylation-specific PCR analysis of *SYK* gene showed hypermethylation of CpG islands within its promoter region. Together with several other DNA-methylated genes, Syk expression in melanoma was restored after treatment with the demethylating agent 5-aza-2-deoxycytidine (Bailet et al. 2009; Muthusamy et al. 2006). However, as shown in breast cancers (Coopman et al. 2000), our study suggested that mechanisms other that promoter methylation might account for loss of Syk expression in certain melanoma cells. Whether Syk is expressed in nevi benign tumors and its silencing can correlate with clinical staging and outcome of the disease remains to be established. Reintroduction of Syk in Syk-deficient melanoma cells dramatically reduced clonogenic survival and three-dimensional tumor spheroid growth and invasion (Bailet et al. 2009). Stable reexpression of Syk in melanoma cell lines did not influence growth speed, but decreased tumor growth and metastatic potential *in vivo* (Hoeller et al. 2005; Muthusamy et al. 2006). Our studies further showed that the growth suppression by Syk in melanoma was mediated through a G0-G1 cell cycle arrest and induction of senescence and did not involve increased cell death (Bailet et al. 2009), in contrast to what is observed in breast carcinomas (Coopman and Mueller 2006). Thus, these observations suggest that reexpression of Syk can orchestrate different tumor suppressive mechanisms in a cell context dependent manner.

Syk and Senescence

Syk-negative melanoma cells genetically modified to express Syk displayed hallmarks of senescent cells, including reduction of proliferative activity and DNA synthesis, large and flat morphology, SA ß-gal activity, and heterochromatic foci. This phenotype was accompanied by c-Jun

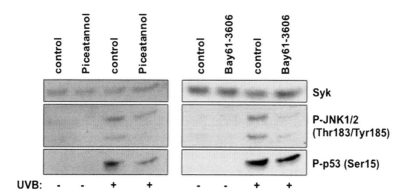

Fig. 22.1 Involvement of Syk in JNK and p53 activation induced by UVB in normal melanocytes. Human primary epidermal melanocytes were isolated from foreskins and cultured as described before (Robert et al. 2006). Cells were treated with control vehicle or with 10 µM of the Syk inhibitors Piceatannol and Bay61-3606 30 min before being subjected or not to UVB irradiation (40 mJ) for 5 min. Cell lysates were then analyzed by immunoblot using antibodies against Syk and phosphorylated forms of JNK1/2 and p53 as indicated

N-terminal kinases (JNK) activation, p53-dependent accumulation of p21Cip1/Waf1, hypophosphorylated RB, and inhibition of E2F1 and E2F1-target genes cyclin A and cyclin B. Together, our study indicated that these biochemical modifications accounted for proliferative arrest and senescence experienced by melanoma cells reexpressing Syk (Bailet et al. 2009). Our findings provided the first evidence that tumor growth inhibition by Syk in melanoma involves phosphorylation on Ser15 and nuclear accumulation of p53 and activation of p53-responsive genes leading to induction of premature senescence. Of note, the senescence program induced by reexpression of Syk is similar to the ones reactivated in melanoma following oncogene inactivation, including MITF, c-Myc, Tbx2 (Kuilman et al. 2010) or expression of the oncosuppressive protein p16INK4a (Gray-Schopfer et al. 2006). However, the mechanisms and signaling pathways through which Syk promotes activation of p53 are still unknown. Because Syk modulates the p53 and the stress-activated JNK pathway, and triggers premature senescence in melanoma cells, we speculate that it may integrate stress responses in melanocytes, thereby mediating senescence induced by oncogenic signaling (e.g. BRAFV600E). In this regard, we observed that pharmacological inhibition of Syk in primary human melanocytes reduced JNK and p53 phosphorylation induced by UVB irradiation (Fig. 22.1) (unpublished observations). These results indicate that Syk contribute to the activation of critical stress response pathways in normal melanocytes, suggesting that loss of Syk in transformed melanocytes may contribute to functional suppression of p53-dependent checkpoints during melanomagenesis.

Discussion and Conclusion

Senescence and cell death are now widely recognized as critical cellular failsafe programs acting together to suppress tumorigenesis. However, the resistance to chemotherapy-induced cell death generally observed in many cancers, including melanomas, has increased the recent interest for a therapeutic manipulation of cellular senescence induction. A better understanding of the molecular pathways controlling senescence is therefore central to future efficient cancer therapies. Establishment of cellular senescence driven by oncogenic or stress signaling is a complex phenomenon (Collado and Serrano 2010). In melanoma cells, our studies have identified the tyrosine kinase Syk as a novel driver of senescence (Bailet et al. 2009). Syk exerts essential activities in the immune and hematopoietic systems, where it drives major biological responses such as innate and adaptative immunity, allergy, inflammation, vascular and bone remodeling (Mocsai et al. 2010). In hematopoietic malignancies and retinoblastomas, increased expression or activity of Syk has been involved in pro-survival programs that are now considered for targeted therapy (Baudot et al. 2009; Friedberg et al. 2009; Zhang et al. 2012). In contrast, loss of Syk expression has been linked to tumor suppression in several other types of solid tumors such as melanoma. We showed that Syk is commonly expressed in normal epidermal melanocytes but absent in or nearly undetectable at the mRNA and protein levels in most human primary and metastatic melanoma cells tested. Transcriptional silencing of Syk in some melanoma cells is associated with aberrant CpG islands methylation of its promoter. Thus, suppression of this tyrosine kinase by melanoma cells predicts a potential role in their transformed phenotype. Importantly, the experimental observations consistent with a tumor suppressive activity of Syk in breast cancer and melanoma have been corroborated by an increasing number of clinical studies showing a correlation between reduced Syk expression and an increased risk for metastasis formation in breast cancer (Coopman and Mueller 2006) and development of invasive melanoma (Bailet et al. 2009; Muthusamy et al. 2006). In addition, epigenetic loss of Syk expression associated with promoter hypermethylation was also found in bladder, hepatocellular, gastric, and ovarian carcinomas (Coopman and Mueller 2006). These observations assign Syk as a potential new prognostic marker and a candidate tumor suppressor gene in different tumor types. However, the mechanisms by which Syk mediates tumor suppression differ between breast carcinomas and melanomas. Whereas sustained expression of Syk in breast cancer cells resulted in Syk accumulation in centrosomes, abnormal cell division and subsequent cell death by mitotic catastrophe (Coopman and Mueller 2006), reexpression of active Syk in melanoma induced a potent cell cycle arrest associated with p53-dependent senescence (Bailet et al. 2009). Thus, the mechanism of tumor suppression by Syk appears to be tissue-specific. In this context, it is interesting to note that human breast cancer cell lines that have lost Syk expression are most often HER2- and E-cadherin-negative and exhibit a mutated p53 status. In contrast, mutations of p53 are rarely found in melanoma, consistent with the idea that Syk mediates senescence or death depending on the genetic background of the tumor.

The p53 pathway is activated by DNA damage response (DDR) during replicative senescence and OIS (Vousden and Lane 2007). Our findings provide evidence that tumor growth inhibition by Syk in melanoma involves activation of p53 and modulation of several p53-responsive genes. Among them, we found that the expression of survivin and caspase 2, two proteins implicated in p53-mediated apoptosis (Vousden and Lane 2007), was suppressed in melanoma upon Syk reexpression. On the other hand, p21Cip1/Waf1 (also known as senescent cell-derived inhibitor), a major effector of p53-induced cell cycle arrest and senescence was strongly induced by Syk in melanoma cells. p21Cip1/Waf1 inhibits cyclin-dependent kinases activity, resulting in hypophosphorylation of RB and repression of E2F-dependent genes required for S phase entry and cell cycle progression. Consistent with this, hypophosphorylation of RB and inhibition of E2F-targets genes (i.e. cyclin A and cyclin B) were observed in cells induced to senesce by Syk (Bailet et al. 2009). Together, these molecular events orchestrated by p53 can account for the proliferative arrest and senescence experienced by melanoma cells reexpressing Syk, rather than cell death. Importantly, we found that whereas the catalytic activity of Syk was required, Syk SH2 domains were dispensable for growth suppression in melanoma and induction of senescence (Bailet et al. 2009). These findings suggest that inhibition of proliferation by Syk reexpression did not require interactions of its SH2 domains with phosphorylated tyrosine or ITAMs on putative receptors or adaptor proteins, leaving open the question of the nature of the mechanisms activating Syk in melanocytic cells. Supporting the idea that this function of Syk is dependant of its enzymatic activity, pharmacological inhibition of Syk impaired p53 phosphorylation in melanocytes exposed to UVB (Fig. 22.1) (unpublished observations).

Inactivation of Syk during melanoma development would have selective advantages for malignant cells. Indeed, besides regulating cell growth and senescence, reexpression of Syk reduced migratory, invasive and metastatic properties of melanoma cells (Bailet et al. 2009; Hoeller et al. 2005), suggesting that Syk acts also as a suppressor of melanoma cell motility. This activity could be related to the regulation of cell-matrix or cell-cell interactions as well as EMT-like process by Syk previously described in mammary epithelial transformation (Coopman and Mueller 2006; Sung et al. 2009). Relative to in breast tumorigenesis where Syk has been described as a regulator of the E-cadherin pathway (Larive et al. 2009; Sung et al. 2009; Zhang et al. 2009), it could be hypothesized that loss of Syk during melanocyte transformation affects cadherin-bound beta-catenin complex, thereby promoting activation of the beta-catenin multifunctional protein implicated in melanocyte senescence bypass and immortalization (Delmas et al. 2007). This hypothesis connecting Syk with the beta-catenin pathway warrants further investigations.

Melanocytes can undergo premature senescence in response to various stresses, including oxidative or oncogenic stress (Kuilman et al. 2010). Because Syk modulates the p53 pathway in melanoma cells, one can speculate that Syk may integrate p53-mediated stress responses in melanocytes. Hence, loss of Syk during *de novo* melanocyte transformation or in preexisting nevi may thus contribute to the senescence escape program during the pathogenesis of melanoma. However, the mechanisms by which Syk promotes activation of the p53 pathway remain ill defined. It is possible that reexpression of Syk activates reactive oxygen species (ROS)-dependent cellular damages resulting in p53 activation and melanoma senescence. The phosphorylation of p53 may also involve the DNA damage kinases ATM and ATR, which have been implicated in melanoma senescence. Indeed, enforced expression of Syk induced phosphorylation of p53 on serine 15 (Bailet et al. 2009), an event generally associated with DDR. However, whether enforced expression of Syk in melanocytic cells can elicit ROS production and

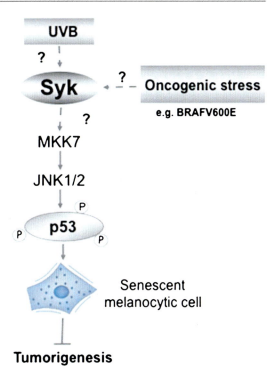

Fig. 22.2 Hypothetic model of the implication of Syk in a p53-dependent pathway to senescence

DDR remain to be tested. Another candidate for the kinase that phosphorylates p53 downstream Syk appears to be the stress-activated JNK that we found activated by Syk in melanoma cells (Bailet et al. 2009). Interestingly, we recently observed that inhibition of Syk by two different pharmacological compounds reduces JNK and p53 phosphorylation in normal melanocytes exposed to UVB (Fig. 22.1) (unpublished observations). Supporting this idea, the JNK1/2 kinase MKK7 has been recently implicated in p53-mediated cell cycle arrest and tumor suppression following oncogenic and genotoxic stress in lung and mammary carcinomas (Schramek et al. 2011). It is therefore tempting to speculate that MMK7 through JNK1/2 and p53 promotes the senescence program evoked by Syk, in melanocytic cells, either downstream UVB exposure or following BRAFV600E expression (Fig. 22.2).

Finally, two important questions regarding Syk function in the melanocytic lineage remain open. What is the physiological role of Syk in melanocyte homeostasis and is Syk a *bona fide* tumor suppressor in melanoma? *In vivo veritas*: answers to these questions will need the generation of melanocyte specific Syk knock-out in the context of the inducible melanocyte specific knock-in BRafV600E mice.

In conclusion, experimental and clinical evidence have shown that spleen tyrosine kinase Syk may function as a tumor suppressor in several carcinomas and skin melanomas. We have provided the first functional link between Syk and the p53-stress pathway in melanocyte and melanoma, therefore identifying Syk as a potential molecular sensor in the control of stress responses involved in senescence, a failsafe program acting as a crucial anti-cancer barrier. How this novel activity of Syk integrates into the contrasting roles of this multifunction protein and what are the clinical implications of these findings require further studies.

Acknowledgements This work was supported by INSERM and research grants from the ARC foundation.

References

Abtahian F, Guerriero A, Sebzda E, Lu MM, Zhou R, Mocsai A, Myers EE, Huang B, Jackson DG, Ferrari VA, Tybulewicz V, Lowell CA, Lepore JJ, Koretzky GA, Kahn ML (2003) Regulation of blood and lymphatic vascular separation by signaling proteins SLP-76 and Syk. Science 299:247–251

Bailet O, Fenouille N, Abbe P, Robert G, Rocchi S, Gonthier N, Denoyelle C, Ticchioni M, Ortonne JP, Ballotti R, Deckert M, Tartare-Deckert S (2009) Spleen tyrosine kinase functions as a tumor suppressor in melanoma cells by inducing senescence-like growth arrest. Cancer Res 69:2748–2756

Baudot AD, Jeandel PY, Mouska X, Maurer U, Tartare-Deckert S, Raynaud SD, Cassuto JP, Ticchioni M, Deckert M (2009) The tyrosine kinase Syk regulates the survival of chronic lymphocytic leukemia B cells through PKCdelta and proteasome-dependent regulation of Mcl-1 expression. Oncogene 28:3261–3273

Bedogni B, Powell MB (2009) Hypoxia, melanocytes and melanoma – survival and tumor development in the permissive microenvironment of the skin. Pigment Cell Melanoma Res 22:166–174

Chen L, Monti S, Juszczynski P, Daley J, Chen W, Witzig TE, Habermann TM, Kutok JL, Shipp MA (2008) SYK-dependent tonic B-cell receptor signaling is a rational treatment target in diffuse large B-cell lymphoma. Blood 111:2230–2237

Chien Y, Scuoppo C, Wang X, Fang X, Balgley B, Bolden JE, Premsrirut P, Luo W, Chicas A, Lee CS, Kogan SC, Lowe SW (2011) Control of the senescence-associated secretory phenotype by NF-kappaB promotes senescence and enhances chemosensitivity. Genes Dev 25:2125–2136

Collado M, Serrano M (2010) Senescence in tumours: evidence from mice and humans. Nat Rev Cancer 10:51–57

Coopman PJ, Mueller SC (2006) The Syk tyrosine kinase: a new negative regulator in tumor growth and progression. Cancer Lett 241:159–173

Coopman PJ, Do MT, Barth M, Bowden ET, Hayes AJ, Basyuk E, Blancato JK, Vezza PR, McLeskey SW, Mangeat PH, Mueller SC (2000) The Syk tyrosine kinase suppresses malignant growth of human breast cancer cells. Nature 406:742–747

de Keizer PL, Packer LM, Szypowska AA, Riedl-Polderman PE, van den Broek NJ, de Bruin A, Dansen TB, Marais R, Brenkman AB, Burgering BM (2010) Activation of forkhead box O transcription factors by oncogenic BRAF promotes p21cip1-dependent senescence. Cancer Res 70:8526–8536

Delmas V, Beermann F, Martinozzi S, Carreira S, Ackermann J, Kumasaka M, Denat L, Goodall J, Luciani F, Viros A, Demirkan N, Bastian BC, Goding CR, Larue L (2007) Beta-catenin induces immortalization of melanocytes by suppressing p16INK4a expression and cooperates with N-Ras in melanoma development. Genes Dev 21:2923–2935

Dhomen N, Reis-Filho JS, da Rocha Dias S, Hayward R, Savage K, Delmas V, Larue L, Pritchard C, Marais R (2009) Oncogenic Braf induces melanocyte senescence and melanoma in mice. Cancer Cell 15:294–303

Fenouille N, Puissant A, Tichet M, Zimniak G, Abbe P, Mallavialle A, Rocchi S, Ortonne JP, Deckert M, Ballotti R, Tartare-Deckert S (2011a) SPARC functions as an anti-stress factor by inactivating p53 through Akt-mediated MDM2 phosphorylation to promote melanoma cell survival. Oncogene 30:4887–4900

Fenouille N, Robert G, Tichet M, Puissant A, Dufies M, Rocchi S, Ortonne JP, Deckert M, Ballotti R, Tartare-Deckert S (2011b) The p53/p21(Cip1/Waf1) pathway mediates the effects of SPARC on melanoma cell cycle progression. Pigment Cell Melanoma Res 24:219–232

Finney BA, Schweighoffer E, Navarro-Nunez L, Benezech C, Barone F, Hughes CE, Langan SA, Lowe KL, Pollitt AY, Mourao-Sa D, Sheardown S, Nash GB, Smithers N, Reis e Sousa C, Tybulewicz VL, Watson SP (2012) CLEC-2 and Syk in the megakaryocytic/platelet lineage are essential for development. Blood 119:1747–1756

Flaherty KT, Hodi FS, Fisher DE (2012) From genes to drugs: targeted strategies for melanoma. Nat Rev Cancer 12:349–361

Friedberg JW, Sharman J, Sweetenham J, Johnston PB, Vose JM, Lacasce A, Schaefer-Cutillo J, De Vos S, Sinha R, Leonard JP, Cripe LD, Gregory SA, Sterba MP, Lowe AM, Levy R, Shipp MA (2009) Inhibition of Syk with fostamatinib disodium has significant clinical activity in non Hodgkin's lymphoma and chronic lymphocytic leukemia. Blood 115:2578–2585

Gembarska A, Luciani F, Fedele C, Russell EA, Dewaele M, Villar S, Zwolinska A, Haupt S, de Lange J, Yip D, Goydos J, Haigh JJ, Haupt Y, Larue L, Jochemsen A, Shi H, Moriceau G, Lo RS, Ghanem G, Shackleton M, Bernal F, Marine JC (2012) MDM4 is a key therapeutic target in cutaneous melanoma. Nat Med 18:1239–1247

Gray-Schopfer VC, Cheong SC, Chong H, Chow J, Moss T, Abdel-Malek ZA, Marais R, Wynford-Thomas D, Bennett DC (2006) Cellular senescence in naevi and immortalisation in melanoma: a role for p16? Br J Cancer 95:496–505

Haferkamp S, Scurr LL, Becker TM, Frausto M, Kefford RF, Rizos H (2009) Oncogene-induced senescence does not require the p16(INK4a) or p14ARF melanoma tumor suppressors. J Invest Dermatol 129:1983–1991

Hoeller C, Thallinger C, Pratscher B, Bister MD, Schicher N, Loewe R, Heere-Ress E, Roka F, Sexl V, Pehamberger H (2005) The non-receptor-associated tyrosine kinase Syk is a regulator of metastatic behavior in human melanoma cells. J Invest Dermatol 124:1293–1299

Kuilman T, Michaloglou C, Mooi WJ, Peeper DS (2010) The essence of senescence. Genes Dev 24:2463–2479

Larive RM, Urbach S, Poncet J, Jouin P, Mascre G, Sahuquet A, Mangeat PH, Coopman PJ, Bettache N (2009) Phosphoproteomic analysis of Syk kinase signaling in human cancer cells reveals its role in cell-cell adhesion. Oncogene 28:2337–2347

Miller AJ, Mihm MC Jr (2006) Melanoma. N Engl J Med 355:51–65

Mocsai A, Ruland J, Tybulewicz VL (2010) The SYK tyrosine kinase: a crucial player in diverse biological functions. Nat Rev Immunol 10:387–402

Muthusamy V, Duraisamy S, Bradbury CM, Hobbs C, Curley DP, Nelson B, Bosenberg M (2006) Epigenetic silencing of novel tumor suppressors in malignant melanoma. Cancer Res 66:11187–11193

Robert G, Gaggioli C, Bailet O, Chavey C, Abbe P, Aberdam E, Sabatie E, Cano A, Garcia de Herreros A, Ballotti R, Tartare-Deckert S (2006) SPARC represses E-cadherin and induces mesenchymal transition during melanoma development. Cancer Res 66:7516–7523

Schramek D, Kotsinas A, Meixner A, Wada T, Elling U, Pospisilik JA, Neely GG, Zwick RH, Sigl V, Forni G, Serrano M, Gorgoulis VG, Penninger JM (2011) The stress kinase MKK7 couples oncogenic stress to p53 stability and tumor suppression. Nat Genet 43:212–219

Scurr LL, Pupo GM, Becker TM, Lai K, Schrama D, Haferkamp S, Irvine M, Scolyer RA, Mann GJ, Becker JC, Kefford RF, Rizos H (2010) IGFBP7 is not required for B-RAF-induced melanocyte senescence. Cell 141:717–727

Soengas MS, Lowe SW (2003) Apoptosis and melanoma chemoresistance. Oncogene 22:3138–3151

Sung YM, Xu X, Sun J, Mueller D, Sentissi K, Johnson P, Urbach E, Seillier-Moiseiwitsch F, Johnson MD, Mueller SC (2009) Tumor suppressor function of Syk in human MCF10A in vitro and normal mouse mammary epithelium in vivo. PLoS One 4:e7445

Vousden KH, Lane DP (2007) p53 in health and disease. Nat Rev Mol Cell Biol 8:275–283

Vredeveld LC, Possik PA, Smit MA, Meissl K, Michaloglou C, Horlings HM, Ajouaou A, Kortman PC, Dankort D, McMahon M, Mooi WJ, Peeper DS (2012) Abrogation of BRAFV600E-induced senescence by PI3K pathway activation contributes to melanomagenesis. Genes Dev 26:1055–1069

Wajapeyee N, Serra RW, Zhu X, Mahalingam M, Green MR (2008) Oncogenic BRAF induces senescence and apoptosis through pathways mediated by the secreted protein IGFBP7. Cell 132:363–374

Weinblatt ME, Kavanaugh A, Genovese MC, Musser TK, Grossbard EB, Magilavy DB (2010) An oral spleen tyrosine kinase (Syk) inhibitor for rheumatoid arthritis. N Engl J Med 363:1303–1312

Zhang X, Shrikhande U, Alicie BM, Zhou Q, Geahlen RL (2009) Role of the protein tyrosine kinase Syk in regulating cell-cell adhesion and motility in breast cancer cells. Mol Cancer Res 7:634–644

Zhang J, Benavente CA, McEvoy J, Flores-Otero J, Ding L, Chen X, Ulyanov A, Wu G, Wilson M, Wang J, Brennan R, Rusch M, Manning AL, Ma J, Easton J, Shurtleff S, Mullighan C, Pounds S, Mukatira S, Gupta P, Neale G, Zhao D, Lu C, Fulton RS, Fulton LL, Hong X, Dooling DJ, Ochoa K, Naeve C, Dyson NJ, Mardis ER, Bahrami A, Ellison D, Wilson RK, Downing JR, Dyer MA (2012) A novel retinoblastoma therapy from genomic and epigenetic analyses. Nature 481:329–334

Micrometastatic Cancer Cells: Role of Tumor Dormancy in Non-small Cell Lung Cancer (NSCLC)

Stefan Werner, Michaela Wrage, and Harriet Wikman

Contents

Abstract 239
Introduction 240
Gene Expression Signatures and Tumor Cell Dissemination 241
Characteristics of DTCs in NSCLC 242
Tumor-Stroma Interactions and Tumor Cell Dormancy 242
Immunosurveillance and Tumor Cell Dormancy 243
Pre-metastatic Niche 244
Quiescence of Dormant Lung Cancer Cells 244
Experimental *In Vivo* Models of Lung Cancer Dormancy 246
Cancer Stem Cells (CSC) and Epithelial-to-Mesenchymal Transition (EMT): Two Features of Dormant Lung Cancer Cells? 247
Role of Metastasis Suppressors in Lung Cancer Dormancy 248
MKK4 and 7 248
Nm23 248
RhoGDI2 248
KISS1 249
PEBP1/RKIP 249
Claudin-1 249
BRMS1 249

References 251

S. Werner • M. Wrage • H. Wikman (✉)
Center of Experimental Medicine, Department of Tumor Biology, University Medical Center Hamburg-Eppendorf, Martinistrasse 52, d-20246 Hamburg, Germany
e-mail: h.wikman@uke.uni-hamburg.de

Abstract

Tumor dormancy is a well known and much studied feature of certain cancers such as breast and prostate cancer, whereas in non-small cell lung cancer (NSCLC) it has not been studied extensively to date. Clinical data have, however, shown that cancer dormancy can also occur in NSCLC, albeit this is only observed in a fairly small subset of NSCLC patients. In this chapter, we summarize the current knowledge about the characteristics and clinical implications of dormant tumor cells in NSCLC. We focus on genetic determinants of NSCLC dissemination and describe the properties of micrometastatic NSCLC cells. Subsequently, we review the different biological features, like tumor-stroma interactions and immunosurveillance, which are known to have vital impact on tumor cell dormancy. Clearly, the ability of a dormant tumor cell to survive and be activated later on is a result of both the interaction with the stroma at the secondary site as well as somatic aberrations and hereditary/susceptibility factors of the dormant tumor cell. We discuss mainly genes and phenomena that have already been reported to be associated with tumor dormancy in other epithelial cancers and are associated with worse prognosis or increased metastatic capacity in NSCLC and thus may very likely play a role in dormancy control of NSCLC cells. Furthermore, we emphasize the few but pioneering experimental lung cancer dormancy models. Detailed and accurate knowledge of the biological characteristics and clinical relevance of dormancy in NSCLC remains limited and further studies are urgently needed to elucidate the underlying biological mechanisms in order to be able to assess the risk profile and to find the optimal treatment for each NSCLC patient.

Keywords

Bone marrow (BM) • Cancer stem cells (CSC) • CD44-HA interactions • Disseminated tumor cells (DTCs) • Epithelial-to-mesenchymal transition (EMT) • Gene expression signatures • Immunosurveillance and tumor cell dormancy • Micrometastatic cancer cells • Non-small cell lung cancer (NSCLC) • Quiescence of dormant lung cancer cells

Introduction

Today it is widely accepted that single disseminated tumor cells (DTCs) are the starting points for later tumor metastases. However, metastatic disease may occur years or decades after successful treatment of the primary tumor by surgery and adjuvant treatment (Wikman et al. 2008). This latency period is due to a clinical phenomenon termed tumor dormancy. Both *in vivo* and *in vitro* experiments have clearly shown that in some cases tumor cells do arrest in a state of dormancy (a reversible growth arrest) which can last very long before they can be reactivated and form a metastatic cell mass. Clinical dormancy can either appear as a micrometastatic mass in which the proliferating population is balanced by a cell division and death or as solitary DTCs. Many, but not all of the biological processes regulating these dormant cells are common to both types of dormancy. One common nominator is the close cross talk with the surrounding microenvironment, which is probably together with intrinsic genetic determinants one of the key components in the dormancy control of tumor cells. Most likely, the ability of a dormant tumor cells to be activated and later to produce a secondary lesion is a function of both: its interaction with the stroma at the secondary site and the combination of somatic aberrations and hereditary/susceptibility factors influencing each process (Rocken 2010; Wikman et al. 2008).

Tumor dormancy is a well known and much studied characteristic of malignant melanoma, prostate, renal cell, and breast cancer. In non-small cell lung cancer (NSCLC) however, tumor dormancy has not been much studied as the majority of patients relapse fairly soon after the primary tumor removal without an obvious dormancy phase. However, a small subset of patients who have a complete resection of the primary tumor do relapse after a long latency phase, even when the primary tumor showed a high proliferation rate (Maeda et al. 2010). Furthermore, lately Demicheli et al. (2012) showed that also resected early-stage NSCLC displays a recurrence dynamics with a specific multipeak pattern of recurrence risk which has already been described for breast cancer. This multipeak pattern of recurrence is independent of anatomical metastatic sites. Importantly, this finding is incompatible with continuous tumor growth, and implies rather that some kind of growth interruption/dormancy of tumors cells occurs during the disease course. Interestingly, the NSCLC recurrence dynamics for women display similarities with the corresponding dynamics described for early breast cancer hazard rates for recurrence, indicating that the recurrence risk pattern results from inherent features of the metastasis development process (Demicheli et al. 2012).

One of the main limitations in dormancy research is the lack of a defined phenotype or molecular profile that identifies a dormant tumor cell. Cellular dormancy, quiescence and even senescence are different nominators for different but partly overlapping cell states of resting cells. Whereas, by definition, dormancy implies an ability to become activated, senescence defines an irreversible state of growth arrest. Therefore, dormancy might be more similar to quiescence than senescence. Quiescent cells are not actively dividing to create new cells, but can re-enter the cell division cycle and proliferate at some later time, which is similar to abrogation of dormancy in case of metastatic outgrowth. Stem cells have been described to be in a state of reversible quiescence. Growing evidence has revealed that also dormant tumor cells may have cancer stem cell-like (CSC) properties that may account for resistance to chemo- and radiation therapy and the ability to generate an overt metastasis upon reactivation (Aguirre-Ghiso 2007; Rocken 2010; Wikman et al. 2008).

The different steps required for the metastatic tumor to arise, including induction of EMT (epithelial-to-mesenchymal transition) and CSC characteristics, are the same for all epithelial tumors and thus not surprising many common genes and pathways have been found to be equally important for the different epithelial tumors. In this chapter, we try to summarize the role of micrometastatic tumor cells and dormancy in NSCLC based mainly on other epithelial cancer models. However, the genes we discuss, have all been reported to show an association with worsen prognosis or association with metastatic capacity in NSCLC.

Gene Expression Signatures and Tumor Cell Dissemination

Numerous gene expression and proteomic studies in NSCLC have been published during the last 10 years trying to define prognostic signatures that could improve therapeutic decision making beyond current practice standards (Subramanian and Simon 2010). In contrast to breast cancer, no gene expression based prognostic signatures in NSCLC have been shown to be a stable signature in different cohorts and thus would be ready for clinical application. Furthermore, the signatures in the different studies have very little in common, in form of genes and even pathways and thus cannot help explaining the underlying biological processes. This is, however, not surprising as different platforms and statistical methods are used with the goal of finding a predictor not a function.

However, in a recent study by Boutros et al. (2009) the authors used a nonlinear semisupervised algorithm and identified a six-gene prognostic signature. This signature was validated in multiple independent validation sets and it was shown to be independent of traditional staging criteria. Interestingly, among the six genes the two genes hypoxia inducible factor 1A (HIF1A) and ring finger protein 5 (RNF5) have been implicated in cancer dissemination and dormancy (Boutros et al. 2009).

One critical mechanism behind tumor dormancy is the ability of the tumor population to induce angiogenesis (angiogenic switch) (Aguirre-Ghiso 2007). The HIF1A transcription factor has been identified as an important transcription factor that mediates the cellular response to hypoxia, promoting both cellular survival and angiogenesis and thus influences metastasis at several levels, also in NSCLC. Upregulation of HIF1A expression is also associated with dissemination of DTC in breast cancer patients (Woelfle et al. 2003). This up-regulation coincided with down-regulation of genes responsible for HIF-1α degradation (e.g., VHL and cullin-2) in BM-positive tumors, which in concert may lead to an accumulation of HIF-1α in tumor cells.

RNF5 modulates cell migration by ubiquitinating paxillin, a focal adhesion molecule. High levels of RNF5 have been associated with decreased survival in human breast cancer. Higher levels of RNF5 were also observed in metastatic melanoma, leukemia, ovarian, and renal tumor-derived cell lines, suggesting that increased RNF5 expression may be a common event during tumor progression (Bromberg et al. 2007).

A few papers have also investigated site specific factors important for the outgrowth of NSCLC cells at a specific site. Vicent et al. (2008) detected a bone signature using a murine lung cancer model. A three gene signature based on TCF4, PRKD and SUSD5 was strongly associated with bone colonization *in vivo*, but was ineffective in promoting local tumor growth and homing of the cells to the bone. Interestingly, these genes interacted with TGFβ and other important regulators of EMT and metastatic processes in various tissues (Vicent et al. 2008).

Characteristics of DTCs in NSCLC

By the time of diagnosis, a large number of NSCLC patients already have DTCs lodged in target organs. These DTCs can be found in secondary organs such as the bone marrow (BM), blood (often referred to as circulating tumor cells; CTC), liver and sentinel lymph nodes (LN) (Wikman et al. 2008).

DTCs are mostly studied in BM or blood aspirates because of the easy access. Interestingly, BM seems to be a common homing organ for DTCs derived from various types of malignant epithelial tumors, including tumors which do not typically form metastases in bone like colon cancer. This suggests that BM might be the preferred reservoir for metastatic tumor cells from where they may re-circulate into other distant organs. However, DTCs most likely occur in multiple organs simultaneously, but analyzing the BM gives a good overview of how these cells function and are characterized in other organs (Wikman et al. 2008).

It is still debated whether the release of cells from the primary tumor is a selective process or rather represents a more unselective shedding of cells into the circulation, and if characterization of these cells thus has any biological significance. We could recently show that tumors from NSCLC patients with DTCs in the BM are associated with a specific molecular signature, suggesting that the dissemination of tumor cells is a specific process driven by a set of distinct genes (Wrage et al. 2009). Especially a deletion on chromosome 4q in primary lung tumors of patients with DTC in the bone marrow as well as in brain metastases suggests that important metastasis suppressing genes are located within this region. Therefore, not only the potential of a tumor to disseminate and form metastasis might be already genetically determined (possibly very early) in primary tumors but also the dormancy control and escape as well as organotropoism of the tumor (Wrage et al. 2009).

Several studies have been conducted on CTCs and DTCs in NSCLC. Most studies have however only enumerated the cells and correlated them to patient outcome (O'Flaherty et al. 2012). Only few studies have tried to characterize the DTCs in order to understand the biology behind the DTCs and dormancy control in NSCLC. However, a few recent papers have shown that there are clearly a subset of CTCs show an expression of EMT markers such as vimentin and N-cadherin. Furthermore, these studies have also shown that there is a large intrapatient heterogeneity in expression of the different epithelial (mainly cytokeratins (CK), E-cadherin and EpCAM) and mesenchymal markers. Interestingly, whereas DTCs found in BM has mostly been described to be in a non-proliferative (negative for Ki67 staining) state, a large fraction of CTCs seems to express Ki67. However, CTC clusters (also called circulating tumor microemboli; CTM) neither show proliferative activity, nor do they contain apoptotic cells, in contrast to solitary CTCs (O'Flaherty et al. 2012).

In conclusion, today a larger number of different techniques exits for detecting CTCs and DTCs. Currently every technique has some weaknesses and a single method might not be enough to detect the very heterogeneous population of CTCs and CTMs. Clearly a pure enumeration of CTC is not enough to understand the mechanisms behind the metastatic processes and thus an exact characterization of the tumor cells is needed, in order to learn more about the biology of metastasis and dormancy control in order to find those patients with highest risk of relapse.

Tumor-Stroma Interactions and Tumor Cell Dormancy

Primary as well as metastatic tumors are composed of both the neoplastic epithelial cells as well as the tumor stroma. The stroma consists of a multitude of different mesenchymal cells, including stromal fibroblasts and endothelial cells as well as all types of immune cells, and extracellular matrix components. Recently studies showed that the tumor-stroma interaction plays a significant role in tumor development and progression (Fridman et al. 2012). In this section we shortly discuss two important aspects of tumor-stroma interaction in relation to tumor cell dormancy: the immunosurveillance and the pre-metastatic niche.

Immunosurveillance and Tumor Cell Dormancy

The influence of the innate and adaptive immune system on tumor growth, dissemination and metastasis is currently highly debated. It has clearly been shown that the immune system does not only act as an important host protection process that inhibits carcinogenesis but also has tumor promoting characteristics. Originally formed by F.M. Burnet and T.R. Prehn tumor immunosurveillance postulated that the majority of malignant tumors are eliminated via the immune system before they become even clinical detectable. Later, it was shown that the adaptive immune system of a naïve mouse had the ability to destroy tumor cells and to develop tumor immunogenicity, and also restrained tumor growth for extended periods of time (immunological equilibrium) (Rocken 2010).

Histopathological analyses of primary lung tumors have shown that the tumors are typically infiltrated by inflammatory and lymphocytic cell, in variable quantities and in different locations within and around the tumor. The presence of high numbers of tumor infiltrating lymphocytes in the primary tumors has been correlated with a good prognosis in lung cancer as well as in melanoma, head and neck, breast and bladder ovarian. And especially high densities of CD8$^+$ effector T cells and CD45RO$^+$ memory T cells as well as T_H1 (helper) cells has been clearly associated with a longer disease-free survival after surgical resection of the primary lung tumor and overall survival. On the other hand, some cell types of the immune system show also tumor promoting functions. The presence of regulatory immune cells (T_{Reg} cells), which can produce immunosuppressive cytokines, as well as T_H17 cells has been shown to have a negative impact on prognosis in lung cancer (Fridman et al. 2012).

The tumor itself can be also very effective in escaping from immune-mediated rejection.

A tumor may directly inhibit antitumor immune responses by multiple mechanisms, e.g. overproducing inhibitors of T cell responses, such as galectin-1, or suppression of induction of proinflammatory danger signals. Some tumor cells showed direct or indirect lesions in antigen processing and presentation pathways that facilitate evasion from adaptive immune recognition, like losses of HLA (human leukocyte antigen) class I proteins. These histocompatibility molecules display antigens on the surface of almost all vertebrate cells and are recognized by CD8$^+$ T cells. Tumor cells without an expression of these glycoproteins are often ignored by CD8$^+$ T cells and the immune system (Fridman et al. 2012).

Cytokines, like interferon-γ (IFNγ) and tumor necrosis factor (TNF), play a critical role in the defense of tumor cells by the immune system and therefore tumor cells have to find alternative ways to escape from these pathways. IFNγ is involved in the regulation of the immune and inflammatory responses. T_H1 cells and natural killer cells (NK) secret IFNγ to recruit leukocytes to the site of infection, which results in increased inflammation. Alteration in the IFNγ receptor signaling pathway in 25 % of the human lung adenocarcinoma cell lines might be a very effective way to escape from the immune system (Fridman et al. 2012). IFN as well as TNF are not only involved in killing and apoptosis, but they also control signaling cascades that regulate cell proliferation, cell cycle progression and the control of signaling molecules that influence cell differentiation. Therefore, immunosurveillance may induce dormancy in single DTCs by arresting their cell cycle through cytokine mediated signals (Rocken 2010).

In a study by Eyles et al. (2010) the authors used a CD8$^+$ T cell depletion in a RET-AAD mice model of melanoma to uncover the role for T cell-dependent tumor immunosurveillance in preventing DTCs from forming metastases. In the presence of CD8$^+$ T cells, the number of Ki67-positive tumor cells is low, whereas the number of Ki67-positive tumor cells is high in the absence of CD8$^+$ T cells. The latter markedly increased the risk of overt metastases. As expression of Ki67 is associated with cell proliferation and cell cycle progression, these data strongly argue that, in addition to cytotoxic effects, a cytostatic mode of tumor silencing may be keeping the DTCs from forming metastases.

Pre-metastatic Niche

The concept of the "stem cell niche" as a distinct specialized microenvironment housing stem cells was first proposed by Schofield and collegians almost 30 years ago. The niche regulates stemness, proliferation and apoptosis resistance of stem cells and is composed of diverse stromal and immune cells, a vascular network, soluble factors and extracellular matrix components which interact with the stem cells via adhesion molecules and paracrine factors (Psaila and Lyden 2009). The hematopoietic stem cells (HSCs) are the best characterized stem cell population and act as precursors of mature blood cells that are defined by their ability to replace the bone marrow system.

The HSC niche in the bone marrow and the microenvironment that constitutes the metastatic tumor niche has been suggested to share many of the same structural and functional components e.g. the osteoblasts, which is important for regulating cellular processes such homing and migration, growth and survival, and quiescence and dormancy. This allows the presumption that the bone marrow microenvironment has special features where DTCs take advantage of niche specific signaling for tumor dormancy and survival as well as evading the immune system (Psaila and Lyden 2009).

The metastatic niche model suggests that the formation of a favorable microenvironment, the so-called pre-metastatic niche, before tumor cells arrive at the metastatic destination is critical for the engraftment of DTCs (Psaila and Lyden 2009). Niche preparation involves stimulation of local fibroblasts by tumor-derived factors and chemokines that attract tumor cells and tumor associated cells. In response to these soluble factors, haematopoietic progenitor cells and macrophages cluster at pre-metastatic niches, creating as initiators an environment to which the tumor cells metastasize. Previously, Kaplan et al. (2005) demonstrated by using a syngeneic Lewis lung cancer mouse model, that HSC that express vascular endothelial growth factor receptor 1 (VEGFR1) are able to travel to tumor-specific pre-metastatic sites and form cellular clusters before the arrival of tumor cells.

Cancer cells and their associated stromal cells secrete a multitude of chemokines that direct the migration, proliferation, and differentiation of the vascular cell network to support the tumor and metastatic microenvironment. In particular CD44, a lymphocyte homing receptor, plays an important role in leukocyte and metastatic cell motility. It is the major marker of breast cancer stem cell (CSC) but has been also described as a NSCLC CSC marker. One of the main properties of CD44 is its binding to hyaluronan (HA), which is accumulated in stem cell niches. This binding plays an important role in HSCs and CSCs homing, EMT and adhesion. Furthermore CD44 is actively involved in the metastatic niche assembly and through its effect on HA secretion and degradation, mainly owing to its activity as a catcher of chemokines, growth factors and matrix-degrading enzymes (Zoller 2011).

CD44-HA interactions are regulated among others by the soluble factors Oncostatin M (OSM) and TGF-β1 (Zoller 2011). OSM is a member of the interleukin (IL)-6 family of cytokines and is involved in several signal transduction pathways, including the stress MAP kinases JNK and p38 pathways. Importantly, the balance between ERK1/2 and p38α/β pathways has been shown to be a key regulator of cancer dormancy control (Sosa et al. 2011). However, whether and how an impaired ERK^{low}/p38^{high} ratio does also influence dormancy of lung cancer cells remains to be determined.

Quiescence of Dormant Lung Cancer Cells

The most obvious feature of single cell dormancy is the lack of proliferative activity. Hence, cellular dormancy is compared to well-studied cell cycle arrest mechanisms like quiescence or senescence.

However, whether the control of cellular dormancy is really equivalent to these biological programs remains unclear. In general, cellular senescence is regarded as a tumor suppressive mechanism that keeps damaged or stressed cells in an irreversible non-proliferative state (Aguirre-Ghiso 2007). On the other hand, in case of metastatic outgrowth tumor dormancy becomes abrogated. Additionally, tumor dormancy occurs in malignant tumor cells that must have already overcome most tumor suppressive mechanisms. Accordingly tumor dormancy and senescence seems to be controlled by at least partly non-overlapping mechanisms.

More parallels are seen between dormant and quiescent cells. Both are not actively dividing and kept in the G0 phase of the cell cycle, which is associated with alteration of the metabolic and transcriptional activity. Also re-entry into the cell cycle in the presence of growth promoting conditions is common for both tumor dormancy and quiescence (Rocken 2010). The ability to appropriately enter and exit the cell cycle during G1 phase is essential for maintaining normal tissue homeostasis and the self-renewal capacity of somatic stem cells. The cycle activity is inversely correlated with the self-renewal capacity in this kind of cells. Progression through the cell cycle is mainly controlled by enzymatic complexes formed by Cyclin-Dependent Kinases (CDKs) and their associated Cyclins. Protein complexes formed by these molecules are antagonized by Cylin-Dependent inhibitors (CDKis) which are viewed as key players in induction and maintenance of quiescence and preservation of homeostasis. Recently, it was shown that expression of CDKN1C/KIP2 (p57) is essential for self-renewal of somatic lung stem cells (Zacharek et al. 2011). In parallel published studies, in which the mechanistic role of p57 in control of HSC quiescence was investigated, it was furthermore shown that the p57 protein interacts with the HSC70 chaperon and prevents CyclinD1 from entering the nucleus. Activating cytokines from the surrounding niche can lead to transcriptional repression and degradation of p57. As a consequence CyclinD1 can enter the nucleus which finally leads to Rb-phosphorylation and abrogation of quiescence (Tesio and Trump 2011).

Since p57 expression is crucial for growth control of both somatic lung as well as hematopoetic stem cells, it is likely that this protein participates in a general regulatory pathway of stem cell quiescence, which might also be important for the intrinsic control of cellular cancer dormancy. The assumption that p57 might play a certain role in the control of lung cancer dormancy is in part supported by the observation that decreased expression levels of p57 in primary lung cancer specimens are associated with poor post-surgical survival time and lymphatic metastasis in lung cancer patients and that low p57 expression is an adverse prognostic factor (Biaoxue et al. 2011).

However, the above mentioned example also emphasizes that the activity of the canonical cell cycle arresting machinery strongly depends on external stimuli, e.g. activating cytokines which were secreted by neighboring cells. These external signals were further processed by a complex network of divers signaling molecules. A good model to study how the tumor cell-microenvironment and distal signaling are controlling tumor cell dormancy is established for head and neck squamous carcinoma. In this model, the metastasis-associated urokinase receptor (uPAR) normally drives tumor progression by interacting with α5β1-integrins which leads to activation if EGFR signaling and transmission of mitogenic signals through the ERK1/2 -pathway. In this model blocking of uPAR, α5β1 or the associated EGF-receptor resulted in suppression of mitogenic ERK1/2 signaling and induction of dormancy. On the other hand, blockage of this adhesion signaling is also leading to an increase of the stress adaptive response hallmarked by p38α/β activation. From there, it is apparent that signaling in dormant tumor cells is characterized by the absence of mitogenic and the presence of stress signals. The $ERK^{low}/p38^{high}$ ratio is a key signaling feature of the studied dormant HEp3 cells that induces quiescence by transcriptional reprogramming. On the other hand proliferation in both metastatic and primary tumor cells is defined by $ERK^{high}/p38^{low}$ ratio (Sosa et al. 2011).

Although there are currently no studies that address mitogenic and stress signaling in dormant lung cancer cells, it has already been shown that impairment of uPAR protein function is also associated with decreased ERK1/2 activation and inhibition of tumor growth in lung cancer cells (Tang et al. 2008). However, as already mentioned before, whether and how an impaired ERK^{low}/p38^{high} ratio does also influence quiescence of lung cancer cells by transcriptional reprogramming remains to be determined.

Experimental *In Vivo* Models of Lung Cancer Dormancy

In spite of several technical difficulties various experimental models have been developed to study tumor dormancy *in vivo*. Among other, this work was mainly focused on lymphoma, melanoma, and breast cancer. These experimental studies led to the discovery of the two different biological types of tumor dormancy, the solitary cell and the micrometastatic dormancy (Rocken 2010). Few but groundbreaking experimental animal models have also been established so far to investigate the dormant phenotype of lung cancer cells. Findings from these models elucidated especially the importance of angiogenic growth to overcome micrometastatic dormancy.

In one of the earlier published dormancy models Lewis lung carcinoma cells were used to study the dormant state of micrometastases under systemic angiogenesis suppression (Holmgren et al. 1995). In this study, the authors inoculated the lung carcinoma cells subcutaneously into C57BL6/J mice. After the tumors reached a given mass they were removed from one half of the animals while the other half underwent a sham operation. Subsequently animals from each group were killed at various time points and the lungs were examined for metastatic growth. Metastases in the lung started to grow rapidly after removal of the primary tumor while in the presence of the primary tumor the disseminated tumor cells remained in a non-growing micrometastatic stage in the examined lungs. Comparison of growing metastases to dormant micrometastases showed high proliferation rates in both groups, whereas determination of the apoptotic index revealed a significant increase of cell death in the non-growing micrometastases. Additionally, metastatic outgrowth correlated with the onset of blood vessel formation in the surrounding areas. It was furthermore shown that outgrowth of dormant micrometastases could be efficiently inhibited by systemic administration of the exogenous angiogenesis inhibitor TNP-470. As possible hypotheses it was strongly supported that metastatic outgrowth is strictly dependent on the availability of oxygen and nutrients and that prevascular micrometastases remain dormant because of balanced tumor cell proliferation and apoptosis. This particular micrometastatic stage is also known as angiogenic dormancy and has become a well established concept in many other tumor types as well (Aguirre-Ghiso 2007).

Another seminal animal model of lung cancer dormancy was published by Kienast et al. (2010), which allows the dissecting of single steps of brain metastasis formation. This novel experimental approach uses multiphoton laser-scanning microscopy to track the fate of individual metastasizing cancer cells in the mouse brain over month in real time. In this study PC14-PE6 lung adenocarcinoma cells, which were initially selected from the parental PC14 cell line from their high metastatic capabilities, were intra-arterially injected into the brains of nude mice. Most importantly the authors pinpointed the ineffectiveness of brain metastasis formation and specified four essential steps for a successful formation of lung cancer cells in the brain: (i) initial arrest of disseminated cells in the brain vessels resulting from size restriction, (ii) dynamic extravasation, (iii) continuous access to oxygen and nutrient supply and (iv) induction of angiogenic growth. Particularly the novel experimental setup enabled to study the fate of solitary dormant cells as well of dormant micrometastatic cell clusters for the duration of several months. While the majority of the tracked cells suffered regression and/or cell death a small subset of 4 % of the PC14-PE6 cells stayed dormant in a single cell stage for up to 51 days and only 2.4 % of the analyzed cells grew out as overt metastases. All the long-term dormant single cells maintained a close association with brain vessels after extravasation and showed no sign of motility. Accordingly, single dormant lung cancer cells may find optimal survival conditions when resting in a strict perivascular position. Additionally, in line with the angiogenic dormancy hypothesis, it was shown that the outgrowth of micrometastatic cell clusters to overt metastasis was essentially dependent on the ability to induce vascular remodeling and angiogenesis. This observation was also confirmed by therapeutically inhibition of VEGF-A which also kept the micrometastases in a state of chronic micrometastatic dormancy.

In conclusion, the few but very thorough animal models in which lung cancer cells were studied clearly support the hypothesis that metastatic outgrowth of dormant cells strongly depends on a switch to the angiogenic phenotype. However, this condition is not unique to lung cancer but thought to be essential for all solid tumor entities (Aguirre-Ghiso 2007). In the above mentioned studies fibrosarcoma as well as melanoma cells were studied in parallel and showed equivalent results.

Cancer Stem Cells (CSC) and Epithelial-to-Mesenchymal Transition (EMT): Two Features of Dormant Lung Cancer Cells?

Both epithelial-to-mesenchymal transition (EMT) cells and cancer stem cells (CSC) have been implicated as important drivers or even prerequisites for the successful dissemination and colonization of DTCs. Whereas, EMT endow the cells a more migratory and invasive activity, CSC cells are identified as a subset of tumor cells, which has the ability to give rise to new tumors when xenografted recapitulating the heterogeneity of the original tumors. Recent findings suggest that DTC in BM and blood in various epithelial cancers do show either CTC and/or EMT properties, as was already discussed in the previous chapter (Bednarz-Knoll et al. 2012; Rocken 2010).

The induction of EMT leads to the loss of epithelial characteristics (such as E-cadherin, claudins, desmoplakin and type IV collagen) and an acquisition of a more mesenchymal phenotype (induction of e.g. vimentin, $\alpha 5\beta 1$ integrin, laminin 5 and fibronectin). The EMT phenotype shares many properties with CSCs, including CD44(high)/CD24(low) expression, induction by hypoxia, RAS, TGFbeta and ZEB1. Both EMT and CTC have also been found to show increased resistance to apoptosis and chemotherapeutic drugs (Bednarz-Knoll et al. 2012).

One of the hallmarks of EMT is the loss of epithelial markers such as E-cadherin. The main transcription factors implicated in E-cadherin down-regulation are Snail, Slug, Zeb1, Zeb2 and TWIST. Interestingly, these transcription factors are also important regulators of the expression of stem cell markers such as aldehyde dehydrogenase 1 (ALDH1) and regulators of the non-proliferative state of cells associated with both CSC and EMT phenotype (Bednarz-Knoll et al. 2012). In a recent paper by Pirozzi et al. (2011) they show that the treatment of the lung cancer cell lines LC31 and A549 with TGFβ-1 resulted in the acquisition of mesenchymal profile (EMT) and in the expression of stem cell markers. These cells showed an increased pneumosphere-forming capacity and tumors-forming ability in NOD/SCID mice, again showing the importance of both EMT and CTC in cancer progression.

Wendt et al. (2011) showed recently that down-regulation of E-cadherin induced by transforming growth factor-β (TGF-β) was sufficient for the outgrowth of the dormant breast cancer cell line D2.OR in three-dimensional (3D) organotypic assays and in the lungs of mice. These intriguing results show that EMT and its down-regulated expression of E-cadherin can circumvent breast cancer dormancy, linking EMT for the first time directly to dormancy control.

In conclusion, CSC and EMT clearly shares many features but whether they are overlapping features of cells at different stages or different cell populations remains to be proven. However, both cell types clearly play an important role in cancer metastasis and dormancy control.

Role of Metastasis Suppressors in Lung Cancer Dormancy

One hypothesis for a late relapse is the existence of cells with prolonged growth arrest with occasional cell division, during which slow accumulation of genetic and/or epigenetic alterations in metastasis suppressor genes (MSG) occurs, giving rise to malignant DTCs and relapse. An activation of dormant cells may also be regulated by MSG that can respond to micro-environmental stress responses. Noteworthy, MSGs are rarely mutated and thus epigenetic events, such as methylation may cause loss of function. Since epigenetic events are reversible, dormancy control by MSGs may be realistic (Hurst and Welch 2011).

MSGs have been shown to play a role in dormancy control and inhibition of metastasis by either inducing DTC growth arrest or by preventing the formation of overt metastases. Today a large fraction of the so far identified MSGs, have been linked with dormancy control or colonization and out growth at secondary sites. Interestingly, in contrast to classical tumor suppressor genes, MSG genes appear to be cancer type specific. Below we discuss briefly those MSGs, which have been implicated in dormancy control and colonization and which also have been shown to play a role in NSCLC (Hurst and Welch 2011).

MKK4 and 7

Mitogen-activated protein kinase 4 (MKK4/MAP4K4) was first described as a metastasis suppressor in prostate cancer. Ectopic expression of MKK4 in AT6.1 prostate cancer cells suppressed the formation of overt lung metastases by >80 % without affecting primary tumor growth or the tumor dissemination. Also ectopic expression of the mitogen-activated protein kinase 7 (MKK7/MAP3K7) in AT6.1 cells, has been shown to suppress the formation of overt lung metastases. MKK4 and 7 are dual-specificity kinases, which can activate both the p38 and JNK MAPK, two important regulators of cancer dormancy control (Haeusgen et al. 2011).

Consistent with its role as a metastasis suppressor, MKK4 down regulation has been found to be associated with poor prognosis in at least pancreatic, breast, ovarian and gastric cancers (Hurst and Welch 2011). In lung adenocarcinoma, MKK4 has also been recently shown to function as a tumor suppressor in the primary tumor and inhibits tumor cell invasion by decreasing PPARγ2 levels (Ahn et al. 2011). The metastasis suppressor activity of MKK7 in lung cancer has not to our knowledge been investigated so far.

Nm23

Non-metastatic protein-23 homolog-1 and 2 (Nm23-H1/NME1 and H2/NME2) were the first metastasis suppressors to be discovered. The Nm23 expression has been extensively studied and shown to play a role in cytoskeletal organizing and signaling pathways. Still it is somewhat unclear, which of these multiple functions plays the critical role in suppressing cancer colonization and metastasis. Generally, with the exception of neuroblastoma, Nm23 expression is inversely correlated with poor survival (Hurst and Welch 2011).

The clinical relevance of Nm23 as a tumor suppressor in NSCLC has been shown in several large studies (Hurst and Welch 2011). In a recent study, 452 stage I NSCLC patients were studied for nm23 expression by immunohistochemistry. Multivariate analysis showed that nm23-H1 down regulation independently predicted shortened survival in NSCLC (Liu et al. 2011). Also in experimental settings nm23 has been shown to influence the lung tumor cell line L9981 ability to metastasize. The expression of epithelial markers such as beta-Catenin, E-Cadherin and TIMP-1 were significantly increased while expression EMT associated markers MMP-2, CD44v6, and VEGF was dramatically decreased in the L9981cell line overexpressing nm23-H1 (Che et al. 2006).

RhoGDI2

Rho GDP dissociation inhibitor 2 (RhoGDI2 or ARHGDIB) was first shown to be a metastasis suppressor in bladder cancers (Gildea et al. 2002).

When RhoGDI2 was transferred back into metastatic bladder cancer cell line T24T that lacked its expression, it suppressed experimental lung metastasis but did not affect *in vitro* growth, or *in vivo* tumorigenicity (Gildea et al. 2002). However, RhoGDI2 has also been shown to promote metastasis in ovarian and breast cancers (Hurst and Welch 2011). In lung cancer a recent study showed that diminished RhoGDI2 expression is associated with higher stage and lymph node metastasis, implying a possible tumor or metastasis suppressor role of RhoGDI2 in lung cancer (Niu et al. 2010).

RhoGDI2 is a regulator of Rho GTPases. One of the key functions of Rho GTPase proteins is to remodel cytoskeleton components thus influence cell mobility. RhoGDI2 is also implicated in acquired resistance to chemotherapeutic agents such as cisplatin (Hurst and Welch 2011).

KISS1

KISS1 is a G-coupled protein receptor ligand. In melanoma cells KISS1 secretion was required for maintenance of disseminated cells in a dormant state. The KISS1 protein has been shown to suppress metastasis of several other tumor entities as well breast, ovarian and pancreatic cancer without blocking orthotopic tumor growth. The exact mode of metastasis inhibition is still unknown but it is though at least partly through the receptor KISSR1 (Hurst and Welch 2011).

Also clinical reports from a variety of tumor entities generally support a positive correlation between KISS1 expression and prognostic factors (Hurst and Welch 2011). In NSCLC the expression of KISS1 has been associated with low TNM stages as well as better prognosis. Also, a negative correlation between KISS1 and MMP-9 protein expression has been found, strongly indicating metastasis suppressive role of KISS1 also in NSCLC (Zheng et al. 2010).

PEBP1/RKIP

Phosphatidylethanolamine binding protein 1 (PEBP1; previously known as RKIP) is though to inhibit nuclear factor (NF)κB and RAF/MEK/ERK signaling pathways. It has been linked to metastasis suppression in malignant melanoma, prostate, colorectal, and breast carcinomas (Hurst and Welch 2011). RKIP blocks activation of Raf-1 by preventing access to activating kinases and phosphorylation of S338 and Y341. RKIP has also been shown to be regulated by Snail signaling, and thus involved in EMT. However, the exact mechanism by which RKIP regulates metastasis is still unclear (Hurst and Welch 2011). In primary NSCLC tumors the expression levels of phospho-RKIP, in contrast to total RKIP, has been shown to be associated with favorable outcome (Huerta-Yepez et al. 2011).

Claudin-1

Claudin-1 (CLDN1) is a transmembrane protein playing a critical role in the maintenance of epithelial and endothelial tight junctions and cell signaling. In breast cancer a highly aggressive claudin-low subtype has been described, which is characterized by EMT and CSC characteristics (Prat et al. 2010).

In a melanoma brain metastasis model the spontaneous brain dormant micrometastatic cells had down regulated CLDN1 (Izraely et al. 2011). In lung adenocarcinoma a low-CLDN1 mRNA and protein expression has been associated with shorter overall survival. Furthermore, overexpression of CLDN1 in NSCLC cell lines inhibited cancer cell dissociation in time-lapse imaging of wound healing, and suppressed cancer cell migration, invasion, and metastasis (Chao et al. 2009).

BRMS1

Breast cancer metastasis suppressor (BRMS1) inhibits the activity of NFκB, and regulates the expression of metastasis-associated microRNA known as metastamir. In breast cancer BRMS1 expression significantly reduces the numbers of solitary single cells that survived after initial arrest within the lung microvasculature, and also inhibits the initiation of growth subsequent to arrest (Hurst and Welch 2011).

Smith et al. (2009) showed that BRMS1 expression in H1299 NSCLC cells significantly decreases both migration and invasion of NSCLC cells *in vitro*. Furthermore, in a xenograft model, BRMS1 suppressed the formation of pulmonary and hepatic metastases but did not significantly affect primary lung tumor growth. Therefore, not surprisingly, BRMS1 expression was associated with improved patient survival. Thus, BRMS1 functions as a metastasis suppressor and may be a prognostic indicator in NSCLC (Smith et al. 2009).

In conclusion, tumor dormancy is a well known and much studied feature of cancers such as breast and prostate where a tumor recurrence might take place even more than a decade after the successful operation of the primary tumor. Also in NSCLC, the phenomenon of tumor dormancy can be observed, even though not as frequent as in e.g. breast cancer and usually associated with shorter relapse times. Clinical data have, however, clearly shown that resected early-stage NSCLC display a multipeak pattern of recurrence, which implies a similar kind of growth kinetic for dormant lung tumor cells, as has been described for breast cancer cells in the course of disease (Demicheli et al. 2012). Although one conclusion from this observation could be that inherent features of the primary tumor are determining the course of the metastasis development process, no specific gene expression signatures in NSCLC have been defined so far, that is today able to predict robustly the course of disease or even improve therapeutic decisions. Clearly a crosstalk between the tumor cell and the environment is crucial in tumor cell survival and dormancy control.

A major limitation for all dormancy research has been the lack of clear markers describing the state of dormancy. Despite this key limitation a substantial number of theories have been generated trying to explain the mechanisms associated with the development, maintenance and breaking of tumor dormancy in cancer in general. Many of these hypotheses are based on several different types of experimental models using also different types of cancers and thus verification of the general applicability is still missing. Also proof of many of these findings in human setting needs still to be proven, especially among NSCLC patients.

Experimental models that address the behavior of dormancy control in NSCLC remain rare. Nonetheless, the few existing animal dormancy models of lung cancer have clearly demonstrated the importance of the angiogenic switch for the outgrowth of lung cancer cells in a new surroundings. Beyond that the ability to image the establishment of brain metastases *in vivo* provides new insights into the evolution and response of dormant lung cancer cells to therapies. In particular the possibility to track the fate of individual long-term dormant tumor cells has revealed further preconditions for metastatic outgrowth.

Dormant cells share several characteristics with quiescent cells, furthermore it is also widely accepted that the activity of the intrinsic cell cycle arresting machinery strongly depends on external stimuli. This suggests that the existence of a favorable microenvironment, in combination with the genetic configuration of DTCs, is critical for their persistence at the metastatic site as well as for metastatic outgrowth. In particular, the bone marrow microenvironment seems to display certain characteristics that favor tumor dormancy and survival of DTCs (Wikman et al. 2008). Recent studies point out that NSCLC patients with DTCs in the bone marrow are associated with a specific molecular signature and that not only the potential of a tumor to disseminate might be already genetically determined in primary tumors but also the configuration that control dormancy as well as organotropoism of DTCs. Besides these factors immunosurveillance seems also to be a vital factor in dormancy control. Immune cells are part of the DTC hosting microenvironment and may act in a pleiotropic manner because in addition to their cytotoxic effects, their secretion of cytokines may contribute to keep DTCs in a cytostatic mode.

The many similarities between somatic stem cells and dormant tumor cells indicates that dormant tumor cells may also possess stem cell-like properties that may account for a long non-proliferative state cell as well as resistance to chemotherapy and radiation. The hematopoietic stem cells found in the bone marrow are known to reside and self-renew in a so called hematopoietic stem-cell niche (HSC), which also protects the stem cells from the surrounding environment and hostile signals. A similar or even overlapping niche has also been described for tumor cells. The metastatic niche model suggests that the formation of a favorable microenvironment in the bone marrow makes it possible for single tumor cells to survive for a long time in foreign and often hostile environments in a dormant or slowly- dividing steady-state condition.

EMT is undoubtedly associated with more aggressive tumor cells with a higher motility and invasion capacity. In addition, EMT and cancer-stem cells also share many features. Whether EMT also plays a role in dormancy control is much more unclear as evidence so far is merely correlative. New experimental approaches are therefore needed to address this question and also needed to reveal whether dormancy of NSCLC cells is also regulated by the known metastasis suppressor genes (MSG).

In addition to the experimental approaches, further studies that address the clinical impact of tumor dormancy in NSCLC should be devised. More knowledge about the biological characteristics and clinical relevance of NSCLC is urgently needed in order to be able to assess the risk profile and to find the right treatment for each cancer patient.

References

Aguirre-Ghiso JA (2007) Models, mechanisms and clinical evidence for cancer dormancy. Nat Rev Cancer 7(11):834–846

Ahn YH, Yang Y, Gibbons DL, Creighton CJ, Yang F, Wistuba II, Lin W, Thilaganathan N, Alvarez CA, Roybal J, Goldsmith EJ, Tournier C, Kurie JM (2011) Map2k4 functions as a tumor suppressor in lung adenocarcinoma and inhibits tumor cell invasion by decreasing peroxisome proliferator-activated receptor gamma2 expression. Mol Cell Biol 31(21):4270–4285

Bednarz-Knoll N, Alix-Panabieres C, Pantel K (2012) Plasticity of disseminating cancer cells in patients with epithelial malignancies. Cancer Metastasis Rev 31(3–4):673–687

Biaoxue R, Xiguang C, Hua L, Hui M, Shuanying Y, Wei Z, Wenli S, Jie D (2011) Decreased expression of decorin and p57(KIP2) correlates with poor survival and lymphatic metastasis in lung cancer patients. Int J Biol Markers 26(1):9–21

Boutros PC, Lau SK, Pintilie M, Liu N, Shepherd FA, Der SD, Tsao MS, Penn LZ, Jurisica I (2009) Prognostic gene signatures for non-small-cell lung cancer. Proc Natl Acad Sci U S A 106(8):2824–2828

Bromberg KD, Kluger HM, Delaunay A, Abbas S, DiVito KA, Krajewski S, Ronai Z (2007) Increased expression of the E3 ubiquitin ligase RNF5 is associated with decreased survival in breast cancer. Cancer Res 67(17):8172–8179

Chao YC, Pan SH, Yang SC, Yu SL, Che TF, Lin CW, Tsai MS, Chang GC, Wu CH, Wu YY, Lee YC, Hong TM, Yang PC (2009) Claudin-1 is a metastasis suppressor and correlates with clinical outcome in lung adenocarcinoma. Am J Respir Crit Care Med 179(2):123–133

Che G, Chen J, Liu L, Wang Y, Li L, Qin Y, Zhou Q (2006) Transfection of nm23-H1 increased expression of beta-Catenin, E-Cadherin and TIMP-1 and decreased the expression of MMP-2, CD44v6 and VEGF and inhibited the metastatic potential of human non-small cell lung cancer cell line L9981. Neoplasma 53(6):530–537

Demicheli R, Fornili M, Ambrogi F, Higgins K, Boyd JA, Biganzoli E, Kelsey CR (2012) Recurrence dynamics for non-small-cell lung cancer: effect of surgery on the development of metastases. J Thorac Oncol 7(4):723–730

Eyles J, Puaux AL, Wang X, Toh B, Prakash C, Hong M, Tan TG, Zheng L, Ong LC, Jin Y, Kato M, Prevost-Blondel A, Chow P, Yang H, Abastado JP (2010) Tumor cells disseminate early, but immunosurveillance limits metastatic outgrowth, in a mouse model of melanoma. J Clin Invest 120(6):2030–2039

Fridman WH, Pages F, Sautes-Fridman C, Galon J (2012) The immune contexture in human tumours: impact on clinical outcome. Nat Rev Cancer 12(4):298–306

Gildea JJ, Seraj MJ, Oxford G, Harding MA, Hampton GM, Moskaluk CA, Frierson HF, Conaway MR, Theodorescu D (2002) RhoGDI2 is an invasion and metastasis suppressor gene in human cancer. Cancer Res 62(22):6418–6423

Haeusgen W, Herdegen T, Waetzig V (2011) The bottleneck of JNK signaling: molecular and functional characteristics of MKK4 and MKK7. Eur J Cell Biol 90(6–7):536–544

Holmgren L, O'Reilly MS, Folkman J (1995) Dormancy of micrometastases: balanced proliferation and apoptosis in the presence of angiogenesis suppression. Nat Med 1(2):149–153

Huerta-Yepez S, Yoon NK, Hernandez-Cueto A, Mah V, Rivera-Pazos CM, Chatterjee D, Vega MI, Maresh EL, Horvath S, Chia D, Bonavida B, Goodglick L (2011) Expression of phosphorylated raf kinase inhibitor protein (pRKIP) is a predictor of lung cancer survival. BMC Cancer 11:259

Hurst DR, Welch DR (2011) Metastasis suppressor genes at the interface between the environment and tumor cell growth. Int Rev Cell Mol Biol 286:107–180

Izraely S, Sagi-Assif O, Klein A, Meshel T, Tsarfaty G, Pasmanik-Chor M, Nahmias C, Couraud PO, Ateh E, Bryant JL, Hoon DS, Witz IP (2011) The metastatic microenvironment: brain-residing melanoma metastasis and dormant micrometastasis. Int J Cancer 131:1071

Kaplan RN, Riba RD, Zacharoulis S, Bramley AH, Vincent L, Costa C, MacDonald DD, Jin DK, Shido K, Kerns SA, Zhu Z, Hicklin D, Wu Y, Port JL, Altorki N, Port ER, Ruggero D, Shmelkov SV, Jensen KK, Rafii S, Lyden D (2005) VEGFR1-positive haematopoietic bone marrow progenitors initiate the pre-metastatic niche. Nature 438(7069):820–827

Kienast Y, von Baumgarten L, Fuhrmann M, Klinkert WE, Goldbrunner R, Herms J, Winkler F (2010) Real-time imaging reveals the single steps of brain metastasis formation. Nat Med 16(1):116–122

Liu C, Liu J, Wang X, Mao W, Jiang L, Ni H, Mo M, Wang W (2011) Prognostic impact of nm23-H1 and PCNA expression in pathologic stage I non-small cell lung cancer. J Surg Oncol 104(2):181–186

Maeda R, Yoshida J, Hishida T, Aokage K, Nishimura M, Nishiwaki Y, Nagai K (2010) Late recurrence of non-small cell lung cancer more than 5 years after complete resection: incidence and clinical implications in patient follow-up. Chest 138(1):145–150

Niu H, Li H, Xu C, He P (2010) Expression profile of RhoGDI2 in lung cancers and role of RhoGDI2 in lung cancer metastasis. Oncol Rep 24(2):465–471

O'Flaherty JD, Gray S, Richard D, Fennell D, O'Leary JJ, Blackhall FH, O'Byrne KJ (2012) Circulating tumour cells, their role in metastasis and their clinical utility in lung cancer. Lung Cancer 76(1):19–25

Pirozzi G, Tirino V, Camerlingo R, Franco R, La Rocca A, Liguori E, Martucci N, Paino F, Normanno N, Rocco G (2011) Epithelial to mesenchymal transition by TGFbeta-1 induction increases stemness characteristics in primary non small cell lung cancer cell line. PLoS One 6(6):e21548

Prat A, Parker JS, Karginova O, Fan C, Livasy C, Herschkowitz JI, He X, Perou CM (2010) Phenotypic and molecular characterization of the claudin-low intrinsic subtype of breast cancer. Breast Cancer Res 12(5):R68

Psaila B, Lyden D (2009) The metastatic niche: adapting the foreign soil. Nat Rev Cancer 9(4):285–293

Rocken M (2010) Early tumor dissemination, but late metastasis: insights into tumor dormancy. J Clin Invest 120(6):1800–1803

Smith PW, Liu Y, Siefert SA, Moskaluk CA, Petroni GR, Jones DR (2009) Breast cancer metastasis suppressor 1 (BRMS1) suppresses metastasis and correlates with improved patient survival in non-small cell lung cancer. Cancer Lett 276(2):196–203

Sosa MS, Avivar-Valderas A, Bragado P, Wen HC, Aguirre-Ghiso JA (2011) ERK1/2 and p38alpha/beta signaling in tumor cell quiescence: opportunities to control dormant residual disease. Clin Cancer Res 17(18):5850–5857

Subramanian J, Simon R (2010) Gene expression-based prognostic signatures in lung cancer: ready for clinical use? J Natl Cancer Inst 102(7):464–474

Tang CH, Hill ML, Brumwell AN, Chapman HA, Wei Y (2008) Signaling through urokinase and urokinase receptor in lung cancer cells requires interactions with beta1 integrins. J Cell Sci 121(Pt 22):3747–3756

Tesio M, Trumpp A (2011) Breaking the cell cycle of HSCs by p57 and friends. Cell Stem Cell 9(3):187–192

Vicent S, Luis-Ravelo D, Anton I, Garcia-Tunon I, Borras-Cuesta F, Dotor J, De Las RJ, Lecanda F (2008) A novel lung cancer signature mediates metastatic bone colonization by a dual mechanism. Cancer Res 68(7):2275–2285

Wendt MK, Taylor MA, Schiemann BJ, Schiemann WP (2011) Down-regulation of epithelial cadherin is required to initiate metastatic outgrowth of breast cancer. Mol Biol Cell 22(14):2423–2435

Wikman H, Vessella R, Pantel K (2008) Cancer micrometastasis and tumour dormancy. APMIS 116(7–8):754–770

Woelfle U, Cloos J, Sauter G, Riethdorf L, Janicke F, van Diest P, Brakenhoff R, Pantel K (2003) Molecular signature associated with bone marrow micrometastasis in human breast cancer. Cancer Res 63(18):5679–5684

Wrage M, Ruosaari S, Eijk PP, Kaifi JT, Hollmen J, Yekebas EF, Izbicki JR, Brakenhoff RH, Streichert T, Riethdorf S, Glatzel M, Ylstra B, Pantel K, Wikman H (2009) Genomic profiles associated with early micrometastasis in lung cancer: relevance of 4q deletion. Clin Cancer Res 15(5):1566–1574

Zacharek SJ, Fillmore CM, Lau AN, Gludish DW, Chou A, Ho JW, Zamponi R, Gazit R, Bock C, Jager N, Smith ZD, Kim TM, Saunders AH, Wong J, Lee JH, Roach RR, Rossi DJ, Meissner A, Gimelbrant AA, Park PJ, Kim CF (2011) Lung stem cell self-renewal relies on BMI1-dependent control of expression at imprinted loci. Cell Stem Cell 9(3):272–281

Zheng S, Chang Y, Hodges KB, Sun Y, Ma X, Xue Y, Williamson SR, Lopez-Beltran A, Montironi R, Cheng L (2010) Expression of KISS1 and MMP-9 in non-small cell lung cancer and their relations to metastasis and survival. Anticancer Res 30(3):713–718

Zoller M (2011) CD44: can a cancer-initiating cell profit from an abundantly expressed molecule? Nat Rev Cancer 11(4):254–267

Quiescent CD4⁺ T Cells Inhibit Multiple Stages of HIV Infection

24

Jerome A. Zack and Dimitrios N. Vatakis

Contents

Abstract	253
Introduction	253
CD4⁺ T Cell Quiescence	254
Quiescent T Cells and HIV Infection	255
Quiescent T Cells Exhibit Multiple Blocks of the HIV Life Cycle	256
Entry, Reverse Transcription and Integration	256
Integration and Viral Expression	257
Restriction Factors	258
Discussion	259
References	260

J.A. Zack
Division of Hematology-Oncology,
David Geffen School of Medicine at UCLA,
Los Angeles, CA 90095, USA

UCLA AIDS Institute, Los Angeles, CA, USA

D.N. Vatakis (✉)
David Geffen School of Medicine at UCLA,
Division of Hematology-Oncology,
Department of Medicine, 615 Charles
E. Young Drive South, BSRB 173,
Mailcode: 736322, Los Angeles, CA 90095, USA
e-mail: dvatakis@ucla.edu

Abstract

Elucidating the block of quiescent CD4⁺ T cells to HIV infection has been an intensely debated issue. Early studies suggested that the virus could not infect this T cell subset; latter studies demonstrated that these cells could inefficiently support HIV infection. The kinetics of infection in quiescent cells was delayed and multiple stages of the viral life cycle were marred by inefficiencies. A number of restriction factors as well as cellular protein have been implicated in the potential block. However, to this date the mechanisms of HIV infection in quiescent cells are still unclear. Further understanding will open the way for better therapeutic approaches and much improved gene therapy protocols using HIV-based vectors.

Keywords

CD4⁺ T cell quiescence • Forkhead Box class O (FOXO) factors • HIV life cycle • Lung Krupple-Like factor (LKLF) • PCR technologies • Polypyrimidine tract binding protein (PTB) • Quiescent T cells and HIV infection • Reverse transcription and integration • RNA and DNA synthesis

Introduction

Some confusion still rules the literature regarding the ability of HIV to infect quiescent cells. This largely stems from major differences between the monocyte/macrophage and T cell lineages in terms of their quiescent status. Studies have tried to elucidate the block of quiescent T cells to HIV infection by either examining the viral life cycle in these cells or by identifying key cellular factors that may have an impact. In this review, we will present the progress made in the field thus far and discuss yet unanswered questions.

CD4+ T Cell Quiescence

Human T lymphocytes, both naive and memory, remain at a quiescent state over long periods of time. The majority of T lymphocytes present in blood, the lymph nodes and spleen are in the G_0 state of the cell cycle (Tzachanis et al. 2001; Tzachanis et al. 2004). Typically following antigen stimulation a small fraction of T cells clonally expand at a high rate (Cotner et al. 1983). After stimulation, the majority of the clonally expanded lymphocytes will undergo apoptosis, leaving behind a small fraction of non-dividing memory cells that live over long periods of time and are recruited only upon antigen re-exposure.

T cell quiescence was largely believed to be a default state. Recent studies, however, have demonstrated that T cell quiescence is an actively maintained state regulated by a plethora of transcription factors (Kuo et al. 1997; Buckley et al. 2001; Tzachanis et al. 2001; Haaland et al. 2005; Yusuf et al. 2008). During quiescence, T cells maintain low levels of cell metabolism, decreased RNA synthesis that is limited to basic housekeeping genes, a small cell size and very long periods of survival (Tzachanis et al. 2004). Therefore, T cell quiescence averts cellular damage from metabolism and prevents inappropriate T cell activation and expansion that could lead to lymphomas, autoimmunity, and lymphopenia due to activation induced cell death (Tzachanis et al. 2004). While the mechanisms of T cell quiescence are still under active investigation, a number of transcription factors have been shown to play a major role in this process.

Lung Krupple-Like factor (LKLF) is one of the first factors identified to regulate T cell quiescence. The protein belongs to the Krupple-Like factors (KLFs) family of proteins. LKLF has been shown to regulate and maintain quiescence (Kuo et al. 1997; Buckley et al. 2001; Haaland et al. 2005). Studies disrupting the expression of the transcription factor LKLF resulted in increased T cell activation, proliferation and cell size (Kuo et al. 1997; Buckley et al. 2001). Furthermore, ectopic expression of the protein in cell lines such as Jurkat T cells induced quiescence (Haaland et al. 2005).

Forkhead Box class O (FOXO) factors have been identified as potential players in cell quiescence and cell death, especially FOXO1, 3 and 4 (Kops et al. 2002; Ouyang et al. 2009; Barnes et al. 2010; Aksoylar et al. 2011). These factors are shown to be active in resting cells and control T cell homeostasis and tolerance, a phenotype reversed by IL-2 mediated activation (Kops et al. 2002). FOXO1 knockout mice demonstrated normal thymopoiesis, elevated numbers of activated T cells and decreased levels of naïve T cells (Ouyang et al. 2009). Furthermore, IL-7 receptor expression was completely abrogated suggesting that this receptor may be a target of FOXO1 (Ouyang et al. 2009). Recent studies have linked Gimap5 to FOXO1, 3, and 4. Gimap5 is a member of the GTPase immunity associated family of proteins (Barnes et al. 2010; Aksoylar et al. 2011). These proteins have been shown to be involved in lymphocyte survival, development, selection and homeostasis. In Gimap5 defective mouse models researchers observed severe intestinal inflammation and lack of tolerance (Barnes et al. 2010; Aksoylar et al. 2011). Gimap5 regulates FOXO 3 and 4 at the protein level by inhibiting their degradation (Aksoylar et al. 2011).

Tob, another nuclear protein, has been recently shown to be involved in cellular quiescence (Tzachanis et al. 2001; Tzachanis and Boussiotis 2009). It belongs to a family of antiproliferative proteins known as APRO (Tzachanis et al. 2001). The protein is expressed in anergic and naïve T cells, is downregulated upon T cell activation and enhances TGF-β signaling pathways, thus, promoting cell quiescence (Tzachanis et al. 2001).

Knockdown of Tob results to increased activation only following only binding to TcR thereby suggesting a role in regulating thresholds of T cell activation (Tzachanis et al. 2001). This was further supported by data showing that ectopic expression of the protein resulted in inhibition of CD3-CD28 mediated proliferation and loss expression of cytokines such as IL-2, IL-4 and IFN-γ (Tzachanis et al. 2001).

Recent studies have implicated Runx1 and Tsc1 as potential regulators of T cell quiescence. Runx1 has been shown to impact various stages of T cell differentiation. However, recently it was shown to be expressed at high levels in naïve T cells but it is downregulated following T cell activation (Wong et al. 2012). Knockdown of Runx1 resulted in severe immune hyperactivation, increased levels of IL-17 and IL-21 production causing fatal autoimmunity (Wong et al. 2012). While the mechanism of action is not yet clear, it is believed that the regulation of pro-inflammatory cytokine production such as IL-17 and IL-21 may be implicated (Wong et al. 2012). Similarly loss of Tsc1 expression results in increased activation, cellularity and eventual cell death. Studies have suggested that Tsc1 may be controlling T cell quiescence by regulating mTORC1, a major protein involved in cell growth, metabolism (Wu et al. 2011; Yang et al. 2011).

In this chapter, we will discuss the potential HIV restriction factors present in quiescent CD4 T cells. Quiescent T cell infection by HIV has been an interesting and controversial subject that has generated a number of high profile studies in the field. While HIV infection does not require mitosis (Weinberg et al. 1991), HIV cannot efficiently infect G_0 cells, as we will describe in the sections to follow. The blocks identified in quiescent cells only partly tell the story of their resistance to HIV. Interestingly, none of the aforementioned transcription factors regulating quiescence are implicated. Understanding these blocks will allow us to develop more effective ways to treat HIV and utilize lentiviral vectors in this cell type with minimal perturbation. The latter can have major implication to lentiviral vector based gene therapy protocols.

Quiescent T Cells and HIV Infection

The infection of quiescent CD4$^+$ T cells by the human immunodeficiency virus (HIV) has been a subject of intense debate. Unlike other retroviruses, HIV replication is not dependent on cell division and is characterized by its ability to infect non-dividing cells and establish a latent infection (Weinberg et al. 1991). Early reports suggested that HIV was able to bind to quiescent T cells but failed to infect them unless they were previously activated (Zagury et al. 1986; Gowda et al. 1989).

Advances in PCR technology demonstrated that quiescent T cells are infectable by HIV (Stevenson et al. 1990; Zack et al. 1990; Spina et al. 1995). However, studies differed on the level as well as the efficiency of infection. Zack et.al demonstrated that the virus could enter quiescent T cells and initiate reverse transcription (Zack et al. 1990). However, the process was not completed resulting in the generation of labile viral cDNA intermediates that degraded over time (Zack et al. 1990). On the other hand, others demonstrated that in HIV infected quiescent T cells there was completion of reverse transcription that resulted in accumulation of viral cDNA in the cytoplasm over prolonged periods of time (Stevenson et al. 1990; Spina et al. 1995). Under this setting, infection was rescued by activation of the infected cells suggesting that there was a defect in the nuclear transport of viral cDNA in quiescent cells (Stevenson et al. 1990; Spina et al. 1995). Further studies by the Vitteta group, focused on the CD25$^+$ and CD25$^-$ T cell populations and their ability to be infected by the virus thereby contrasting activated versus non-activated T cells (Ramilo et al. 1993; Borvak et al. 1995; Chou et al. 1997). In a series of studies, they showed that the CD25$^-$ T cells, representing non-activated T cells, were not infectable by HIV while the CD25$^+$ T cells were able to support infection in the absence of any stimulation. However, when total human peripheral blood mononuclear cells were infected, the non-activated cells had copies of viral DNA. This was the first evidence of cell-to-cell mediated infection of quiescent cells.

Based on these early studies, it was evident that the life cycle of HIV in quiescent CD4 T cells was quite distinct from that of activated T cells and warranted further investigation. Subsequent studies using a flow cytometry based cell cycle progression assay that assessed the levels of both cellular RNA and DNA synthesis were able to distinguish non-dividing T cells into two categories: (i) cells in the G_0/G_{1a} phase which is characterized by undetectable levels of DNA and RNA synthesis and (ii) the G_{1b} phase which is characterized by high levels of RNA expression in the absence of DNA synthesis (Korin and Zack 1998). Cells in the G_{1b} stage of the cell cycle, while non-dividing and thus resting, were permissive to infection. On the other hand, the $G_{0/1a}$ cells were deemed as truly quiescent and they were resistant to HIV infection (Korin and Zack 1998). Therefore, the above data provided a foundation for the discrepancies seen among the different groups in terms of the permissiveness of quiescent T cells to HIV infection. Furthermore, it underscored the importance of distinguishing truly quiescent cells from resting but activated T cells. However, it was not yet clear what was behind the block presented to HIV by quiescent cells and what stage of the viral life cycle was impacted.

Quiescent T Cells Exhibit Multiple Blocks of the HIV Life Cycle

Entry, Reverse Transcription and Integration

Studies assessing the efficiency of HIV entry in quiescent T cells have been limited, especially in comparison to activated cells. However, based on the early work on quiescent cells, it was clear that the virus could enter quiescent CD4 T cells efficiently. Our group did compare HIV entry between quiescent and activated cells and found no significant differences between the two groups (Vatakis et al. 2007). Similarly, Agosto et.al demonstrated that the use of CXCR4-expressing envelope was more efficient than VSV-g in transducing quiescent cells (Agosto et al. 2009).

As entry seemed unaffected in quiescent T cells, the next stages of the viral cycle impacted are viral uncoating and reverse transcription. The study of viral uncoating is quite challenging in primary cells such as quiescent T cells. This is further complicated by the fact that quiescent T cells are very small in size with very limited cytoplasm. The majority of studies looking at this stage of the viral life cycle have been carried out in cell lines. However, one study using lysates from activated and quiescent cells showed that in contrast to that of quiescent cells lysates from activated cells resulted in efficient uncoating of HIV virions. While the authors were able to fractionate potential factors involved in uncoating they did not further identify them (Auewarakul et al. 2005).

The development of more sensitive PCR technologies as well as cell purification protocols allowed for a better characterization of HIV reverse transcription and integration in quiescent T cells. A series of elegant studies by the Siliciano group shed more light on the infection of quiescent T cells by HIV. Utilizing a linker-mediated PCR assay, they measured the rate of reverse transcription and degradation of the non-integrated linear viral DNA (Pierson et al. 2002). They showed that reverse transcription in quiescent T cells was completed in 2–3 days, whereas in activated cells it was finished within 4–6 h. The linear piece of DNA had a half-life of about 1 day. The slow rate of reverse transcription and the labile nature of the newly synthesized cDNA severely compromised the ability of the virus to establish a productive infection. These observations were supported by a follow up study showing that the linear non-integrated viral DNA was integration competent (Zhou et al. 2005). These studies pointed to a potential block at the early stages of HIV infection.

Furthermore, Swiggard et.al showed that while reverse transcription was decreased in quiescent T cells, full length HIV cDNA accumulated over time, was stable for approximately 3 days, and partial viral reverse transcripts were degraded (Swiggard et al. 2004; Swiggard et al. 2005). However, the use of an alternative method of infection, spinoculation, raised the possibility of partial cell activation, which may have improved the stability of the full-length viral cDNA.

We took a more detailed look at the kinetics of HIV infection in quiescent T cells and compared this against that of stimulated cells. In our studies, not only did we see delays in reverse transcription but also significantly lower levels of reverse transcription in quiescent T cells (Vatakis et al. 2007). More specifically, initiation of reverse transcription was 30-fold lower in quiescent T cells. There was some completion of reverse transcription which was delayed by 16 h compared to that seen in activated cells. Interestingly, this inefficient infection process of quiescent T cells was not rescued with immediate stimulation (Vatakis et al. 2007). Thus, all the studies point to a strong early block to infection.

Integration and Viral Expression

The inefficiencies seen in reverse transcription did impact downstream events of the HIV viral life cycle. The development of a sensitive and quantitative assay allowed detection of low levels of integration in HIV infected cells, which proved very useful in the studies outlined below. Using this assay and spinoculation as the infection method, the O'Doherty group detected integrated virus in quiescent cells (Swiggard et al. 2005). Furthermore, viral expression was induced following stimulation of infected quiescent cells with IL-7 or anti-CD3/anti-CD28 co-stimulation. The results from this study demonstrated that a latent infection could be established in quiescent CD4 T cells. However, these studies did not reveal any deficiencies that may arise following reverse transcription, such as nuclear import of viral cDNA.

In our studies, the delays seen in quiescent T cells impacted integration as this process was delayed by 24 h in quiescent CD4 T cells. However, the efficiency at which the viral cDNA was integrated was two to threefold less than that seen in activated cells. However, this difference was within the limits of standard error and thus not significant. More interestingly, quiescent T cells expressed 48 h post-infection multiply spliced viral RNA at significantly lower levels than stimulated cells. Despite the expression of viral mRNA, there was no detectable Gag protein expression. Furthermore, the levels of integrated HIV copies remained unchanged for 5 days suggesting that these cells did not die after infection but were latently non-productively infected (Vatakis et al. 2007). Therefore, these data suggested that once reverse transcription took place, the viral cDNA was efficiently transported to the nucleus and integrated. Furthermore the provirus expressed low levels of viral mRNA, but the lack of viral protein suggested a potential post-transcriptional block.

Furthermore, the above studies raised further questions regarding HIV integration site selection and viral expression. Integrated virus was found in resting cells of HIV infected individuals but this was attributed to infection of previously activated T cells that return to quiescence (Han et al. 2004). There was no indication that these cells were infected while quiescent. Furthermore, the presence of viral mRNA but the lack of detectable viral protein in quiescent T cells was quite intriguing (Vatakis et al. 2007). This raised the question of whether integration in quiescent T cells is distinct from that seen in their stimulated counterparts. As HIV preferentially integrates into actively transcribing genes and T cell quiescence is an actively maintained state, it could be inferred that a distinct distribution of integration sites could explain our observations. Others and we examined the distribution of integration sites in quiescent CD4 T cells (Brady et al. 2009; Vatakis et al. 2009). Based on our data, integration in both activated and quiescent CD4 T cells occurred mainly in transcriptionally active genes that were not affected by the T cell activation state such as housekeeping genes (Vatakis et al. 2009). However, a screening of the LTR ends in both the provirus and the 2-LTR circles revealed some interesting patterns. In quiescent CD4 T cells we observed elevated numbers of abnormal LTR-host DNA junctions. Furthermore, the levels of 2-LTR circles with abnormal LTR junctions was also higher in quiescent cells; the abnormalities were primarily extensive deletions (Vatakis et al. 2009). It is suggested that the delays in reverse transcription had a detrimental effect on the ends of the viral cDNA. Similar patterns of integration were seen in studies by Brady et.al. However, in their studies, HIV integration patterns were somewhat different between stimulated and quiescent T cells (Brady et al. 2009). HIV appeared to integrate in less transcriptionally active regions in quiescent cells when compared to stimulated cells, but the observed differences were modest.

To this date, there have been no studies showing viral release from HIV infected quiescent CD4 T cells in the absence of any stimulation. This can be proven quite important since in densely packed lymphoid tissues very low level viral release can be sufficient to support ongoing replication. Recent work in the SIV rhesus macaque model suggested that resting cells can spontaneously release virions (Nishimura et al. 2009). However, it is not clear if these resting cells are at the G_{1b} stage of the cell cycle. If they are, it is expected that they will make virus as they are transcriptionally active. Additional studies have shown that the lack of the polypyrimidine tract binding protein (PTB) in resting cells results in nuclear retention of multiply spliced viral RNA limiting the production of virions (Lassen et al. 2006). However, it is unclear what the mechanism of action is and, thus, if this is an added block to infection.

Restriction Factors

As stated earlier, quiescent cells are characterized by low metabolic rates and low levels of cellular RNA transcription. HIV, like any other virus, relies on cellular resources to replicate efficiently. Thus, it is very reasonable to infer that the lack of cellular substrates or raw materials can have a negative impact on viral replication. One such limited resource is the total nucleotide pool as it is important for viral reverse transcription. Pretreatment of quiescent T cells with exogenous nucleosides improved reverse transcription, increased the levels of integrated provirus and 2-LTR circles but still failed to rescue infection (Korin and Zack 1999; Plesa et al. 2007). Furthermore, the low metabolic rates of quiescent T cells would result in decreased amounts of available ATP. As the processes of reverse transcription and integration are energy dependent, it is expected that this could limit infection. However, there have not been studies looking into the impact of ATP levels on the efficiency of HIV infection. This suggested that the presence of inhibitory factors or the absence of supportive processes were responsible for this phenotype.

siRNA technology has enabled researchers to identify the role of a number of genes using siRNA-mediated knockdown. This has also been used to study the effect of cellular factors on HIV replication. Ganesh et.al used this technology to identify a potential HIV restriction factor in quiescent T cells (Ganesh et al. 2003). Murr1 is a protein involved in copper regulation and inhibition of NFκB activity through inhibition of IκB degradation by the proteasome (Ganesh et al. 2003). When knocked down in cell lines, it decreased levels of IκB-α and enhanced NFκB activity. Furthermore, Murr1 is highly expressed in T cells. When the authors knocked down Murr1 using siRNA, they observed enhanced infection of quiescent T cells demonstrated by increased Gag expression. This suggested Murr1 may regulate HIV infection in quiescent CD4 T cells (Ganesh et al. 2003). The authors used nucleofection as means to introduce the siRNA. While the expression of CD25, CD69 and HLA-DR T cell activation markers was not observed, the process of nucleofection may have had an impact on the infection. However, no additional studies were performed to further elucidate the role of this protein.

Apolipoprotein B mRNA editing catalytic peptide like 3G (APOBEC3G) has been shown to have strong anti-viral activity and is classified as an innate antiviral factor. APOBEC3G (A3G) is a cellular cytidine deaminase expressed in T cells and was initially found to be a potent antiviral factor against *vif* deficient HIV-1 (Sheehy et al. 2002; Mangeat et al. 2003; Zhang et al. 2003). Studies showed that in the absence of APOBEC3G causes severe hypermutation of the viral genome resulting resulting in non-productive infection.

Vif sequesters APOBEC3G mediating its degradation by the proteasome (Sheehy et al. 2002). Additional studies showed that A3G is found in a catalytically active low molecular mass form in quiescent T cells and in an inactive high molecular mass form in activated T cells, thus impacting permissiveness to infection between cellular activation states (Chiu et al. 2005).

A3G knockdown in quiescent T cells initially indicated that the restriction to infection in quiescent T cells is abrogated in the absence of A3G. By nucleofecting an siRNA against A3G into quiescent T-cells, Chiu et al. demonstrated that these cells could be infected once A3G is eliminated even though there was no apparent indication of cellular activation (Chiu et al. 2005). However, two groups have independently published results that would indicate otherwise. Using identical techniques with the same, and two additional, siRNAs against A3G, Kamata et al. could not reproduce the results of the initial experiments (Kamata et al. 2009). Further, Santoni et al. knocked down A3G using both stable shRNA and ectopic *vif* expression in activated T-cells then allowed them to return to a resting state before infecting them with HIV-1. They found no difference in infection between cells with, and those without, A3G, nor did they find a correlation between viral restriction and high vs. low molecular mass A3G complexes (Santoni de Sio and Trono 2009). Thus, while A3G remains a major antiviral factor, its impact on quiescent T cell resistance to HIV does not seem to be major.

Manganaro and colleagues shifted the focus on the lack of a cellular protein as a factor for the block to productive infection in quiescent T cells. In recent studies they demonstrated that c-Jun N-terminal kinase (JNK) phosphorylates viral integrase, which in turn interacts with the peptidylprolyl-isomerase enzyme Pin1 causing a conformational change in integrase (Manganaro et al. 2010). This results in increased integrase stability and completion of viral integration. Since quiescent T cells do not express JNK, they argued that the process is not efficient in quiescent cells (Manganaro et al. 2010). These findings do provide support for the earlier studies that suggested the presence of a preintegrated viral DNA in resting cells that acts as an inducible reservoir (Stevenson et al. 1990; Spina et al. 1995). However, they do not explain the recent findings by our group as well as others that show integration occurring in quiescent T cells and that the defects seen in these cells are prior to that event in the viral life cycle (Swiggard et al. 2004; Vatakis et al. 2007; Brady et al. 2009; Vatakis et al. 2009). In addition, recent work by the Chow group (Briones et al. 2010), has implicated integrase HIV viral core stability. It is possible that JNK and Pin1 may be improving core stability, thus positively impacting reverse transcription. If this is the case, then JNK and Pin1 will have an impact on the early rather than later events of HIV life cycle.

Finally, Glut1 has recently been added to the list of potential cellular factors that may facilitate HIV infection. Like JNK and Pin1, the absence of this cellular factor seems to impact HIV infection (Loisel-Meyer et al. 2012). More specifically, Glut1 is a major glucose transporter in T cells both mature T cells and thymocytes (Loisel-Meyer et al. 2012). The expression of the protein is upregulated following exposure to IL-7 and conventional T cell activation. When knocked out, HIV infection was abrogated in activated T cells (Loisel-Meyer et al. 2012). Furthermore, its expression was linked to increased permissiveness to HIV, since double positive thymocytes expressing high levels of Glut1 were more likely to be infected by HIV than their low Glut1 expressing counterparts. This is the first study linking T cell metabolism to productive HIV infection.

In summary, while it is accepted that there is a plethora of cellular factors that can enhance or restrict HIV replication in quiescent CD4 T cells, it is becoming more evident that the concerted action of multiple events is responsible for the lack of productive infection seen in quiescent CD4 T cells.

Discussion

The permissiveness of quiescent CD4 T cells to HIV infection has been quite controversial. However, the development of more quantitative and sensitive techniques has provided us with some concrete answers as to the nature of HIV infection in this cell type. It is now clear that quiescent T cells can become infected very inefficiently by HIV. It seems that the blocks present are focused on the early stages of infection. However, blocks in post-transcriptional events may also be implicated. While there has been major progress, it is still unclear how quiescent T cells block infection. A number of restriction factors have been presented as a possible explanation but their role is at best incremental. Similarly, the lack of key cellular factors and/or raw materials while important does not seem to tell the whole story. The major hurdle for the virus seems to be localized at the early stages of the life cycle, uncoating and reverse transcription. Poor uncoating in quiescent cells can prove detrimental to infection but studies thus far are very limited due to the inherent difficulties studying the viral capsid (Auewarakul et al. 2005). Moreover, quiescent T cell physiology is quite intriguing. They are characterized by very small size, a small cytoplasm and a very large nucleus. Thus, a more careful examination of quiescent cell physiology may yield important answers.

At the post-integration level, quiescent cells still pose a very important enigma. Are they capable of making virus or is there another block? The defect we saw in LTR ends of infected quiescent cells were mainly the result of inefficiencies in reverse transcription. Epigenetic studies on these cells are imperative as they can shed some light on the state of the provirus. In addition, since we have detected viral RNA, the potential of posttranscriptional events such as the lack of PTB may pose a late stage block to infection.

In summary, identifying the blocks in HIV infection of quiescent T cells remains quite important as increased knowledge can have major benefits in the area of study; one is the development of improved means to block infection especially at the early stages of the viral life cycle. Second, as HIV-based vectors are becoming more prevalent in gene therapy protocols, understanding fully the infection processes will allow for the development of more effective gene therapy vectors. Finally, HIV infected quiescent T cells can shed some more light as to the potential of our immune system to detect and eliminate non-activated cells as targets. Are dormant cells whether viral infected or cancerous effective at evading our immune system?

Acknowledgements This work was supported by in part by NIH/NIAID AI 070010-06A1, NIH Martin Delaney Collaboratory(to J.A.Z.), UCLA Center for AIDS Research NIH/National Institute of Allergy and Infectious Diseases Grant AI028697, NIH/NIDA R21 DA031036-01A1 (D.N.V.).

References

Agosto LM, Yu JJ, Liszewski MK, Baytop C, Korokhov N, Humeau LM, O'Doherty U (2009) The CXCR4-tropic human immunodeficiency virus envelope promotes more-efficient gene delivery to resting CD4+ T cells than the vesicular stomatitis virus glycoprotein G envelope. J Virol 83(16):8153–8162

Aksoylar HI, Lampe K, Barnes MJ, Plas DR, Hoebe K (2011) Loss of immunological tolerance in Gimap5-deficient mice is associated with loss of foxo in CD4+ T cells. J Immunol 188(1):146–154

Auewarakul P, Wacharapornin P, Srichatrapimuk S, Chutipongtanate S, Puthavathana P (2005) Uncoating of HIV-1 requires cellular activation. Virology 337(1): 93–101

Barnes MJ, Aksoylar H, Krebs P, Bourdeau T, Arnold CN, Xia Y, Khovananth K, Engel I, Sovath S, Lampe K, Laws E, Saunders A, Butcher GW, Kronenberg M, Steinbrecher K, Hildeman D, Grimes HL, Beutler B, Hoebe K (2010) Loss of T cell and B cell quiescence precedes the onset of microbial flora-dependent wasting disease and intestinal inflammation in Gimap5-deficient mice. J Immunol 184(7):3743–3754

Borvak J, Chou CS, Bell K, Van Dyke G, Zola H, Ramilo O, Vitetta ES (1995) Expression of CD25 defines peripheral blood mononuclear cells with productive versus latent HIV infection. J Immunol 155(6):3196–3204

Brady T, Agosto LM, Malani N, Berry CC, O'Doherty U, Bushman F (2009) HIV integration site distributions in resting and activated CD4+ T cells infected in culture. AIDS 23(12):1461–1471

Briones MS, Dobard CW, Chow SA (2010) Role of human immunodeficiency virus type 1 integrase in uncoating of the viral core. J Virol 84(10):5181–5190

Buckley AF, Kuo CT, Leiden JM (2001) Transcription factor LKLF is sufficient to program T cell quiescence via a c-Myc–dependent pathway. Nat Immunol 2(8): 698–704

Chiu YL, Soros VB, Kreisberg JF, Stopak K, Yonemoto W, Greene WC (2005) Cellular APOBEC3G restricts HIV-1 infection in resting CD4+ T cells. Nature 435(7038):108–114

Chou CS, Ramilo O, Vitetta ES (1997) Highly purified CD25- resting T cells cannot be infected de novo with HIV-1. Proc Natl Acad Sci U S A 94(4):1361–1365

Cotner T, Williams JM, Christenson L, Shapiro HM, Strom TB, Strominger J (1983) Simultaneous flow cytometric analysis of human T cell activation antigen expression and DNA content. J Exp Med 157(2): 461–472

Ganesh L, Burstein E, Guha-Niyogi A, Louder MK, Mascola JR, Klomp LW, Wijmenga C, Duckett CS, Nabel GJ (2003) The gene product Murr1 restricts HIV-1 replication in resting CD4+ lymphocytes. Nature 426(6968):853–857

Gowda SD, Stein BS, Mohagheghpour N, Benike CJ, Engleman EG (1989) Evidence that T cell activation is required for HIV-1 entry in CD4+ lymphocytes. J Immunol 142(3):773–780

Haaland RE, Yu W, Rice AP (2005) Identification of LKLF-regulated genes in quiescent CD4+ T lymphocytes. Mol Immunol 42(5):627–641

Han Y, Lassen K, Monie D, Sedaghat AR, Shimoji S, Liu X, Pierson TC, Margolick JB, Siliciano RF, Siliciano JD (2004) Resting CD4+ T cells from human immunodeficiency virus type 1 (HIV-1)-infected individuals carry integrated HIV-1 genomes within actively transcribed host genes. J Virol 78(12): 6122–6133

Kamata M, Nagaoka Y, Chen IS (2009) Reassessing the role of APOBEC3G in human immunodeficiency virus type 1 infection of quiescent CD4+ T-cells. PLoS Pathog 5(3):e1000342

Kops GJ, Dansen TB, Polderman PE, Saarloos I, Wirtz KW, Coffer PJ, Huang TT, Bos JL, Medema RH, Burgering BM (2002) Forkhead transcription factor FOXO3a protects quiescent cells from oxidative stress. Nature 419(6904):316–321

Korin YD, Zack JA (1998) Progression to the G1b phase of the cell cycle is required for completion of human immunodeficiency virus type 1 reverse transcription in T cells. J Virol 72(4):3161–3168

Korin YD, Zack JA (1999) Nonproductive human immunodeficiency virus type 1 infection in nucleoside-treated G0 lymphocytes. J Virol 73(8):6526–6532

Kuo CT, Veselits ML, Leiden JM (1997) LKLF: a transcriptional regulator of single-positive T cell quiescence and survival. Science 277(5334):1986–1990

Lassen KG, Ramyar KX, Bailey JR, Zhou Y, Siliciano RF (2006) Nuclear retention of multiply spliced HIV-1 RNA in resting CD4+ T cells. PLoS Pathog 2(7):e68

Loisel-Meyer S, Swainson L, Craveiro M, Oburoglu L, Mongellaz C, Costa C, Martinez M, Cosset FL, Battini JL, Herzenberg LA, Atkuri KR, Sitbon M, Kinet S, Verhoeyen E, Taylor N (2012) Glut1-mediated glucose transport regulates HIV infection. Proc Natl Acad Sci U S A 109(7):2549–2554

Manganaro L, Lusic M, Gutierrez MI, Cereseto A, Del Sal G, Giacca M (2010) Concerted action of cellular JNK and Pin1 restricts HIV-1 genome integration to activated CD4+ T lymphocytes. Nat Med 16(3): 329–333

Mangeat B, Turelli P, Caron G, Friedli M, Perrin L, Trono D (2003) Broad antiretroviral defence by human APOBEC3G through lethal editing of nascent reverse transcripts. Nature 424(6944):99–103

Nishimura Y, Sadjadpour R, Mattapallil JJ, Igarashi T, Lee W, Buckler-White A, Roederer M, Chun TW, Martin MA (2009) High frequencies of resting CD4+ T cells containing integrated viral DNA are found in rhesus macaques during acute lentivirus infections. Proc Natl Acad Sci U S A 106(19):8015–8020

Ouyang W, Beckett O, Flavell RA, Li MO (2009) An essential role of the Forkhead-box transcription factor Foxo1 in control of T cell homeostasis and tolerance. Immunity 30(3):358–371

Pierson TC, Zhou Y, Kieffer TL, Ruff CT, Buck C, Siliciano RF (2002) Molecular characterization of pre-integration latency in human immunodeficiency virus type 1 infection. J Virol 76(17):8518–8531

Plesa G, Dai J, Baytop C, Riley JL, June CH, O'Doherty U (2007) Addition of deoxynucleosides enhances human immunodeficiency virus type 1 integration and 2LTR formation in resting CD4+ T cells. J Virol 81(24):13938–13942

Ramilo O, Bell KD, Uhr JW, Vitetta ES (1993) Role of CD25+ and CD25- T cells in acute HIV infection in vitro. J Immunol 150(11):5202–5208

Santoni de Sio FR, Trono D (2009) APOBEC3G-depleted resting CD4+ T cells remain refractory to HIV1 infection. PLoS ONE 4(8):e6571

Sheehy AM, Gaddis NC, Choi JD, Malim MH (2002) Isolation of a human gene that inhibits HIV-1 infection and is suppressed by the viral Vif protein. Nature 418(6898):646–650

Spina CA, Guatelli JC, Richman DD (1995) Establishment of a stable, inducible form of human immunodeficiency virus type 1 DNA in quiescent CD4 lymphocytes in vitro. J Virol 69(5):2977–2988

Stevenson M, Stanwick TL, Dempsey MP, Lamonica CA (1990) HIV-1 replication is controlled at the level of T cell activation and proviral integration. Embo J 9(5): 1551–1560

Swiggard WJ, O'Doherty U, McGain D, Jeyakumar D, Malim MH (2004) Long HIV type 1 reverse transcripts can accumulate stably within resting CD4+ T cells while short ones are degraded. AIDS Res Hum Retroviruses 20(3):285–295

Swiggard WJ, Baytop C, Yu JJ, Dai J, Li C, Schretzenmair R, Theodosopoulos T, O'Doherty U (2005) Human immunodeficiency virus type 1 can establish latent infection in resting CD4+ T cells in the absence of activating stimuli. J Virol 79(22): 14179–14188

Tzachanis D, Boussiotis VA (2009) Tob, a member of the APRO family, regulates immunological quiescence and tumor suppression. Cell Cycle 8(7): 1019–1025

Tzachanis D, Freeman GJ, Hirano N, van Puijenbroek AA, Delfs MW, Berezovskaya A, Nadler LM, Boussiotis VA (2001) Tob is a negative regulator of activation that is expressed in anergic and quiescent T cells. Nat Immunol 2(12):1174–1182

Tzachanis D, Lafuente EM, Li L, Boussiotis VA (2004) Intrinsic and extrinsic regulation of T lymphocyte quiescence. Leuk Lymphoma 45(10):1959–1967

Vatakis DN, Bristol G, Wilkinson TA, Chow SA, Zack JA (2007) Immediate activation fails to rescue efficient human immunodeficiency virus replication in quiescent CD4+ T cells. J Virol 81(7):3574–3582

Vatakis DN, Kim S, Kim N, Chow SA, Zack JA (2009) HIV integration efficiency and site selection in quiescent CD4+ T cells. J Virol. doi:10.1128/JVI.00356-09

Weinberg JB, Matthews TJ, Cullen BR, Malim MH (1991) Productive human immunodeficiency virus type 1 (HIV-1) infection of nonproliferating human monocytes. J Exp Med 174(6):1477–1482

Wong WF, Kohu K, Nakamura A, Ebina M, Kikuchi T, Tazawa R, Tanaka K, Kon S, Funaki T, Sugahara-Tobinai A, Looi CY, Endo S, Funayama R, Kurokawa M, Habu S, Ishii N, Fukumoto M, Nakata K, Takai T, Satake M (2012) Runx1 deficiency in CD4+ T cells causes fatal autoimmune inflammatory lung disease due to spontaneous hyperactivation of cells. J Immunol 188(11):5408–5420

Wu Q, Liu Y, Chen C, Ikenoue T, Qiao Y, Li CS, Li W, Guan KL, Zheng P (2011) The tuberous sclerosis complex-mammalian target of rapamycin pathway maintains the quiescence and survival of naive T cells. J Immunol 187(3):1106–1112

Yang K, Neale G, Green DR, He W, Chi H (2011) The tumor suppressor Tsc1 enforces quiescence of naive T cells to promote immune homeostasis and function. Nat Immunol 12(9):888–897

Yusuf I, Kharas MG, Chen J, Peralta RQ, Maruniak A, Sareen P, Yang VW, Kaestner KH, Fruman DA (2008) KLF4 is a FOXO target gene that suppresses B cell proliferation. Int Immunol 20(5):671–681

Zack JA, Arrigo SJ, Weitsman SR, Go AS, Haislip A, Chen IS (1990) HIV-1 entry into quiescent primary lymphocytes: molecular analysis reveals a labile, latent viral structure. Cell 61(2):213–222

Zagury D, Bernard J, Leonard R, Cheynier R, Feldman M, Sarin PS, Gallo RC (1986) Long-term cultures of HTLV-III–infected T cells: a model of cytopathology of T-cell depletion in AIDS. Science 231(4740):850–853

Zhang H, Yang B, Pomerantz RJ, Zhang C, Arunachalam SC, Gao L (2003) The cytidine deaminase CEM15 induces hypermutation in newly synthesized HIV-1 DNA. Nature 424(6944):94–98

Zhou Y, Zhang H, Siliciano JD, Siliciano RF (2005) Kinetics of human immunodeficiency virus type 1 decay following entry into resting CD4+ T cells. J Virol 79(4):2199–2210

Part III
Stem Cells and Cancer Stem Cells

Senescent-Derived Pluripotent Stem Cells Are Able to Redifferentiate into Fully Rejuvenated Cells

25

Ollivier Milhavet and Jean-Marc Lemaitre

Contents

Abstract	265
Introduction	266
Cellular Senescence in Cell Reprogramming	266
Reprogramming Senescent Cells	269
Generation of iPSCs from Senescent Cells	269
Rejuvenation of Cells from Senescent Derived iPSCs	271
Discussion	272
Adding LIN28 and NANOG to the Yamanaka's Four Genes Cocktail, a Key to Reprogram Senescent Cells	273
References	275

O. Milhavet (✉) • J.-M. Lemaitre (✉)
Laboratory of Plasticity of the Genome and Aging,
Institute of Functional Genomics, 141 Rue de la
Cardonille, 34094 Montpellier Cedex 05, France
e-mail: ollivier.milhavet@igf.cnrs.fr;
jean-marc.lemaitre@igf.cnrs.fr

Abstract

Direct reprogramming of human somatic cells into induced pluripotent stem cells (iPSCs) provides a unique opportunity to derive patient-specific stem cells with potential application in cell therapies and without the ethical problems concerning the use of human embryonic stem cells (hESCs). However, cellular senescence, which contributes to aging and restricted longevity, has for long been described as a barrier to the derivation of iPSCs. However, recently, we demonstrated, using a specific protocol, that cellular senescence is not a limit to cell reprogramming and that age-related cellular physiology could be reversed. We showed that iPSCs generated from senescent and centenarian cells are indistinguishable from hESCs and have been fully rejuvenated as shown by their ability to re-differentiate into young cells.

In this chapter, we will present the initial results that described senescence as a barrier to cell reprogramming before detailing the experiments that allowed us to derive iPSCs from senescent cells and show that these cells were reverted to a younger state. We will finally discuss the molecular mechanistic possibly involved in cell reprogramming of aged cells and propose a model that could provide new insights into iPSCs technology and pave the way for regenerative medicine for aged patients.

Keywords

Cell reprogramming • Embryonic stem cells (ESCs) • Generation of iPSCs • NANOG and LIN28 • OCT4 • SOX2 • KLF4 and c-MYC • Oxidative stress and mitochondrial dysfunction • Short hairpin RNA (shRNA) • Telomerase • Undifferentiated embryonic cell transcription factor 1(UTF1) • Vitamin C (VitC)

Introduction

The discovery of iPSCs by Yamanaka's group and the rapid progress that have followed have opened up a new era in autologous regenerative medicine whereby patient-specific pluripotent cells could potentially be derived from adult somatic cells. In many laboratories, iPSCs have been reproducibly obtained from different cell types by forced expression of the four transcription factors OCT4, SOX2, KLF4 and c-MYC (Takahashi et al. 2007), or by an alternative combination, substituting KLF4 and c-MYC by NANOG and LIN28 (Yu et al. 2007). However, cell reprogramming still suffers from several hurdles that have to be overcome before clinical applications (Hanna et al. 2010). Among them, cellular senescence that increases in elderly was thought to be a critical issue. Senescence is linked to physiological aging, and is characterized by an irreversible cell cycle arrest in response to various forms of stress, including activation of oncogenes, shortened telomeres (replicative senescence), DNA damage, oxidative stress and mitochondrial dysfunction (Campisi and d'Adda di Fagagna 2007; Collado et al. 2007). A common feature is the activation of the $p53/p21^{CIP1}$ and $p16^{INK4A}/pRb$ pathways that play key roles both in the initiation of the senescence program and in its maintenance, alteration of cell morphology and metabolism, epigenetic changes, increase in senescence-associated β-galactosidase activity and formation of senescence-associated heterochromatic foci, so called SAHF (Narita et al. 2003; Zhang et al. 2007).

After the original discovery by Yamanaka's group, reprogramming into iPSCs became a major area of research. However, efficiency of this strategy and sometimes lack of quality, in terms of genomic integrity and similarity to bona fide embryonic stem cells (ESCs), appeared rapidly as two major issues that could potentially limit translating iPSC-based reprogramming into clinical application. Notably, only a small proportion of cells become pluripotent cells (less than 1%), suggesting the possible need of additional factors to increase the efficiency of reprogramming and increase the stability of derived cells during this multi step process. Thus many strategies have early been developed to achieve high level reprogramming that basically means a high number of colonies without chromosomal aberration. Such strategies involved different sets of factors, diverse ways of delivery, varying the cell of origin, changing the stoichiometry of the factors or even the order in which they are delivered (for example see (Feng et al. 2009; Robinton and Daley 2012)). Actually the potential understanding of the roadblocks to cell reprogramming has rapidly became focused in order to decipher the deepest mechanisms of cell reprogramming. Several hypothesis have been formulated and investigated to explain the low efficiency of reprogramming and among them a possible explanation, as reported by several groups, might be that cellular senescence is a barrier to reprogramming due to up-regulation of p53, $p16^{INK4A}$ and $p21^{CIP1}$, which are also induced by some of the iPSC reprogramming factors (Banito et al. 2009; Esteban et al. 2010; Hong et al. 2009; Kawamura et al. 2009; Li et al. 2009; Marion et al. 2009a; Utikal et al. 2009). Some of the important results, which were decisive for our understanding of cell reprogramming mechanisms were obtained while exploring this hypothesis.

Cellular Senescence in Cell Reprogramming

In the heat of Yamanaka and Thomson's discoveries, many researches focused on ways to improve cell reprogramming efficiency. When screening for a large panel of factors potentially able to facilitate cell reprogramming from human adult fibroblasts, Zhao et al. (2008) discovered that disrupting expression of the tumor suppressor protein gene TP53, using RNA silencing with concomitant introduction of UTF1 gene (Undifferentiated embryonic cell Transcription Factor 1) enhanced the efficiency of iPSC generation. This phenomenon was pretty powerful since it was observed even when *c-MYC* oncogene was removed from the combination, thus using a three factors set only. Notably, if the maximal effect was observed when p53 was silenced and UTF1 overexpressed at the same time while using a four factor reprogramming set containing *OCT4, SOX2, KLF4,* and *c-MYC*, the only addition of a siRNA against p53 to the original Yamaka four factors cocktail was able to increase cell reprogramming efficiency, as denoted by the increase number of colonies generated. However, although these results were the first to our knowledge to experimentally point out the effect of p53 expression on cell reprogramming leading to the hypothesis, that overcoming cellular senescence might also be required, the specificity and the mechanisms of such importance remained poorly understood (Zhao and Daley 2008).

Few months later, a decisive step was accomplished in our understanding of the role of cellular senescence pathways in cell reprogramming. From an historical point of view it is pretty astonishing to notice that five decisive publications on this concept were published in the very same issue of Nature in august 2009.

In direct lineage of the study revealing the importance of p53, Hong et al. (2009) used fibroblasts from mice deleted for the p53 gene to demonstrate that these cells were tremendously easier to be converted into an embryonic state even without using MYC in the reprogramming cocktail. Moreover, iPSCs were generated from terminally differentiated cells, namely T lymphocytes previously activated in proliferation, obtained from these same mice. They also confirmed that, in human cells, downregulation of the expression of p53 by either dominant negative mutant or short hairpin RNA (shRNA) directed against p53, increased the efficiency of human iPSC generation. Strikingly, they noted that suppression of the retinoblastoma protein (RB), a tumor suppressor protein, whose inhibition is involved in senescence induction (Narita et al. 2003), did not enhance iPSC generation. To reinforce their demonstration they performed DNA microarray analyses and identified common p53-regulated genes, similarly modified in expression both in mouse and human fibroblasts. They ultimately concluded that the p53–p21 pathway serves as a barrier in iPSC generation. Thus, they identified a signaling pathway involved in cellular senescence, but also known to be implicated in DNA damage response, as a major brake to cell reprogramming.

These results were strongly supported by the work of Kawamura et al. (2009) who first observed that the cell reprogramming process by itself increases the level and activity of p53 protein and thus the expression of p21 in mouse embryonic fibroblasts. As expected, inhibition of p53 levels using shRNA or ablation of the protein by homologous recombination resulted in an increased reprogramming efficiency in producing iPSC colonies. However, they also reported that acting downstream of p53, for example by inhibiting p21, p19^{Arf} or p16^{Ink4a} expression, also increased reprogramming efficiency as well as reducing the activity of p53 by genetic modulation of Mdmx a critical negative regulator of the p53. They, finally reported that it was possible to fully reprogram mouse embryonic fibroblasts to pluripotency by the forced expression of only two factors, Oct4 and Sox2, when p53 levels are reduced. Finally, they obtained similar results demonstrating that downregulation of p53 activity increases reprogramming efficiency using human somatic cells.

The tumor suppressor protein p53 has a crucial involvement in preventing the propagation of DNA-damaged cells including those containing short or dysfunctional telomeres (Collado et al. 2007). Interestingly, it was previously demonstrated that efficient production of iPSCs, as measured by the capacity to produce viable chimeric mice, using cells from engineered telomerase deficient mice requires active telomerase and that telomere shortening represents a potent barrier for iPSC generation (Marion et al. 2009b). Moreover, reintroduction of telomerase in deficient cells allows the re-elongation of short telomeres in derived iPSC to a state similar to that observed in ESCs associated to the acquisition of specific telomere epigenetic marks of ESCs. This was even observed in iPSCs generated from cells from old donor mice. This initial demonstration led this same group to determine if further explanations may be related to the presence of DNA damage in mouse cells undergoing reprogramming. They demonstrated that iPSCs could be more efficiently generated from embryonic fibroblasts with short telomeres derived from telomerase-deficient mice when expression of p53 was abrogated using shRNA. Similar results were obtained when cells were exposed to various sources of DNA damage. This study confirmed the increased efficiency of human iPSC generation when the expression of p53 was inhibited with shRNA. Thus, they concluded that p53 limits reprogramming of mouse and human cells by restricting conversion of DNA-damaged cells into iPSCs. Actually, the authors established that p53 activity induces apoptosis of suboptimal damaged cells very early in the time frame of iPSC generation, between 9 and 13 days after infection that is the time of pluripotency induction well before typical iPSC colonies could be observed. However, when p53 was silenced in cells, the resulting iPSCs generated carried persistent DNA damage and chromosomal aberrations (Marion et al. 2009a). Hence, this study further emphasized the central function of p53 in cell reprogramming.

If the first studies we discussed in this chapter mainly focused on p53, another approach was to look at upstream signaling pathways. Li et al. (2009) thus focused on the *INK4a/ARF* locus which encodes three important tumor suppressor proteins (p16^{Ink4a} and p19^{Arf} from *Cdkn2a*, and p15^{Ink4b} from *Cdkn2b*) that activate two critical anti-proliferative pathways, namely, the Rb and p53 pathways. Activation of these well known pathways prevents the propagation of aberrant cells, by induction of either apoptosis or senescence. The authors showed that the locus is completely silenced in iPCSs, as well as in ESCs, although it could be reactivated after differentiation. They first observed that reprogramming mouse fibroblasts is a process that naturally includes the silencing of the *INK4a/ARF* locus and is not the result of a selective process to select rare cells. Thus they could observe that genetic inhibition of this locus by shRNA increased the efficiency of iPSC generation as measured both by kinetic and quantitative parameters. Consequently, they identified the *INK4a/ARF* locus as a main barrier to reprogramming by activation of p53 and p21 although some mechanistic differences were apparent when comparing mouse and human cells. Indeed, it finally emerged that while in murine cells, *ARF* rather than *INK4a* is the main barrier to reprogramming, in human fibroblasts, *INK4a* is more important than *ARF*. Interestingly, they determined that aging in mice upregulates the *INK4a/ARF* locus leading to less efficient reprogramming in cells from old organisms, but this defect can be rescued by silencing *INK4a/ARF*.

Permanently inhibiting p53 introduce strong chromosomal aberration (Marion et al. 2009a), but another way to bypass the senescence roadblock would be to immortalize cells. This approach was investigated by Utikal et al. (2009). Their work first confirmed that the reprogramming potential into iPSCs was decreased when senescence was induced in primary murine fibroblasts after serial passaging. Again, this is consistent with the notion that loss of replicative potential provides a barrier for reprogramming. They also noticed that growth of MEFs in low oxygen (4%) can counteract culture-induced upregulation of senescence associated factors (p16^{Ink4a}, p19^{Arf} and activation of p53), thereby extending replicative potential. Thus, low oxygen culture conditions could increase reprogramming efficiency, in agreement with the notion that p16^{Ink4a} and activated p53 inhibit reprogramming. They could obtain similar results showing that low expression of proteins encoded by the *INK4a/ARF* locus promote cell reprogramming, but also that using a spontaneously immortalized cell line, due to deficiency in components of the Arf-p53 pathway, eliminated a roadblock during the reprogramming of somatic cells into iPSCs, leading to an increased reprogramming efficiency of these cells. Here again, when p53 was silenced using shRNA technology, cells that were not able to produce iPSCs could eventually be reprogrammed to an embryonic state. Thus, this work showed that after the acquisition of immortality, in that case obtained by epigenetic silencing of the *INK4a/ARF* locus, almost every somatic cells can potentially generate iPSCs since all the roadblocks on their way to an ES-like state seem to have been removed.

Indeed, these five major publications clearly point out the major role of cellular senescence in strongly inhibiting cell reprogramming with the central role of p53 through the *INK4a/ARF* pathway. They also demonstrate the impact of DNA damage and its relation with telomere shortening. Consequently, they provided very strong arguments to show that senescence induction pathways is an impassable barrier to reprogramming.

The direct relationship between the process of iPSCs reprogramming and senescence was further investigated in mice cells by Banito et al. who observed that the reprogramming process itself could initially induces senescence. Indeed, they could demonstrate that, few days after expression of the four reprogramming factors, some cells might present a senescent-like morphology with an enlarged cytoplasm and displayed senescent-associated β-galactosidase activity, and senescence-associated heterochromatic foci (SAHF). Molecular analysis showed that not only senescence-associated proteins like p53, $p16^{INK4a}$, and $p21^{CIP1}$ were upregulated after induction of cell reprogramming but also that molecular actors of the DNA damage cascade response (DDR), like increased phosphorylation of ATM and ATR targets or of the histone variant H2A.X. Increased oxydative stress markers, like 8-Oxoguanine, were also detected. Furthermore, they observed chromatin remodeling of the *INK4a/ARF* locus leading to its upregulation that involved the histone demethylase JMJD3 known as a key contributor for the remodeling of the *INK4a/ARF* locus during senescence (Banito et al. 2009). Here again, inhibition of senescence by silencing expression of $p16^{INK4a}$, $p21^{CIP1}$, or p53 in human fibroblasts with shRNAs improved reprogramming efficiency. Thus, factors and processes involved in reprogramming evoke cellular senescence adding to its role as a barrier and thus limiting cell reprogramming efficiency.

Interestingly a very recent study unexpectedly corroborated the role of senescence as a barrier for cell reprogramming. In the hunt for small molecules able to increase the efficacy of reprogramming it was demonstrated that vitamin C (VitC) enhanced iPSC generation from both mouse and human somatic cells. This vitamin acts, at least to some extent, by alleviating cell senescence, but interestingly not in relation to its antioxidant properties, since various compounds tested with well known antioxidant activities did not have any noticeable effect. Actually, the authors could demonstrate that VitC induces a small but significant reduction in p53 and p21 protein levels in VitC treated cells (Esteban et al. 2010). Further, the same group, in a more recent study, dissected the mechanism of action of VitC demonstrated that it induces H3K36me2/3 demethylation in mouse embryonic fibroblasts and also promotes the Jumonji histone demethylase activities of Jhdm1a/1b, already been shown to demethylate H3K36me2/3 (He et al. 2008). Actually, removal of H3K36me2/3 by Jhdm1a/1b was observed in the *INK4a/ARF* locus reinforcing the role of this locus and thus of senescence in blocking cell reprogramming (Wang et al. 2011).

All these essential results suggested at that time that cellular aging might be an important limitation for the derivation of iPSCs for therapeutic purposes from elderly individuals. It was thus essential to determine if it was possible to efficiently reverse cellular senescence and reprogram somatic cells into iPSCs, and that such reprogrammed iPSCs could not be distinguished from ESCs. Indeed, as we are going to detail now, our laboratory could obtain such results and could demonstrate that re-differentiation, led to rejuvenated cells with a reset cellular physiology, defining a new paradigm for cell rejuvenation.

Reprogramming Senescent Cells

Generation of iPSCs from Senescent Cells

Because iPSC generation from senescent cells was only assayed using the initial four gene set described by Yamanaka's group, we decided to investigate the effect of a different gene combination of six factors. This novelty came from the demonstration by Thomson's group that OCT4 and SOX2 in conjunction with NANOG and LIN28 could also induce cell reprogramming of human somatic cells replacing c-MYC and KLF4 (Yu et al. 2007). Then, the idea that each combination of four factors might undergo slightly different roads for the same final effect of reprogramming into iPSCs, led us to investigate whether both combination could have a synergic effect, increasing "reprogramming pressure" able to override cellular senescence. Moreover, the specific nature of LIN28 and NANOG argues for their addition to a cell reprogramming cocktail. Indeed, NANOG over-expression has been described to facilitate reprogramming in a predominantly cell-division-rate-independent manner and LIN28 over-expression, similarly to the inhibition of the $p53/p21^{CIP1}$ pathway, increased the cell division rate resulting in accelerated kinetics of iPSCs production (Hanna et al. 2009; Yu et al. 2009). Thus, proliferative human diploid fibroblasts from a 74-year-old donor were induced into replicative senescence by serial passaging. Senescence was confirmed by FACS analysis showing a permanent cell cycle arrest, increase in senescence-associated β-galactosidase activity, up-regulation of $p16^{INK4A}$ and $p21^{CIP1}$ and formation of SAHF (Narita et al. 2003). These cells were then infected with individual lentiviruses carrying each of the six genes OCT4, SOX2, KLF4, c-MYC, NANOG and LIN28 (Lapasset et al. 2011). Few days after lentiviral infection, we first observed disappearance of SAHFs in infected senescent cells, and then a restored proliferation after three weeks. Colonies resembling hESCs appeared at 35–40 days post-infection, with a mean reprogramming efficiency similar to proliferative fibroblasts infected under the same conditions and were successfully maintained in culture during more than 35 passages on either MEF feeders or in feeder-free culture conditions. We further characterized iPSC colonies from proliferative and senescent fibroblasts to assess successful reprogramming by confirming the continued presence of stem cell surface markers SSEA-4 and TRA-1-60, as well as re-expression of endogenous OCT4, SOX2, NANOG and REX1 pluripotent marker genes, compared to the parental fibroblasts and to control hESCs and iPSCs from the Thomson's laboratory (Yu et al. 2007). Reactivation of endogenous pluripotency genes either in iPSCs from senescent or proliferative cells was also confirmed by DNA demethylation of the *OCT4* and *NANOG* promoters that are usually highly methylated in fibroblasts. To exclude any cell type-specific effects, we also achieved efficient reprogramming using another cell type IMR-90 induced or not into replicative senescence and thus, demonstrated our ability to consistently reprogram senescent cells with the six-factor gene cocktail.

Next we assessed the pluripotency of the derived iPSCs by evaluating their ability to differentiate into the three embryonic lineages. All iPSC clones generated from the senescence-triggered or proliferative fibroblasts were able to differentiate efficiently into endoderm, ectoderm and mesoderm in vitro, as shown by immunodetection of specific lineage markers after differentiation, but also in vivo, after injection in mice and analysis of resulting teratomas. These experiments clearly demonstrated that iPSCs derived from senescent cells could regain pluripotency similar to hESCs.

It has to be noted that in our original study (Lapasset et al. 2011) we failed, despite all our efforts, to achieve reprogramming of senescent cells using the two described four factors cocktail (Takahashi et al. 2007; Yu et al. 2007) even when using additional treatments described to enhance reprogramming efficiency (Feng et al. 2009). Thus, these results suggested that the unique combination of the six transcription factors is a key determinant for a successful reprogramming reversing cellular senescence to derive iPSCs, without any direct suppression of senescence effectors contributing as safeguards of the genome.

Finally, we demonstrated that our protocol can efficiently reinstates self-renewal capacity and pluripotency from centenarian fibroblasts and thus that cellular aging and senescence are definitely not a barrier to reprogramming.

Rejuvenation of Cells from Senescent Derived iPSCs

This first part of our demonstration allowed us to establish a new paradigm and reach a new step in cell reprogramming. However, if the cell program had been successfully reset it remained to determine if iPSCs derived from senescent fibroblasts and aged donors had maintained or lost their aged characteristics. Moreover, an important question to address was to determine if senescent-derived iPSCs were able to redifferentiate into fully rejuvenated cells.

Thus, in our original study (Lapasset et al. 2011), we showed that, similar to hESC lines, all iPSCs generated from replicative senescent or proliferative fibroblasts from old donors exhibited down-regulated expression of $p16^{INK4A}$ and $p21^{CIP1}$ that are common hallmarks of senescence and aging.

We then focused on telomeres length since, in humans, progressive telomere shortening is thought to be one of the mechanisms underlying organism aging (Sahin and Depinho 2010) and the previous demonstration, already discussed, that telomere shortening represents a potent barrier for iPSC generation in engineered telomerase-deficient cells (Marion et al. 2009b). Moreover, although iPSCs, created by current protocols, generally exhibit an initial increased telomere size compared to the parental differentiated cells, prematurely aged (shortened) telomeres appear to be a common feature of these iPSC and their cell progeny (Feng et al. 2010; Suhr et al. 2009; Vaziri et al. 2010). However, all these iPSCs were not produced using a six factor cocktail. Indeed, when we measured the telomere length of iPSCs generated from senescent or centenarian cells, we found an increased mean size. Moreover, we were able to maintain all iPSC lines for more than a hundred population doubling without any decrease in telomere length or loss of self-renewal and pluripotency properties. Interestingly, in some iPSC clones, telomeres were even longer than in control hESCs, suggesting possible additional improvements in iPSC generation due to our original cocktail for increased proliferation capacity of re-differentiated cells.

To confirm the disappearance of senescence markers in senescent cell derived iPSCs we also performed transcriptomic analysis of selected iPSCs (Lapasset et al. 2011). When compared to their parental counterparts and with hESCs and iPSCs generated with a four factor cocktail from different laboratories we confirmed that the specific pluripotency genes were expressed in our iPSCs at a similar level to hESCs and iPSCs. Moreover, it was striking to note that our derived iPSCs obtained using six factors clustered with hESCs, separating them from previously described iPSCs derived with only four factors, although the parental aged and senescent fibroblasts gene expression profiles clustered together compared to embryonic and post-natal fibroblasts sharing a general common aging signature. Looking back at telomeres, these transcriptomic analysis also demonstrated that genes involved in telomere metabolism and maintenance described previously (Vaziri et al. 2010) were precisely reset in our iPSCs, which clustered with hESCs and not with iPSCs generated by four factors and, that iPSCs expressed higher levels of telomerase transcripts compared to hESCs concomitantly with longer telomeres measured by FISH, demonstrating that our six factor-based strategy was highly efficient for telomere function resetting.

Since oxidative stress and mitochondrial dysfunction are well described common features of senescence and aging (Moiseeva et al. 2009; Passos et al. 2007), we wondered whether these functions were also specifically reprogrammed from senescent and aged cells. Clustering of transcriptomes using a specific subset of genes previously described for their involvement in these regulatory pathways (Armstrong et al. 2010; Prigione et al. 2010) confirmed that our iPSCs have reset these functions to an embryonic-like state. Moreover, the overall mitochondrial activity was measured by evaluating the trans-membrane potential and, we found an increased membrane potential in iPSCs to a level similar to that in hESCs, confirming the reprogramming of the altered mitochondrial activity from old and senescent fibroblasts to an hESC-like metabolism. Analysis of mitochondrial properties in iPSCs illustrates how nuclear reprogramming, in resetting gene expression programs to pluripotency, might also reprogram cellular organelle function.

Thus, these last results indicate collectively that our six factor-based reprogramming strategy generates iPSCs indistinguishable from hESCs. However, this gave rise to a major remaining question of whether senescent-derived iPSCs are able to produce young re-differentiated cells.

To answer to this last question, we first established that fibroblasts derived from iPSCs generated from proliferative, senescent or aged cells could proliferate similarly to young fibroblasts and did not enter prematurely into senescence (Lapasset et al. 2011), as opposed to previous results for re-differentiated iPSCs generated from different healthy proliferating cells or from premature aging syndromes (Feng et al. 2010; Liu et al. 2010). We demonstrated that these re-differentiated fibroblasts could re-enter in replicative senescence, indicating they did not acquire during reprogramming process from senescence-arrested cells a non reversible immortalization state and that they retained intact the senescence induction pathways. This was demonstrated by senescence specific markers and re-shortened telomere size, but with an increased number of population doubling before replicative senescence, similar to young proliferative fibroblasts. That latter experiment demonstrated a rejuvenated proliferation potential of aged and senescent cells induced by a six factors-based strategy.

To confirm and assess rejuvenation of the global cellular physiology, we performed transcriptomic analysis of the re-differentiated cells and, indeed, the global gene expression profiles of re-differentiated fibroblasts from our iPSCs actually clusters with young proliferative embryonic fibroblasts derived from hESCs, whereas their parental counterparts share a common aging signature, separating them from post natal fibroblasts. Similar results were obtained after gene expression profile clustering of genes associated with oxidative stress and mitochondrial activity. All together, we definitely demonstrated a rejuvenated physiology of senescent and aged cells after the six factor-based reprogramming through the pluripotent state (Lapasset et al. 2011).

We would like to further stress the importance of the use of six factors, and we believe that additional demonstrations are needed to clearly decipher the mechanisms involved. An interesting argument for this point is the work of Prigione et al. (2011) that have achieved cell reprogramming of fibroblasts from a single 84-year-old donor using the original four factors cocktail from Yamanaka, and demonstrated that the resulting iPSCs harbored genomic alterations in the cells obtained that not only preclude their theoretical use for clinical application but also establish that the protocol used was not optimal.

Discussion

Our work has unquestionably set a new mark for iPSCs research and translation of this technology into therapy applications by providing a new paradigm for cell rejuvenation. We have established that it is possible, using a pertinent reprogramming strategy, to efficiently reprogram senescent cells and cells derived from very old individuals into iPSCs and that cellular senescence and aging are not a barrier to reprogramming towards pluripotency. After cell reprogramming, pluripotent cells re-differentiated into fibroblasts have rejuvenated extended lifespan and characteristics of young proliferative embryonic fibroblasts that means that these cells have been completely clear of their former aged cellular phenotype. Our study demonstrates the importance of epigenetic modifications in aging to reverse major hallmarks of the cellular aging physiology. It is noteworthy to remark that previous studies have actually bypassed senescence by preventing cells to reach this state in order to induce reprogramming of cells heading to senescence and have not, as we accomplished ourselves, reverted cells from a senescent state to a rejuvenated one through reprogramming. Although the mechanistic reasons for our successful results remain unclear, several rational hypothesis can be formulated.

Adding LIN28 and NANOG to the Yamanaka's Four Genes Cocktail, a Key to Reprogram Senescent Cells

It sounds obvious from all the results reported in this chapter that silencing p53 gene while an efficient way of action to reprogram senescent cells, is not a viable strategy since it will induce genomic alterations. From our own results and according to current knowledge, it sounds that efficient cell reprogramming of senescent cells requires at least the use of a six factors cocktail. We have provided evidences, as reported here, that only the combination of OCT4, SOX2, KLF4, c-MYC, NANOG and LIN28 permits to reset a senescent phenotype to obtain rejuvenated cells, whereas the four Yamanaka factors are inefficient to reprogram senescent cells. This might restrict the use of Yamanaka's cocktail to non aged people while the elderly population is mostly concerned by regenerative medicine. Thus, we believe that the differences with the original set described by Yamanaka is a critical issue that has to be further studied.

NANOG plays a central role in the maintenance of ESC identity in tight relation with the two other transcription factors OCT4 and SOX2 (Chambers et al. 2003, 2007; Chambers and Tomlinson 2009; Mitsui et al. 2003). Moreover, in the context of senescent cells reprogramming, it has been shown that p53 directly inhibits NANOG expression in both mice and humans (Lin et al. 2005; Qin et al. 2007) and one can imagine that overexpression of NANOG during the reprogramming process allows to counterbalance the negative action of p53. We could also speculate that ectopic expression of NANOG might titrate p53, eventually capturing it in the cytoplasm to allow reprogramming by the Four Yamanaka factors. In this scenario, it appears that the addition of NANOG to the original cocktail could greatly enhance the reprogramming efficiency of senescent/aged cells.

The importance of the addition of LIN28 seems more appealing to us but might have a synergic effect when combined with NANOG addition. LIN28 has the ability to repress let-7 accumulation. This microRNA, originally discovered in the nematode Caenorhabditis elegans, regulates cell proliferation and differentiation and both its sequence and function are highly conserved in mammals (Roush and Slack 2008). LIN28 by inhibiting the production of mature let-7 favors proliferation, and in contrast, when high levels of mature let-7 are produced this inhibits the expression of proteins like MYC or HMGA2. If the impact on MYC expression is easily interpretable, knowing it is one of the important factor from the Yamanaka's cocktail that has been shown to increase cell reprogramming efficiency, the action on HMGA2 is more subtle. In fact, to our point of view, HMGA2, a non-histone chromatin proteins that could alters chromatin structure, has a main contribution to cell reprogramming of senescent cells. Indeed, it has been shown that HMGA2 could negatively regulate the *INK4a/ARF* locus in mouse stem cells (Nishino et al. 2008) and since the *INK4a/ARF* locus is one of the main barrier to iPSC reprogramming, acting on this locus should be beneficial and potentiated by the action of Nanog on p53.

These observations about the putative roles of NANOG and LIN28 when added to the original Yamanaka's factors lead us to propose a mechanistic model related to the synergistic action of the six factors in the cocktail that we established as crucial to efficiently reprogram senescent/aged cells (Fig. 25.1).

However, other ways to improve cell reprogramming of senescent/aged cells have to be evaluated. Among them, the use of small molecules might be of great interest. In particular molecules that were already demonstrated to increase cell reprogramming efficiency by acting at the chromatin level like DNA methyltransferase inhibitor, like 5-aza-cytidine (AZA), or histone deacetylase inhibitors, like valproic acid (VPA) or trichostatin A (TSA) (Feng et al. 2009). Since cell reprogramming requires a large remodeling it is highly pertinent to test such epigenetic modulators although other small molecules could be efficient.

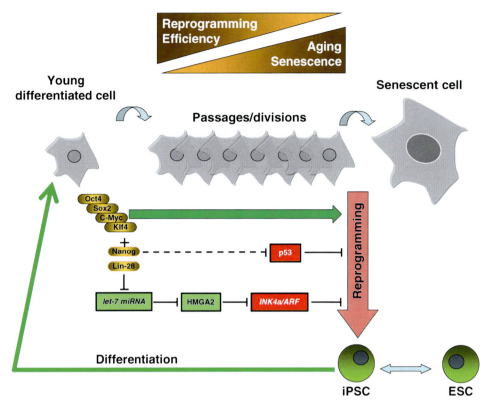

Fig. 25.1 Model depicting the synergistic action of the six factors in the cocktail to efficiently reprogram senescent/aged cells. The efficiency of reprogramming, commonly performed by using Yamanaka's cocktail containing OCT4, SOX2, C-MYC and KLF-4, was described as inversely proportional to the age of cells and blocked by senescence. This inhibition is mainly due to the action of p53 protein and *INK4a/ARF* locus. The addition of NANOG to the four original factors allows to counteract the negative action of p53 on the expression of endogenous NANOG. We believe that LIN28 is highly important to the cocktail through its signaling pathway going from inhibition of let-7 microRNA and ultimately leading to inhibition of the *INK4a/ARF* locus. However, it is not excluded that NANOG and LIN28 also have direct effect by activating the pluripotency program

Another approach would be to assess other factors that might be valuable to increase cell reprogramming. In the publication from Yamanaka's group describing for the first time cell reprogramming in mouse, the authors initially tested 24 factors before restricting their cocktail to only four (Takahashi and Yamanaka 2006). We believe that some of these factors should be reassessed for their propensity to ameliorate cell reprogramming of senescent/aged cells. For example, among these factors is UTF1 that we already discussed and was demonstrated to enhance the efficiency of iPSC generation (Zhao et al. 2008) but others might be beneficial.

Micro RNA also constitutes a very interesting area of investigation to increase reprogramming efficiency of cells from aged patient. It might be a subtle way to fine tune gene expression during cell reprogramming. It is especially relevant since highly efficient miRNA-mediated reprogramming of mouse and human somatic cells was achieved (Anokye-Danso et al. 2011). Moreover, in one of the article previously discussed, Banito et al. also demonstrated that ectopic expression of the miR-302 cluster inhibited growth arrest and up-regulation of p21[CIP1] and p130 observed during reprogramming induced senescence (Banito et al. 2009). Thus exploration of miRNA based alternatives should be undertaken.

Last but not least is the amelioration of cell culture conditions in order to diminish general stress conditions like oxidative stress that might increase replicative stress favoring senescence. In particular we think about low oxygen conditions that are particularly important not only to decrease oxidative stress but also diminish activation of p53 and upregulation of p16^{Ink4a} and p19^{Arf} as previously demonstrated (Utikal et al. 2009). Low oxygen conditions are the golden standard for culturing hESCs and hypoxia inducible factors have been shown to regulate pluripotency and proliferation of hESCs cultured at reduced oxygen tensions in part by upregulating pluripotency genes like OCT4, NANOG or SOX2 (Forristal et al. 2010). Thus, conducting cell reprogramming in reduced oxygen environment sounds obvious but it is also probable that other factors that could improve cell culture conditions remain to be discovered.

Although all these hypothesis remain to be confirmed, it opens very interesting new investigation areas in order to establish optimal strategies for developing cell-based therapies for aged patients. Moreover, the demonstration that a six factor-based reprogramming could bypass the senescence barrier and reverse the aged altered cellular physiology has establish a new point of reference for the study of aging and cell reprogramming.

References

Anokye-Danso F, Trivedi CM, Juhr D, Gupta M, Cui Z, Tian Y, Zhang Y, Yang W, Gruber PJ, Epstein JA et al (2011) Highly efficient miRNA-mediated reprogramming of mouse and human somatic cells to pluripotency. Cell Stem Cell 8:376–388

Armstrong L, Tilgner K, Saretzki G, Atkinson SP, Stojkovic M, Moreno R, Przyborski S, Lako M (2010) Human induced pluripotent stem cell lines show stress defense mechanisms and mitochondrial regulation similar to those of human embryonic stem cells. Stem Cells 28:661–673

Banito A, Rashid ST, Acosta JC, Li S, Pereira CF, Geti I, Pinho S, Silva JC, Azuara V, Walsh M et al (2009) Senescence impairs successful reprogramming to pluripotent stem cells. Genes Dev 23:2134–2139

Campisi J, d'Adda di Fagagna F (2007) Cellular senescence: when bad things happen to good cells. Nat Rev Mol Cell Biol 8:729–740

Chambers I, Tomlinson SR (2009) The transcriptional foundation of pluripotency. Development 136: 2311–2322

Chambers I, Colby D, Robertson M, Nichols J, Lee S, Tweedie S, Smith A (2003) Functional expression cloning of Nanog, a pluripotency sustaining factor in embryonic stem cells. Cell 113:643–655

Chambers I, Silva J, Colby D, Nichols J, Nijmeijer B, Robertson M, Vrana J, Jones K, Grotewold L, Smith A (2007) Nanog safeguards pluripotency and mediates germline development. Nature 450:1230–1234

Collado M, Blasco MA, Serrano M (2007) Cellular senescence in cancer and aging. Cell 130:223–233

Esteban MA, Wang T, Qin B, Yang J, Qin D, Cai J, Li W, Weng Z, Chen J, Ni S et al (2010) Vitamin C enhances the generation of mouse and human induced pluripotent stem cells. Cell Stem Cell 6:71–79

Feng B, Ng JH, Heng JC, Ng HH (2009) Molecules that promote or enhance reprogramming of somatic cells to induced pluripotent stem cells. Cell Stem Cell 4:301–312

Feng Q, Lu SJ, Klimanskaya I, Gomes I, Kim D, Chung Y, Honig GR, Kim KS, Lanza R (2010) Hemangioblastic derivatives from human induced pluripotent stem cells exhibit limited expansion and early senescence. Stem Cells 28:704–712

Forristal CE, Wright KL, Hanley NA, Oreffo RO, Houghton FD (2010) Hypoxia inducible factors regulate pluripotency and proliferation in human embryonic stem cells cultured at reduced oxygen tensions. Reproduction 139:85–97

Hanna J, Saha K, Pando B, van Zon J, Lengner CJ, Creyghton MP, van Oudenaarden A, Jaenisch R (2009) Direct cell reprogramming is a stochastic process amenable to acceleration. Nature 462:595–601

Hanna JH, Saha K, Jaenisch R (2010) Pluripotency and cellular reprogramming: facts, hypotheses, unresolved issues. Cell 143:508–525

He J, Kallin EM, Tsukada Y, Zhang Y (2008) The H3K36 demethylase Jhdm1b/Kdm2b regulates cell proliferation and senescence through p15(Ink4b). Nat Struct Mol Biol 15:1169–1175

Hong H, Takahashi K, Ichisaka T, Aoi T, Kanagawa O, Nakagawa M, Okita K, Yamanaka S (2009) Suppression of induced pluripotent stem cell generation by the p53-p21 pathway. Nature 460:1132–1135

Kawamura T, Suzuki J, Wang YV, Menendez S, Morera LB, Raya A, Wahl GM, Belmonte JC (2009) Linking the p53 tumour suppressor pathway to somatic cell reprogramming. Nature 460:1140–1144

Lapasset L, Milhavet O, Prieur A, Besnard E, Babled A, Aït-Hamou N, Leschik J, Pellestor F, Ramirez JM, De Vos J et al (2011) Rejuvenating senescent and centenarian human cells by reprogramming through the pluripotent state. Genes Dev 25:2248–2253

Li H, Collado M, Villasante A, Strati K, Ortega S, Canamero M, Blasco MA, Serrano M (2009) The

Ink4/Arf locus is a barrier for iPS cell reprogramming. Nature 460:1136–1139

Lin T, Chao C, Saito S, Mazur SJ, Murphy ME, Appella E, Xu Y (2005) p53 Induces differentiation of mouse embryonic stem cells by suppressing Nanog expression. Nat Cell Biol 7:165–171

Liu GH, Barkho BZ, Ruiz S, Diep D, Qu J, Yang SL, Panopoulos AD, Suzuki K, Kurian L, Walsh C et al (2011) Recapitulation of premature ageing with iPSCs from Hutchinson-Gilford progeria syndrome. Nature 472:221–225

Marion RM, Strati K, Li H, Murga M, Blanco R, Ortega S, Fernandez-Capetillo O, Serrano M, Blasco MA (2009a) A p53-mediated DNA damage response limits reprogramming to ensure iPS cell genomic integrity. Nature 460:1149–1153

Marion RM, Strati K, Li H, Tejera A, Schoeftner S, Ortega S, Serrano M, Blasco MA (2009b) Telomeres acquire embryonic stem cell characteristics in induced pluripotent stem cells. Cell Stem Cell 4:141–154

Mitsui K, Tokuzawa Y, Itoh H, Segawa K, Murakami M, Takahashi K, Maruyama M, Maeda M, Yamanaka S (2003) The homeoprotein Nanog is required for maintenance of pluripotency in mouse epiblast and ES cells. Cell 113:631–642

Moiseeva O, Bourdeau V, Roux A, Deschenes-Simard X, Ferbeyre G (2009) Mitochondrial dysfunction contributes to oncogene-induced senescence. Mol Cell Biol 29:4495–4507

Narita M, Nunez S, Heard E, Narita M, Lin AW, Hearn SA, Spector DL, Hannon GJ, Lowe SW (2003) Rb-mediated heterochromatin formation and silencing of E2F target genes during cellular senescence. Cell 113:703–716

Nishino J, Kim I, Chada K, Morrison SJ (2008) Hmga2 promotes neural stem cell self-renewal in young but not old mice by reducing p16Ink4a and p19Arf expression. Cell 135:227–239

Passos JF, Saretzki G, Ahmed S, Nelson G, Richter T, Peters H, Wappler I, Birket MJ, Harold G, Schaeuble K et al (2007) Mitochondrial dysfunction accounts for the stochastic heterogeneity in telomere-dependent senescence. PLoS Biol 5:e110

Prigione A, Fauler B, Lurz R, Lehrach H, Adjaye J (2010) The senescence-related mitochondrial/oxidative stress pathway is repressed in human induced pluripotent stem cells. Stem Cells 28:721–733

Prigione A, Hossini AM, Lichtner B, Serin A, Fauler B, Megges M, Lurz R, Lehrach H, Makrantonaki E, Zouboulis CC et al (2011) Mitochondrial-associated cell death mechanisms are reset to an embryonic-like state in aged donor-derived iPS cells harboring chromosomal aberrations. PLoS One 6:e27352

Qin H, Yu T, Qing T, Liu Y, Zhao Y, Cai J, Li J, Song Z, Qu X, Zhou P et al (2007) Regulation of apoptosis and differentiation by p53 in human embryonic stem cells. J Biol Chem 282:5842–5852

Robinton DA, Daley GQ (2012) The promise of induced pluripotent stem cells in research and therapy. Nature 481:295–305

Roush S, Slack FJ (2008) The let-7 family of microRNAs. Trends Cell Biol 18:505–516

Sahin E, Depinho RA (2010) Linking functional decline of telomeres, mitochondria and stem cells during ageing. Nature 464:520–528

Suhr ST, Chang EA, Rodriguez RM, Wang K, Ross PJ, Beyhan Z, Murthy S, Cibelli JB (2009) Telomere dynamics in human cells reprogrammed to pluripotency. PLoS One 4:e8124

Takahashi K, Yamanaka S (2006) Induction of pluripotent stem cells from mouse embryonic and adult fibroblast cultures by defined factors. Cell 126:663–676

Takahashi K, Tanabe K, Ohnuki M, Narita M, Ichisaka T, Tomoda K, Yamanaka S (2007) Induction of pluripotent stem cells from adult human fibroblasts by defined factors. Cell 131:861–872

Utikal J, Polo JM, Stadtfeld M, Maherali N, Kulalert W, Walsh RM, Khalil A, Rheinwald JG, Hochedlinger K (2009) Immortalization eliminates a roadblock during cellular reprogramming into iPS cells. Nature 460: 1145–1148

Vaziri H, Chapman KB, Guigova A, Teichroeb J, Lacher MD, Sternberg H, Singec I, Briggs L, Wheeler J, Sampathkumar J et al (2010) Spontaneous reversal of the developmental aging of normal human cells following transcriptional reprogramming. Regen Med 5:345–363

Wang T, Chen K, Zeng X, Yang J, Wu Y, Shi X, Qin B, Zeng L, Esteban MA, Pan G et al (2011) The histone demethylases Jhdm1a/1b enhance somatic cell reprogramming in a vitamin-C-dependent manner. Cell Stem Cell 9:575–587

Yu J, Vodyanik MA, Smuga-Otto K, Antosiewicz-Bourget J, Frane JL, Tian S, Nie J, Jonsdottir GA, Ruotti V, Stewart R et al (2007) Induced pluripotent stem cell lines derived from human somatic cells. Science 318:1917–1920

Yu J, Hu K, Smuga-Otto K, Tian S, Stewart R, Slukvin II, Thomson JA (2009) Human induced pluripotent stem cells free of vector and transgene sequences. Science 324:797–801

Zhang R, Chen W, Adams PD (2007) Molecular dissection of formation of senescence-associated heterochromatin foci. Mol Cell Biol 27:2343–2358

Zhao R, Daley GQ (2008) From fibroblasts to iPS cells: induced pluripotency by defined factors. J Cell Biochem 105:949–955

Zhao Y, Yin X, Qin H, Zhu F, Liu H, Yang W, Zhang Q, Xiang C, Hou P, Song Z et al (2008) Two supporting factors greatly improve the efficiency of human iPSC generation. Cell Stem Cell 3:475–479

The Transcription Factor GATA2 Regulates Quiescence in Haematopoietic Stem and Progenitor Cells

26

Neil P. Rodrigues and Alex J. Tipping

Contents

Abstract	277
Introduction	278
Cellular and Molecular Control of Quiescence in Haematopoietic Stem Cells	278
GATA2 and Proliferative Control	279
Enforced GATA2 Expression Confers Quiescence	279
Quiescence Conferred by GATA2 Is Level-Dependent	281
Effects of Enforced GATA2 Expression on Haematopoietic Reconstitution	281
GATA2 Target Genes in Cell Cycle Control	282
GATA2 Modulates Cytokine Receptor Expression	284
Haploinsufficiency of Gata2 Also Impacts Quiescence Program	285
Discussion	285
References	287

N.P. Rodrigues
National Institutes of Health, Center for Biomedical Research Excellence in Stem Cell Biology,
Roger Williams Medical Center, Boston University of Medicine, Providence, RI 02908, USA

Department of Dermatology, Boston University School of Medicine, Boston, MA 02118, USA

Center for Regenerative Medicine, Boston University School of Medicine, Boston, MA 02118, USA

A.J. Tipping (✉)
University College London (UCL), Cancer Institute, Paul O'Gorman Building, 72 Huntley Street, London, WC1E 6DD, UK
e-mail: a.tipping@ucl.ac.uk

Abstract

Control of haematopoietic stem cell (HSC) proliferation is critical in preventing bone marrow failure or haematological malignancy. Understanding the mechanisms that balance the requirement to restrain excessive HSC proliferation while allowing for production of blood cells and maintenance of the HSC pool is therefore of substantial clinical interest. Herein we discuss the nature of HSC quiescence and the role of the zinc finger transcription factor GATA2 in regulating a gene expression program which reversibly confers quiescence on HSCs and committed progenitors. We present data extending previous observations of reduced HSC and progenitor functionality in the context of enforced *GATA2* expression, and begin to demonstrate the molecular mechanisms by which the GATA2 program appears to function in restraining HSC and progenitor cell proliferation. Conversely, we also show that Gata2 haploinsufficiency impacts the quiescent program of HSCs and committed progenitors, demonstrating that HSC proliferation is exquisitely responsive to either up or down-regulation of GATA2 level. Finally, we discuss the clinical manifestations of loss-of-function *GATA2* mutations and high *GATA2* expression in the pre-malignant myelodysplastic syndromes and myeloid leukaemia.

Keywords

CD34+ CD38-cells • CD34+ cord blood cells • Chemokine and cytokine signals • Common myeloid progenitors (CMPs) • GATA2 expression • Granulocyte-macrophage progenitor (GMP) lineage • Haematopoietic stem cell (HSC) • Haploinsufficiency of GATA2 • Kinases CDK4 and CDK6 • Quiescence-associated transcription factor MEF/ELF4

Introduction

Appropriate regulation of cell division in adult tissue-specific stem and progenitor cells is essential for maintenance of tissue turnover and integrity. In the blood system, the most primitive haematopoietic stem cells (HSCs) in residence in the bone marrow (BM) are faced with a requirement to periodically divide, to replenish the HSC pool via self renewal, and to generate more mature progenitor cells which in turn produce large numbers of functional blood and immune cells. Entry into and exit from cycle, and the nature of the division (self-renewing versus differentiative) must be tightly controlled, in order to prevent exhaustion of the HSC pool which can lead to bone marrow failure syndromes, or corruption of the HSC pool that can lead to haematological malignancy. The mechanisms by which this is achieved are beginning to be understood.

Cellular and Molecular Control of Quiescence in Haematopoietic Stem Cells

The state of dormancy within which HSCs largely reside is referred to as quiescence. Operationally, this involves reversible exit from cell cycle despite exposure to mitogenic signals. Cells can persist in this state for some time, dividing relatively infrequently; BrdU retention assays suggest that HSCs cycle once on average over a timescale measured in weeks. Modelling analyses of H2B-GFP retention suggest that between 1 and 11% of HSCs divide per day; modelling in humans suggests that each HSC divides around once every 40 weeks. Stress haematopoiesis (modelled by 5-FU treatment to kill all cycling cells) triggers a large proportion of quiescent HSCs to enter cycle; successive 5-FU treatments are eventually fatal, via HSC exhaustion and pancytopenia (Cheng et al. 2000).

Cells in quiescence (G_0) have a 2n DNA content, and are distinguishable from G_1 counterparts via reduced expression of the nuclear antigen Ki-67, or in living cells by reduced staining of polysomal RNA with Pyronin Y. Previous studies (Venezia et al. 2004) have examined the transcriptional program of quiescent haematopoietic stem (HSCs) and progenitor cells (HPCs) and shown that they significantly differ from their cycling counterparts. Although these data need to be interpreted carefully, they do suggest that exit from cycle is associated with transcriptional changes, including reduced metabolism, as might be expected from their reduced polysomal RNA content.

It has been suggested that HSCs are induced to enter quiescence by their exposure to physical and biochemical signals in the BM, at endosteal niches. Residency in some regions of the BM has been proposed to promote quiescence and maintenance of primitivity, whereas activation of cycling/differentiation correlates with movement into a distinct vascular niche. Indeed, cells isolated from the lumen of the BM are less able to reconstitute murine BM than immunophenotypically identical cells from the endosteum (Grassinger et al. 2010). However, a variant model posits the existence of distinct HSC niches which drive dormancy or regenerative self-renewal divisions respectively (Trumpp et al. 2010). Presumably the latter niche is more akin to that which allows differentiation of HSCs into less mature progenitor cells.

Recent evidence suggests that the regulation underlying this control is complex and redundant, and involves HSC-supportive contributions from multiple cell types which varies by location (Ehninger and Trumpp 2011). It appears that supporting cells (whether perivascular mesenchymal stem cells, osteoblasts or other cell types) produce chemokine and cytokine signals which somehow induce quiescence in HSCs and physically retain them. Candidate molecules produced by such cells include CXCL12, ANG1, and TPO (Li 2011) as well as structural proteins which are contacted by HSCs; presumably HSCs exposed to the correct factor(s) execute a gene expression program which drives cell cycle exit. Some of these interactions, or the signalling pathways they stimulate, may be druggable targets for clinical intervention. For example, use of the CXCR4inhibitor Plerixafor inhibitsCXCL12 stimulation of this pathway and hence exerts its effects in freeing HSCs from their niche. Conditional deletion of *Cxcr4*in mice is associated with mobilisation, decreased quiescence and increased cycling of HSCs (Flt3- LSK (Lin- Sca + Kit+)) in a cell-intrinsic manner which may involve reduced expression of the CDKI*p57* and induction of *Ccnd1*(Nie et al. 2008). Importantly, Plerixafor is now being utilised clinically to mobilise HSCs for transplantation.

Ultimately however, these complex extrinsic signals are transduced into the nucleus and integrated at the level of transcription factor activity. Identifying the relevant transcription factors in this setting is therefore critical to understanding the mechanisms by which cells can transition between cycling and dormancy in HSCs.

GATA2 and Proliferative Control

Early observations of transcription factor expression in murine HSCs showed high levels of expression of *Gata2* (Orlic et al. 1995). Indeed, high expression of *Gata2* has been proposed to mark dormant HSCs in the BM (Persons et al. 1999). The functional basis of the correlation between high GATA2 expression and HSC function had earlier been assessed in two seminal papers. In one, GATA2 expression was enforced by retroviral transduction in primary murine bone marrow cells, with the effect of reducing proliferation and output in functional assays of HSCs/HPCs (Persons et al. 1999). This reduced proliferation was not directly associated with altered cell cycle distribution or increased apoptosis but suppressed haematopoietic output in bone marrow transplantation experiments that assay HSC activity. In another study, it was demonstrated that activation of the estradiol-receptor fusion GATA2/ER in the murine multi-potent FDCP-mix model conferred growth arrest and accumulation in G_0/G_1 *in vitro*, although this was associated with spontaneous differentiation into myelomonocytic cells (Heyworth et al. 1999). Together, these data confirmed that GATA2 activity was relevant to growth control of HSCs/HPCs, but did not clarify whether these effects on cell growth were directly related to quiescence.

Enforced GATA2 Expression Confers Quiescence

Previous observations of high *Gata2* expression in largely quiescent murine HSCs led us to examine the association between quiescence and *GATA2* expression in human HSCs/HPCs, as defined by surface expression of CD34 but not CD38 (CD34+ CD38-). Defining quiescent HSCs/HPCs as being within the G_0/G_1 DNA peak, but with low staining for Pyronin Y, we sorted freshly-thawed human cord blood cells together with their cycling Pyronin-high counterparts. Quantitative RT-PCR analysis confirmed that quiescent HSCs/HPCs expressed roughly 2.5× more *GATA2* than cycling cells. Given this observation, we elected to investigate the effect of overexpressing *GATA2* in human cord blood HSCs/HPCs. Lentiviral transduction with a bicistronic h*GATA2*-IRES-eGFP construct conferred quiescence as measured by reduction of both Pyronin staining (Fig. 26.1a, b), and *in vitro* proliferation (Fig. 26.1c).

Reduced proliferative potential was mirrored by reduced output in both CFC and LTC-IC assays. The former *in vitro* assay measures the ability of HPCs to form colonies of mature cells in response to differentiative cytokine signals; the latter measures HSC frequency via limiting dilution, with HSCs able to self-renew in contact with bone marrow stromal cultures, such that after 6 weeks of *in vitro* culture HPCs which are present must have been generated as the progeny of a plated HSC. In CFC assays CD34+ CD38- cells with enforced expression of *GATA2* produced

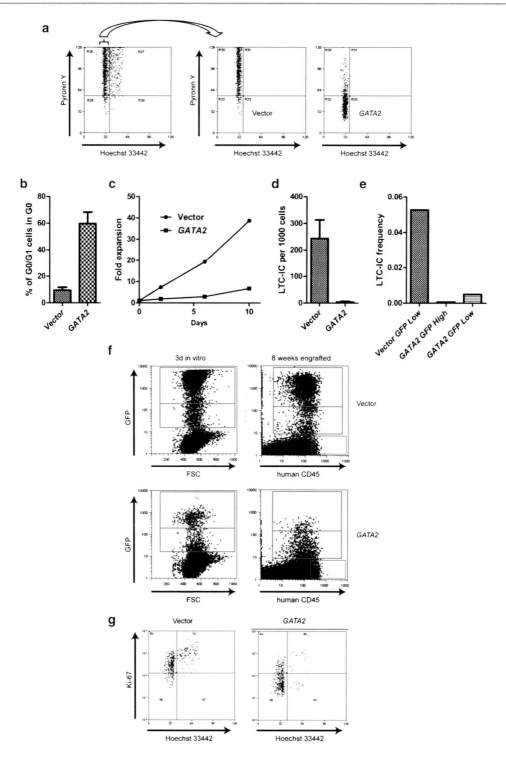

Fig. 26.1 (a) Lentiviral transduction of CD34+ CD38- cord blood mononuclear cells with a bicistronic h*GATA2*-IRES-eGFP construct confers quiescence as measured by reduction of Pyronin staining. These data are collated in (b) as means +SEM. (c) Quiescence is associated with a proliferative defect in CD34+ CD38- cord blood mononuclear cells cultured in SCF, TPO and Flt3L, in the absence of enhanced apoptosis (not shown). (d) Long-

fewer colonies, and colonies which did form were smaller. Frequency of all colony types was reduced, with erythroid colony frequency least affected. In LTC-IC assays with CD34+ CD38- cells, the compound effect of reduced proliferation during long-term culture combined with reduced CFC potential served to almost abrogate LTC-IC frequency (Fig. 26.1d). This last observation needs placing in context, as the LTC-IC assay involves a long-term culture phase on supportive stroma, during which HSCs/HPCs proliferate and produce progenitors with CFC activity that are detected after 6 weeks culture much as for normal CFC assays. Plating equal numbers showed that many fewer cells were produced from *GATA2*-transduced cells during the long-term culture phase than from control vector cells; thus reduced HSC proliferation yields fewer progenitors with CFC potential, and as already described, CFC activity was reduced also. Taken together, enforced expression of *GATA2* confers a functional defect on HSCs and HPCs which is intimately tied to increased quiescence and restrained proliferation.

Quiescence Conferred by GATA2 Is Level-Dependent

We further investigated the relationship between high *GATA2* expression and quiescence by sorting and experimentally interrogating cells with high and low GFP expression; GFP expression from bicistronic constructs is proportional to *GATA2* transgene expression. We found clear level-dependent inhibition of cell performance in these assays. Low-level overexpression of *GATA2* (~2-fold) in CD34+ cells was associated with minimal inhibition of *in vitro* proliferation, whereas higher levels (~9-fold) almost totally inhibited proliferation. Similarly, low-level overexpression barely affected performance of CD34+ cells in CFC assays, whereas high-level expression strongly inhibited CFC formation. In LTC-IC assays of the more primitive CD34+ CD38- fraction, a ~2-fold increase in *GATA2* expression effectively reduced LTC-IC frequency by ~10-fold, whereas ~9-fold overexpression almost completely abrogated LTC-IC frequency (Fig. 26.1e). These data together demonstrate that enforced *GATA2* expression exerts level-dependent effects on HSC/HPC performance *in vitro*.

Effects of Enforced GATA2 Expression on Haematopoietic Reconstitution

We therefore predicted that xenograft transplantation assays in NODSCID mice would show limited engraftment and restricted haematopoiesis from *GATA2*-transduced cells, with low-level overexpressing cells performing better than high-level cells. At 8 weeks after transplantation, sacrificed animals overexpressing *GATA2* in their grafts showed reduced GFP expression when compared with the GFP profile of the initial graft. Vector-transduced cells did not show any reduction in GFP expression profile (Fig. 26.1f). These data were strongly supportive of a level-dependent inhibition of haematopoiesis conferred by *GATA2*. These effects persisted for 12 weeks post-transplantation, suggesting that these data are due to effects of GATA2 on CB HSCs. Typically, and as seen in animals receiving a vector-transduced graft, the majority of human

Fig. 26.1 (continued) term culture-initiating cell (LTC-IC) assays show that enforced *GATA2* expression in CD34+ CD38- cord blood cells abrogates performance in this assay, as discussed in the text. (**e**) This effect is dependent on the level of *GATA2* expression which is enforced in the cells, but cells with the lowest *GATA2* expression are still inhibited in their LTC-IC frequency. (**f**) Mouse xenotransplantation of CD34+ CD38- cells with enforced expression of *GATA2* in NOD-SCID animals. Transduced CD34+ CD38- cord blood cells were injected into the tail veins of sublethally-irradiated mice, with an aliquot reserved for *in vitro* culture to reveal the GFP (and hence *GATA2*) expression profile in the absence of selective pressure. 8 weeks after transplantation, the reduced GFP profile revealed a failure of *GATA2*-high cells to engraft and repopulate the bone marrow, presumably due to increased quiescence. *GATA2*-low cells were more capable of engraftment and reconstitution. (**g**) Human cells recovered from engrafted NOD-SCID showed reduced *in vivo* expression of the proliferation marker Ki-67 upon enforced *GATA2* expression

cells in the BM are lymphoid. In animals receiving the *GATA2*-transduced graft however, most of the human cells were myeloid. It may be that reduced proliferative capacity in the myeloid progenitor fraction allows them to persist until 8 weeks, or alternatively enforced *GATA2* expression may favour the production or survival of myeloid rather than lymphoid cells. Given the implication of reduced cell output *in vivo*, and previous cell cycle data, we isolated human cells from xenografted animals and examined their cell cycle profile by staining for Ki-67 and found that the *GATA2*-conferred cell cycle defect was still present *in vivo* (Fig. 26.1g).

GATA2 Target Genes in Cell Cycle Control

We explored the effects of *GATA2* overexpression on expression of genes encoding key cell cycle regulatory proteins, and cell cycle inhibitors that regulate HSCs/HPC quiescence (Passegue et al. 2005). We first asked whether *GATA2*-conferred quiescence was due to induction of cyclin-dependent kinase inhibitors. By quantitative RT-PCR, we found that enforced *GATA2* overexpression in CD34+ CD38- cord blood cells did not induce $p16^{INK4A}$, $p19^{INK4D}$, $p21^{CIP1}$, $p27^{KIP1}$, or $p57^{KIP2}$ (data not shown). Enforcing *GATA2* expression in *p21−/−* LSKcells conferred quiescence to an equivalent degree as in wild-type LSKcells (Fig. 26.2a), and in these *p21−/−* LSKcells p27 protein was not induced by either a compensatory or parallel mechanism (Fig. 26.2b). Thus quiescence appears to be conferred by high GATA2 expression independently of $p21^{CIP1}$ and $p27^{KIP1}$.

We next examined expression of genes encoding key regulatory proteins in control of cell cycle in HSCs/HPCs(Passegue et al. 2005), concentrating on D-type cyclins and cyclin-dependent kinases relevant to the transition from quiescence into cycle. *CCND3* has been described as being expressed at low levels in quiescent murine LT-HSCs and repressed in T lymphoma cells induced into quiescence by glucocorticoid treatment (Rhee et al. 1995). Moreover, mice lacking all D-type cyclins demonstrated a severe haematopoietic defect in spite of other aspects of embryogenesis proceeding relatively normally; this is consistent with a role for D-type cyclins in restraining exhaustion of specifically haematopoietic stem and progenitor cells. We found *CCND3* to be repressed by enforced *GATA2* expression, whereas *CCND1* and *CCNE1* were not affected (Fig. 26.2c). In addition, expression of the kinases *CDK4* and *CDK6* was reduced (Fig. 26.2d). Chromatin immunoprecipitation analysis shows strong binding of GATA2 to the loci of *CCND3*, *CDK4* and *CDK6*, suggesting their regulation by GATA2 might be direct. CDK4 and CDK6 are the catalytic subunits of D-type cyclins and together act early in G_1; thus, reduced expression of several players in this axis might be responsible for the effects of high GATA2activity on reducing cell cycle entry, although further experiments are required to investigate this. Elevated expression of the Notch target gene *HES1* was observed in response to *GATA2* (Fig. 26.2e), a result which fits well with the proposed role for this TF in inhibiting progenitor cell cycling (Yu et al. 2006), Reduced expression of the quiescence-associated transcription factor *MEF/ELF4* (Lacorazza et al. 2006) was observed in response to GATA2 overexpression (Fig. 26.3a), and direct binding of GATA2 to the *MEF* locus was demonstrated by chromatin immunoprecipitation in FDCP-mix murine progenitor cells. Taken together, this suggested that repression of *MEF* might be, in part, responsible for the induction of quiescence mediated by *GATA2*. To test this we simultaneously enforced expression of both *GATA2* and *MEF* from lentiviral vectors marked with GFP and LNGFR respectively. *GATA2* conferred quiescence as measured by Hoechst/Pyronin staining to an equivalent degree in cells with enforced expression of *MEF* as it did in cells which did not overexpress *MEF* (Fig. 26.3b). This indicates either that MEF is not a point of control in GATA2-conferred quiescence, or at least that GATA2-conferred quiescence mechanisms are pleiotropic.

Given the highly synchronous activation of GATA2 afforded by the GATA2/ER fusion protein model, and thus its utility in rapidly inducing

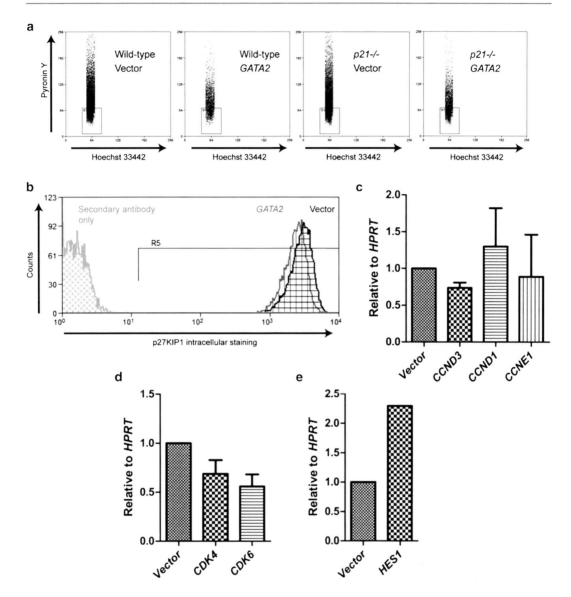

Fig. 26.2 (a) Enforced *GATA2* expression in LSK (Lin- Sca+ Kit+) cells from mice deleted for *p21* confers quiescence as measured by Hoechst/Pyronin just as in wild-type cells, confirming that *p21* is not a critical regulator of quiescence caused by *GATA2*. (b) These LSK cells do not show increased expression of p27 protein by intracellular flow cytometry, suggesting that p27 cannot substitute for p21 in quiescence conferred by GATA2, and is similarly uninvolved. (c) Quantitative RT-PCR analysis of cyclins in cord blood CD34+ CD38- cells. Expression relative to HPRT is shown as mean +SEM relative to vector-transduced cells. Only cyclin D3 was significantly repressed by enforced GATA2 expression. (d) In addition, cyclin-dependent kinases 4 and 6 were significantly repressed in a manner confirmed by Western blotting. (e) Quantitative RT-PCR analysis showing *GATA2* induces expression of the Notch target *HES1* in cord blood CD34+ CD38- cells. Expression relative to HPRT is shown as mean +SEM relative to vector-transduced cells

quiescence on a timescale measured in hours, Tariq Enver's laboratory generated stable clones of the factor-dependent multipotent cell line FDCP mix expressing GATA2/ER. These cells grew well *in vitro*, implying largely normal cell cycle characteristics. Microarray time course analysis of these cells after activation of the ER moiety with ligand showed a highly significant

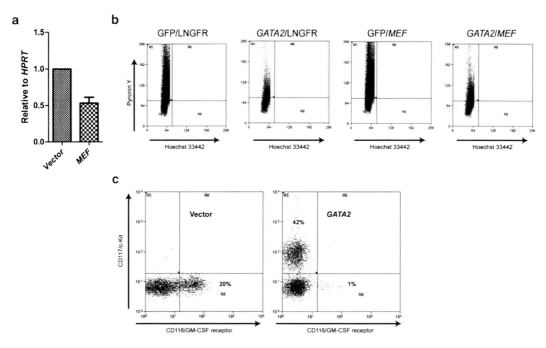

Fig. 26.3 (**a**) Quantitative RT-PCR analysis of *MEF/ELF4* in cord blood CD34+ CD38- cells. Expression relative to HPRT is shown as mean +SEM relative to vector-transduced cells. (**b**) Enforcing expression of both *GATA2* and its repressive target *MEF* in cord blood CD34+ cells confers quiescence similarly to *GATA2* expression alone, suggesting *GATA2* is not conferring quiescence solely via reduced *MEF* expression. (**c**) Expression of GM-CSF receptor (CD116) and c-Kit (CD117) are altered by enforced *GATA2* expression in myeloid differentiation conditions in a manner predicted to make cells less sensitive to GM-CSF and more sensitive to SCF

modulation of genes involved in cell cycle, including regulation of almost all *Mcm* (minichromosome maintenance) genes which together form part of the origin recognition complex (ORC) which licenses DNA replication origins in early G_1 (Tariq Enver, unpublished data). These data are also notable for showing the modulation of multiple amino-acyl biosynthetic proteins, perhaps revealing a role for GATA2 in modulating biosynthetic processes in cells which are preparing for a period of cellular dormancy.

GATA2 Modulates Cytokine Receptor Expression

We also examined the effect of enforced *GATA2* expression in CD34+ cord blood cells induced to differentiate towards terminal myeloid fates by a combination of Flt3L, SCF, IL3, G-CSF, M-CSF and GM-CSF. We found that enforcing *GATA2* expression caused changes in expression of cytokine receptors relevant to this process. After 12 days in myeloid differentiation conditions, *GATA2*-transduced cells did not express the receptor for GM-CSF, whereas around 25% of control cultures showed GM-CSFr expression. Failure to express GM-CSFr was correlated with induction of c-Kit expression consistent with the induction of c-Kit by GATA2 proposed in mast cells (Maeda et al. 2010), although these preliminary data do not imply simple conversion from a GM-CSFr+ fate to a c-Kit+ fate, as a higher proportion of cells become c-Kit+ than are otherwise GM-CSFr+ (Fig. 26.3c). It is tempting to speculate that part of a putative quiescence program in HSCs, apart from simply reducing metabolic activity, slowing cell cycling must be mediated in terms of enforcing non-responsiveness to signals in the milieu. In support of this idea, a link between *GATA2* expression and c-Kit signaling has been proposed in the *Gata-2* knockout model

(Tsai et al. 1994). Down-regulating surface receptors and signaling molecules via transcription factors such as GATA2 may silence responses to extrinsic cues in HSCs. Together, these data suggest that target genes of GATA2 constitute at least part of a quiescence program which can be reversibly activated in primitive haematopoietic cells in order to restrain their proliferation despite exposure to proliferative signals. Given our data showing effects on expression of cytokine receptors, together with direct regulators of cycle, GATA2 appears to regulate genes at multiple levels through which cytokine signals are connected to control of cell cycle.

Haploinsufficiency of Gata2 Also Impacts Quiescence Program

As high expression of *GATA2* enforces quiescence in HSCs, it is reasonable to conversely predict that reducing *GATA2* would attenuate quiescence. We directly assessed this hypothesis in *Gata2+/–* mice which provide an amenable model to assess the impact of reduced GATA2 level on HSC function. *Gata2+/–* adult marrow was diminished in both number and function of HSCs, which was unexpectedly associated with increased quiescence (Rodrigues et al. 2005). Thus, both up- and down-regulation of GATA2 regulates adult HSC quiescence (Rodrigues et al. 2005; Tipping et al. 2009). While our studies in the enforced expression setting have identified critical regulatory components that regulate quiescence(Tipping et al. 2009), the molecular basis for increased quiescence in HSCs from *Gata2+/–* mice remains unclear and awaits further investigation. Hence, it is not currently known whether lower and higher levels of GATA2 regulate HSC quiescence through the same partner proteins and target genes. For multiple reasons, our contention is that quiescence in these two settings is regulated by largely distinct mechanisms. Firstly, in addition to increased quiescence in HSCs from *Gata2+/–* mice, increased HSC apoptosis (Rodrigues et al. 2005) was observed together with decreased expression of anti-apoptotic BCL-XL. A mechanistic link between quiescence and apoptosis pathways, including those related to the BCL-XL family (Janumyan et al. 2003), has been demonstrated and it is possible that GATA2 and BCL-XL jointly regulate quiescence and apoptosis in *Gata2+/–* mice. In marked contrast, high GATA2 expression enforces quiescence without any effect on apoptosis. Secondly, it is entirely possible that the quiescent phenotype of HSCs from *Gata2-+/–* mice was a secondary protective mechanism, acting to prevent the erosion of HSC numbers caused by enhanced HSC apoptosis. Comparatively, the predominant, primary phenotype of enforcing GATA2 expression in HSCs was increased quiescence. Finally, reducing *Gata2* level in murine HSCs elicited a milder increase in quiescence than increasing it by the same factor (2-fold); this suggests that enforced expression of *GATA2* imposes a broader, stronger shutdown of the cell cycle machinery.

Further examination of bone marrow from *Gata2+/–* mice revealed that committed HPCs of the granulocyte-macrophage progenitor (GMP) lineage were attenuated in number and less able to self-renew *in vitro*, or produce mature GM progeny while other committed HPCs, such as common myeloid progenitors (CMPs) and lymphoid progenitors, remained unchanged by *Gata2* deficiency (Rodrigues et al. 2008). The origin of this functional defect was mapped to the quiescent fraction of *Gata2+/–* GMPs and at the molecular level to reduced HES-1 expression, as alluded to above a critical regulator of cell cycling in HSCs/HPCs. Thus, in *Gata2+/–* mice, GATA2 regulates quiescence at differentiation stage-specific levels of the haematopoietic hierarchy, at the HSC and GMP level.

Discussion

Previous studies had suggested a possible role for GATA2 in regulation of stem and progenitor cell growth and survival, although a formal causative role in regulation of HSC quiescence had not been demonstrated unambiguously. Enforcing *GATA2* expression in murine BM was previously shown to inhibit haematopoiesis *in vivo* (Persons et al. 1999), a result we confirmed and extended in human umbilical cord blood HSCs, yet the molecular mechanisms involved were not addressed. In addition, the effects of enforcing *GATA2* activity have been controversial, with different variants of *GATA2* having opposite effects on function and distinct molecular interactions depending on cell context (Kitajima et al. 2002). For example, in ES cells, enforced GATA2 activity is associated with increased clonogenic cell output rather than inhibition(Kitajima et al. 2002); however, in the same study, GATA2/ER activity repressed clonogenic cell output in a manner consistent with the data reported here for GATA2. Although these data raise concerns about the relative biological activities of the two proteins, investigated at the level of interactions with other TFs, they are broadly consistent with the known roles of GATA2.

The molecular regulator(s) of *GATA2*itself are unclear, with many suggested inducers of *GATA2* expression proposed. The correlative observations of high *GATA2*expressionin quiescent versus cycling HSCs from mouse BM (Venezia et al. 2004) and human cord blood(Tipping et al. 2009) suggest that niche interactions or other stimuli drive high *GATA2* expression in a cellular context whereby quiescence is the result. The specific regulators of*GATA2* are largely undescribed, but has been proposed to include C-MYB (Lorenzo et al. 2011); indeed the *c-myb–/–* phenotype strongly resembles the *Gata2–/–* model (Sumner et al. 2000). Other putative *GATA2* inducers include cytokine and integrin stimuli relevant to the niche (Dao and Nolta 2007), SCL (Dey et al. 2010), FLI1 (Liu et al. 2008) and EVI1 (Yuasa et al. 2005). GATA2 is also proposed to positively autoregulate, and for its activity to be modulated by post-translational modifications (Rodrigues et al. 2012). Although our data demonstrate a role for GATA2 in regulating gene expression programs associated with cellular quiescence, the precise molecular mechanism(s) by which cell cycle is arrested or extended in response to GATA2 are poorly-elucidated and require more experimentation. Determining global DNA binding sites contacted by GATA2 in quiescent versus cycling cells will be important, as well as investigations of GATA2 partner proteins relevant in HSCs to either regulating quiescence-effecting target genes or controlling cell cycle directly. A detailed dissection of the interaction partners of GATA2 in multipotent cells will be important. Indeed, the known rapid turnover of GATA2protein (Minegishi et al. 2005) may have relevance to the mechanisms by which quiescent cells re-enter cycle in the absence of *GATA2*-inductive signals. Better data correlating the transience of HSC quiescence with GATA2 level in a physiological, homeostatic setting is required to properly address this question.

As an important level dependent regulator of cell cycle and cell fate in HSCs/HPCs, one might expect GATA2 activity to be relevant to clinical haematology. Indeed, a slew of recent publications have tabled *GATA2* as a target for somatic mutation in predisposition to myelodysplasia (MDS), chronic myeloid leukaemia and acute myeloid leukaemia (Rodrigues et al. 2012). In several of these studies *GATA2* mutations liable to affect DNA binding have associated with similar but variable clinical phenotypes giving rise to MDS or leukaemia. This implies either mutation-specific clinical effects, or that these mutations act similarly but are extensively modified by co-operating genetic or environmental events. Nonetheless, these loss-of-function mutations alter the ability of GATA2 to interact with its target genes and is associated with predisposition to MDS/leukaemia in a manner predictable from its role in restraining HSC and progenitor cell proliferation. In addition, high *GATA2* expression in leukaemia has been associated with poor prognosis (Vicente et al. 2012) in a manner consistent with the conferral of quiescence in malignant HSCs/HPCs or leukaemia propagating cells and thus insensitivity to chemotherapeutic drugs, or alternatively via conferral of an anti-apoptotic gene expression program.

Identifying and understanding the target genes of GATA2 which comprise the GATA2-conferred quiescence gene expression program in normal and malignant haematopoiesis, and the niche signals which induce *GATA2* expression is the next challenge. These data will yield important insights that may enable specific therapeutic targeting of quiescent tumour cells which are resistant to chemotherapy by virtue of their lack of division. Separately, efforts to manipulate HSCs in HSC expansion and transplantation protocols have been limited by the relative quiescence of HSCs and identifying GATA2 target genes regulating the quiescence program in normal HSCs may facilitate such efforts.

Acknowledgement This work was supported by Leukaemia Lymphoma Research, the Medical Research Council, and EuroSyStem. Funding to Neil Rodrigues was provided by National Center for Research Resources (5P20RR018757-10) and the National Institute of General Medical Sciences (8 P20 GM103414-10) from the National Institutes of Health, BD Biosciences and the Rhode Island Foundation. We acknowledge our collaborators within the laboratories of Tariq Enver and Sten Eirik Jacobsen and the Weather all Institute of Molecular Medicine, Oxford; and thank Lorenza Lazzari (Cell Factory, Milan, Italy).

References

Cheng T, Rodrigues N, Shen H, Yang Y, Dombkowski D, Sykes M, Scadden DT (2000) Hematopoietic stem cell quiescence maintained by p21cip1/waf1. Science 287:1804–1808

Dao MA, Nolta JA (2007) Cytokine and integrin stimulation synergize to promote higher levels of GATA-2, c-myb, and CD34 protein in primary human hematopoietic progenitors from bone marrow. Blood 109:2373–2379

Dey S, Curtis DJ, Jane SM, Brandt SJ (2010) The TAL1/SCL transcription factor regulates cell cycle progression and proliferation in differentiating murine bone marrow monocyte precursors. Mol Cell Biol 30:2181–2192

Ehninger A, Trumpp A (2011) The bone marrow stem cell niche grows up: mesenchymal stem cells and macrophages move in. J Exp Med 208:421–428

Grassinger J, Haylock DN, Williams B, Olsen GH, Nilsson SK (2010) Phenotypically identical hemopoietic stem cells isolated from different regions of bone marrow have different biologic potential. Blood 116:3185–3196

Heyworth C, Gale K, Dexter M, May G, Enver T (1999) A GATA-2/estrogen receptor chimera functions as a ligand-dependent negative regulator of self-renewal. Genes Dev 13:1847–1860

Janumyan YM, Sansam CG, Chattopadhyay A, Cheng N, Soucie EL, Penn LZ, Andrews D, Knudson CM, Yang E (2003) Bcl-xL/Bcl-2 coordinately regulates apoptosis, cell cycle arrest and cell cycle entry. EMBO J 22:5459–5470

Kitajima K, Masuhara M, Era T, Enver T, Nakano T (2002) GATA-2 and GATA-2/ER display opposing activities in the development and differentiation of blood progenitors. EMBO J 21:3060–3069

Lacorazza HD, Yamada T, Liu Y, Miyata Y, Sivina M, Nunes J, Nimer SD (2006) The transcription factor MEF/ELF4 regulates the quiescence of primitive hematopoietic cells. Cancer Cell 9:175–187

Li J (2011) Quiescence regulators for hematopoietic stem cell. Exp Hematol 39:511–520

Liu F, Walmsley M, Rodaway A, Patient R (2008) Fli1 acts at the top of the transcriptional network driving blood and endothelial development. Curr Biol 18:1234–1240

Lorenzo PI, Brendeford EM, Gilfillan S, Gavrilov AA, Leedsak M, Razin SV, Eskeland R, Saether T, Gabrielsen OS (2011) Identification of c-Myb target genes in K562 cells reveals a role for c-Myb as a master regulator. Genes Cancer 2:805–817

Maeda K, Nishiyama C, Ogawa H, Okumura K (2010) GATA2 and Sp1 positively regulate the c-kit promoter in mast cells. J Immunol 185:4252–4260

Minegishi N, Suzuki K, Kawatani Y, Shimizu R, Yamamoto M (2005) Rapid turnover of GATA-2 via ubiquitin-proteasome protein degradation pathway. Genes Cells 10:693–704

Nie Y, Han YC, Zou YR (2008) CXCR4 is required for the quiescence of primitive hematopoietic cells. J Exp Med 205:777–783

Orlic D, Anderson S, Biesecker LG, Sorrentino BP, Bodine DM (1995) Pluripotent hematopoietic stem cells contain high levels of mRNA for c-kit, GATA-2, p45 NF-E2, and c-myb and low levels or no mRNA for c-fms and the receptors for granulocyte colony-stimulating factor and interleukins 5 and 7. Proc Natl Acad Sci U S A 92:4601–4605

Passegue E, Wagers AJ, Giuriato S, Anderson WC, Weissman IL (2005) Global analysis of proliferation and cell cycle gene expression in the regulation of hematopoietic stem and progenitor cell fates. J Exp Med 202:1599–1611

Persons DA, Allay JA, Allay ER, Ashmun RA, Orlic D, Jane SM, Cunningham JM, Nienhuis AW (1999) Enforced expression of the GATA-2 transcription factor blocks normal hematopoiesis. Blood 93:488–499

Rhee K, Bresnahan W, Hirai A, Hirai M, Thompson EA (1995) c-Myc and cyclin D3 CcnD3. genes are independent targets for glucocorticoid inhibition of lymphoid cell proliferation. Cancer Res 55:4188–4195

Rodrigues NP, Boyd AS, Fugazza C, May GE, Guo Y, Tipping AJ, Scadden DT, Vyas P, Enver T (2008) GATA-2 regulates granulocyte-macrophage progenitor cell function. Blood 112:4862–4873

Rodrigues NP, Janzen V, Forkert R, Dombkowski DM, Boyd AS, Orkin SH, Enver T, Vyas P, Scadden DT (2005) Haploinsufficiency of GATA-2 perturbs adult hematopoietic stem-cell homeostasis. Blood 106:477–484

Rodrigues NP, Tipping AJ, Wang Z, Enver T (2012) GATA-2 mediated regulation of normal hematopoietic stem/progenitor cell function, myelodysplasia and myeloid leukemia. Int J Biochem Cell Biol 44:457–460

Sumner R, Crawford A, Mucenski M, Frampton J (2000) Initiation of adult myelopoiesis can occur in the absence of c-Myb whereas subsequent development is strictly dependent on the transcription factor. Oncogene 19:3335–3342

Tipping AJ, Pina C, Castor A, Hong D, Rodrigues NP, Lazzari L, May GE, Jacobsen SE, Enver T (2009) High GATA-2 expression inhibits human hematopoietic stem and progenitor cell function by effects on cell cycle. Blood 113:2661–2672

Trumpp A, Essers M, Wilson A (2010) Awakening dormant haematopoietic stem cells. Nat Rev Immunol 10:201–209

Tsai FY, Keller G, Kuo FC, Weiss M, Chen J, Rosenblatt M, Alt FW, Orkin SH (1994) An early haematopoietic defect in mice lacking the transcription factor GATA-2. Nature 371:221–226

Venezia TA, Merchant AA, Ramos CA, Whitehouse NL, Young AS, Shaw CA, Goodell MA (2004) Molecular signatures of proliferation and quiescence in hematopoietic stem cells. PLoS Biol 2:e301

Vicente C, Vazquez I, Conchillo A, Garcia-Sanchez MA, Marcotegui N, Fuster O, Gonzalez M, Calasanz MJ, Lahortiga I, Odero MD (2012) Overexpression of GATA2 predicts an adverse prognosis for patients with acute myeloid leukemia and it is associated with distinct molecular abnormalities. Leukemia 26:550–554

Yu X, Alder JK, Chun JH, Friedman AD, Heimfeld S, Cheng L, Civin CI (2006) HES1 inhibits cycling of hematopoietic progenitor cells via DNA binding. Stem Cells 24:876–888

Yuasa H, Oike Y, Iwama A, Nishikata I, Sugiyama D, Perkins A, Mucenski ML, Suda T, Morishita K (2005) Oncogenic transcription factor Evi1 regulates hematopoietic stem cell proliferation through GATA-2 expression. EMBO J 24:1976–1987

Dormancy and Recurrence of Cancer Stem Cells in Bone: Role of Bone Morphogenetic Proteins

Sambad Sharma, Fei Xing, and Kounosuke Watabe

Contents

Abstract	289
Introduction	290
Clinical Significance of Dormancy	291
Molecular Mechanisms of Tumor Dormancy	292
Cancer Stem Cell and Dormancy	293
Tumor Dormancy in the Bone	294
Roles of BMP7 in Dormancy in the Bone	295
Conclusions and Future Directions	297
References	297

S. Sharma • F. Xing • K. Watabe (✉)
Cancer Institute, University of Mississippi
Medical Center, Jackson, MS, USA
e-mail: kwatabe@siumed.edu

Abstract

Metastatic disease is the leading cause of death in cancer patients. Despite significant advancement in understanding the pathological process at cellular and molecular level, there is no effective cure to this devastating disease till date. Dissemination of cancer cell into the blood is considered to occur even at an early stage in cancer progression that eventually leads to formation of secondary tumor or metastasis after remaining dormant for a long or short period of time. This variation in time period can be attributed to the properties of both disseminated tumor cells and the new homing microenvironment at the metastatic sites. Increasing line of evidences indicate that the disseminated cells possess cancer stem cell like properties that render them with the ability to disseminate from primary tumor and remain dormant in the new microenvironment at the distant organs. However, it is not clear about how and when these cancer stem-like cells (CSCs) are able to switch into proliferative state to manifest as recurrent disease. CSCs must adopt different genetic and epigenetic changes to potentiate themselves with proliferative phenotype leading to recurred tumor. On the other hand, immediate microenvironment of CSCs exerts restrictive barrier for growth of these cells or even induce cell death. To evade from the inhibitory signaling from the microenvironment, these disseminated CSCs switch to dormant phenotype. It has been recently found that bone morphogenetic proteins (BMPs) released by microenvironmental cells plays a critical role in inducing dormancy and deprivation of BMPs leads to reversal of dormancy and increase in proliferation of dormant cells. Role of BMPs in the dormancy-recurrence axis may offer a therapeutic approach to retain the disseminated CSCs in perpetual dormant state. Therefore, comprehensive understanding of the role of BMPs in various microenvironment of disseminated tumor cell is considered to open a new avenue for the treatment of cancer metastasis.

Keywords

Bone marrow and trastuzumab • Bone morphogenetic protein 7 (BMP7) • Cancer stem cell and dormancy • Circulating tumor cells (CTC) • Disseminated tumor cells (DTC) • Mesenchymal to epithelial transition (MET) • Minimal residual disease (MRD) • Molecular mechanisms of tumor dormancy • Tumor dormancy in bone • Vascular cell adhesion molecule-1 (VCAM-1)

Introduction

Cancer dormancy is referred to the period when primary tumor has already been diagnosed, surgically resected and patient is clinically disease-free before relapse as recurred tumor on different organs (Sosa et al. 2011). During this period, tumor stays asymptomatic and occult. This period is also termed as minimal residual diseased-state. After years of dormancy, these cancer cells may start to grow and eventually form secondary tumor (Aguirre-Ghiso 2007). While cancer progresses, tumor cells are disseminated from primary tissue to different organs and these cells are known as disseminated tumor cells (DTC) that can be identified by immunocytochemical staining (Allgayer and Aguirre-Ghiso 2008). Tumor cells are also found to circulate in the bloodstream and are termed as circulating tumor cells (CTC). Because CTC potentially seed to different organs in the body to establish metastatic tumor, the number of CTCs can be used as diagnostic and prognostic marker (Slade et al. 2009). Previously, dissemination of tumor cells was thought to occur only at a later stage of cancer progression. However, recent evidences indicate that dissemination is one of the earliest steps, even before primary tumor is diagnosed (Huseman et al. 2008). These disseminated tumor cells must acquire ability to survive immune surveillance before their seeding to different parts of the body as outlined in Fig. 27.1.

In the process of dissemination, immune surveillance leads to apoptosis of the DTC; however, proliferating cells compensate for apoptosis leading to dynamic equilibrium of these cells. This state of dynamic equilibrium is also known as micrometastasis and has latency and survival potential through crosstalk between cancer cell and immune system via specific secretory factors (Aguirre-Ghiso 2007; Mackie et al. 2003). Micrometastasis can also occur due to a lack of signaling for inducing angiogenesis (Aguirre-Ghiso 2007). Similarly, immediate tumor microenvironment may inhibit angiogenic signal, further contributing to the state of micrometastasis. By overcoming all these barriers, some tumor cells grow as a recurred tumor years or even decades after dissemination from their primary organ. These cells must acquire genetic or epigenetic changes or gain a highly favorable environment to relapse. Therefore, both genetic reprogramming and environmental suitability is equally important in deciding the fate of these disseminated tumor cells, and understanding the biology of dormant cells may reveal the exact mechanism behind the dormancy-recurrence loop. Recent evidences also suggest that dormant cells must retain cancer stem like properties to disseminate from the primary tumor, reside in distant organs as dormant tumor cell and eventually relapse into secondary tumor (Paez et al. 2011). Identifying the mechanism that leads to switch from dormancy to recurrence can offer a therapeutic target either by preventing recurrence with inhibition of factors responsible for dissemination or by expression of factors that retains these cells in a dormant phase. There are multiple mechanisms that have been proposed in the regulation of

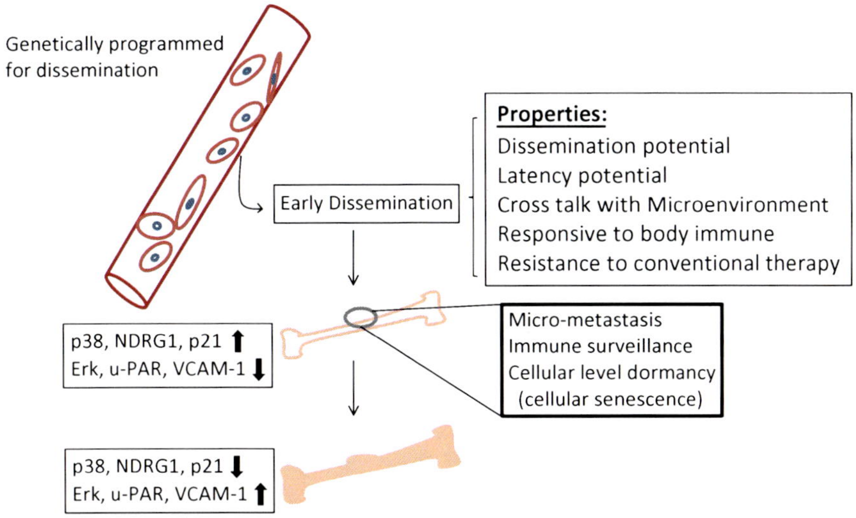

27 Dormancy and Recurrence of Cancer Stem Cells in Bone: Role of Bone Morphogenetic Proteins

Fig. 27.1 Properties of dormant disseminated cells that lead to dormancy and factors responsible for recurrence: Cells disseminate from primary tumor into the bloodstream and reside in distant organs (only bone shown in figure). Being genetically programmed, these cells possess the ability to survive and remain dormant. Dormancy occurs via mechanism of micro-metastasis, immune surveillance and cellular level dormancy. Various factors play roles during maintenance of dormancy or relapse into recurrent tumor

dormant-recurrence switch. One of the mechanisms includes reversible senescence of CSCs (Kobayashi et al. 2011). Disseminated tumor cells residing in different organs undergo senescence as a means to survive within that environment. By reversing back from the senescence, CSCs start proliferating and thus overcome the dormant phase, and this reversal was shown to be inhibited by Bone Morphogenetic protein (BMP). This chapter focuses on the roles of BMPs in dormancy and mechanisms behind the reversible senescence.

Clinical Significance of Dormancy

Surgical resection of the primary tumor is an inevitable step in cancer treatment for most cases. Usually, surgery is followed by chemotherapy or radiotherapy; however, dormant cells are known to be resistant to chemotherapy (Naumov et al. 2003; Aguirre-Ghiso et al. 1999). Although the adjuvant therapies are generally inefficient in eliminating cytokeratin positive disseminated cells in the bone marrow, recent findings suggest that drugs like Trastuzumab can decrease clinical relapse in breast cancer patients by decreasing chemo-resistant CK-19 positive CTC (Braun et al. 2000; Georgoulias et al. 2012). Therapeutic efficiency of this drug is not yet validated at a clinical level, however, these lines of evidences provide new hope in the treatment of cancer dormancy. Early identification of CK positive dormant cells may increase efficacy of treatment in cancer patients. While CTC is readily detectable, dormant tumor cells are clinically unidentifiable and the significance of these dormant tumor cells is highlighted only after recurrence. Time required and number of patients identified for recurrence varies and highly depends on types, and stage of cancer. Besides patients' health status also makes cancer recurrence an unpredictable phenomenon. In one case of breast cancer, circulating tumor cells were found in patients' blood for 22 years (Goss and Chambers 2010). What differentiates these tumor cells in the time they require to recur is not known. Currently, identification of dormant tumor cell includes isolation of DTC from the bone marrow of patients or isolation of CTC from peripheral blood, which has not been able to reveal the fact behind the time difference for recurrence

(Husemann et al. 2008); however, these studies relate dissemination as a very early step during cancer progression. In one study of prostate cancer, 70 % of men who had undergone radical prostatectomy were detected with DTC in their bone marrow prior to surgery, suggesting early escape of DTC (Morgan et al. 2009). Importantly, 5–40 % of patients harbored micrometastatic disease even when the disease was pathologically confined to the organ at the time of surgery (Ellis et al. 2003), suggesting that very few genetic changes are required for dissemination to occur. Bone marrow samples were even found to be DTC-positive in patients for 9 years after surgery of primary prostate tumor which provides evidence to the role of microenvironment in modulating the state of dormancy of disseminated tumor cells (Weckermann et al. 2009). Understanding the mechanism by which these disseminated cells become dormant may provide more detailed answers to variation in time for recurrence. In one study comprising 800 patients, presence of DTC was found to be related to significantly higher tumor recurrence and death of patients. However, this was not the case in ovarian cancer. Although high percentage of patients showed presence of DTC in the bone, they rarely developed bone metastasis (Allgayer and Aguirre-Ghiso 2008). This adds controversy to whether presence of DTC in the bone marrow actually relates to recurrence or not. Recent study on copy number variation analysis of circulating free DNA isolated from plasma of breast cancer patients has established one more way to distinguish dormant cells at clinical level. Shaw et al. (2012) have recently shown that genomic profile of circulating free DNA represented a set of copy number variation associated with minimal residual diseased state of the patients. Similarly, Schwarzenbach et al. (2009) identified cell-free tumor DNA from the plasma of 81 prostate cancer patients which served as a marker for metastatic progression of the cancer.

The results of clinical studies on dormancy also offer a tremendous challenge to researchers. Although there have been significant improvements, more detailed findings that unravels the mechanism of each type and subtypes of cancers are required to target the dormant cells at a clinical level. Drugs like Trastuzumab can be promising in breast cancers although it may not be equally effective in the case of other cancers; however, more decisive findings in the pathological mechanism that can be translated into the clinical setting are needed for therapeutic application. Pre-clinical studies must be robust to support clinical trials for significant advances in the treatment of dormancy. Retention of the tumor cell in a dormant state may serve as a therapeutic option so that cancer can be treated as a chronic disease.

Molecular Mechanisms of Tumor Dormancy

In general, tumor dormancy ensues when proliferation is counter balanced by apoptosis due to (i) insufficiency of angiogenesis, (ii) regulation by microenvironmental factors and (iii) immunosurveillance (Aguirre-Ghiso 2007). This check and balance between growing tumor cells that inhibits proliferation of disseminated cells in different organs is also known as micrometastasis; however, tumor cells can also transit to dormant phase at a cellular level due to cell cycle arrest. Several different molecular signaling has been identified that play roles in inducing dormancy. One of the prominent findings relates dormancy and recurrence process to Erk1/2-p38α/β axis (Sosa et al. 2011). Erk- extracellular regulated kinase and p38 kinases belong to Mitogen activated protein kinases (MAPK) family, and alternate ratio of Erk1/2 and p38α/β were found during cancer progression. During early dissemination of tumor cells, Erk expression was found to be high whereas p38 expression was found to be low. However, during dormancy, lower expression of Erk and higher expression of p38 expression inhibited the growth of the tumor (Aguirre-Ghiso et al. 2003). The new environment of the disseminated cells imposed restrictive effect for tumor growth that lead to difference in expression of Erk and p38 kinases, and this differential expression attributed to dormancy within that microenvironment. Another finding relates dormancy to genetic programming of the tumor cells that renders them with the potential to survive immune surveillance by acquiring genetic and epigenetic changes enough to seed and survive in different organs as disseminated dormant cells; however, additional changes or genetic reprogramming were still required in order to switch from dormant to proliferative state (Bragado et al. 2012).

Cellular dormancy represents cell cycle arrest that occurs on a cellular level due to expression of cell cycle inhibitor proteins such as p21, p27 and p16 or loss of pro-proliferative signals (Aguirre-Ghiso et al. 1999; Sosa et al. 2011). Together with expression of cell cycle inhibitors, dormant tumor cells can also undergo a phase of senescence at the cellular level in order to evade restrictive barriers, when they disseminate. Cellular senescence, which is the phase of growth arrest, is known to occur to limit excessive growth of cancer. Although senescence has long been considered to be irreversible, this process was recently found to be reversible during dormancy through intricate interaction with microenvironmental cells (Kobayashi et al. 2011)., suggesting that signaling between cancer cells and its immediate microenvironment plays a role in inhibiting cells from proliferation and induces dormancy. Figure 27.1 shows overview of molecular signaling involved in dissemination and cellular senescence. Whenever cells are under restrictive signaling or incapacitated with enough changes to proliferate, it undergoes cellular senescence. As soon as it gains supportive signaling, it then starts to proliferate, reversing the stage of senescence. It is also documented that dormancy and cellular senescence are induced by the expression of tumor and metastasis suppressor genes (Kobayashi et al. 2011).

Cancer Stem Cell and Dormancy

Recent stem cell theory suggests that cancer stem-like cells (CSCs) are tumor-initiating cells that have self-renewal and differentiation capabilities and can drive cancer progression. One of the important properties of CSCs includes epithelial to mesenchymal transition (EMT). CSCs are also known to be resistant to radio- and chemotherapy (Pantel and Alix-Panabieres 2007; Polyak and Weinberg 2009). The fact that dormant cells can survive conventional therapy provided clue that dormant cells may possess CSCs properties (Naumov et al. 2003). According to the cancer stem cell theory, a dormant and recurrent tumor cell must retain stem cell–like properties. On the other hand, properties of DTC such as resistance to chemo-therapy and their restrictive environment, potential to relapse in suitable microenvironment and staining for cytokeratin clearly shows that DTC are dormant forms of tumor initiating cells. Moreover, CTC has previously been shown to possess EMT and stem cell characteristics in majority of metastatic breast cancer patient (Aktas et al. 2009), which further documented the CSC characteristics of dormant cells. However, EMT is not a necessary and sufficient condition for metastatic relapse. Muller et al. (2010) has hypothesized that mesenchymal to epithelial transition (MET) should also take place for formation of overt metastasis. EMT leads to dissemination of primary tumor cells to target organs; however, after homing to the secondary organ such as bone marrow, these DTC undergo MET before metastatic relapse. This hypothesis explains the finding of high incidence of DTC compared to CTC in blood and bone marrow sample from same patients when assayed based on the cytokeratin marker. There might be even a possibility that MET occurs before homing, thereby endowing the capability of extravasation to disseminated cells.

Recent findings show increasing evidences in the role of BMPs (Bone Morphogenetic proteins) in inducing differentiation and reducing features of CSCs in cancers related to colon, brain, hepatic and breast (Gao et al. 2012; Zhang et al. 2012). Interestingly, an inhibitor of BMPs, Coco, was shown to enhance the features of CSC and induced lung metastasis of breast cancer cells. BMP-induced signaling from the stroma initially inhibited metastasis in lung and induced dormancy; however this dormant nature was reversed by the Coco expression in the lung leading to formation of macrometastasis. Furthermore, BMP was shown to be able to inhibit features of CSCs such as tumor initiation and self-renewal *in vivo* and *in vitro*, respectively. Coco also promoted expression of the transcription factors related to embryonic stem cells including Nanog, Sox-2, Oct-4 and Hippo transducer Taz that would otherwise be inhibited by BMPs (Gao et al. 2012). Similarly, signaling from BMP4 was found to induce differentiation of CD133-positive cells in hepatocellular carcinoma. It is noteworthy that Varga and Wrano (2005) have also shown the role of BMP related pathway in the promotion of differentiation and inhibition of self-renewal in pluripotent and adult stem cells. Pretreatment of mice with BMP was indeed found to inhibit colonization of metastatic breast cancer in the bone (Buijs et al. 2007). These mounting evidences clearly indicate that BMP signaling inhibits successful metastasis by inhibiting the CSC traits or by promoting differentiation. Therefore, disseminated cancer cells can be maintained in a dormant state by these signaling while reversal of this signaling can be induced by factors such as Coco to reactivate the CSC properties of the cells leading to recur into secondary tumor.

Tumor Dormancy in the Bone

Bone is one of the major metastatic sites for many types of tumors. Cancer cells are known to induce osteoclastic bone resorption by secreting certain growth factors or cytokines. Bone resorption further releases growth factors from bone matrix which in turn enhances cancer cell proliferation and bone metastasis. However, DTC reside as Minimal Residual Disease (MRD) in bone and follows this vicious cycle only after a long period of time with transition from the dormant state to proliferative stage (Andrews 2012). Genetic determinants of these DTC may explain the lengthening of occurrence of the vicious cycle. Aguirre-Ghiso et al. (1999) has recently shown the role of u-PAR receptor in regulating a shift between single cell tumor dormancy and proliferation. Binding of u-PAR ligand activates plasmin that degrades extracellular matrix and basement membrane components of bone leading to invasion, extravasation and metastasis, and therefore, regulates the switch to proliferative state. More recently, Lu et al. (2011) identified VCAM-1 (Vascular cell adhesion molecule-1) from non-proliferative indolent cells that were initially isolated from bone metastasizing cell line. However, these cells were able to metastasize via the expression of VCAM-1 and played a critical role in the escape of indolent phase of the cancer cell to aggressive state. Indeed, breast cancer patients with high expression of VCAM-1 relapse earlier compared to patients with low VCAM-1. These results suggest that expression of molecules such as u-PAR and VCAM-1 may provide a switch in re-activating the dormant state of cancer cell (Fig. 27.1); however, what regulates their expressions of these genes during dormancy and recurrence is still virtually unknown.

Bone marrow is known to be the favorite place of homing for breast, prostate, colon, lung and ovarian cancers. Bone microenvironment is primarily composed of bone-derived mesenchymal cells that are far unlikely to be present in the primary tumor stroma (Pedersen et al. 2012). Correlation between the presence of DTC in the bone of patient to bone metastasis varies between types of cancers. In the case of prostate and breast cancers, bone is the most common site for metastasis, and this can be attributed to both physiochemical properties of the bone and certain chemokine signaling specially mediated by CXCR4, CXCR6, and CXCR7 (Thobe et al. 2011). It is still controversial whether DTC present in the bone correlates with minimally residual disease in these cancers; however, quantitative development of DTC in bone has association with bone recurrence in patients (Allgayer and Aguirre-Ghiso 2008). Recently, it has been shown that DTC of prostate cancer actually targets hematopoietic stem cell (HSC) niche in the bone and replaces it before residing there (Pedersen et al. 2012). In the case of breast cancer, disseminated cells were found to exist near endosteum of the bone, and these cells form gap junctional intercellular communication (GJIC) with supporting hematopoietic cell and stromas (Kiel and Morrison 2008). Interestingly, Lim et al. (2011) has recently identified certain microRNAs (miRNAs) that transported from bone marrow stroma to breast cancer leading to tumor cell quiescence. These miRNAs crossed GJIC via gap junction and exosomes and inhibited tumor cell proliferation by reducing CXCL12 expression. Early dissemination of cancer cells to bone has also been noted in lung and colon cancer (Pantel et al. 2003; Allgayer et al. 1997). These cancers normally do not metastasize to bone; however, DTC, identified through its staining for certain cytokeratin indicated poor prognosis for colon cancer.

Disseminated tumor cells in the bone have to encounter the suppressive signaling from the bone microenvironment either via direct cell-cell interaction, cytokines or exosomal miRNA. Therefore, targeting HSC niche may be one of the steps in the process of acclimatization to suppressive signaling. Tumor cells must breakdown this restrictive effect to grow as a metastatic tumor. Expression of receptor like uPAR or protein like VCAM-1 may be used as such a mechanism to reactivate the dormant cells, and changes in the expression of these and other unidentified factors are crucial in deciding the fate of disseminated tumor cells.

Roles of BMP7 in Dormancy in the Bone

Emerging evidences on the role of microenvironment in dormancy-recurrence loop explore the promising end in the therapeutic treatment of dormancy, and cellular senescence has been recently found to be one of the mechanisms responsible for inducing dormancy. Various mechanisms have been known to trigger cellular senescence which limits excess or aberrant cellular proliferation, and therefore, senescence provides an anti-carcinogenic programming to normal cells thus preventing normal cell transformation into a malignant one (Collado et al. 2007). In the cancer cells, it has been known that the expression of senescent-inducing proteins such as cell cycle inhibiting such as p53, p21, and p16 are indicative of tumor suppressive effect. Induction of senescence was also found to occur in response to chemotherapy and radiation treatment (Roninson 2012). In cancer, oncogene induced senescence has been one of the significantly noticed event. Collado et al. (2005) showed that premalignant lesions but not adenocarcinomas of lung were abundantly positive for senescent cells in K-Ras conditional knockin mouse model verifying its tumor suppressive role during early cancer progression. Thus, senescence has a critical role to play in both normal and cancer cells, and it was found to be a process to induce dormancy in prostate cancer stem-like cells (Fig. 27.2).

Kobayashi et al. (2011) have recently shown the role of Bone Morphogenetic protein 7 (BMP7) in inducing dormancy of CSCs in prostate cancer. They have shown that secretory factors from stromal cells were able to suppress the growth of prostate cancer cells in-vitro. In an approach to identify the factor involved in this suppression, BMP7 was found to induce senescence in prostate cancer stem-like cells, and BMP7 was indeed expressed significantly higher in bone stromal cells than prostate cancer stem-like cells. BMP7, a member of TGF-β family of proteins known for its role in bone morphogenesis, was found to be secreted from various bone stromal cells. This protein was previously known to inhibit bone metastasis *in vivo*; however, the exact mechanism underlying this inhibition was unclear (Buijs et al. 2007). Importantly, senescence induced by BMP7 in CSCs was shown to be reversible and CSCs regained its ability to proliferate immediately after withdrawal of BMP7 both *in vitro* and *in vivo*. Therefore, key players in reversing this senescence could be potential therapeutic targets for inhibiting relapse in bone. The authors have also shown that BMP7 activates p38 MAPK, cell cycle inhibitor, p21 and tumor metastasis suppressor gene, NDRG1 (N-Myc downstream regulated gene) in CSCs. NDRG1 is previously known to suppress tumor metastasis in prostate, breast and colon cancers. The expression of NDRG1 also has been shown to have inverse correlation with bone metastasis in prostate cancer patients (Bandhopadhyay et al. 2004). As mentioned earlier in this chapter, high p38 expression dictates early dissemination and

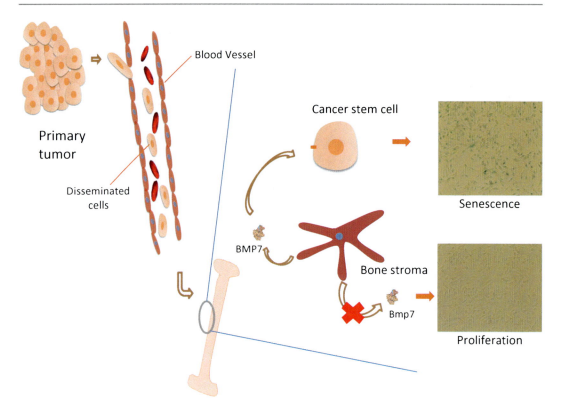

Fig. 27.2 A primary tumor cell disseminates into the bloodstream reaching distant organs including bone. Breast and prostate cancer are known to frequently metastasize to the bone. The stroma of the bone secretes BMP7 in response to CSCs. BMP7 further inhibit proliferation of CSCs by inducing cellular senescence (β-galactosidase stained cells). Breakdown of Bmp7-induced-senescence of CSCsl leads to reversal of senescence and enhances cell proliferation

dormancy whereas high Erk phosphorylation promotes proliferation and metastasis. It was found that, in the presence of BMP7, phosphorylation of Erk was greatly suppressed whereas BMP7 withdrawal reversed this phosphorylation. Being stimulated by BMP7, NDRG1, in a p38 and p21 dependent manner further induced reversible cellular senescence in CSCs via G1 arrest (Kobayashi et al. 2011). In this context, it should be noted that cell cycle arrest at G1 phase is known to be important for disseminated cancer cells in evading immune recognition to remain in sub-clinical disease state (Aguire-Ghiso 2007).

Interestingly, analysis of expression profiles of BMP receptors using existing database showed expression of BMPR2 (Bone Morphogenetic protein Receptor 2) to be positively correlated with recurrence-free survival of patients. In addition, knockdown of BMPR2 in cancer cell abrogated BMP7-induced dormancy *in vivo*, indicating the immense role of this receptor during dormancy. The effect of BMP7 on recurrent growth was also found to be independent of Androgen Receptor (AR) status, and both AR independent and AR dependent cell lines showed senescence phenotype after addition of BMP7 (Kobayashi et al. 2011). Accordingly, it was suggested that BMP7 functioned in an osteoinductive manner to suppress bone tumor by inhibiting osteolysis which is in accordance with the previous notion that increase in osteolytic type lesion is a precursor of relapse of prostate cancer (Shimazaki et al. 1992).

As mentioned above, early dissemination of cancer cells from their primary environment is insufficient for metastasis; however, these early disseminated cells may represent certain subset of cancer stem-cell that possesses the capability to disseminate and reside in different body organs, although they lack mechanisms for proliferation. Secretion of BMP7 is one of the mechanisms to inhibit proliferation of the CSC within bone which would otherwise have successfully grown into secondary tumor. In a clinical setting, CSCs within the bone must essentially overcome restrictions from its immediate environment, with some genetic and epigenetic changes, and proliferate to form metastasized tumor. One potential model of recurrence can be adaptation and integration of disseminated CSC within niche of hematopoietic stem cell rendering them with the potential to revert from state of MET to EMT leading to cell cycle progression. Senescent state induced by BMP7 would then be reversed by utilizing favorable factors from hematopoietic stem cell niche for tumor growth. Based on the result of this study and considering the role of BMP7-p38-NDRG1 axis, a molecule that mimics BMP7 in combination with drug targeting EMT can potentially inhibit bone recurrence and can be used to treat the patients who are diagnosed with disseminated cells.

Conclusions and Future Directions

Overall, evidences presented above show the critical roles of bone microenvironment during dormancy of disseminated tumor cells. Various line of evidences indicate that DTC, more specifically a disseminated tumor stem cell, remains dormant in the bone for a certain time period before its recurrence whereas in some cases the bone was used only as a transit point to metastasize to different organs especially in the case of gastric cancer. Dormant DTC persist either by withdrawing itself from the cell cycle or by balancing the rate of proliferation with the rate of cell death. Identification of DTC and CTC has been a major breakthrough for the study of disseminated cells in the bone and blood. Unraveling the precise mechanism behind dormancy by using DTC and CTC as a tools can render new direction for finding a therapeutic treatment for cancer. Approaches can be made either by targeting the dormant cell itself or targeting factors responsible for recurrence to prevent relapse. Similarly, cell surface markers or antigens of DTC associated with dormant phenotype of cancer cells at particular site can be functionally characterized. The fact that dormant tumors are asymptomatic and undetected due to its small size can be overcome by improvisation in isolation techniques to increase the efficiency of DTC or CTC isolation. The increase in quantity of isolation of disseminated cells can also be helpful in further exploration of properties and characteristics of these cells. Moreover, establishment of model(s), whether in-vivo or in-vitro, that recapitulates dormancy has become a necessity to decode unexplored mechanisms involved in dormancy. There is also an immense need of a system that can recapitulate microenvironment of various organs as change in factor(s) within the microenvironment has been shown to be important in the regulation of stemness of organ residing CSCs. Comprehensive characterization of DTC and CTC can contribute to establish prominent understanding of these cells and to expand knowledge behind its regulation. Research identifying the molecular signatures induced by each process of dormancy (immune-surveillance, cellular, micro-environment and angiogenesis) can open new horizon in the field of dormancy. Robust targeted pre-clinical finding to block the early stage dissemination can be promising but is only possible after comprehensive characterization of disseminated cells and proper identification of mechanisms and regulations involved in dormancy of CSCs.

Acknowledgement This work was supported by the National Institutes of Health (grants R01CA124650 and R01CA129000 to K. Watabe) and the US Department of Defense (grant PC101369 to K. Watabe and grant PC094294 to F. Xing).

References

Aguirre-Ghiso JA (2007) Models, mechanisms and clinical evidence for cancer dormancy. Nat Rev Cancer 7(11):834–846

Aguirre-Ghiso JA, Kovalski K, Ossowski L (1999) Tumor dormancy induced by downregulation of urokinase receptor in human carcinoma involves integrin and MAPK signaling. J Cell Biol 147:89–104

Aguirre-Ghiso JA, Estrada Y, Liu D, Ossowski L (2003) ERK (MAPK) activity as a determinant of tumor growth and dormancy; regulation by p38 (SAPK). Cancer Res 63:1684–1695

Aktas B, Tewes M, Fehm T, Hauch S, Kimmig R, Kasimir-Bauer S (2009) Stem cell and epithelial-mesenchymal transition markers are frequently overexpressed in circulating tumor cells of metastatic breast cancer patients. Breast Cancer Res 11(4):R46

Allgayer H, Aguirre-Ghiso JA (2008) The urokinase receptor (u-PAR)—a link between tumor cell dormancy and minimal residual disease in bone marrow? APIMS 116(7–8):602–614

Allgayer H, Heiss MM, Riesenberg R, Grutzner KU, Tarabichi A, Babic R, Schildberg FW (1997) Urokinase Plasminogen Activator Receptor (uPA-R): one potential characteristic of metastatic phenotypes in minimal residual tumor disease. Cancer Res 57:1394–1399

Andrews AN (2012) From tumor dormancy to the vicious cycle of bone metastasis. IBMS Bonekey 9:110

Bandhopadhyay S, Pai SK, Hirota S, Hosobe S, Takano Y, Saito K, Piquemal D, Commes T, Watabe M, Gross SC, Wang Y, Ran S, Watabe K (2004) Role of putative tumor metastasis suppressor gene Drg-1 in breast cancer progression. Oncogene 23:5675–5681

Bragado P, Sosa MS, Keely P, Condeelis J, Aguirre-Ghiso JA (2012) Microenvironment dictating tumor cell dormancy. Minim Residual Dis Circ Tumors Cells Breast Cancer 195:25–39

Braun S, Kentenich C, Janni W, Hepp F, Waal JD, Willgeroth F, Sommer H, Pantel K (2000) Lack of effect of adjuvant chemotherapy on the elimination of single dormant tumor cells in bone marrow of high-risk breast cancer patients. J Clin Oncol 1:80–86

Buijs JT, Rentsch CA, van der Horst G, van Overveld PGM, Wetterwald A, Schwaninger R, Henriquez NV, Dijke PT, Borovecki F, Markwalder R, Thalmann GN, Papapoulos SE, Pelger RCM, Vukicevic S, Cecchini MG, Lowik CWGM, van der Pluiim G (2007) BMP7, a putative regulator of epithelial homeostasis in the human prostate, is a potent inhibitor of prostate cancer bone metastasis in vivo. Am J Pathol 171(3):1047–1057. doi:10.2353%2Fajpath.2007.070168#pmc_ext

Collado M, Gil J, Efeyan A, Guerra C, Schuhmacher AJ, Barradas M, Benguria A, Zaballos A, Flores JM, Barbacid M et al (2005) Tumour biology: senescence in premalignant tumours. Nature 436(7051):642

Collado M, Blasco MA, Serrano M (2007) Cellular senescence in cancer and aging. Cell 130(2):223–233

Ellis WJ, Pfitzenmaier J, Colli J, Arfman E, Lange PH, Vessella RL (2003) Detection and isolation of prostate cancer cells from peripheral blood and bone marrow. Urology 61(2):277–281

Gao H, Chakraborty G, Lee-Lim AP, Mo Q, Decker M, Vonica A, Shen R, Brogi E, Brivanlou AH, Giancotti FG (2012) The BMP inhibitor coco reactivates breast cancer cells at lung metastatic sites. Cell 150(4):764–779

Georgoulias V, Bozionelou V, Agelaki S, Perraki M, Apostolaki S, Kallergim G, Kalbakis K, Xyrafas A, Mavroudis D (2012) Trastuzumab decreases the incidence of clinical relapses in patients with early breast cancer presenting chemotherapy-resistant CK-19mRNA-positive circulating tumor cells: results of a randomized phase II study. Ann Oncol 23:1744–1750

Goss PE, Chambers AF (2010) Does tumour dormancy offer a therapeutic target? Nat Rev Cancer 10(12):871–877

Husemann Y, Geigl JB, Schubert F, Meyer M, Burghart E, Forni G, Eils R, Fehm T, Riethmuller G, Klein CA (2008) Systemic spread is an early step in breast cancer. Cancer Cell 13:58–68

Kiel MJ, Morrison SJ (2008) Uncertainty in the niches that maintain haematopoietic stem cells. Nat Rev Immunol 8(4):290–301

Kobayashi A, Okuda H, Xing F, Pandey PR, Watabe M, Hirota S, Pai SK, Liu W, Fukuda K, Chambers C, Wilber A, Watabe K (2011) Bone morphogenetic protein 7 in dormancy and metastasis of prostate cancer stem-like cells in bone. J Exp Med 208(13):2641–2655

Lim PK, Bliss SA, Patel SA, Taborga M, Dave MA, Gregory LA, Greco SJ, Bryan M, Patel PS, Rameshwar P (2011) Gap junction-mediated import of microRNA from bone marrow stromal cells can elicit cell cycle quiescence in breast cancer cells. Cancer Res 71(5):1550–1560

Lu X, Mu E, Wei Y, Riethdorf S, Yang Q, Yuan M, Yan J, Hua Y, Tiede BJ, Lu X, Haffty BG, Pantel K, Massague J, Kang Y (2011) VCAM-1 promotes osteolytic expansion of bone micrometastasis of breast cancer by engaging α4β1-positive osteoclast progenitors. Cancer Cell 20(6):701–714. doi:10.1016/j.ccr.2011.11.002#doilink

MacKie RM, Reid R, Junor B (2003) Fatal melanoma transferred in a donated kidney 16 years after melanoma surgery. N Engl J Med 348:567–568

Morgan TM, Lange PH, Porter MP, Lin DW, Ellis WJ, Gallaher IS, Vessella RL (2009) Disseminated tumor cells in prostate cancer patients after radical prostatectomy and without evidence of disease predicts biochemical recurrence. Clin Cancer Res 15(2):677–683

Muller V, Alix-Panabieres C, Pantel K (2010) Insights into minimal residual disease in cancer patients: implications for anti-cancer therapies. Eur J Cancer 46(7):1189–1197

Naumov GN, Townson JL, MacDonald IC, Wilson SM, Bramwell VH, Groom AC, Chambers AF (2003) Ineffectiveness of doxorubicin treatment on solitary dormant mammary carcinoma cells or late developing metastases. Breast Cancer Res Treat 82:199–206

Paez D, Labonte MJ, Bohanes P, Zhang W, Benhanim L, Ning Y, Wakatsuki T, Loupakis F, Lenz H (2011) Cancer dormancy: a model of early dissemination and late recurrence. Clin Cancer Res 18:645–653

Pantel K, Alix-Panabieres (2007) The clinical significance of circulating tumor cells. Nat Clin Pract Oncol 4:62–63

Pantel K, Izbicki J, Passlick B, Angstwurm M, Haussinger K, Thetter O, Riethmuller G (2003) Frequency and prognostic significance of isolated tumour cells in bone marrow of patients with non-small-cell lung cancer without overt metastases. Lancet 347(9002):649–653

Pedersen EA, Shiozawa Y, Pienta KJ, Taichman RS (2012) The prostate cancer bone marrow niche: more than just 'fertile soil'. Asian J Androl 3:423–427

Polyak K, Weinberg RA (2009) Transitions between epithelial and mesenchymal states: acquisition of malignant and stem cell traits. Nat Rev Cancer 9(4):265–273

Roninson IB (2012) Tumor cell senescence in cancer treatment. Cancer Res 63:2705–2715

Schwarzenbach H, Alix-Panabieres C, Muller I, Letang N, Vendrell JP, Rebillard X, Pantel K (2009) Cell-free tumor DNA in blood plasma as a marker for circulating tumor cells in prostate cancer. Clin Cancer Res 15(3):1032–1038

Shaw JA, Page K, Blighe K, Hava N, Guttery D, Ward B, Brown J, Ruangpratheep C, Stebbing J, Payne R, Palmieri C, Cleator S, Walker RA, Coomber RC (2012) Genomic analysis of circulating cell-free DNA infers breast cancer dormancy. Genome Res 22(2):220–231

Shimazaki J, Higa T, Akimoto S, Masai M, Isaka S (1992) Clinical course of bone metastasis from prostatic cancer following endocrine therapy: examination with bone x-ray. Adv Exp Med Biol 324:269–275

Slade MJ, Payne R, Riethdorf S, Ward B, Zaidi SAA, Stebbing J, Palmieri C, Sinnett HD, Kulinskaya E, Pitfield T, McCormack RT, Pantel K, Coombes RC (2009) Comparison of bone marrow, disseminated tumour cells and blood-circulating tumour cells in breast cancer patients after primary treatment. Br J Cancer 100:160–166

Sosa MS, Avivar-Valderas A, Bragado P, HC W, Aguirre-Ghiso JA (2011) ERK1/2 and p38α/β signaling in tumor cell quiescence: opportunities to control dormant residual disease. Clin Cancer Res 17(18):5850–5857

Thobe MN, Clark RJ, Bainer RO, Prasad SM, Rinker-Schaeffer CW (2011) From prostate to bone: key players in prostate cancer bone metastasis. Cancers 3(1):478–493

Varga AC, Wrano JL (2005) The disparate role of BMP in stem cell biology. Oncogene 24(37):5713–5721

Weckermann D, Polzer B, Ragg T, Blana A, Schlimok G, Arnholdt H, Bertz S, Harzmann R, Klein CA (2009) Perioperative activation of disseminated tumor cells in bone marrow of patients with prostate cancer. J Clin Oncol 27:1549–1556

Zhang L, Sun H, Zhao F, Lu P, Ge C, Li H, Hou H, Yan M, Chen T, Jiang G, Xie H, Cui Y, Huang X, Fan J, Yao M, Li J (2012) BMP4 administration induces differentiation of CD133$^+$ hepatic cancer stem cells, blocking their contributions to hepatocellular carcinoma. Cancer Res 72(16):4276–4285

Role of Microenvironment in Regulating Stem Cell and Tumor Initiating Cancer Cell Behavior and Its Potential Therapeutic Implications

Ana Krtolica

Contents

Abstract 301
Introduction 302
Stem Cells' Behavior Is Controlled by Their Niches: Examples of Epidermal Stem Cell Niche and of Hypoxia 302
Disruption of Niche-Stem Cell Interactions Induces Pathological States: Premature Aging and Tumorigenesis 304
Tumor-Initiating and Cancer Stem Cell-Like State Is Affected by Niche/Microenvironment 305
Modulating the Effects of Microenvironment: Examples of Escaping Immunosurveillance and of Promoting Epithelial-to-Mesenchymal Transition 306
Therapeutic Implications 309
References 310

A. Krtolica (✉)
Life Sciences Division, Lawrence Berkeley National Laboratory, Berkeley, CA 94720, USA
e-mail: akrtolica@sllsciences.com

Abstract

Microenvironment plays a key role in controlling stem cell fate and thereby regulating tissue homeostasis and repair. It consists of acellular and cellular components that interact with stem cells and their progenitors and through signaling cascades influence balance between self-renewal, differentiation and dormancy. Under pathological conditions, disruptions in microenvironment can generate signals that stimulate untimely or aberrant stem cell differentiation or self-renewal, or activate de-differentiation of progenitor cells, leading to diseased states such as cancer. However, while unaltered microenvironment can restrain transformed cell behavior inhibiting malignant phenotypes, transformed cancer cells that exhibit resistance to conventional therapies and tumor initiating capacity are capable of inducing more permissive and immunotolerant microenvironment that promotes tumor growth and metastasis. Better understanding of their behavior and interactions with microenvironment opens up novel avenues for devising more efficacious cancer therapies.

Keywords

Antigen-processing machinery (APM) • Cancer stem cells (CSC) • Epidermal stem cell niche • Hypoxia • Immunosurveillance and anti-cancer therapy • Major histocompatibility complex (MHC) • Mammalian target of rapamycin (mTOR) • Mesenchymal stem cells (MSC) • Mouse mammary tumor virus (MMTV) promoter • T cell receptors (TCR)

Introduction

Stem cells represent populations of cells within organism that are capable of self-renewal and differentiation into one or more cell types that form a specific tissue (uni- or multi- potent stem cells, respectively) or the whole body (pluripotent or omnipotent stem cells). Orderly development and tissue repair require tight balance between arrested stem cell state (dormancy), stem cell self-renewal and differentiation, and in case of multi-potent and pluripotent stem cells, between multiple differentiation pathways. This is achieved by stringent control of stem cell behavior exerted through signals from their microenvironment (niche). For instance, signals stemming from cell-cell and cell-matrix interactions, changes in oxygen concentration (i.e., hypoxia) and nutrient availability, all affect stem cell state and phenotype. Escape from and/or deregulation of the signaling cues present in microenvironment loosens control over stem cell function and results in disruption of tissue homeostasis leading to pathological states. Understanding the mechanisms involved in the cross talk between acellular and cellular elements of the microenvironment and their effects on normal and cancer stem cell behavior is crucial for the development of more efficacious treatments for pathological states associated with aberrant stem and/or progenitor cell phenotypes such as cancer.

Stem Cells' Behavior Is Controlled by Their Niches: Examples of Epidermal Stem Cell Niche and of Hypoxia

Embryonic development and multiple physiologic processes depend on timely and regulated activation and differentiation of stem cells, processes controlled by signals from the niche. Typically, the stem cell niche is a defined anatomical compartment that includes cellular and acellular components that integrate both systemic and local cues to regulate the biology of stem cells. Cells, blood vessels, matrix glycoproteins, and the three-dimensional space that is formed from this architecture provide a highly specialized microenvironment for a stem cell. Stem cell niches facilitate the interaction between the stem cell and surrounding cells in a spatially and temporally defined manner that maintains tissue homeostasis. Niche includes extracellular components as well as diffusible factors to provide the proper regulation of the stem cell.

Epidermal stem cell niche. We will illustrate the role of the niche with an example of interfollicular epidermal stem cells in human skin. In humans large portions of the skin lack hair folicules and stem cells appear to be dispersed along the basal compartment of the interfollicular epidermis. While here, specific niche is difficult to define in morphological terms, patterns or gradients in structural elements and/or positive and negative signals generate niches that enable maintenance and functionality of the stem cell population. Even in the absence of morphological separation present in more commonly described epidermal stem cell hair buldge niche of the mouse, it still incorporates all the key interactions present in other stem cell niches. In both cases, epidermal stem cells reside in fine-tuned microenvironments that are controlled by constant cell–cell and cell–matrix interactions. They sit on and, through integrin and other receptors interact with, a specialized extracellular matrix layer called basement membrane that provides an interface between the epidermis and the underlying dermis. It is made from a complex network of extracellular matrix molecules, including several laminins, type IV collagen, nidogen and perlecan, all of which are necessary for its native structure and epidermal tissue formation. Cellular constituents of the niche include fibroblasts, endothelial cells and inflammatory cells and, presumably, also neighboring keratinocytes and melanocytes, Merkel cells and Langerhans cells. When three-dimensional in vitro model of human skin (dermal equivalent), was modified in a way that allowed for the formation of an authentic dermis-like matrix, it enabled long-term regeneration of the epidermis (Boehnke et al. 2007). These studies also confirmed the major impact of keratinocytes on ECM assembly and maturation in the dermis and clearly underlined the relevance of mutual epithelial–mesenchymal interaction for establishing a proper stem cell niche.

Epidermal stem cell maintenance is likely regulated by Notch ligand Delta1 whose expression is confined to the basal layer of human epidermis, with highest levels in regions presumably harboring stem cells. Furthermore, epidermal stem cell differentiation and maintenance of interfollicular epidermis are determined by a fine-tuned balance between intracellular levels of Notch and p63 (Watt et al. 2008). Another crucial regulator of homeostasis in the interfollicular epidermis is the EGFR with its multiple mediators such as amphiregulin, epiregulin, heparin binding-epidermal growth factor and transforming growth factor (TGF)-α, all acting in an autocrine and paracrine manner. The role for several major EGFR-activated pathways is described including MAPK, PI3K/AKT, JAK/STAT and PKC cascades. Antagonizing interaction of EGFR and Notch family members in the differentiating epidermis resulted in their mutual down-regulation (Fuchs 2008) strongly arguing for their essential functional interdependence.

Role of hypoxia. Stem cell fate is further regulated by conditions in the microenvironment. For example, low oxygen tension (hypoxia) maintains undifferentiated states of embryonic, hematopoietic, mesenchymal, and neural stem cell phenotypes and also influences their proliferation and cell-fate commitment. It has been hypothesized that the presence of low oxygen tension in stem cell niches offers a selective advantage that is well suited to their particular biological role. That is, essentially all cells that undergo aerobic metabolism are subject to some degree of oxidative stress through the generation of reactive oxygen species that can damage DNA. This effect is demonstrated by the fact that mouse embryonic fibroblasts accumulate more mutations and senesce faster when cultured under 20 % oxygen than cells cultured under 3 % oxygen (Parrinello et al. 2003). By residing in anatomical compartments that experience relatively low oxygen tensions (in the range of 1–9 %), stem cells may escape this damage and maintain low proliferation rate. In addition, hypoxia has been shown in multiple stem cell systems to activate molecular pathways that control Oct4 and Notch signaling, two important regulators of stemness. Indeed, human ESC derivation from single embryonic cells (blastomeres) has been enhanced under mildly hypoxic conditions (8 %) and eliminated the need for serum, essential ingredient for blastomere derivation under 20 % oxygen. This gives further support to the notion that lower oxygen tension promotes better survival and self-renewal of pluripotent ESC (Ilic et al. 2009). Finally, oxygen tensions as low as 1 % appear to decrease proliferation and maintain ESC pluripotency, while higher oxygen tensions (3–5 %) appear to maintain pluripotency with no effect on proliferation (Ezashi et al. 2005). These results suggest that proliferation and perhaps even stem cell dormancy may be regulated by gradients of oxygen tension supplied by stem cells' local niche.

Hypoxia appears to maintain an immature blast-like quality in mouse hematopoietic stem cells (HSC) with a primitive phenotype and enhanced engraftment capabilities (Eliasson and Jonsson 2010). Several investigators have demonstrated that slow-cycling HSC are more likely to localize in the low oxygen areas of the marrow, away from blood vessels, whereas fast cycling early hematopoietic progenitors with limited capacity for self-renewal reside in areas much closer to vasculature (Kubota et al. 2008). Although hypoxic cultivation of bone marrow cells has been shown to increase their ability to repopulate and engraft, it is still unclear whether these effects are due to direct action on HSCs or other stromal elements, as many of these experiments were performed with whole marrow or partially purified cell populations (Eliasson and Jonsson 2010). HSCs present in the hypoxic niche express higher levels of Notch-1, telomerase, and the cell-cycle inhibitor p21 than cells closer to the vasculature (Jang and Sharkis 2007). Remarkably, extremely low oxygen tensions (0.1 %) push CD34+ cells into an essentially quiescent state (Hermitte et al. 2006). HIF-1 has emerged as a likely candidate for this regulatory mechanism, as several groups have demonstrated that HIF can mediate cell-cycle arrest in multiple cell lines (Koshiji et al. 2004). In addition, mice with defective HIF signaling exhibit numerous hematopoietic pathologies with prominent defects in hematopoiesis that are embryonically lethal (Eliasson and Jonsson 2010). Collectively, this evidence suggests that hypoxia is a critical component of the HSC niche, and exposure of HSC to elevated oxygen tensions negatively affects their self-renewal while promoting cell-cycle entry and differentiation.

Hypoxia also likely plays a role in a neural stem cell (NSC) maintenance. In the human brain, partial oxygen pressure (pO_2) varies from approximately 3 % (23 mmHg) to 4 % (33 mmHg), demonstrating a physiological oxygen gradient that is the highest in the alveolar space and the lowest in tissues where NSC likely reside. It is thus, likely that NSCs in the brain are located in a relatively hypoxic environment. Thus, in addition to the intercellular signals, soluble factors, blood vessels, and the extracellular matrix proteins found in neurogenic niches, oxygen tension may be an additional important component of the neural stem cell niche. In vitro, hypoxia is able to promote an undifferentiated state in neural crest stem cells and NSC (Morrison et al. 2000). Observations regarding an enhancement in survival and proliferation of NSCs in mild hypoxic conditions have also been made (Morrison et al. 2000; Pistollato et al. 2007). It has been shown that p53 phosphorylation increases in cultures maintained at 20 % oxygen resulting in cell-cycle arrest, decreased proliferation, and differentiation of NSCs toward the glial lineage (Pistollato et al. 2007). This finding suggests that oxygen tensions in the environment influence both NSC stemness (or the maintenance of an undifferentiated state) by inhibiting their differentiation and their specific fate by modulating important intracellular pathways such as p53 and Notch signaling (Gustafsson et al. 2005; Pistollato et al. 2007). Therefore, oxygen tension in the neural niche functions to maintain stem cell self-renewal and an undifferentiated state. Although direct measurements of oxygen tension of the subventricular zone (SVZ) have not been made, our current understanding of the cytoarchitecture and its relation to adjacent blood cells suggests that oxygen can be limiting near the ependymal surface where the neural stem cells reside. In conclusion, a hypoxic microenvironment facilitates stemness and prevents NSC from differentiating. Changes in redox status or other local cues mobilize the NSC population to proliferate, differentiate, or migrate.

Other factors such as nutrient availability also affect stem cell behavior. The mammalian target of rapamycin (mTOR) seems to play a key role within a cell in integrating a myriad of external and internal signals including niche oxygen levels and nutrient availability, cell energy status, presence of cellular stressors, and growth factors. The finely tuned response of mTOR to these stimuli results in alterations in stem cell metabolism, differentiation and cell growth, playing a major role in stem cell homeostasis and lifespan determination (reviewed in Russell et al. 2011).

Disruption of Niche-Stem Cell Interactions Induces Pathological States: Premature Aging and Tumorigenesis

Disruptions/failures in stem cell regulation lead to pathological states such as premature aging and cancer. For example, untimely and altered differentiation of mesenchymal stem cells (MSC) by upregulated downstream Notch signaling causes premature-aging disease Hutchinson–Gilford Progeria Syndrome (HGPS). This is due to a mutant version of lamin A protein called progerin that increases availability of the SKIP, nuclear matrix-associated co-activator of the Notch targets' transcription. Significantly, activation of Notch pathway not only induced premature differentiation of MSC, but also enhanced osteogenesis while inhibiting differentiation of hMSCs into adipose tissue altering a balance between multiple differentiation pathways (Scaffidi and Misteli 2008). Alterations in stem cell regulation can also be caused by external signals and insults from the microenvironment such as ionizing radiation. For example, ionizing radiation-induced premature differentiation of melanocyte stem cells and resulting melanocyte stem cells depletion lead to irreversible hair graying (Inomata et al. 2009).

The altered niche environment can also induce aberrant activation of stem cell phenotype and self-renewal, the phenomenon that may lead to tumorigenesis. There is strong evidence to support the role of permissive microenvironment in promoting tumorigenesis both at premalignant and malignant stages. Multiple studies have shown that tumor-associated stroma can promote tumorigenesis by creating pro-inflammatory microenvironment. For example, tumor associated-fibroblasts isolated from the tumor stroma produce plethora of pro-inflammatory cytokines and growth factors and stimulate malignant transformation of multiple epithelial cell types including breast and prostate (Olumi et al. 1999; Aboussekhra 2011). Pro-inflammatory secretome including IL-1, IL-6, IL-8 and GROα, has also been shown to contribute to pro-carcinogenic microenvironment associated with aging and stress-induced cell senescence (Coppe et al. 2010). Moreover, these cytokines have been shown to play key role in supporting cancer stem cell phenotypes and stem cell self-renewal (Krtolica et al. 2011; Korkaya et al. 2012). Indeed, recent evidence suggests that secretion of IL-1 by carcinoma cells attracts MSC to tumor-associated stroma and via prostaglandin E2 signaling induces MSC to generate pro-inflammatory cytokines that promote β catenin activation and cancer stem cell phenotype (Li et al. 2012).

However, microenvironment can also restrict cell behavior and, in some instances, restrain frankly malignant state through direct control of growth and invasiveness. For example, Weaver and colleagues have shown that malignant phenotype of breast tumor cells can be reversed in three-dimensional culture and in vivo by integrin β1 blocking antibodies which induced them to form polarized acini and cease growth (Weaver et al. 1997). Additionally, microenvironment can exert control through immunosurveillance immune system-mediated tumor cell recognition and consequent destruction (see discussion below; also reviewed in Qi et al. 2012).

Tumor-Initiating and Cancer Stem Cell-Like State Is Affected by Niche/Microenvironment

There is currently growing evidence for the presence of cancer stem-like or tumor initiating cells in multiple tumor type – both hematological malignancies and solid cancers (Bonnet and Dick 1997; Reya et al. 2001; Lathia et al. 2011). Cancer stem cells (CSC) or tumor initiating cells (TIC) are functionally defined by their potential to recapitulate tumor from which they were isolated at the single cell level. To this end, they are usually identified using serial transplantation experiments where limited numbers of cells isolated from the original tumor are transplanted into recipient animals and, once tumor is formed, this procedure is repeated multiple times with additional animals demonstrating CSC/TIC ability to initiate new tumors. New evidence demonstrates CSC/TIC clonal capacity even within natural tumor niches (Chen et al. 2012; Driessens et al. 2012; Schepers et al. 2012). These characteristics imply an unlimited proliferative capacity and also an ability to differentiate into all cell types present in the given tumor. What makes CSC/TIC-like populations within tumor even more therapeutically relevant is that their phenotype is often associated with high resistance to common therapeutic modalities of cancer treatment: chemotherapy and ionizing radiation. It is therefore, hypothesized, that they may be the major source of tumor re-growth and patient relapse after therapy. Indeed, there is growing clinical evidence that this may be the case (Li et al. 2008).

The ability of cancer cells to establish themselves in a foreign cellular environment is an essential characteristic of successful metastasis and a defining characteristic of CSC/TICs. Furthermore, there is some evidence that CSC/TIC phenotype may significantly overlap with phenotype of cells undergoing epithelial to mesenchymal transition (EMT), a phenomenon associated with increased tumor aggressiveness and metastasis. EMT is driven by transcription factors, including SNAIL1/2, ZEB1/2, or TWIST1/2, which increase the invasiveness of epithelial cells. In several studies, the induction of EMT has been shown to enhance self-renewal and the acquisition of CSC characteristics (Ansieau et al. 2008; Mani et al. 2008). In contrast, some studies demonstrate that tumor cells with an epithelial, not mesenchymal, phenotype survive in the circulation and form distant metastasis (Tsuji et al. 2008; Celia-Terrassa et al. 2012). For example, in prostate cancer cell lines, subpopulations with a strong epithelial gene program were enriched in highly metastatic CSC/TIC, whereas mesenchymal subpopulations showed reduced numbers of CSC/TIC. Collectively, these studies illustrate cancer stem cell plasticity and the fact that cell-type specific characteristics govern their self-renewal and mesenchymal gene interactions (Celia-Terrassa et al. 2012). Nevertheless, these data taken together with CSC/TIC chemo- and radiation-therapy resistance and capacity to form new tumors, suggest that it is quite likely that the cells with CSC/TIC characteristics may be the main sources of metastasis.

Importantly, every aspect of CSC/TIC behavior is under influence of microenvironment. For example, presence of drugs in tumor microenvironment and circulation may support survival and provide growth advantage to drug resistant tumor cells such as those expressing CSC/TIC phenotypes. Another example are hypoxic tumor microenvironments that often harbor quiescent (non-dividing) and tumor initiating cell populations. Hypoxia maintains the stem-like phenotype and prevents differentiation of CSC/TIC. It has been shown to promote self-renewal of glioblastoma and colorectal cancer stem cell-like CSC/TIC populations by inducing PI(3)K and ERK 1/2 pathways and regulating CDX1 and Notch1, respectively (Soeda et al. 2009; Yeung et al. 2011). While the degree to which quiescent and CSC/TIC populations overlap in hypoxic regions remains to be elucidated and may vary between different stages and types of tumors, it is clear that hypoxic tumor cells exhibit high resistance to common chemotherapeutic agents and are thus, likely responsible for tumor reoccurrence and, potentially, metastasis. Consequently, tumor hypoxia has been shown to correlate with poorer patient outcomes in multiple cancer types including colon, breast, prostate, and brain cancer.

Yet another effect of microenvironment is exerted through provisional extracellular matrix laid down by tumor-associated fibroblasts. It may promote migration and invasiveness thru integrin-fibronectin interactions and support survival of cells that have undergone EMT and are capable of metastasizing from the primary tumor. Tumor-associated fibroblasts and MSC present in tumor-activated stroma can also secrete a number of cytokines and growth factors creating a pro-inflammatory environment. This, in turn, can both directly affect CSC/TIC proliferation and motility (see discussion above) and also, modulate immune response (see next section).

Modulating the Effects of Microenvironment: Examples of Escaping Immunosurveillance and of Promoting Epithelial-to-Mesenchymal Transition

Tumor cell microenvironment contains, and is altered by, various components of both innate and adaptive immune system and their products. Depending on tumor immunogenicity, both initial tumor formation and progression of the disease can be significantly affected by host anti-tumor responses. By targeting premalignant and malignant cells in a process called immunosurveillance, immune responses can eliminate cancer cells prior to tumors becoming clinically apparent or can attenuate tumor growth and progression. However, mounting evidence shows that CSC/TIC effectively evade host immunosurveillance through multiple mechanisms including altered immunogenicity, production of immunomodulatory factors and direct interactions with tumor-infiltrating immune cells. We will discuss these different mechanisms in the following paragraphs.

MHC class I molecules are one of two primary classes of major histocompatibility complex (MHC) molecules and are found on every nucleated cell of the body. Their function is to present fragments of proteasome-degraded cellular and external proteins to cytotoxic T cells (CTL, CD8+). T cell receptors (TCR) and CD8 co-receptors on CTL plasma membrane interact with MHC I molecules that present the foreign protein fragments (antigens) on the surface of the affected cell. This interaction activates CTL to lyse the foreign antigen presenting cell.

MHC class I molecules play a key role in the immune recognition of transformed cells. It was recently reported that CSC/TIC may predominantly lose MHC class I molecules, and selectively silence the expression of tumor-associated antigens (TAAs), whose presence is associated with differentiated state, leading to resistance to immune rejection. Indeed, selective or general downregulation of MHC class I molecules may suppress the ability of class I MHC-restricted CTL to lyse CSC/TIC.

Consequently, flow cytometry analysis of glioblastoma multiforme (GBM) and astrocytoma CSC/TIC-enriched CD133+ cell fractions revealed that the majority of CD133+ cells did not express detectable MHC class I molecules or natural killer (NK) cell activating ligands on their cell surface (Wu et al. 2007). This may render them resistant to adaptive and innate immune surveillance. In addition, defects were found in the expression of antigen-processing machinery (APM) components in the cultured population of GBM-derived CSC/TIC (GSC; (Di Tomaso et al. 2010)). APM molecules included MHC class I molecules and their heavy chains (i.e., A-HC), β2-microglobulin, constitutive proteasome subunits (δ, MB1, and Z), immunoproteasome (LMP2, LMP7, and LMP10), transporter molecules (TAP), and chaperon molecules (tapasin, calnexin, calreticulin, and ERp57). While low levels of expression were also detected in corresponding fetal bovine serum-derived tumor cell lines in most cases expression was higher in tumor cell lines than in CSC/TIC (CSC/TIC are typically isolated under serum-free conditions, since serum constituents induce their differentiation into progenitor tumor cell types that have lower tumorigenic capacity). These results are in line with the previously reported decreased expression of MHC and APM molecules in a variety of human tumors and derived cell lines (Johnsen et al. 1998) and suggest that in CSC/TIC isolated from these tumor types there is low efficiency in antigen processing and presentation. Therefore, CSC/TIC may display unique immunophenotypes, such as downregulation of MHC class I molecules, differentiation-associated TAAs and APM components, that enable them to effectively evade host immunosurveillance.

Another mechanism, by which CSC/TIC may avoid the attack of immune system, is by inducing immunogenic tolerance through functional inactivation of antigen-reactive T lymphocytes or activation of regulatory T cells (Treg). Lymphocyte tolerance or anergy is likely induced by direct T cell inhibition via secretion of immunosuppressive cytokines, including IL-4, IL-10 and TGF-β. For example, researchers detected high levels of IL-4 and IL-4R in CD133+ stem-like cells in colon cancer. IL-4 has previously been reported to suppress apoptosis by enhancing the expression of anti-apoptotic proteins cFLIP, Bcl-xL, and PED in many tumor cell lines, including chronic lymphocytic leukemia B cells (Dancescu et al. 1992), as well as prostate, breast, and bladder tumor cell lines (Conticello et al. 2004). Additionally, IL-4 has the capability of inhibiting the proliferation and immune responses of helper T (Th, CD4+) cells, and also exhibits immunoregulatory effect on B cells, mastocytes, and macrophages. Recent evidence shows that GSC can prevent CTL mediated specific immune cytotoxicity (Trapani and Sutton 2003), and GSC can strongly inhibit the proliferation of Th cells. Effects of CSC/TIC IL-4 signaling may include autocrine inhibition of the apoptosis and induction of immune tolerance. In addition, TGF-β signaling pathway is specifically activated in the CSC/TIC fraction of breast cancers (Shipitsin et al. 2007), and secreted morphogens in the TGF-β super family as well as their receptors are preferentially expressed by CD133+ brain tumor CSC/TIC (Piccirillo et al. 2006) and by ABCB5+ malignant melanoma CSC/TIC (Schatton et al. 2008). It was shown that TGF-β negatively influences antitumor capabilities of host CTL. This activity is multi-directional and is based on the impairment of Fas-mediated apoptosis of tumor cells, downregulation of IFN-γ secretion and disturbed expression of perforin and granzymes by CTL (Jarnicki et al. 2006). Indeed, T cell-specific blockade of TGF-β signaling was found to enhance the antitumor immune response in mice challenged with live tumor cells (Gorelik and Flavell 2001). Moreover, CSC/TIC may induce high levels of Treg cells to suppress the antitumor immune response and ultimately promote tumorigenic growth (Qi et al. 2012).

Immunogenic tolerance may also be achieved through clonal anergy of macrophages and dendritic cells (DC). Tumor associated macrophages (TAM) constitute one of the major immune cell populations responsible for both tumor rejection and promotion. The high expression of CD47 on leukemia stem cells (LSC) of AML patients can reduce the macrophage-induced phagocytosis of LSC and thus, decrease LSC clearance by innate immune system (Jaiswal et al. 2009). CD47, also known as integrin-associated protein (IAP), can inhibit macrophage-mediated phagocytosis by binding signal-regulatory protein-α (SIRPα) on their surface (Barclay and Brown 2006). Disruption of the CD47–SIRPα interaction with a monoclonal antibody against CD47 preferentially enabled the phagocytosis of AML LSC by human macrophages (Majeti et al. 2009). In addition, CD47–SIRPα interaction between CD47 expressed by LSC and SIRPα present on DC surface can also inhibit DC activation (Barclay and Brown 2006). Moreover, functional inactivation of DC, a major type of antigen-presenting cells, can affect the activation of initial T cells and inhibit the adaptive immune response.

Expression of another immunosuppressive protein, CD200, was significantly higher in the CSC/TIC relative to differentiated tumor cell fractions isolated from prostate, breast, colon and brain tumors. Additionally, CD200 was co-expressed with CSC/TIC markers (Kawasaki et al. 2007). CD200 is a type Ia membrane protein which exerts suppressive effects through binding to its receptor, CD200R. CD200R is present on the surface of myeloid DCs, monocyte/macrophage lineage and on T lymphocytes. It was shown that the stimulation of CD200R on DCs triggered tumor supporting reactions mediated by Th2 cytokines and increased Treg activity, thought to hamper tumor-specific effector T cell immunity (Curiel et al. 2004). Conversely, blockade of CD200/CD200R interactions with monoclonal anti-CD200 antibodies resulted in a shift towards Th1 activity and attenuated immune tolerance (Kretz-Rommel et al. 2007; Rygiel et al. 2012).

While attenuation of the immune response may promote carcinogenesis, the activity of immune system itself can also promote tumor development. For example, chronic inflammatory responses mediated by activated B cells and associated antibodies have been directly shown to be critical in the initiation of skin cancer in K14-HPV16 mice (de Visser et al. 2005). Furthermore, tumor growth could be promoted by TAM (Mantovani et al. 1992). TAM can contribute to either a pro-tumorigenic or anti-tumorigenic environment depending on their capacity to present antigens, produce inflammatory cytokines, stimulate angiogenesis, and enable cytotoxic activity. While tumors evade macrophage phagocytosis through the expression of anti-phagocytic signals, including CD200 and CD47 as discussed above, cytokine production and antigen presentation by macrophages have also been shown to directly impact tumor growth (Jaiswal et al. 2009).

Moreover, the evidence suggests that immune effectors can induce EMT following an acute or chronic inflammatory response. Likely, CTL cells trafficking into the tumor microenvironment can produce direct mediators of EMT, or alternatively, can produce other cytokines or chemokines (e.g., MCP-1), which can attract additional immune effectors (e.g., macrophages) that provide the stimuli. When epithelial tumors from neu-transgenic mice, that express the cell surface neu-oncogene under control of the mammary epithelial cell-specific mouse mammary tumor virus (MMTV) promoter, were transplanted into nontransgenic syngeneic mice, a T-cell-dependent rejection occurred. However, the mice subsequently relapsed with tumors enriched in neu-negative variant tumor cells that had a mesenchymal phenotype. CTL cells were required for outgrowth of the neu-negative mesenchymal variants suggesting local induction of

EMT (Santisteban et al. 2009). Furthermore, CTL cells isolated from mice primed with neupositive tumor cells were able to induce antigen loss when co-cultured with neu-positive tumor cells. Tumor cells isolated from relapsed mice showed that these tumors had characteristics of breast cancer stem cells (BCSC), as indicated by the cell surfaceCD24−/loCD44+ marker profile, enhanced mammosphere formation and tumorigenicity, elevated expression of drug transporters, DNA repair enzymes, and enhanced resistance to chemotherapy and radiation. In accordance with characteristics of true CSC/TIC, BCSC gave rise in vivo to tumors with a heterogeneous tumor population consisting of both CD24− and CD24+ tumor cells with predominant neu-positive epithelial phenotype, suggesting that the immune induced EMT was fully reversible. Thus, in contrast to their typically ascribed protective role, CTL cells were capable of inducing tumors to undergo EMT and to acquire BCSC properties and a more aggressive tumor phenotype.

Therapeutic Implications

In devising anti-cancer therapy, it is important to take into account the effects of microenvironment. Probably best known example and most used cancer treatment that relies on altering microenvironment, is inhibition of angiogenesis by anti-VEGF antibody (bevacizumab/avastin) and thus, deprivation of tumors of their oxygen and nutrition supply. Another example is inhibition of hedgehog pathway through inactivation of smoothened (Saridegib/IPI-926, GDC-0449/vismodegib, LDE-225/erismodegib) that may act to eliminate fibrous tissue that hinders drugs from reaching the cancer, while also directly affecting TIC/CSC growth. Tumor site allografts of healthy endothelial cells embedded in polymer matrix (PVS-30200) delivered at the time of surgical tumor removal to block tumor growth have also shown promise in preclinical studies.

There are multiple other novel approaches that may be tackled and are at different stages of preclinical/clinical development. For example, eliminating CSC/TIC through stimulating external signals that activate their differentiation may serve to sensitize tumor cells to standard therapy. Oxygenating hypoxic regions of tumors has the potential to promote cell cycle entry of quiescent tumor cells and to induce differentiation of hypoxic niche-dependent CSC/TIC, reducing resistance to antineoplastic therapies. Inhibiting promotion of EMT and lowering chronic inflammation while activating anti-tumor immune responses could provide additional approaches. Chronic inflammation is often associated with increased cancer risk in humans: patients with inflammatory bowel disease have an increased risk of colorectal cancer; Helicobacter pylori infection is associated with gastric adenocarcinomas and mucosa associated lymphomas; and chronic hepatitis is associated with hepatocellular carcinoma (Lu et al. 2006). For patients with chronic inflammatory conditions, therefore, suppressing the immune response can reduce subsequent cancer development. Considering the complexity of the disease, most likely the best ways for treating cancer patients are going to be individualized combination therapies based on well stratified patient populations and may include one or multiple of aforementioned modalities in conjunction with more traditional therapies.

A series of therapeutic strategies targeting CSC/TIC have been developed, such as inhibiting proliferation, promoting differentiation, inducing apoptosis, and enhancing the sensitivity of chemo radiation. Preliminary experimental results indicate that these strategies can target CSC/TIC and inhibit their functions albeit so far with limited success. Therefore, there is an urgent need for further in-depth investigations of the mechanisms that may lead to rational basis for treatment development.

Although identification of therapies that selectively target CSC/TIC is an important goal, a parallel and perhaps equally efficacious approach would be to target the mechanisms of plasticity that generate and maintain the CSC/TIC population in tumors, namely elements of their microenvironment. These include both extracellular factors controlled on systemic and local levels such as hypoxia, cytokines, growth factors and extracellular matrix, as well as different cellular components, including niche and stromal cells and various constituents of the immune system that contribute to tumor milieu. For example, it has been suggested by Reiman and colleagues (Reiman et al. 2010) that because immunity is able to induce BCSCs, one approach would be to define and to target the specific immune effectors of this process. Although activated CTL cells are the critical effectors of EMT in mouse mammary tumor cells, it is not possible to generally target CTL cells given their important role in protection against infection. Skewing of the macrophage response within the tumor microenvironment from a M2 (tumor-promoting) to M1(tumor-eradicating) phenotype may have the potential to reduce tumor invasion and metastases. Having a Th2 or Treg response may promote breast cancer metastases, suggesting that agents (e.g., vaccines) that shift from Th2/Treg to an antitumor Th1 response may be useful.

Another approach to inhibit EMT-associated tumor progression is to target the pathways involved in the induction of EMT that specifically lead to the acquisition of CSC/TIC characteristics, as recently shown for inhibitors of EMT mediated by TGFβ or Hedgehog pathways (Wan et al. 2008; Gupta et al. 2009). The immune system has long been viewed as a co-conspirator with developing tumors; more recent data have shown that it can also selectively target tumor cells at early stages of cancer progression. An important goal now is to identify how to reduce the tumor promotion abilities of the immune system while preserving or increasing its ability to eliminate tumor cells.

In conclusion, while tumor cell populations undergoing malignant transformation may not in some cases represent CSC or TIC (and in many cases may not arise from normal stem cells), they inevitably harbor within their cells that share with CSC/TIC some of their key properties: the ability to self-renew and to give rise to tumors and, often, also the capacity to differentiate into multiple tumor-associated cell types. It is these characteristics that may be selected for and/or supported by permissive microenvironment and are thus, important to be studied in that context. The underlying mechanisms promise to open up novel approaches to developing drug targets and therapies that may lead to increased disease-free survival and reduction in metastatic disease.

References

Aboussekhra A (2011) Role of cancer-associated fibroblasts in breast cancer development and prognosis. Int J Dev Biol 55:841–849

Ansieau S, Bastid J, Doreau A, Morel AP, Bouchet BP, Thomas C, Fauvet F, Puisieux I, Doglioni C, Piccinin S, Maestro R, Voeltzel T, Selmi A, Valsesia-Wittmann S, Caron de Fromentel C, Puisieux A (2008) Induction of EMT by twist proteins as a collateral effect of tumor-promoting inactivation of premature senescence. Cancer Cell 14:79–89

Barclay AN, Brown MH (2006) The SIRP family of receptors and immune regulation. Nat Rev Immunol 6:457–464

Boehnke K, Mirancea N, Pavesio A, Fusenig NE, Boukamp P, Stark HJ (2007) Effects of fibroblasts and microenvironment on epidermal regeneration and tissue function in long-term skin equivalents. Eur J Cell Biol 86:731–746

Bonnet D, Dick JE (1997) Human acute myeloid leukemia is organized as a hierarchy that originates from a primitive hematopoietic cell. Nat Med 3:730–737

Celia-Terrassa T, Meca-Cortes O, Mateo F, de Paz AM, Rubio N, Arnal-Estape A, Ell BJ, Bermudo R, Diaz A, Guerra-Rebollo M, Lozano JJ, Estaras C, Ulloa C, Alvarez-Simon D, Mila J, Vilella R, Paciucci R, Martinez-Balbas M, de Herreros AG, Gomis RR, Kang Y, Blanco J, Fernandez PL, Thomson TM (2012) Epithelial-mesenchymal transition can suppress major attributes of human epithelial tumor-initiating cells. J Clin Invest 122:1849–1868

Chen J, Li Y, Yu TS, McKay RM, Burns DK, Kernie SG, Parada LF (2012) A restricted cell population propagates glioblastoma growth after chemotherapy. Nature 488:522–526

Conticello C, Pedini F, Zeuner A, Patti M, Zerilli M, Stassi G, Messina A, Peschle C, De Maria R (2004) IL-4 protects tumor cells from anti-CD95 and chemotherapeutic agents via up-regulation of antiapoptotic proteins. J Immunol 172:5467–5477

Coppe JP, Desprez PY, Krtolica A, Campisi J (2010) The senescence-associated secretory phenotype: the dark side of tumor suppression. Annu Rev Pathol 5:99–118

Curiel TJ, Coukos G, Zou L, Alvarez X, Cheng P, Mottram P, Evdemon-Hogan M, Conejo-Garcia JR, Zhang L, Burow M, Zhu Y, Wei S, Kryczek I, Daniel B, Gordon A, Myers L, Lackner A, Disis ML, Knutson KL, Chen L, Zou W (2004) Specific recruitment of regulatory T cells in ovarian carcinoma fosters immune privilege and predicts reduced survival. Nat Med 10:942–949

Dancescu M, Rubio-Trujillo M, Biron G, Bron D, Delespesse G, Sarfati M (1992) Interleukin 4 protects chronic lymphocytic leukemic B cells from death by apoptosis and upregulates Bcl-2 expression. J Exp Med 176:1319–1326

de Visser KE, Korets LV, Coussens LM (2005) De novo carcinogenesis promoted by chronic inflammation is B lymphocyte dependent. Cancer Cell 7:411–423

Di Tomaso T, Mazzoleni S, Wang E, Sovena G, Clavenna D, Franzin A, Mortini P, Ferrone S, Doglioni C, Marincola FM, Galli R, Parmiani G, Maccalli C (2010) Immunobiological characterization of cancer stem cells isolated from glioblastoma patients. Clin Cancer Res 16:800–813

Driessens G, Beck B, Caauwe A, Simons BD, Blanpain C (2012) Defining the mode of tumour growth by clonal analysis. Nature 488:527–530

Eliasson P, Jonsson JI (2010) The hematopoietic stem cell niche: low in oxygen but a nice place to be. J Cell Physiol 222:17–22

Ezashi T, Das P, Roberts RM (2005) Low O_2 tensions and the prevention of differentiation of hES cells. Proc Natl Acad Sci U S A 102:4783–4788

Fuchs E (2008) Skin stem cells: rising to the surface. J Cell Biol 180:273–284

Gorelik L, Flavell RA (2001) Immune-mediated eradication of tumors through the blockade of transforming growth factor-beta signaling in T cells. Nat Med 7:1118–1122

Gupta PB, Onder TT, Jiang G, Tao K, Kuperwasser C, Weinberg RA, Lander ES (2009) Identification of selective inhibitors of cancer stem cells by high-throughput screening. Cell 138:645–659

Gustafsson MV, Zheng X, Pereira T, Gradin K, Jin S, Lundkvist J, Ruas JL, Poellinger L, Lendahl U, Bondesson M (2005) Hypoxia requires notch signaling to maintain the undifferentiated cell state. Dev Cell 9:617–628

Hermitte F, Brunet de la Grange P, Belloc F, Praloran V, Ivanovic Z (2006) Very low O_2 concentration (0.1%) favors G0 return of dividing CD34+ cells. Stem Cells 24:65–73

Ilic D, Giritharan G, Zdravkovic T, Caceres E, Genbacev O, Fisher SJ, Krtolica A (2009) Derivation of human embryonic stem cell lines from biopsied blastomeres on human feeders with minimal exposure to xenomaterials. Stem Cells Dev 18:1343–1350

Inomata K, Aoto T, Binh NT, Okamoto N, Tanimura S, Wakayama T, Iseki S, Hara E, Masunaga T, Shimizu H, Nishimura EK (2009) Genotoxic stress abrogates renewal of melanocyte stem cells by triggering their differentiation. Cell 137:1088–1099

Jaiswal S, Jamieson CH, Pang WW, Park CY, Chao MP, Majeti R, Traver D, van Rooijen N, Weissman IL (2009) CD47 is upregulated on circulating hematopoietic stem cells and leukemia cells to avoid phagocytosis. Cell 138:271–285

Jang YY, Sharkis SJ (2007) A low level of reactive oxygen species selects for primitive hematopoietic stem cells that may reside in the low-oxygenic niche. Blood 110:3056–3063

Jarnicki AG, Lysaght J, Todryk S, Mills KH (2006) Suppression of antitumor immunity by IL-10 and TGF-beta-producing T cells infiltrating the growing tumor: influence of tumor environment on the induction of CD4+ and CD8+ regulatory T cells. J Immunol 177:896–904

Johnsen A, France J, Sy MS, Harding CV (1998) Down-regulation of the transporter for antigen presentation, proteasome subunits, and class I major histocompatibility complex in tumor cell lines. Cancer Res 58:3660–3667

Kawasaki BT, Mistree T, Hurt EM, Kalathur M, Farrar WL (2007) Co-expression of the toleragenic glycoprotein, CD200, with markers for cancer stem cells. Biochem Biophys Res Commun 364:778–782

Korkaya H, Kim GI, Davis A, Malik F, Henry NL, Ithimakin S, Quraishi AA, Tawakkol N, D'Angelo R, Paulson AK, Chung S, Luther T, Paholak HJ, Liu S, Hassan KA, Zen Q, Clouthier SG, Wicha MS (2012) Activation of an IL6 inflammatory loop mediates trastuzumab resistance in HER2+ breast cancer by expanding the cancer stem cell population. Mol Cell 47:570–584

Koshiji M, Kageyama Y, Pete EA, Horikawa I, Barrett JC, Huang LE (2004) HIF-1alpha induces cell cycle arrest by functionally counteracting Myc. Embo J 23:1949–1956

Kretz-Rommel A, Qin F, Dakappagari N, Ravey EP, McWhirter J, Oltean D, Frederickson S, Maruyama T, Wild MA, Nolan MJ, Wu D, Springhorn J, Bowdish KS (2007) CD200 expression on tumor cells suppresses antitumor immunity: new approaches to cancer immunotherapy. J Immunol 178:5595–5605

Krtolica A, Larocque N, Genbacev O, Ilic D, Coppe JP, Patil CK, Zdravkovic T, McMaster M, Campisi J, Fisher SJ (2011) GROalpha regulates human embryonic stem cell self-renewal or adoption of a neuronal fate. Differentiation 81:222–232

Kubota Y, Takubo K, Suda T (2008) Bone marrow long label-retaining cells reside in the sinusoidal hypoxic niche. Biochem Biophys Res Commun 366:335–339

Lathia JD, Gallagher J, Myers JT, Li M, Vasanji A, McLendon RE, Hjelmeland AB, Huang AY, Rich JN (2011) Direct in vivo evidence for tumor propagation by glioblastoma cancer stem cells. PLoS ONE 6:e24807

Li X, Lewis MT, Huang J, Gutierrez C, Osborne CK, Wu MF, Hilsenbeck SG, Pavlick A, Zhang X, Chamness GC, Wong H, Rosen J, Chang JC (2008) Intrinsic resistance of tumorigenic breast cancer cells to chemotherapy. J Natl Cancer Inst 100:672–679

Li HJ, Reinhardt F, Herschman HR, Weinberg RA (2012) Cancer-stimulated mesenchymal stem cells create a carcinoma stem cell niche via prostaglandin E2 signaling. Cancer Discov 2:840–855

Lu H, Ouyang W, Huang C (2006) Inflammation, a key event in cancer development. Mol Cancer Res 4:221–233

Majeti R, Chao MP, Alizadeh AA, Pang WW, Jaiswal S, Gibbs KD Jr, van Rooijen N, Weissman IL (2009) CD47 is an adverse prognostic factor and therapeutic antibody target on human acute myeloid leukemia stem cells. Cell 138:286–299

Mani SA, Guo W, Liao MJ, Eaton EN, Ayyanan A, Zhou AY, Brooks M, Reinhard F, Zhang CC, Shipitsin M, Campbell LL, Polyak K, Brisken C, Yang J, Weinberg RA (2008) The epithelial-mesenchymal transition generates cells with properties of stem cells. Cell 133:704–715

Mantovani A, Bottazzi B, Colotta F, Sozzani S, Ruco L (1992) The origin and function of tumor-associated macrophages. Immunol Today 13:265–270

Morrison SJ, Csete M, Groves AK, Melega W, Wold B, Anderson DJ (2000) Culture in reduced levels of oxygen promotes clonogenic sympathoadrenal differentiation by isolated neural crest stem cells. J Neurosci 20:7370–7376

Olumi AF, Grossfeld GD, Hayward SW, Carroll PR, Tlsty TD, Cunha GR (1999) Carcinoma-associated fibroblasts direct tumor progression of initiated human prostatic epithelium. Cancer Res 59:5002–5011

Parrinello S, Samper E, Krtolica A, Goldstein J, Melov S, Campisi J (2003) Oxygen sensitivity severely limits the replicative lifespan of murine fibroblasts. Nat Cell Biol 5:741–747

Piccirillo SG, Reynolds BA, Zanetti N, Lamorte G, Binda E, Broggi G, Brem H, Olivi A, Dimeco F, Vescovi AL (2006) Bone morphogenetic proteins inhibit the tumorigenic potential of human brain tumour-initiating cells. Nature 444:761–765

Pistollato F, Chen HL, Schwartz PH, Basso G, Panchision DM (2007) Oxygen tension controls the expansion of human CNS precursors and the generation of astrocytes and oligodendrocytes. Mol Cell Neurosci 35:424–435

Qi Y, Li RM, Kong FM, Li H, Yu JP, Ren XB (2012) How do tumor stem cells actively escape from host immunosurveillance? Biochem Biophys Res Commun 420:699–703

Reiman JM, Knutson KL, Radisky DC (2010) Immune promotion of epithelial-mesenchymal transition and generation of breast cancer stem cells. Cancer Res 70:3005–3008

Reya T, Morrison SJ, Clarke MF, Weissman IL (2001) Stem cells, cancer, and cancer stem cells. Nature 414:105–111

Russell RC, Fang C, Guan KL (2011) An emerging role for TOR signaling in mammalian tissue and stem cell physiology. Development 138:3343–3356

Rygiel TP, Karnam G, Goverse G, van der Marel AP, Greuter MJ, van Schaarenburg RA, Visser WF, Brenkman AB, Molenaar R, Hoek RM, Mebius RE, Meyaard L (2012) CD200-CD200R signaling suppresses anti-tumor responses independently of CD200 expression on the tumor. Oncogene 31:2979–2988

Santisteban M, Reiman JM, Asiedu MK, Behrens MD, Nassar A, Kalli KR, Haluska P, Ingle JN, Hartmann LC, Manjili MH, Radisky DC, Ferrone S, Knutson KL (2009) Immune-induced epithelial to mesenchymal transition in vivo generates breast cancer stem cells. Cancer Res 69:2887–2895

Scaffidi P, Misteli T (2008) Lamin A-dependent misregulation of adult stem cells associated with accelerated ageing. Nat Cell Biol 10:452–459

Schatton T, Murphy GF, Frank NY, Yamaura K, Waaga-Gasser AM, Gasser M, Zhan Q, Jordan S, Duncan LM, Weishaupt C, Fuhlbrigge RC, Kupper TS, Sayegh MH, Frank MH (2008) Identification of cells initiating human melanomas. Nature 451:345–349

Schepers AG, Snippert HJ, Stange DE, van den Born M, van Es JH, van de Wetering M, Clevers H (2012) Lineage tracing reveals Lgr5+ stem cell activity in mouse intestinal adenomas. Science 337:730–735

Shipitsin M, Campbell LL, Argani P, Weremowicz S, Bloushtain-Qimron N, Yao J, Nikolskaya T, Serebryiskaya T, Beroukhim R, Hu M, Halushka MK, Sukumar S, Parker LM, Anderson KS, Harris LN, Garber JE, Richardson AL, Schnitt SJ, Nikolsky Y, Gelman RS, Polyak K (2007) Molecular definition of breast tumor heterogeneity. Cancer Cell 11:259–273

Soeda A, Park M, Lee D, Mintz A, Androutsellis-Theotokis A, McKay RD, Engh J, Iwama T, Kunisada T, Kassam AB, Pollack IF, Park DM (2009) Hypoxia promotes expansion of the CD133-positive glioma stem cells through activation of HIF-1alpha. Oncogene 28:3949–3959

Trapani JA, Sutton VR (2003) Granzyme B: pro-apoptotic, antiviral and antitumor functions. Curr Opin Immunol 15:533–543

Tsuji T, Ibaragi S, Shima K, Hu MG, Katsurano M, Sasaki A, Hu GF (2008) Epithelial-mesenchymal transition induced by growth suppressor p12CDK2-AP1 promotes tumor cell local invasion but suppresses distant colony growth. Cancer Res 68:10377–10386

Wan XB, Long ZJ, Yan M, Xu J, Xia LP, Liu L, Zhao Y, Huang XF, Wang XR, Zhu XF, Hong MH, Liu Q (2008) Inhibition of Aurora-A suppresses epithelial-mesenchymal transition and invasion by downregulating MAPK in nasopharyngeal carcinoma cells. Carcinogenesis 29:1930–1937

Watt FM, Estrach S, Ambler CA (2008) Epidermal Notch signalling: differentiation, cancer and adhesion. Curr Opin Cell Biol 20:171–179

Weaver VM, Petersen OW, Wang F, Larabell CA, Briand P, Damsky C, Bissell MJ (1997) Reversion of the malignant phenotype of human breast cells in three-dimensional culture and in vivo by integrin blocking antibodies. J Cell Biol 137:231–245

Wu A, Wiesner S, Xiao J, Ericson K, Chen W, Hall WA, Low WC, Ohlfest JR (2007) Expression of MHC I and NK ligands on human CD133+ glioma cells: possible targets of immunotherapy. J Neurooncol 83:121–131

Yeung TM, Gandhi SC, Bodmer WF (2011) Hypoxia and lineage specification of cell line-derived colorectal cancer stem cells. Proc Natl Acad Sci U S A 108:4382–4387

Index

A

α2M transcription enhancement element (ATEE), 78–79
Accelerated cellular senescence
 DDAH enzymes, 11–14
 nitric oxide synthesis, 10, 13
 ROS formation, 12, 13
 SAH, 10
 telomerase, 8–10, 12
Acute cutaneous wounds, 90
Acute myeloid leukemia (AML), 133
ADMA. *See* Asymmetric dimethylarginine (ADMA)
Alanine-glyoxylate aminotransferase-2 (AGXT2), 7–8
Alpha-2-macroglobulin (α2M)
 Alzheimer β protein, 73
 ATEE, 78–79
 extra-cellular pan-proteinase inhibitor, 73
 gene expression level of, 77
 human plasma, 73
 NGF-β, 73
 p16INK4a, 77, 78
 PZP, 73
 replicative senescence (*see* Replicative senescence)
 senescence biomarker, 74
 senescent cells, 78
 single-copy gene, 73
 TGF-β, 73
 transcription regulatory mechanism, 78
 up-regulated expression, 77
Alzheimer β-protein, 73
Anaphase-promoting complex (APC/C)
 Cdh1/Cdc20, 209
 cell cycle progression, 208–209
 G1/S and G2/M, 209
 regulation of Cdh1
  MEFs, 212–213
  multiple mechanisms, 210–212
  phosphorylation and dephosphorylation events, 210, 211
  PTEN, 211–212
  RB/E2F1 and CLASPIN/CHK1/P53 pathways, primary human fibroblasts, 213–215
 SCF E3 biquitin ligase complexes, 208–209
Angiogenic dormancy, 173
Antigen-processing machinery (APM), 307

APC/cyclosome. *See* Anaphase-promoting complex (APC/C)
Apolipoprotein B mRNA editing catalytic peptide like 3G (APOBEC3G), 258–259
ARF/P53-dependent senescence pathway, 36–37
ARF/P53-indepenent senescence pathway
 cellular senescence, 39–40
 F-Box protein family, 38
 proteolytic and non-proteolytic ubiquitination, 38–39
Asymmetric dimethylarginine (ADMA)
 accelerated cellular senescence (*see* Accelerated cellular senescence)
 aging, 4
 cardiovascular diseases, 5
 cellular senescence and elderly, 7
 eukaryotic nuclear proteins, 5
 human endothelial cells, 4
 human somatic cells, 4
 L-NMMA and SDMA, 6
 metabolism, 6–8
 nitric oxide (NO), 5
 PRMTs, 5, 6
 SA β-gal, 5
Ataxia-telangiectasia mutated (ATM) protein, 198
ATR-X syndrome, 148
Atypical adenomatous hyperplasia (AAH), 198

B

BCCC. *See* Branchial cleft cyst carcinoma (BCCC)
Bone marrow-derived mesenchymal stem cells (BM-MSC), 97–98
Bone morphogenetic protein 7 (BMP7), 295–297
Brahma (BRM), 104
Branchial arches, 166
Branchial cleft cyst carcinoma (BCCC), 166
 aetiopathogenesis, 168, 169
 cholesterol crystals, 167
 contrast-enhanced CT scan, 167
 diagnostic criteria, 168
 embryonic development, 167
 facial nerve, 167
 fine-needle aspiration, 167
 histopathology, 169
 lateral neck masses, 168

Branchial cleft cyst carcinoma (BCCC) (cont.)
    MRI, 167
    neck mass, 167
    origin of, 167
    sternocleidomastoid muscle, 167
    surgical excision, 167
    TSC populations, 176
    type II lesions, 167
    type I lesions, 167
    ultrasonography, 167
Breast cancer
    activated stroma, 158
    breast density, 157
    CAF signature, 159
    Cav-1 deficient stroma, 159
    co-conspirator theory, 159
    collagen-I, 158
    dormancy, 162
    dynamic interaction, 156
    epithelial-stromal interactions, 157
    extracellular matrix, 156, 158
    glandular epithelial tissues, 156
    initiation and progression, 159
    LOX, 159
    mammary density, 158
    metastatic phenotype, 156
    metastatic site, 159–160
    non-neoplastic tissues, 158
    tamoxifen treatment, 158
Breast cancer metastasis suppressor (BRMS1), 249–251
Breast cancer stem cells (BCSC), 309

C
Cancer associated fibroblasts (CAFs), 159, 169
Cancer of unknown primary (CUP), 176
Cancer stem cells (CSC), 305–306
    BCSC, 309
    bone morphogenetic proteins, 293–294
    CD200, 308
    and dormancy, 293–294
    embryologic developmental programs, 171
    and EMT, 247
    immunogenic tolerance, 307
    MHC class I molecules, 307
    SCC, 171
    TGF-β signaling pathway, 307
    therapeutic implications, 309–310
Caveolin-1, 123
CD4+ T cell quiescence
    FOXO, 254
    and HIV infection
        APOBEC3G, 258–259
        entry, reverse transcription and integration, 256–257
        Glut1, 259
        integration and viral expression, 257–258
        LTR, 257–258
        Murr1, 258
        NFκB activity, 258
        PCR technology, 255
        restriction factors, 258–259
        RNA and DNA synthesis, 256
        siRNA technology, 258
        viral cDNA, 255
    LKLF, 254
    Runx1 and Tsc1, 255
CD47-SIRPα interaction, 308
Cdk inhibitors (CdkIs), 208, 209
Cellular senescence, 54, 197–198
    and aging, 112
    Arf/p53/p21 tumor suppressor pathway, 135
    biomarkers, 46
    and cancer, 112
    Caveolin-1, 123
    cell reprogramming
        H3K36me2/3, 269
        INK4a/ARF locus, 268, 269
        p53 expression, 267, 268
        shRNA, 267, 268
        telomerase, 267–268
        UTF1, 267
    characteristic features, 28, 112
    chemotherapy-induced senescence, 204
    clinical implications, 204–205
    cyclin D1-induced senescence, 202, 203
    definition, 197
    DNA damaging agents, 204
    Hayflick limit, 112
    HNRNP (see Heterogeneous nuclear ribonucleoproteins (HNRNP))
    inducers of, 198–199
    INGs (see INhibitors of Growth (INGs) proteins)
    irreversible cell cycle, 112
    irreversible growth arrest, 134
    JDP2 (see Jun dimerization protein 2 (JDP2))
    malignant transformation, 201–202
    markers and characteristics, senescent cells, 199–200
    melanocyte transformation, 229–230
    MNK1 roles, 117
    nuclear lamina (see Nuclear lamina)
    OIS, 35–36
    p16/pRb tumor suppressor pathway, 135
    p53 proteins, 135, 200, 201
    pathways in, 200–201
    premalignant lesions, tumor progression, 202–203
    premature senescence, Cdh1 loss
        in MEFs, 212–213
        RB/E2F1 and CLASPIN/CHK1/P53 pathways, primary human fibroblasts, 213–215
    pro-senescence therapy, 40–41
    pten-loss induced senescence, 36
    RB, 200, 201
    Rb proteins, 135
    replicative senescence (see Replicative senescence)
    stress-activated MAPKS, 112–113
    TGF-β (see Transforming growth factor-beta (TGF-β))
    tumorigenesis and anti-cancer therapy, 28–29
    tumor promoting factor, 203–204

Ceramide
    activated protein kinase, 84
    biology, 83–84
    de novo synthesis, 86
    glucosylceramide synthase pathway, 86
    mechanisms of action, 84–85
    senescent cells, 85–86
    sphingomyelinase, 86
Chromatin
    ATP-dependent remodelers, 144, 145
    CpG islands, 145
    DNA damaging agents, 144
    DNA methylation, 145, 146, 150–152
    DNA replication process, 144
    DNA-templated processes, 144
    DNMT3A and DNMT3B, 145
    gene expression, 144
    HAT
        and ageing, 149–150
        and cancer, 149
        CBP, 149
        multisubunit complexes, 149
    hetero-and eu-chromatin, 144, 145
    nucleosomes, 144
    primary protein components, 144
    remodeling complexes
        ATR-X syndrome, 148
        Juberg Marsidi syndrome, 148
        Schimke immuno-osseous dysplasia, 148
        senescence and cancer, 146–148
        Smith Fineman Myers syndrome, 148
        Sutherland-Haan syndrome, 148
        Williams syndrome, 148
Circulating tumor cells (CTCs), 242, 290, 291
c-Jun N-terminal kinase (JNK), 233, 235, 244, 259
Claudin-1 (CLDN1), 249
Collagen-I expression, 162
Common myeloid progenitors (CMPs), 285
Connective tissue growth factor (CTGF)
CREB binding protein (CBP), 149
CSC. *See* Cancer stem cells (CSC)
Cullin-RING E3 ligase (CRL), 29–30
Cutaneous regeneration
    dermal papilla (DP) cells, 97
    hair follicle stem cells, 95–97
    mesenchymal stem cells, 97–98
Cyclin-dependent kinase inhibitors (CDKi), 189, 245
Cyclin-dependent kinases (CDKs), 208–210, 215, 222, 245
Cytokines. *See* Interleukin-6 (IL-6)

D
Dendritic cells (DC), 308
Dermal-condensation (DC), 90
Dermal papilla (DP) cells, 97
Dihydroceramide, 82, 84
Dimethylaminohydrolase (DDAH) enzymes, 6–7, 11–14
Disseminated tumor cells (DTCs), 172–175, 290–292
    in bone marrow, 242
    and CTCs, 242

NSCLC
    characteristics of, 242
    and gene expression signatures, 241
DNA damage response (DDR), 198–199, 234

E
Edar signaling, 93, 94
Electrophoretic mobility shift assays (EMSA), 76–78
Embryonic day (ED), 90
Embryonic hair follicle
    dermal-condensation (DC), 90
    growth, regression and rest, 90
    induction and morphogenesis
        BMPs, 93–94
        dermal condensate formation, 92
        Edar signaling, 93
        epithelial-mesenchymal signaling interaction, 92
        "first dermal signal," 90
        heparanase, 93
        Notch signaling, 92
        PDGF-A, 92
        "second dermal signal" regulation, 92
        Wnt/ β-catenin signaling, 90–91
    pattern formation, 94–95
    quiescence and activation, 91, 95
    Wnt signaling, 90
Embryonic stem cells (ESCs), 266, 268
Endosomal sorting complex required for transport (ESCRT), 21
Endothelial progenitor cells (EPCs)
    exosomes and microvesicles, 21
    flow cytometry, 18
    hematopoietic stem cell lineage, 18
    IRI, 23–24
    pro-angiogenic and anti-apoptotic effects, 19
    repair process, 19
    soluble angiogenic factors, 19
    vascular homeostasis and repair, 19
    vascular injury, 18
Epidermal growth factor (EGF) signal, 94–95
Epithelial-to-mesenchymal transition (EMT)
    cancer stem cell, 293
    CSC, 247
    immune effectors, 308
    inhibiting promotion of, 309
    primary tumor, 306
    TGF-β, 224
    transcription factors, 305–306
Euchromatin, 144
Exosome biogenesis, 20–21
Extracellular matrix (ECM), 174
Extracellular signal-regulated kinase (ERK), 174
Extracellular vesicles (EVs)
    bioactive molecules, 20
    biogenesis of, 20–21
    cell-to-cell communication, 19
    classification and nomenclature, 20
    IRI, 23–24
    in vitro cell cultures, 19

Extracellular vesicles (EVs) (cont.)
    pleiotropic functions, 21
    prokaryotes and eukaryotes, 19
    receptor-ligand interactions, 20
    tumor neoangiogenesis, 21–22

F
Fibrosis
    breast cancer (see Breast cancer)
    collagen-I expression, 162
    lungs, 161–162
    mechanisms of, 156–157
Fluorescence intensity, 66
Fluorescent recovery after photobleaching (FRAP), 66
5-Fluorouracil (5-FU), 175
Forkhead Box class O (FOXO), 254
Functional promoter region determination, 74–75

G
Gap junctional intercellular communication (GJIC), 294, 295
GATA2 expression
    BCL-XL family, 285
    CD34+ CD38-cells, 279–281
    cell cycle control, target genes in
        CCND1, 282
        CCND3, 282
        CCNE1, 282
        CDK4 and CDK6, 282
        D-type cyclins, 282
        HES1, 282, 283
        LSKcells, 282, 283
        MEF/ELF4, 282, 284
        quantitative RT-PCR analysis, 284
    cytokine receptor expression, 284–285
    GFP expression, 281
    GM-CSFr+, 284
    on haematopoietic reconstitution, 280–282
    haploinsufficiency of, 285
    LTC-IC assays, 280–281
    molecular regulator, 286
    and proliferative control, 279
    Pyronin staining, 279, 280
Glioblastoma multiforme (GBM), 307
Glucosylceramide synthesis, 82
GP130 and signaling molecules, 55
Granulocyte-macrophage progenitor (GMP), 285

H
Haematopoietic stem cells (HSCs)
    and CSCs, 244
    hypoxic niche, 303
    metastasis, 172
    quiescence, 245
        cellular and molecular control, 278–279
        GATA2 (see GATA2)
    VEGFR1, 244

Hair follicle. See Embryonic hair follicle
Herpesvirus-associated ubiquitin-specific protease (HAUSP), 185, 187
Heterogeneous nuclear ribonucleoproteins (HNRNP)
    A1 protein
        MNKL regulation, subcellular distribution, 115–116
        molecular mechanisms, 114
        p38 MAPK inhibition, 114–115
        phosphorylation level, 114
    A/B family proteins, 114
    cytoplasmic accumulation, 114
    pre-mRNA processing, 113
    RNA binding proteins, 113
    stress-induced phosphorylation sites, 114
Histone acetyltransferases (HATs), 148–149, 181, 182
Histone deacetylases (HDACs), 148–149, 181–183
HNRNP. See Heterogeneous nuclear ribonucleoproteins (HNRNP)
Homologous recombination (HR), 39
HPV. See Human papilloma virus (HPV)
HSCs. See Haematopoietic stem cells (HSCs)
Human diploid fibroblasts (HDF), 85
Human mammary epithelial cells (HMECs), 222, 223
Human mesenchymal stem cells (hMSCs), 63
Human papilloma virus (HPV), 175–176
Human T lymphotropic virus type 1 (HTLV-1) Tax protein, 39
Human umbilical vein endothelial cells (HUVEC), 8, 59, 85
Hutchinson-Gilford Progeria Syndrome (HGPS), 62, 190, 304
Hyperplastic adenomas, 196
Hypoxia inducible factor (HIF), 160, 185
Hypoxia inducible factor 1A (HIF1A), 241

I
IL-6. See Interleukin-6 (IL-6)
Immunodeficiency, Centromere instability and Facial anomalies syndrome (ICF syndrome), 151
Immunoreceptor tyrosine-based activation motifs (ITAMs), 231
Immunosurveillance, 306, 307
Induced pluripotent stem cells (iPSCs)
    and hESCs, 271, 272
    mitochondrial properties, 271–272
    senescent cells
        c-MYC, 273, 274
        endogenous pluripotency genes, 270
        HMGA2, 273
        INK4a/ARF locus, 273
        KLF-4, 273, 274
        LIN28, 270, 273, 274
        microRNA, 273–275
        NANOG, 270, 273–275
        OCT4, 270, 273–275
        SOX2, 270, 273–275
        telomeres, 271
    VitC, 269

INhibitors of Growth (INGs) proteins
  in aging
    oncogene induced senescence, 188–189
    SIPS, 188
    telomeres and replicative senescence, 187–188
  in cancer
    apoptosis, 185–186
    cell cycle regulation, 184–185
    cell motility and angiogenesis, 185
    DNA damage response and genomic stability, 185
    ing genes, mutation of, 186
    p53 tumour suppressor, 186–187
  chromatin regulators, 182–183
  discovery of, 180
  post-translational modification, 182
  senescence, changes with
    epigenetic changes, 190
    genetic changes, 189–190
    morphological changes, 189
    senescence markers, 190–191
  short non-coding RNAs, 184
  structural components of, 180–182
  tumour suppressor networks and senescence
    ARF-p53-p21 pathway, 191
    ING-HAT/HDAC pathway, 192
    p16-pRb-E2F pathway, 191–192
Inner nuclear membrane (INM), 62
Insulin-like growth factor-binding protein 5 (IGFBP5), 53–59
Integrin-associated protein (IAP), 308
Interferon-γ (IFNγ), 243
Interleukin-6 (IL-6)
  gp130 receptor molecule, 55
  premature senescence
    chains and constitutively activated STAT3, 56–57
    DDR, 57
    IGFBP5, 56, 58
    p53 levels, 57
    rheumatoid arthritis, 55
    ROS, 56–57
    SA-β gal activity, 56
    SASP/SMS, 54
    signal transducing receptor, 56
    STAT3-IGFBP5 axis, 57–58
    tumorigenesis, 55
  signal transduction pathways, 55, 56
  specific IL-6 receptor, 55
Ischemia-reperfusion injury (IRI), 18, 23–24

J
Juberg Marsidi syndrome, 148
Jun dimerization protein 2 (JDP2)
  AP-1 family, 103
  cell cycle progression, 103
  heterodimers, 103
  histone chaperone, 102, 103
  INK4a/ARF locus, 106–108
  p300/CBP-associated factor, 103
  PRCs, 102
  transcription factor, 102

K
Keratinocyte growth factor (KGF), 94–95
Krupple-Like factors (KLFs), 254

L
Label-retaining cells (LRCs), 95
Latency associated peptide (LAP), 161
Leucine zipper-like (LZL), 181, 186
Leukemia stem cells (LSC), 308
Lin-Sca+ Kit+ (LSK) cells, 282, 283
Lipidomics profiles, 87
Long-term culture-initiating cell (LTC-IC) assays, 280–281
Loss of heterozygosity (LOH), 186
Lrig1-expressing cells, 96
Lung cancer dormancy
  CSC, 247
  EMT, 247
  ERK^{high}/p38^{low} ratio, 245, 246
  experimental *in vivo* models, 246–247
  metastasis suppressors
    BRMS1, 249–251
    CLDN1, 249
    KISS1, 249
    MKK4 and 7, 248
    Nm23, 248
    PEBP1/RKIP, 249
    RhoGDI2, 248–249
  quiescence of, 244–246
Lung Krupple-Like factor (LKLF), 254
Lungs fibrosis, 161–162
Lysosomal dysfunction, 124
Lysyl oxidase (LOX), 159

M
Major histocompatibility complex (MHC) molecules, 306, 307
Mammalian target of rapamycin (mTOR), 304
Manganese superoxide dismutase (MnSOD), 46
Mechano-transduction process, 156
Meiomitosis, 170
Melanoma
  BRAF-induced senescence, 134
  BRAF mutations, 228, 229
  ERK, 134
  NRAS mutations, 228, 229
  oncogenic events, 228
  pathogenesis of, 228–229
  PIR expression, 135–136
  PTEN, 229
  radial-growth phase (RGP), 134
  Ras-Raf-MEK-ERK signalling pathway, 134
  senescence bypass, melanocyte, 230
  skin cancer, 134
  Syk (*see* Spleen tyrosine kinase (Syk))
  TP53 mutation rate, 135
Mesenchymal stem cells (MSC), 97–98, 152, 304, 305
Mesenchymal to epithelial transition (MET), 293

Metastasis suppressor genes (MSG), 248, 251
Micrometastatic cancer cells. *See* Non-small cell lung cancer (NSCLC)
Microvesicle biogenesis, 20
Minimal Residual Disease (MRD), 294
Mitochondrial dysfunction
  energy metabolism
    ATP, ADP, and AMP, 47
    cancer cells, 47
    senescence, 47–48
  hydrogen peroxide, 46
  intracellular organelles, 46
  mtDNA, 46, 47
  p53 role, 49
  RISP, 49–50
  ROS, 46
  shRNA, 49–50
Mitochondrial Free Radicals Theory of Aging, 49
Mitochondrial single-stranded DNA-binding protein (mtSSB), 49
Mitogen-activated protein kinase 4 (MKK4/MAP4K4), 248
Mitogen-activated protein kinase 7 (MKK7/MAP3K7), 248
Mitogen activated protein kinases (MAPK) family, 292
Mouse embryonic fibroblasts (MEFs), 210, 212–213
Mouse mammary tumor virus (MMTV) promoter, 308
Multinucleate/polyploid giant cells (MN/PGs), 169
Multiple tumor suppressor-1 (MTS1), 191
Multivesicular bodies (MVBs), 20
Myelodysplasia (MDS), 286

N
NEDD8 activating enzyme (NAE) inhibitor MLN4924
  cancer cells growth, 30
  irreversible cellular senescence, 30
  p21-dependent senescence, 30–31
Neddylation, 29–30
Nemosis, 169
Neoangiogenesis
  definition, 18
  extracellular vesicles, 21–22
  IRI, 18
  paracrine mechanisms, 19
  physiological process, 18
  tumor vascularization, 18
  vascular homeostasis and repair, 19
Neosis, 169–170
Neubauer cell chamber, 9
Neural stem cell (NSC), 304
N-monomethyl-L-arginine (L-NMMA), 6
N-Myc downstream regulated gene (NDRG1), 295–297
Noncoding RNAs (ncRNAs), 184
Non-metastatic protein-23 (Nm23), 248
Non-small cell lung cancer (NSCLC)
  dormant lung cancer cells (*see* Lung cancer dormancy)
  DTCs
    characteristics of, 242
    gene expression signatures, 241
    tumor-stroma interactions and tumor cell dormancy, 242
    immunosurveillance, 243–244
    pre-metastatic niche, 244
Notch intracellular domain (NICD), 92
Nuclear envelope (NE), 62, 63
Nuclear lamina
  biophysical properties, 64–66
  cell classification, 66, 67
  cytoplasmic functions, 68
  detection, 63, 65
  HGPS patients, 62
  hMSCs, 63
  INM via farnesyl hydrophobic group, 62
  intranuclear structures, 63, 64
  lamin A-and lamin B-encoding genes, 62, 68
  lamin proteins, 62
  mammalian cells, 63
  nuclear activities, 66
  plasma membrane, 68
  progerin accumulation, 62
  structural changes, 63, 68
  structural deformation, 63
  structure of, 68
Nuclear localization signal (NLS), 180, 182, 187, 211

O
Oncogene Bcr-Abl, 50
Oncogene-induced senescence (OIS), 35–36, 230
Oncostatin M (OSM), 244
Outer nuclear membrane (ONM), 62
Oxidative phosphorylation (OXPHOS), 47, 49–50

P
P38 MAP kinase pathway, 114–115
Pancreatic intraductal neoplasias (PIN), 198
Partial bromodomain (PBD), 180–181
PCNA-interacting-protein (PIP), 180–181
Peptidylarginene deminase 4 (PAD4), 187
Pharyngeal fistulas. *See* Branchial cleft cyst carcinoma (BCCC)
Pharyngeal pouches, 166
Phosphatase and tensin homolog (PTEN)-loss induced senescence, 36
Phosphatidylethanolamine binding protein 1 (PEBP1/RKIP), 249
Pirin (PIR) protein
  biological function, 132
  in cancer, 133–134
  melanoma
    BRAF-induced senescence, 138
    chemotherapy, 138
    colony formation assays, 136
    growth curve experiments, 136
    knock-down approach, 136, 138
    molecular mechanism, 138
    OHT, 138
    SASP, 136
    senescent phenotype, 138

tumor model, 136
  WM266-4 metastatic, 136, 137
 molecular pathways, 132–133
 senescence and melanoma progression, 139, 140
 structure, 132
Plant homeodomain (PHD), 180–182
Platelet-derived growth factor-A (PDGF-A), 92
Pleiotropic functions, 21
Polybasic region (PBR), 181
Polycomb repressive complexes (PRCs), 102
Polypyrimidine tract binding protein (PTB), 258
Positive element identification, 75–77
Pregnancy zone protein (PZP), 73
Premalignant lesions, 196–197, 202–203
Premature cell senescence
 Cdh1 loss
  in MEFs, 212–213
  RB/E2F1 and CLASPIN/CHK1/P53 pathways, primary human fibroblasts, 213–215
 collagen XVIII and endostatin, 126
 interleukin-6 (IL-6)
  chains and constitutively activated STAT3, 56–57
  DDR, 57
  IGFBP5, 56, 58
  p53 levels, 57
  rheumatoid arthritis, 55
  ROS, 56–57
  SA-β gal activity, 56
  SASP/SMS, 54
  signal transducing receptor, 56
  STAT3-IGFBP5 axis, 57–58
  tumorigenesis, 55
 lysosomal dysfunction, 124, 125
 molecular signatures of, 122–124
 SIPS (see Stress-induced premature senescence (SIPS))
 sirtuins-cancer, 124–126
Proliferating cell nuclear antigen (PCNA), 180–181, 185, 186
Pro-senescence therapy, 40–41
Protein arginine methyltransferases (PRMTs), 5
Protein phosphatases, 84
Proteolytic and non-proteolytic ubiquitination, 38–39
Pyrrollidine dithiocarbamate (PDTC), 5, 6

R
Ras-Raf-MEK-ERK signalling pathway, 134
Rb and p53 tumor suppressor pathways, 102
Reactive oxygen species (ROS), 12, 57
Replicative senescence, 24, 46
 α2M
  constant expression level of, 74
  cultured fibroblasts, 73
  functional promoter region determination, 74–75
  HeLa cells, 74
  human diploid fibroblasts, 77
  mRNA level of, 73
  positive element identification, 75–77
  RNA samples, 73
 p38 pathway, 113
Reporter gene assay, 77

Retinoblastoma (RB), 200–203
Rieske iron sulfur protein (RISP), 49–50
Ring finger protein 5 (RNF5), 241
Rubinstein Taybi syndrome, 149

S
SA-β-galactosidase activity, 201, 202
SAHF. See Senescence associated heterochromatin foci (SAHF)
Schimke immuno-osseous dysplasia, 148
Secondary glioblastoma multiforme, 196
Secreted Protein Acidic and Rich in Cysteine (SPARC), 229, 230
Senescence. See Cellular senescence
Senescence-associated β-galactosidase (SA β-gal), 5
Senescence associated heterochromatin foci (SAHF), 46, 199–203, 266
 ATP-dependent chromatin-remodeling, 104
 chromatin, 102
 chromosome condensation, 103
 chromosome regions, 102
 heterochromatin, 102
 HIRA and ASF1a, 103
 linker histone H1, 102
 and pRb and p53 Pathways, 105
 PRC1 and PRC2 silencing, 105–106
 proliferation-promoting genes, 102
Senescence-associated secretary phenotype (SASP), 46, 54, 136
Senescence-like growth arrest (SLGA), 222, 223
Senescence-messaging secretome (SMS), 54
Short hairpin RNA (shRNA), 267, 268
Signal transducer and activator of transcription 3 (STAT3), 56–58
SIPS. See Stress-induced premature senescence (SIPS)
Sirtuins, 124–126
Skp1-Cullin1-F-box complex (SCF), 208–209
Skp2 regulation
 cellular senescence, 39–40
 F-Box protein family, 38
 proteolytic and non-proteolytic ubiquitination, 38–39
Smith Fineman Myers syndrome, 148
Soluble IL-6R (sIL-6R), 55–57
Sphingolipids
 biology, 82
 cancer, 82–83
 ceramide (see Ceramide)
 implications, 86–87
Spleen tyrosine kinase (Syk)
 epigenetic loss, 234
 hematopoietic malignancies and retinoblastomas, 234
 hematopoietic signaling and diseases, 231
 hypothetic model, 235
 inactivation of, 235
 in leukocytes, 231
 in melanomagenesis
  epidermal melanocytes and melanoma cells, 232
  and senescence, 232–233
 in non-hematopoietic cancers, 231–232
 p21Cip1/Waf1, 234

Spleen tyrosine kinase (Syk) (*cont.*)
  p53-dependent pathway, 235
  role of, 232
  ROS, 235
Squamous cell carcinoma (SCC), 168
  CSC hypothesis, 171
  HPV, 175
Stem cells, 96, 177
  CSCs, 171
  environmental signaling, 171
  epidermal stem cell niche, 302–303
  HSC, 172
  hypoxia, role of, 303–304
  malignant transformation
    apoptosis, 169
    epigenetics, 169
    meiomitosis, 170
    nemosis, 169
    neosis, 169–170
    senescence, 169
  microenvironmental control, 172
  premature aging and tumorigenesis, 304–305
  SCC, 171–172
  stem cell niche, 172
  TSCs, 172
Stress-activated MAPKS, 112–113
Stress-induced premature senescence (SIPS), 188
  affecting tissue functions, 126
  cancer therapy, 127
  cell stress, 124, 125
  endothelial cells, 124
  lysosomal permeabilization, 124
  molecular mechanisms, 122–123
  tumor epithelial cells, 126
Subventricular zone (SVZ), 304
Sutherland-Haan syndrome, 148
Syk. *See* Spleen tyrosine kinase (Syk)
Symmetrical dimethylarginine (SDMA), 6

T
T cell receptors (TCR), 307
Telomeres, 187–188
Trans-acting factors, 76
Transforming growth factor-beta (TGF-β), 93
  and cellular senescence, 222–223
  EMT, 224
  in human carcinomas, 221–222
  MH1 and MH2 domains, 221
  RI, RII and RIII, 220

R-Smads and Smad4, 220–221
TGFBR2 mutation, 221
in tumor progression
  in early stage, 223
  in later stage, 223–224
Trichostatin A (TSA), 183
Tumor-associated antigens (TAAs), 307
Tumor associated macrophages (TAM), 308
Tumor development, 47
Tumor dormancy, 172
  angiogenic dormancy, 173
  in bone, 294–295
  cellular dormancy-autophagy, 174
  clinical significance of, 291–292
  dominant tumor dormancy escape, 174–175
  Erk1/2 and p38α/β, 292
  immunesurveillance, 173–174
  molecular mechanisms, 292–293
  roles of BMP7, 295–297
Tumorigenesis and anti-cancer therapy, 28–29
Tumor initiating cells (TIC), 305–306
Tumor necrosis factor (TNF), 243
Tumor neoangiogenesis. *See* Neoangiogenesis
Tumors suppressor protein 53 (TP53) proteins, 200

U
Ubiquitin-binding domain (UBD), 181
Undifferentiated embryonic cell Transcription Factor 1 (UTF1), 267
Urokinase plasminogen activator (Upa), 162

V
Vascular cell adhesion molecule-1 (VCAM-1), 294, 295
Vascular endothelial growth factor receptor 1 (VEGFR1), 244
Vascular endothelial growth factor receptor 1+ (VEGFR1+), 160
Vitamin C (VitC), 269

W
Warburg effect, 47
Williams syndrome, 148
Wnt pathway, 123

X
Xeroderma pigmentosum group A protein (XPA), 185

Printed by Publishers' Graphics LLC
LMO131224.15.18.324